DEVELOPMENT IN MAMMALS
VOLUME 3

DEVELOPMENT IN MAMMALS

VOLUME 3

Editor
Martin H. Johnson
Director of Studies in Physiology and Fellow Christ's College, Cambridge

1978

NORTH-HOLLAND PUBLISHING COMPANY
AMSTERDAM · NEW YORK · OXFORD

ISBN North-Holland for the series: 0 7204 0632 3
ISBN North-Holland for this volume: 0 7204 0663 3

Published by:

Elsevier/North-Holland Biomedical Press
335 Jan van Galenstraat, P.O.Box 211,
Amsterdam, The Netherlands.

Sole distributors for the U.S.A. and Canada:

Elsevier/North-Holland Inc.
52 Vanderbilt Avenue
New York, N.Y. 10017.

Library of Congress Cataloging in Publication Data (Revised)
Main entry under title:

Development in mammals.

Includes bibliographies and indexes.
1. Embryology--Mammals. I. Johnson, M. H.
QL959.D36 599'.03'3 77-5894
ISBN 0-7204-0632-3 (Series)
ISBN 0-7204-0663-3 (Volume)

Printed in the Netherlands.

PREFACE

Volume 3 of Development in Mammals shows some alterations in style, but has adopted the same approach to content as did Volumes 1 and 2. Authors have again been encouraged to develop new ideas and to be provocative, and to use unpublished data where these would support an argument. Within this flexible frame-work, a number of contributions have been received relating to four broad categories of topic.

Two contributions deal with the acrosome reaction, one approach biochemical and the other biophysical. Two contributions are concerned specifically with models for cell:cell interaction in development, either by viral particles or through localised fluidity changes in cell membranes. Five contributions deal with the relationships between stem cells, pluripotentiality and the transmission of the germ line. Finally, three contributions are concerned with the fundamental aspects of morphogenesis - proliferation, migration and clonal growth.

Volume 4 of Development in Mammals, which will appear in about fifteen months time, will pursue some of these themes further. In particular, we hope to include contributions on the relationship between current ideas on morphogens and gradients and the established ideas about inducers. Scientists interested in making a contribution to, or suggesting contributors for, such a topic should contact the editor.

I wish to thank Shirley French, Raith Overhill and Chris Burton for their help in preparing this volume, and the contributors for their stimulating and prompt manuscripts.

23rd March, 1978.

Martin H. Johnson,
Department of Anatomy,
Cambridge.

CONTENTS

THE MAMMALIAN SPERM ACROSOME REACTION, A BIOCHEMICAL APPROACH

Stanley Meizel

Department of Human Anatomy
School of Medicine
University of California, Davis
Davis, California 95616
U.S.A.

The mammalian sperm acrosome reaction, a membrane fusion and vesiculation involving the acrosome (a sperm head organelle) and the overlying plasma membrane, is essential to fertilization. In this chapter, the major aims are to present a critical review of methods and results of studies of the biochemical factors which influence the *in vitro* acrosome reaction and to suggest possible biochemical mechanisms for the acrosome reaction. In addition, discussion of other topics relevant to these aims (e.g. morphology of the acrosome reaction), and of the possible relationships of *in vitro* studies to *in vivo* situations will be included. The reader is also directed to useful reviews by Gwatkin (1976), Yanagimachi (1977) and Bedford and Cooper (1978) concerned with mammalian gametes and fertilization.

THE SIGNIFICANCE OF THE ACROSOME REACTION

The acrosome reaction apparently allows certain hydrolytic enzymes such as hyaluronidase to be released from the acrosome and exposes other membrane-bound hydrolytic enzymes such as acrosin (see Allison and Hartree, 1970; Srivastava *et al.*, 1974 and reviews by Morton, 1976 and Hartree, 1977). Such enzymes might then aid in sperm penetration of the egg investments and plasma membrane. The former include: the cumulus oophorus, a multilayer of retained ovarian granulosa cells surrounding the zona pellucida and embedded loosely in a matrix containing hyaluronic acid; the corona radiata, part of the cumulus closest to the zona pellucida and bound to the zona in a particularly tight manner through microvilli (in some species the corona does not remain tightly bound to the zona after ovulation); and the zona pellucida, a protein-polysaccharide envelope surrounding the egg plasma membrane (see Zamboni, 1970 and Franchi and Baker, 1973, for morphological details of the egg investments).

The occurrence of the acrosome reaction does appear to be essential for sperm penetration of the cumulus oophorus and the zona pellucida (see section 5). In addition, Yanagimachi and Noda (1970a) and Noda and Yanagimachi (1976) have suggested (based on their experimental results) that the occurrence of the acrosome reaction is also essential for sperm-egg fusion.

THE RELATIONSHIP OF CAPACITATION TO THE ACROSOME REACTION

Bedford (1970) has suggested that the acrosome reaction should be viewed as an event which occurs only to sperm which have been previously capacitated. This description involves redefinition of the term "capacitation" which had originally been used by Chang and Austin as an operational definition to collectively describe all of the changes to sperm occurring in the female tract or *in vitro* and which gave the sperm capacity to fertilize (see reviews by Austin, 1975 and Chang and Hunter, 1975, for a more complete historical review of capacitation studies). In this chapter, the term capacitation is still an operational definition, but is limited to those events which prepare the sperm for acrosome reaction.

The events which have been suggested to occur during capacitation and which might therefore play a role in allowing the acrosome reaction to occur include: the loss of sperm surface stabilizing proteins (Oliphant and Brackett, 1973); changes in membrane fluidity (O'Rand, 1977) and subtle ultrastructural changes in membrane components(Koehler, 1976). Bedford and Cooper (1978) have critically reviewed the information available concerning the nature of changes occurring in sperm membranes during sperm maturation in the male reproductive tract and during capacitation in the female tract or *in vitro*.

THE ACROSOME

The mammalian sperm acrosome, a cap-like, membrane-limited organelle located in the anterior two-thirds of the sperm head contains hydrolytic enzymes which are important in fertilization (Figure 1). This organelle has various sizes and shapes in different animals, but always consists of the following (Austin, 1968; Franklin *et al.*,1970; Fawcett, 1970): a continuous limiting membrane composed of an outer region (that region in

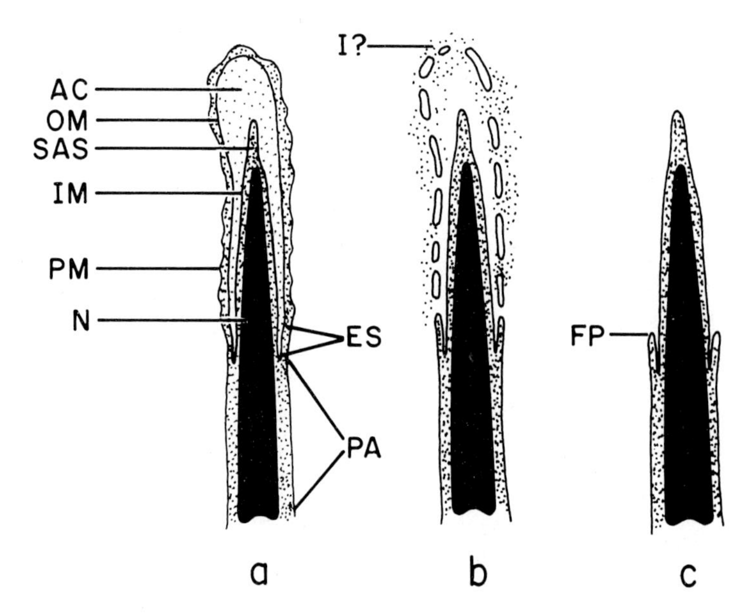

Figure 1: Mammalian sperm acrosome reaction (modified from Figure 1 of Bedford, 1968). (a) Rabbit sperm head: The cap-like structure of the acrosome is evident here. AC - those acrosomal contents not membrane bound; OM - outer acrosomal membrane; IM - inner acrosomal membrane; SAS - subacrosomal space (also known as perforatorium when large and electron dense, as in rabbit); PM - plasma membrane; N - nucleus (nuclear membrane not shown); ES - equatorial segment (narrower caudal region of the acrosome also known as acrosomal collar); PA - postacrosomal sheath (previously known as postnuclear cap). The PA is a specialized membrane region underlain by a dense layer not shown here. (b) Acrosome reaction. The acrosome reaction may be initiated in the apical region of the acrosome. I? - represents one such apical initiation site. The fusion and vesiculation occurring between outer acrosomal membrane and overlying plasma membrane allows the "soluble" contents of the acrosome including any unbound enzymes to leave the acrosome. (c) After the acrosome reaction. Vesicles are shed, thus exposing any remaining enzymes bound to inner acrosomal membrane. FP - fusion point of the outer acrosomal membrane with the plasma membrane along the anterior limit of the equatorial segment. This stable fusion enables the sperm to remain surrounded by a single continuous membrane.

close proximity to the overlying plasma membrane), an inner region (the infolded surface of the acrosomal cap which lies in close apposition to the nuclear membrane); the equatorial segment or acrosomal collar (the particularly narrow caudal region of the cap where the outer and inner acrosomal membranes join); the subacrosomal space, in some cases known as the perforatorium (an electron dense region, in some cases rod-like, located in a space between the anterior end of the nucleus and the acrosome); and the "soluble" contents of the acrosome (electron-dense granular contents not bound to membranes).

The hydrolytic enzyme content of the acrosome and its origin from the Golgi apparatus during spermatogenesis have resulted in the acrosome being compared to a "specialized lysosome" (Allison and Hartree, 1970; Morton, 1976). Friend (1977) has suggested that the acrosome may be more similar to a secretory granule. Both descriptions may be valid. The pH of the hamster sperm acrosome is 5 or less, which resembles the lysosomal pH (Meizel and Deamer, 1978), and the acrosome reaction appears to resemble the exocytosis seen in secretory granules (Friend *et al.*,1977).

THE "TRUE" ACROSOME REACTION

MORPHOLOGY

The morphological events considered characteristic of the acrosome reaction of a mammalian sperm capable of fertilization have been observed in rabbits, rodents, pig and man (see Bedford and Cooper, 1978, for literature). Transmission electron microscopic studies indicate that the "true" acrosome reaction in mammalian sperm is an organized, progressive membrane fusion and vesiculation involving the outer acrosomal membrane and its overlying sperm head plasma membrane (Figure 1) which results in the loss of the "soluble" contents of the acrosome (Franklin *et al.*,1970; Bedford, 1970). An important electron microscopic criterion of this "true" acrosome reaction is the fusion of remaining outer membrane and plasma membrane at the anterior margin of the equatorial segment, thus allowing for a continuity of sperm membranes after the acrosome reaction (Bedford, 1968; Franklin *et al.*,1970). The acrosome reaction may begin at the apical border of the sperm head (Bedford, 1974; Friend, 1977). Freeze-fracture studies of the *in vitro* guinea pig sperm (Friend *et al.*,1977) appear to demonstrate some of the more subtle morphological events occurring during the acrosome reaction.

There are also "false" acrosome reactions which occur in dead or dying sperm (Bedford, 1970). The random nature of these membrane changes can be detected by transmission electron microscopy. At the light microscopic level (phase contrast, interference contrast or anoptral contrast), the loss or partial loss of the acrosome in strongly motile sperm is generally taken as representing a completed or still in progress normal acrosome reaction (Austin, 1975). In such observations, it is the absence or partial loss of the cap-like structure of the acrosome which is detected (see Austin and Bishop, 1958 for photographs demonstrating light microscopic morphology of motile sperm which have undergone the acrosome reaction).

In light microscopic studies, a whiplash-like flagella movement called activation is also seen in capacitated guinea pig and hamster sperm (Yanagimachi, 1970a; Yanagimachi and Usui, 1974), and the occurrence of a high percentage of acrosome reactions in motile sperm along with a high percentage of activation is generally considered an indication of capacitation and physiological acrosome reactions (Austin, 1975).

DETECTION OF THE "TRUE" ACROSOME REACTION

Obviously one cannot rule out the possibility that some of these acrosome reactions occurring in motile sperm and viewed by light microscopy may be abnormal, but the criteria based on using motile sperm is far superior to the use of stained sperm smears to count acrosome reactions (the absence of the cap-like structure of the acrosome). The major problem, of course, in the latter technique is that it cannot distinguish between sperm which were dead prior to fixation and which had undergone a "false" acrosome reaction and those which had undergone "true" reaction prior to fixation. There are always dead sperm present in *in vitro* incubations. In addition, the danger exists in sperm smear studies that the sperm were damaged in some way by the incubation prior to fixation and that the acrosome reactions detected by staining represent a population of damaged sperm acrosomes particularly fragile and sensitive to fixation.

Although the ideal combination of studies would combine counts at light microscope level of acrosome reactions in motile sperm with various kinds of electron microscopic studies and with fertilization studies, this ideal is not easy to attain. Transmission electron microscopic observations can be difficult because it takes numerous hours to find enough sperm sections oriented in such a way as to be able to distinguish whether the "true" acrosome reaction occurred in more than a few sperm. Freeze-fracture studies may provide new insights into the morphology of the "true" acrosome reaction (Friend *et al.*,1977), but the interpretation of results obtained by this method has certain difficulties (Friend *et al.*,1977). Even fertilization does not prove that many "true" acrosome reactions occurred because even a small percentage of normal acrosome-reacted sperm can still yield relatively high percentages of fertilization when the total sperm number incubated is high (as in most *in vitro* studies). In any case, it can at least be strongly suggested that in the future biochemical studies more than one method will be used to determine whether "true" acrosome reactions have occurred.

Small sizes of sperm acrosomes in animals such as the mouse and human make it difficult to detect the loss or presence of the acrosomes at the light microscopic level. However, Iwamatsu and Chang (1969) used phase contrast microscopy to detect the acrosome reaction in motile mouse sperm slowed down by the addition of egg white to the medium. The present author in a limited number of observations has been able to detect the acrosome loss in weakly motile mouse sperm using interference contrast microscopy. In addition, the absence of the acrosome in motile rabbit sperm can be detected by phase contrast microscopy (Overstreet and Cooper, manuscript in preparation). Improved interference and phase contrast techniques and vital staining techniques should be developed to encourage the use of these species with relatively small acrosomes in acrosome reaction studies. There are also animals, such as guinea pig, hamster, bull and ram, in which sperm acrosomes are large enough to be detected easily in motile sperm at the light microscope level. However, most *in vitro* acrosome reaction studies have been done with the guinea pig and hamster because of the lack of systems for inducing capacitation and the acrosome reaction *in vitro* for the ram, bull and other sperm with large acrosomes. The development of new incubation systems for such sperm should be encouraged. The use of sperm, other than hamster and guinea pig, for biochemical studies of the acrosome reaction is important because comparison between animals is essential to determine the "universality" of any suggested mechanisms.

THE QUESTION OF THE "PHYSIOLOGICAL" SITE OF THE ACROSOME REACTION

Bedford (1970) has pointed out that the morphology of the "true" acrosome reaction has been seen on rare occasions in testicular, epididymal, ejaculate and uterine sperm. Obviously the presence of an acrosome reaction in the male reproductive tract has no bearing on our attempt to understand the site of the "physiological" acrosome reaction of the fertilizing sperm. Notwithstanding the fact that uterine fluid can induce a true acrosome reaction in bovine sperm (Wooding, 1975), the uterus is a site which cannot be considered seriously as the site for acrosome reaction induction in the fertilizing sperm. Bedford (1970) has discussed the evidence which shows that rabbit sperm recovered from rabbit uterus 15 hours after insemination do not show a "true" acrosome reaction. It would seem, in any event, that the uterus would be too early a site for the acrosome reaction of the fertilizing sperm. Such an early acrosome reaction might allow sperm enzymes such as hyaluronidase, to leave the sperm head earlier than needed for fertilization. Thus, these enzymes would be wasted. Also any polypeptide enzyme inhibitors present in the uterus could gain access to enzymes remaining bound to the sperm after the acrosome reaction. If, for example, acrosin is essential for hydrolytic penetration of the zona pellucida as suggested by the results of Stambaugh *et al.* (1969), then uterine acrosin inhibitors (Fritz *et al.*,1975a) might bind to acrosin and block fertilization at a later time in the oviduct (see also review by Hartree, 1977).

If we then assume that the "physiological" site of the acrosome reaction is somewhere in the oviduct, we can then ask whether that oviductal site is within the lumen of a particular oviductal segment or perhaps, more specifically, while sperm are very near or within an egg investment. In most mammals, the usual, but not sole, site of fertilization is the ampulla of the oviduct (Blandau, 1973).

Yanagimachi and Noda (1972) and Bedford (1972) have discussed the difficulties in identifying and studying the "fertilizing" sperm in *in vitro* and *in vivo* studies. Those authors suggested that most of the supernumerary acrosome-reacted sperm closely associated with the egg and its investments, as observed in both the *in vitro* and *in vivo* studies, were probably capable of fertilization. Therefore, it is not unreasonable to accept the conclusions of *in vitro* experiments with hamster sperm (Yanagimachi and Noda, 1970b), and *in vivo* experiments with guinea pig (Yanagimachi and Mahi, 1976) and rabbit (Bedford, 1968, 1972), that for most sperm the acrosome reaction begins prior to or early during sperm penetration through the cumulus matrix.

Release of acrosomal hyaluronidase is probably involved in sperm penetration of the cumulus oophorus and, thus, should occur close to the cumulus during the acrosome reaction (which presumably should also occur close to the cumulus). Such a role for hyaluronidase is supported by the report that antibodies to rabbit hyaluronidase inhibit *in vitro* rabbit sperm penetration of the cumulus (Dunbar *et al.*,1976), and the observation that guinea pig sperm release their hyaluronidase activity *in vitro* in synchrony with the acrosome reaction (Talbot and Franklin 1974a; Rogers and Yanagimachi, 1975a). However, Talbot and Franklin (1974a,b) have also reported that there was no synchrony of hyaluronidase release and the acrosome reaction in hamster sperm *in vitro*.

Further complications in accepting the cumulus as a "physiological" site for acrosome reactions arise from several older observations which suggest that in certain ungulates, the cumulus oophorus may be "sparse or absent" prior to *in vivo* fertilization (see Blandau, 1961, for literature). In addition, Bedford (1968) has observed that at least a few rabbit sperm appeared to penetrate the cumulus *in vivo* without their having undergone an obvious acrosome reaction. Moore and Bedford (1977) have reported preliminary evidence which demonstrates that cumulus-free hamster eggs are fertilized *in vivo* and Harper (1970) and Overstreet and Bedford (1974) have reported that cumulus-free rabbit eggs can be fertilized *in vivo*. Moreover, Gwatkin *et al.* (1976) have reported *in vitro* results which suggested to them that the acrosome reaction of the "fertilizing" hamster sperm must occur on the zona pellucida (see section 7 for further discussion). However, Austin and Bishop (1958) reported the occurrence of acrosome reactions of motile sperm in the oviductal fluid of several naturally mated rodents. Overstreet and Cooper (manuscript in preparation) have observed the absence of acrosomes in motile rabbit sperm flushed from the isthmus of the rabbit oviduct (in the absence of the products of ovulation). Yanagimachi and Mahi (1976) observed very few such sperm flushed from the "distal" guinea pig oviduct prior to ovulation, but did observe many more flushed from the "proximal" portion of the oviduct after ovulation.

The reports of oviductal acrosome reactions occurring in sperm not near the cumulus lead to the question of whether or not the acrosin of sperm with such "early" acrosome reactions would be inhibited and thus unable to aid in zona penetration. It is, therefore, of interest that Stambaugh *et al.* (1974) have reported that rhesus monkey oviductal polypeptide inhibitors of acrosin are lowest in concentration around the time of ovulation. Thus, the acrosin of these sperm which had undergone acrosome reactions around the time of ovulation (whether in close proximity to oviductal ovulation products or not) might be less readily inhibited than at other times.

In summary, at the present time the only valid general statement that can be made about the "physiological" site of the acrosome reaction of the "fertilizing sperm" is that it is somewhere in the oviduct. Even this statement is complicated by the possibility that there appears to be more than one possible "physiological" site for the acrosome reaction.

IN VITRO HAMSTER AND GUINEA PIG SPERM ACROSOME REACTION INDUCING SYSTEMS

Since most biochemical studies of the acrosome reaction have been done with cauda epididymal hamster and guinea pig sperm, it seems appropriate to discuss the conditions used for incubation of guinea pig and hamster sperm during these studies. One aspect of these conditions, the metabolites required, will be discussed in a later section.

SYNCHRONOUS AND NON-SYNCHRONOUS SYSTEMS

Most biochemical studies of the hamster (e.g. Meizel and Lui, 1976) and guinea pig acrosome reaction (e.g. Rogers *et al.*, 1977) were carried out in non-synchronized systems where the percentage of acrosome reactions continued to increase for about 1 to 1.5 hours after detection of initial reactions. Yanagimachi and Usui (1974) have been able to develop a more synchronous system for the induction of the guinea pig acrosome reaction. In their method, sperm were capacitated by pre-incubation for 10 hours in a Ca^{2+} free medium containing serum albumin, and then 40-80% of the sperm underwent acrosome reactions 10 minutes after the addition of Ca^{2+}. Until recently, such a system did not exist for hamster sperm. However, Lui and Meizel (manuscript in prepation) have developed a capacitation system for hamster sperm which also permits relatively synchronous induction of the acrosome reaction.

"DEFINED" AND "UNDEFINED" INCUBATION SYSTEMS

Guinea pig sperm have been capacitated and acrosome reactions induced in media consisting of buffered salt solutions in which all of the added salts and metabolites are known, and which either did or did not contain serum albumin (Yanagimachi, 1972; Barros *et al.*,1973; Rogers and Yanagimachi, 1975b; Yanagimachi and Usui, 1974; Noda and Yanagimachi, 1976). In most biochemical experiments, cauda epididymal guinea pig sperm were washed in order to reduce contamination by epididymal fluid molecules.

Biochemical studies of hamster sperm acrosome reaction induction have been carried out in bicarbonate buffered Tyrode's solution mixed with bovine serum albumin or other protein and a dialyzable or protein-free "motility factor". The motility factor was derived from a number of sources including follicular fluid (Yanagimachi, 1970b; Lui *et al.*, 1977), blood (Yanagimachi, 1970c; Morton and Bavister, 1974), adrenal glands (Bavister *et al.*,1976), and sperm (Bavister and Yanagimachi, 1977). Since the composition of the motility factor is not completely known (see section 8 for further discussion), one can say that hamster systems are less defined than the guinea pig incubation system.

It should also be pointed out that only recently have washed hamster sperm been used for biochemical studies of the acrosome reaction (Lui and Meizel, 1977; Lui and Meizel, manuscript in preparation). Such hamster sperm washed free of epididymal fluid by a method similar to that of Bavister and Yanagimachi (1977), and incubated with bovine serum albumin and a protein-free motility factor from bovine adrenals underwent "true" acrosome reactions (Friend and Meizel, unpublished electron microscopic observations), and were able to fertilize hamster eggs (Cornett, Bavister and Meizel, work in progress). In future biochemical sudies of the acrosome reaction in various species, washed sperm should be used in order to reduce the massive contamination of epididymal fluid or seminal plasma molecules.

Although it is true that some guinea pig incubation systems are more defined than those of the hamster, the former are not really completely defined systems. In the first place, the presence of albumin in the media in some guinea pig experiments (or hamster experiments) actually detracts from the defined nature of the medium. Albumin is known to bind many ligands (Peters, 1975). Lui and Meizel (1977) have shown that a high fatty acid content in serum albumin can inhibit acrosome reaction induction by albumin. Although we do not yet know whether serum albumin ligands play a positive role in the acrosome reaction, we cannot be sure that only serum albumin (and not its ligand content) was important in the *in vitro* induction of guinea pig or hamster acrosome reactions (see also chapter by Pratt in this volume). In addition, even the crystalline serum albumin used in many experiments contains serum protein contaminants, but it is true that these contaminants do not appear to be involved in the acrosome reaction (Lui *et al.*,1977). The fact remains, however, that the use of serum albumin does not ensure a defined system. Finally, not all samples of serum albumin will induce acrosome reactions.

Another difficulty that exists in all acrosome reaction studies, including those with guinea pig sperm, is that a certain percentage of sperm die during the incubations. Although under some conditions as many as 70% of the sperm can remain alive after 4 hours of incubation, in many incubation systems no more than 50% are alive after 4 hours of incubation. When sperm die, they can release proteins and small molecules into the incubation medium. There is no evidence that such molecules could play a role in stimulating or inhibiting capacitation or the acrosome reaction, but such a possibility must at least be considered when evaluating biochemical influences on the acrosome reaction (particularly when drawing conclusions based on the use of "defined" incubation media).

OTHER INCUBATION FACTORS WHICH CAN INFLUENCE THE ACROSOME REACTION

The pH of an incubation can affect the number of acrosome reactions occurring in hamster sperm (Mahi and Yanagimachi, 1973). Incubations of guinea pig sperm even in bicarbonate media are usually done in air. The pH of such incubations can vary during the experiment by at least 0.4 pH units (e.g. Rogers and Yanagimachi, 1975b). Some biochemical factors might change the pH even more under certain conditions. Although a number of hamster sperm studies have also been done in air using a bicarbonate medium, many of the recent studies have been done in CO_2 incubators with a mixture of 95% air and 5% CO_2. This gas mixture keeps the incubation pH constant ±0.1 pH unit (e.g. 7.5-7.6 in 25mM bicarbonate).

In this regard, it is of interest that Mahi and Yanagimachi (1973) reported that a pH range of 7.2-7.8 was maximal for the hamster sperm *in vitro* acrosome reaction induced by detoxified blood serum. Bavister (1969) obtained the highest amount of hamster egg fertilization *in vitro* at pH 7.6-7.8. It is also of interest that in a preliminary report, Maas *et al.* (1977) observed an increase in pH of the rhesus monkey oviductal lumen to pH 7.5-7.8 at the time of ovulation.

Another variable which must be considered is that the sperm number used by different laboratories can vary by as much as an order of magnitude. It is known that the sperm number can influence the number of hamster sperm capacitated and stimulated to undergo acrosome reactions (Talbot *et al.*, 1974). Perhaps a particular ratio of sperm and initiator molecules are required for optimal results.

The osmotic pressure of the medium and the incubation temperature have been shown to influence hamster sperm capacitation and acrosome reaction induction (Mahi and Yanagimachi, 1973). Most studies are done at 37°C. Bavister (1974) has observed that an incubation medium with osmotic pressure 20% lower than 290 mosmol results in leakage of hamster sperm factors which then act to stimulate sperm motility. However, most experiments with guinea pig and hamster are done with solutions 280-310 mosmol (this range does not affect the results).

SOURCES OF REPRODUCTIVE TRACT MOLECULE(S) WHICH MIGHT BE THE INITIATOR(S) OF THE ACROSOME REACTION IN VITRO

In this section I will discuss *in vivo* and *in vitro* evidence which is related to whether or not various female reproductive tract cells, cell products and fluids provide the molecule(s) which initiate the acrosome reaction of the "fertilizing" sperm under normal *in vivo* conditions.

The word "trigger" has often been used when discussing initiation of the acrosome reaction in invertebrate sperm by the "fertilizin" of egg investments, and initiation of the mammalian sperm acrosome reaction by exogenous factors. However, the term "initiator" is more appropriate when referring to the former than the latter. The use of the word trigger could be taken to imply that a mammalian sperm acrosome reaction occurs within a second as is the case in sea urchin sperm (Tegner and Epel, 1973). Although the mammalian sperm acrosome reaction might prove to be instantaneous *in vivo*, evidence from *in vitro* studies have suggested that the acrosome reaction takes 3-6 minutes in guinea pig sperm (Yanagimachi and Usui, 1974) and 10-20 minutes in the hamster sperm (Talbot and Franklin, 1976; Meizel and Lui, unpublished).

The "physiological" initiator might turn out to be a complex molecule such as a specific protein or a simpler molecule such as a particular ion. It may also be that the concentration of an initiator at a particular site in the oviduct or *in vitro* in a particular egg investment could be as important as the chemical nature of the initiator. It is also important to consider the possibility that more than one type of initiator molecule may be required.

OVIDUCTAL FLUID AS A SOURCE OF INITIATOR MOLECULE(S)

Barros and Austin (1967) reported that acrosome reactions could be induced in motile hamster sperm *in vitro* by incubation of the sperm with oviductal fluid from a unilaterally ovariectomized hamster. Iwamatsu and Chang (1972) have observed acrosome reactions in motile hamster sperm induced by rabbit oviductal fluid. Mukherjee and Leppes (1972) reported that human postovulatory oviductal fluid stimulated the acrosome reaction (assayed by stained sperm smears) of human, rat and mouse sperm, but mention the fact that such a fluid is partially derived from follicular fluid.

However, as has been discussed above, motile sperm have been observed to undergo acrosome reactions in the oviduct in the absence of the products of ovulation. In this regard, it is of interest that serum albumin is present in rabbit and human oviductal fluid in a concentration as high as that in the serum (Moghissi, 1970; Beier, 1974) and, as will be discussed, serum albumin is involved in acrosome reaction induction *in vitro*. In addition, Ca^{2+} (essential for the *in vitro* acrosome reaction) and pyruvate and lactate (metabolites required for optimal *in vitro* acrosome reaction induction) are present in relatively high concentrations in the oviduct (see Rogers and Yanagimachi, 1975b; 1976).

FOLLICULAR FLUID AS A SOURCE OF INITIATOR MOLECULE(S)

Follicular fluid proteins are nearly all derived from blood serum (Edwards, 1974). Yanagimachi (1969) has shown that untreated follicular fluid from hamster, rat, and mouse (but not rabbit) would induce capacitation and the acrosome reaction of hamster sperm. Yanagimachi (1970b,c) has also shown that bovine follicular fluid and blood serum of several animals could induce the acrosome reaction in the hamster sperm if the fluid and blood were first detoxified by heating to 56°C. Similar results were obtained (Yanagimachi, 1970c) if sperm were incubated with a heat-stable, dialyzable factor from blood together with certain samples of serum albumin.

Bavister and Morton (1974) have also shown that certain samples of serum albumin, but not all, could induce capacitation and the acrosome reaction when incubated with sperm and a low molecular weight motility factor from blood (similar to Yanagimachi dialyzable factor, but in this case isolated by gel chromatography). Bavister and Morton (1974) also reported that the apparent molecular weights of the proteins in blood which induced the highest percentage of acrosome reactions and activated sperm were greater than 80,000, but that certain samples of serum albumin (65,000 molecular weight) could replace the blood proteins. All of these *in vitro* experiments certainly suggested that serum albumin was at least one of the proteins in blood responsible for inducing the acrosome reaction *in vitro*.

Since it is also known that serum albumin is present in high concentrations in follicular fluid (Edwards, 1974), it was not surprising that Lui *et al.*(1977) demonstrated that serum albumin was the bovine follicular fluid protein involved in inducing the acrosome reaction in hamster sperm incubated with a protein-free motility factor from follicular fluid. Mukherjee and Leppes (1972) used stained sperm smears to determine that human follicular fluid stimulated the acrosome reaction of rat, mouse and human sperm. Bedford (1969) reported preliminary results showing that the loss of acrosomes from capacitated rabbit sperm could be stimulated by untreated rabbit follicular fluid, but stated that it was not clear whether these reactions were "true" acrosome reactions.

Oliphant *et al.*(1977) used a stained sperm smear technique and electron microscopy in studying the acrosome reactions induced by rabbit, cat, pig, human and bovine follicular fluid in precapacitated rabbit sperm. The acrosome reaction inducing activity of bovine follicular fluid was destroyed by proteases, and serum albumin failed to replace follicular fluid in acrosome reaction induction experiments. Therefore, Oliphant *et al.* (1977) concluded that the acrosome reaction inducing factor of follicular fluid was a protein, but not albumin. They also reported that the acrosome reaction inducing activity of human, pig and cow was reduced by at least 50% when heated to 56°C (the effect of heating cat and rabbit follicular fluid was not tested). Yanagimachi (1969) observed that unheated rabbit bovine follicular fluid killed hamster sperm unless first heated at 56°C (Yanagimachi, 1970b). Thus, the results observed by Oliphant *et al.*(1977) may have been due, at least in part, to acrosomal damage, perhaps by the complement system (which is inactivated at 56°C). Some "true" acrosome reactions must have occurred because Oliphant (1976) reported that 16 of 41 denuded rabbit eggs were fertilized by rabbit sperm which had been precapacitated and then incubated with unheated bovine follicular fluid (see also section 8).

Although follicular fluid can induce acrosome reactions *in vitro*, rabbit eggs free of cumulus oophorus and corona radiata could be fertilized *in vivo* in the absence of the products of ovulation (including follicular fluid) (Harper, 1970; Overstreet and Bedford, 1974), and similar results have been obtained with hamster eggs (Moore and Bedford, 1977).

In summary, there can be no doubt that follicular fluid can induce a "true" acrosome reaction *in vitro*, but *in vivo* studies certainly show that follicular fluid molecules are not "essential" for the acrosome reaction. However, still unanswered are the following questions concerning those animals in which the fluid is normally present in the oviduct after ovulation: Does a follicular fluid molecule normally play an initiator role in such animals? Does the cumulus, as hypothesized by Lui and Meizel (1977), bind follicular fluid serum albumin, thus presenting sperm at or near the cumulus in the oviduct with a high local concentration of that acrosome reaction-inducing protein?

THE CUMULUS OOPHORUS AND/OR CORONA RADIATA AS SOURCES OF INITIATOR MOLECULE(S)

Austin and Bishop (1958) observed acrosome reactions in sperm within the cumulus of eggs of several naturally mated rodents. Gwatkin (1976) has reviewed the evidence from his laboratory which suggests that cumulus cells are involved in capacitation, but in those studies effects on acrosome reactions were not investigated. Gwatkin and Anderson (1973) reported that certain glycosidase inhibitors could inhibit hamster sperm capacitation by cumulus cells. Gwatkin *et al.* (1972) reported that both the cumulus cells and their intercellular matrix were required for hamster sperm capacitation *in vitro*. Gwatkin *et al.* reported that the matrix component was "at least partially inorganic". In view of recent work by Yanagimachi and Usui (1974) showing the importance of Ca^{2+} to the acrosome reaction, one wonders whether the matrix component of Gwatkin *et al.* was Ca^{2+}. Gwatkin *et al.* also found that washing the cumulus cells removed an essential component which was regenerated during reincubation of cells (see section 8).

Yanagimachi (1969) reported that hamster cumulus washed free of follicular fluid could not capacitate hamster sperm nor induce acrosome reactions. Perhaps the contradictory results of Yanagimachi and Gwatkin *et al.* (1972) were due to some important factor being washed out of the cumulus in Yanagimachi's experiments, or perhaps the difference in incubation buffer composition in some way accounts for the results.

Soupart and Morgenstern (1973) and Soupart and Strong (1974) have reported that gonadotropin-stimulated corona cells were apparently involved in the *in vitro* capacitation of human sperm. Penetration of zona pellucida only occurred in those eggs surrounded by cumulus or at least corona cells, and exogenous gonadotropins increased the incidence of zona penetration (Soupart and Morgenstern, 1973). Acrosome reactions occurred in the presence of the corona cells but no comparison was made of the number occurring in the presence and absence of exogenous hormones.

Austin *et al.*(1973) and Bavister (1973) described *in vitro* experiments with hamster sperm which definitely suggested that the cumulus induced acrosome reactions in hamster sperm. Austin *et al.*(1973) also treated hamster cumulus and eggs with trypsin for hours until the zonae were digested and the cumulus were largely dispersed. Hamster sperm still underwent acrosome reactions and fertilized the eggs when incubated with trypsin-treated cumulus-egg. Austin *et al.* concluded from those various results that a cumulus protein was probably not involved in acrosome reaction induction, but that those results would not eliminate the possibility of polypeptide hydrolysis fragments or protein-bound molecules, such as steroids being involved in the acrosome reaction. (See section 8 for a discussion of experiments relevant to these points).

A number of studies have demonstrated that fertilization and thus presumably the acrosome reaction, occurs in several animals *in vitro* and *in vivo* in the absence of the cumulus oophorus (see Bedford and Cooper, 1978, for literature and section 5 of this chapter for further discussion). Bedford and Cooper point out that these experiments prove that the cumulus is not "essential" for the induction of the acrosome reaction. Although that conclusion is certainly valid, acrosome reactions might have occurred in those *in vitro* studies during sperm preincubation (as was the case for human sperm, described by Yanagimachi *et al.*,1976). Therefore, although the studies reviewed by Bedford and Cooper (1978) prove that the cumulus is not essential, they do not prove that the cumulus cannot be a source of initiator when present *in vitro*. Moreover, the *in vivo* studies do not really settle the question of whether or not the cumulus plays an *in vivo* role in the induction of the acrosome reaction in those animals, such as the hamster, guinea pig, rabbit and human, where the cumulus is normally present during fertilization (see also section 7).

ZONA PELLUCIDA OR OOCYTE AS SOURCES OF INITIATOR MOLECULES(S)

Overstreet and Bedford (1975) reported that treatment of rabbit eggs with neuraminidase, trypsin, chymotrypsin or antibody to progesterone failed to inhibit the occurrence of acrosome reactions in capacitated sperm incubated with these eggs. Austin *et al.* (1973) dissolved the zonae and dispersed the cumulus of hamster eggs with trypsin and were able to observe acrosome reactions in hamster sperm after addition to the treated eggs. Although such biochemical treatments of egg investments must be attempted, negative results (such as those with rabbit and hamster) do not prove that the zona is not the source of an initiator since the hypothetical zona molecule may be resistant to the particular treatments used. Also, even if the proteases partially hydrolyzed a zona molecule which was involved in the acrosome reaction, the fragments might still be effective (see section 8).

Wolf (1977) reported that the penetration of zona-free mouse eggs by precapacitated mouse sperm was inhibited by trypsin inhibitors. Nicosia *et al.* (1977) suggested that such results might indicate a role for an egg cortical granule trypsin-like enzyme in the acrosome reaction. However, other explanations are possible (as discussed in section 8).

Gwatkin *et al.* (1976) have reported results which infer the existence of a zona factor for the acrosome reaction. In their experiments, cumulus-free hamster eggs added to sperm preincubated with dispersed cumulus cells were penetrated prior to the time when most "free" sperm had undergone an acrosome reaction. In addition, these authors reported that the "fertilizing" sperm appears to undergo the acrosome reaction while associated with isolated zonae, and suggested that penetration of the isolated zonae did not occur because of the absence of an oocyte factor. Gwatkin *et al.* concluded that the acrosome reaction of the "fertilizing" sperm occurs on the surface of the zona; that the acrosomal cap does not vesiculate "extensively"; and also that the cap is not immediately lost because it is required for binding to the zona. Although the scanning electron micrographs used to support those conclusions may not be able to resolve vesicles if present, they do show many acrosomal caps bound to the zona. Unfortunately, it is not possible to be certain that the acrosome reactions which lead to the loss of these caps occurred while on the zona. The fact that preincubated sperm penetrated eggs prior to a high percentage of acrosome loss, might have been due to the high sperm numbers used. The low percentages of acrosome reactions which were detected in free sperm during the time of optimal fertilization would still result in several hundred acrosome reacted sperm per egg. Thus, the fertilizing sperm may have undergone acrosome reactions while unbound. It is difficult to judge the validity of the conclusion concerning the infertility of the free sperm which underwent acrosome reactions later in the incubation because no estimates of motility or activation were given. Although the results of Gwatkin are certainly provocative, further studies will be necessary to confirm all the conclusions drawn by those authors.

SOMATIC CELLS (OTHER THAN CUMULUS CELLS) PRESENT IN THE REPRODUCTIVE TRACT AS POSSIBLE SOURCES OF INITIATOR MOLECULE(S)

There is no evidence which demonstrates that cells, other than those of the cumulus, might be involved in the acrosome reaction. However, Barros (1974) has reviewed evidence in support of or against the possibility that contact of sperm with uterine or oviductal cells is necessary for *in vivo* capacitation. In addition, Soupart (1970) has suggested that uterine leukocytes may be involved in capacitation of rabbit sperm, and Ericsson (1969) has reported *in vitro* capacitation of rabbit sperm by mule eosinophils.

THE INFLUENCE OF VARIOUS MOLECULES ON THE OCCURRENCE OF ACROSOME REACTION

This section will discuss several exogenous molecules which were tested *in vitro;* those molecules within sperm which evidence suggests may play a role *in vitro;* certain exogenous and endogenous molecules which appear to influence capacitation *in vitro,* but which need to be investigated in relationship to the acrosome reaction.

One important question which must be kept in mind when these experiments are evaluated is whether a molecule, which is shown to be effective, stimulated capacitation, the acrosome reaction or both. It is obvious that a factor which increases the number of capacitated sperm may also increase the number of acrosome reacted sperm, provided that an acrosome reaction stimulus is present. In such studies in our laboratory, we always eventually try to determine whether a factor was influencing capacitation or, more directly, the acrosome reaction. This distinction is made by utilizing: activation as a probable marker for hamster sperm capacitation; synchronous capacitation systems so that many acrosome reactions can be induced by potential initiators within a few minutes in already capacitated sperm; time course studies of potential stimulators or inhibitors so that it may be determined if these compounds affect sperm at a time too early to be during the acrosome reaction or too late to be during capacitation.

Another question which should be asked in evaluating the overall results is whether there is any interrelationship between all the molecules which can influence the acrosome reaction (see section 9). A final question which is perhaps most important of all is whether any of these effective *in vitro* factors can play a role in the *in vivo* acrosome reaction. The questions raised above will be considered throughout the remainder of this chapter.

PROTEINS EXOGENOUS TO SPERM

Serum Albumin

We have previously mentioned the fact that not all samples of serum albumin are capable of inducing high percentages of acrosome reaction (see section 7). What is the explanation for this? Johnson (1975) has suggested that the fatty acid binding ability of serum albumin might prove to be important in acrosome reaction induction by that protein. Experiments by Lui and Meizel (1977) and Lui *et al.* (1977) have shown that the fatty acid content of serum albumin does greatly influence its ability to induce acrosome reactions. When fatty acids are removed from serum albumin, the serum albumin becomes 5-10 fold better in inducing acrosome reactions, and when fatty acids are added back to the albumin molecule at high levels the albumin loses that acrosome reaction inducing ability acquired when fatty acids were removed (see also Pratt, this volume).

The removal of all fatty acids from serum albumin was carried out by treatment of serum albumin with charcoal at acid pH. Removal of two-thirds of the original fatty acid content could be carried out at the same time as a more complete purification of serum albumin by precipitation of the protein with trichloroacetic acid and resolubilization in ethanol. Both the fatty acid-free and low fatty acid serum albumin were excellent inducers of acrosome reactions. Moreover, such a more highly purified albumin (immunoelectrophoretically homogeneous) had not been used for acrosome reaction induction until those experiments of Lui *et al.*(1977) and Lui and Meizel (1977).

Davis (1976a) had previously reported that the amount of fatty acids associated with serum albumin can influence the capacitation of unwashed rat sperm. He found that a high fatty acid content inhibited capacitation, but acrosome reactions were not studied. Davis did not attempt to add fatty acids back to albumin to determine their effect, but did add cholesterol and found that the cholesterol-saturated albumin inhibited capacitation. From these and other results (Davis, 1976b), he concluded that the removal of sperm membrane cholesterol might be an important event in capacitation which could lead to membrane labilization. However, in his experiments, sperm motility was also inhibited by the cholesterol-saturated albumin solutions used. Thus, fertilization (his measure of capacitation) might have been inhibited, at least in part, because of poor motility.

As yet, we have no evidence to prove how the presence of fatty acids on serum albumin influences the ability of serum albumin to induce acrosome reactions in hamster sperm. This point is discussed to some extent by Lui and Meizel (1977). It should also be mentioned that most of these experiments reported by Lui and Meizel were carried out with the sperm incubated in the presence of epididymal fluid, but fatty acid-containing albumin was also a poor inducer of acrosome reactions compared to fatty acid-free albumin when incubated with washed hamster sperm.

My laboratory has also recently attempted to answer two questions concerning the role of serum albumin in acrosome reaction induction: is the effect of albumin on capacitation, the acrosome reaction directly, or both;and is the intact albumin molecule necessary for its effect on acrosome reaction induction in hamster sperm? Results reported by Bavister (1974) have strongly suggested that serum albumin does play a role in capacitation of hamster sperm. Bavister was able to show that hamster sperm preincubated with serum albumin could then undergo acrosome reactions when incubated with the products of ovulation.

Lui and Meizel (1977) have shown that if hamster sperm were incubated for 4½ hours with protein-free motility factor(s) and a sample of serum albumin, known to be a poor inducer of acrosome reactions, only 13-20% acrosome reactions occurred. However, an additional 25% of the sperm underwent acrosome reactions within 15 minutes after addition (at 4½ hours) of a sample of serum albumin which had been previously shown to be capable of inducing a high percentage of acrosome reactions. These results suggest that serum albumin also plays a more direct role in the acrosome reaction induction of hamster sperm.

Lui and Meizel (1977) have hydrolyzed bovine serum albumin with pepsin so that the remaining large and small molecular weight polypeptide fragments did not give an agar immunodiffusion plate precipitation line with antibodies to the bovine serum albumin. The fraction of hydrolyzed albumin preparation containing fragments of 30,000 and above still induced acrosome reactions in hamster sperm. However, in these experiments the incubation medium had to include an intact serum albumin not capable of inducing a high percentage of acrosome reactions, as well as a motility factor. It would appear that the intact molecule is not necessary for acrosome reaction induction, but may be necessary for capacitation and/or sperm viability. It must be pointed out, however, that these experiments of Lui and Meizel (1977) do not prove or disprove a requirement in the acrosome reaction for albumin ligands or long amino acid sequences in the large remaining polypeptide fragments of the pepsin hydrolysate.

Blank *et al.*(1976) have claimed that bovine serum albumin adsorbs to about one half of the rabbit sperm surface and have postulated that this adsorption would result in physical changes in the membrane which are important for the acrosome reaction. However, the polarographic measurements used for the detection of adsorption were not (as those authors claimed) made with sperm which had been incubated in conditions approximating *in vitro* rabbit sperm capacitation.

If serum albumin is involved in the *in vivo* acrosome reaction, the particular ligands (such as fatty acids) carried by it could influence the protein's acrosome reaction inducing activity (see section 9). In addition, Lui *et al.*(1977) have shown that the concentration of the follicular fluid serum albumin also influences the protein's effectiveness in acrosome reaction induction. Follicular fluid serum albumin bound to the cumulus might provide such a high, local concentration (see also section 7).

Exogenous proteins other than serum albumin

Bavister and Morton (1974) reported that proteins of between 80,000 and 180,000 apparent molecular weight were the most effective of the human blood serum proteins in inducing acrosome reactions of hamster sperm when incubated with the sperm and a protein-free motility stimulating serum factor. Cohn fraction V bovine serum albumin and Cohn fraction V fatty acid-free human serum albumin were able to replace the serum proteins in the induction of acrosome reactions in that system, but crystalline bovine serum albumin, crystalline chicken ovalbumin, Cohn fraction V bovine α-globulin and Cohn fraction III bovine β-globulin were unable to stimulate acrosome reactions. All of these protein preparations were able to maintain sperm viability in the presence of the motility stimulating fraction, except β-globulin which killed the hamster sperm.

Reports that various enzymes, such as glucuronidase and trypsin, can capacitate sperm *in vitro* have been reviewed by Gwatkin (1976). However, Yanagimachi (1975) reported that the following exogenous enzymes did not induce the acrosome reaction and activation in guinea pig sperm: trypsin, α-chymotrypsin, papain, bromelain, neuraminidase, β-glucuronidase, lipase, phospholipase C, α-amylase, β-amylase, cellulase, α-glucosidase and β-glucosidase, β-galactosidase, amyloglucosidase and cytochrome C. If enzymes such as glucuronidase and trypsin can capacitate sperm as reviewed by Gwatkin (1976), why then did they not increase acrosome reactions in the experiments of Yanagimachi? The answer, of course, is that although sperm must first be capacitated in order to undergo acrosome reactions, agents which can capacitate do not necessarily also have the ability to initiate the acrosome reaction.

Yanagimachi (1975) did observe, however, that pronase induced activation of all sperm and that occasionally 1-15% of activated sperm showed an acrosome reaction if examined within one hour after the start of incubation. However, it should be noted that the pronase preparation was not completely purified (see Calbiochem catalogue). He also reported that unheated blood sera of rabbit and guinea pig could induce the acrosome reaction and activation at low concentrations (too high a concentration of unheated sera killed the sperm), and that sera treated at 56°C did not kill the sperm but would not induce the acrosome reaction and activation unless complement from rabbit or guinea pig was added to the heat treated sera. The result obtained with complement is difficult to explain because it is known that treatment at 56°C inactivates the complement system, and Yanagimachi (1970c) had previously reported that hamster sperm were killed by unheated blood sera. Perhaps certain levels of complement such as those in diluted blood sera can induce acrosome reactions without killing the sperm as long as a sufficient concentration of serum albumin is present (as in the heat inactivated sera) to help maintain sperm viability.

Meizel and Lui (unpublished) have observed that human β-lactoglobulin (known to be a fatty acid binding protein; Peters, 1975) was capable of inducing acrosome reactions in hamster sperm (serum albumin and motility factor had to be present to maintain the sperm viability). I have already discussed the fact that serum albumin fatty acid content influences that protein's acrosome reaction inducing activity. The fact that addition of fatty acids to serum albumin inhibited its acrosome reaction inducing activity (see earlier) taken in combination with the β-lactoglobulin result does suggest that a possible role for serum albumin in the acrosome reaction might be in binding fatty acids (this point will be discussed later in the chapter - and see also Pratt, this volume).

THE EFFECT OF IONS (EXTERNAL AND INTERNAL)

Calcium ions

Although the requirement for calcium in the invertebrate acrosome reaction has long been known (Dan, 1956), it was not until the work of Yanagimachi and Usui (1974) that calcium was shown to be essential for the acrosome reactions of mammalian sperm *in vitro*. Yanagimachi and Usui found that the presence of calcium ions in the medium was necessary for the acrosome reaction and for activation of guinea pig sperm. In their experiments, acrosome reactions could be initiated in many sperm within a few minutes when calcium was added to guinea pig sperm which had been previously preincubated in a calcium-free medium containing albumin. It is of interest that although capacitation must have occurred during the preincubation,

Ca^{2+} was still required for the expression of the activated motility characteristic of capacitated guinea pig and hamster sperm. Experiments by Summers *et al.* (1976) and Talbot *et al.* (1976) utilized the calcium ionophore A23187 to demonstrate a role for calcium in the guinea pig and hamster acrosome reaction. The ionophore A23187 increases the sperm membrane permeability to calcium, and thus the uptake of Ca^{2+} appears to be necessary for the acrosome reaction. The results of those experiments with ionophores suggest that the acrosome reaction did not occur in the presence of the ionophore until capacitation had been achieved and that the ionophore might have reduced the time required for capacitation. A nearly synchronous hamster sperm system (Lui and Meizel, manuscript in preparation; see later) has also been developed in which most acrosome reactions are not initiated until the calcium ionophore A23187 was added. Somewhat more direct evidence for Ca^{2+} uptake during capacitation has been obtained by Babcock and co-workers (manuscript in preparation) who have observed a phase of ^{45}Ca uptake by guinea pig sperm which "parallels or slightly precedes the time course of the acrosome reaction". Those results did not include intracellular localization of the Ca^{2+} uptake.

Yanagimachi and Usui (1974) suggested several possible mechanisms in which Ca^{2+} could have a role in the acrosome reaction. For example, Ca^{2+} might neutralize negative charges on membranes so that they would come together to fuse. Another possible mechanism suggested was that Ca^{2+} might bind to membrane phospholipids and by somehow inducing changes in permeability, would increase water influx into the acrosome. This water influx would swell the acrosome and bring membranes closer together in order to fuse.

Poste and Allison (1973) suggested that a general mechanism for membrane fusions included the "close approximation of the membranes". Yanagimachi and Usui (1974) may be correct in suggesting that the swelling of the acrosomal cap observed by them brings the outer acrosomal membrane closer to the plasma membrane, but the mechanism for the remainder of the acrosome reaction (including the fusion initiation step) is still not obvious.

Yanagimachi and Usui (1974) and Meizel and Lui (1976) have suggested that Ca^{2+} might influence the acrosome reaction through its effect on sperm enzymes. These suggestions will be discussed further later in this section.

Friend (1977) has used a cytochemical method to detect specific localizations of calcium ions in the acrosomes of uncapacitated guinea pig sperm. Calcium appeared to be highly localized in an apical site and near-apical site within the sperm head. Friend also suggested that the acrosome reaction vesiculation stage appears to begin at the apex of the sperm head. Yanagimachi and Usui (1974) and Bedford (1974) have also suggested that the acrosome reaction appears to begin in the anterior part of the sperm head.

Other divalent cations

Rogers and Yanagimachi (1976) reported that Mg^{2+} could inhibit the calcium initiated acrosome reaction in guinea pig sperm *in vitro*. They also suggested that it is possible that the relative amounts of Mg^{2+} and Ca^{2+} in the male and female reproductive tracts may play a role in controlling the occurrence of the acrosome reaction *in vivo*. Although it is true that there is more Ca^{2+} than Mg^{2+} in the oviductal fluid of several mammals, the Mg^{2+} to Ca^{2+} ratios vary quite a bit in the male reproductive tract (as reviewed by Rogers and Yanagimachi). However, it is still possible that such Mg^{2+} could play a role in controlling the acrosome reaction as suggested by those authors.

Yanagimachi and Usui (1974) have shown that Sr^{2+}, but not Mg^{2+}, could replace Ca^{2+} in the medium as a stimulus for the induction of the acrosome reaction. As Johnson (1975) has discussed, such results could mean that the contraction of microfilaments is involved in the acrosome reaction. However, Johnson (1975) also reported preliminary studies which showed that an inhibitor of microfilament contraction cytochalasin B did not inhibit the effect of Ca^{2+} on guinea pig sperm acrosome reactions.

Meizel and Lui (1976) have reported that the presence of Zn^{2+} in the medium could inhibit the acrosome reaction of hamster sperm induced by bovine follicular fluid and that addition of extra Ca^{2+} to the medium could overcome that inhibition. The possible mechanisms involved will be discussed in section 9.

Monovalent cations

A preliminary report by Schackmann *et al.*(1977) and unpublished experiments discussed by Collins and Epel (1977) indicate that H^+ is released and Ca^{2+} taken up during the sea urchin sperm acrosome reaction. Schackmann *et al.* also studied the effects of several ionophores and concluded that the sea urchin acrosome reaction does not necessarily occur when Ca^{2+} uptake and H^+ release are detected, and that K^+ flux may be an important part of the acrosome reaction. Tilney (1977) has reviewed evidence from his laboratory which indicates that a rise in acrosomal pH is essential for the polymerization of actin during development of the acrosomal process, an event characteristic of the echinoderm sperm acrosome reaction. Although similar experiments with mammalian sperm have not yet been reported, two other reports are relevant to the sea urchin results. Meizel and Deamer (1978) have shown that the pH within the hamster sperm acrosome is pH 5 or less (see later). Rosado *et al.*(1977) reported that K^+ was very high in human sperm head fractions assayed by atomic absorption spectroscopy. Also, Toyoda and Chang (1974) have reported that rat sperm capacitation was increased by preincubation of the sperm in high K^+/Na^+ ratio (the effect on acrosome reactions was not investigated).

SPERM ENZYMES WHICH MAY BE INVOLVED IN THE ACROSOME REACTION

Evidence for the involvement of sperm trypsin-like enzyme, probably acrosin, in the acrosome reaction

The idea that trypsin-like enzymes may have a role in membrane fusion events has been discussed by Schuel *et al.*(1973). Gordon (1973) has hypothesized a role for sperm proteases such as acrosin in the vesiculation part of the acrosome reaction. Yanagimachi and Usui (1974) have even suggested that possibly the acrosome reaction is not a true membrane fusion, but is a "local" disintegration of membranes which may involve activation of acrosomal enzymes by water and Ca^{2+} influx.

However, the first evidence to suggest the involvement of a trypsin-like enzyme in the acrosome reaction was reported by Meizel and Lui (1976). They found that the low molecular weight (< 1,000) synthetic trypsin inhibitors TLCK, NPGB, benzamidine, and p-aminobenzamidine all inhibited the acrosome reactions of hamster sperm without appreciably inhibiting sperm activation in an incubation system containing bovine follicular fluid.

However, the larger molecular weight polypeptide trypsin inhibitors (> 20,000) from soybean, lima bean, and egg white did not inhibit acrosome reactions even at concentrations of 1 mg/ml. Zinc ions known to inhibit several enzymes, including acrosin, inhibited the acrosome reaction, and Ca^{2+} known to stimulate a number of enzymes, including acrosin, overcame the inhibition due to Zn^{2+}. The low molecular weight synthetic chymotrypsin inhibitor TPCK did not inhibit the acrosome reaction. The absence of inhibition of activation by the synthetic trypsin inhibitors suggested that capacitation was not inhibited in these experiments. It was suggested by Meizel and Lui (1976) that the effective inhibitors inhibited proacrosin activation and/or acrosin activity. It is important to mention here that in my laboratory's studies of the effect of trypsin inhibitors on the acrosome reaction, each inhibitor had its own narrow concentration range at which it was effective (Meizel and Lui, unpublished experiments). Too low a concentration would not inhibit and too high a concentration inhibited activation and/or motility. Furthermore, we have also observed that the effective concentrations may be modified by the incubation conditions used (washed sperm, follicular fluid, etc.).

More recent inhibitor studies by Lui and Meizel (manuscript in preparation) have utilized washed sperm incubated with purified serum albumin and a protein-free ultrafiltrate containing hamster adrenal motility factor(s). The synthetic inhibitors NPGB, TLCK and benzamidine inhibited acrosome reactions under these conditions, but soybean trypsin inhibitor did not. For the sake of argument, it can be suggested that the effective trypsin inhibitors could be inhibiting some non-trypsin-like enzyme. It is true that, for example, TLCK can inhibit papain (Whitaker and Perez-Villasenor, 1968). However, TPCK which also inhibits papain (Whitaker and Perez-Villasenor, 1968) did not inhibit acrosome reactions (Meizel and Lui, 1976). In addition, the effective inhibitors all function by different mechanisms(Walsh, 1970): TLCK alkylates histidine residues on the active site; NPGB titrates the serine groups at the active site; and benzamidine acts as a competitive inhibitor. Thus, it is highly unlikely that a nontrypsin-like enzyme present in sperm would be inhibited by all three of these trypsin inhibitors.

Studies were also undertaken in my laboratory (Lui and Meizel, manuscript in preparation) to determine whether or not the trypsin-like activity involved in the acrosome reaction influenced capacitation or more directly the acrosome reaction. For these experiments, a relatively synchronous incubation system was used. In this system, washed sperm were incubated for several hours with a protein-free ultrafiltrate containing hamster adrenal gland motility factor(s) and a serum albumin which is capable of inducing only a low percentage of acrosome reactions in that time. After incubation for the several hour period, ionophore A23187 was added and as many as 50% of the sperm underwent acrosome reactions within 10-15 minutes. However, when the trypsin inhibitors TLCK, NPGB, or benzamidine were added a few minutes prior to the calcium ionophore, most of the acrosome reactions which would have been induced by the ionophore did not occur. Soybean inhibitor (1 mg/ml) did not inhibit acrosome reactions in this system even if added one hour prior to ionophore. In addition, the low molecular weight (< 700) inhibitor of thermolysin-like enzymes, phosphoramidon (Aoyagi and Umezawa, 1975) did not inhibit acrosome reactions (even at 300 mg/ml). The results obtained with TLCK, NPGR and benzamidine strongly suggest that it is the acrosome reaction rather than earlier capacitation steps which is being influenced by the trypsin-like activity, but the possibility that sperm trypsin-like activity can also influence some aspect of capacitation cannot be ruled out by these experiments.

Another question which can be asked about the inhibitor results is whether the trypsin-like enzyme being inhibited is present within the sperm or is exogenous to the sperm. In a preliminary report, Green (1976) observed that induction of the guinea pig acrosome reaction *in vitro* by the calcium ionophore (A23187) was "delayed" by a synthetic trypsin inhibitor, but not by a protein trypsin inhibitor (the inhibitors were not further identified). The calcium ionophore, of course, increases membrane permeability to Ca^{2+}, and since Ca^{2+} is essential for the acrosome reaction Green concluded that the inhibition of trypsin activity must have been internal. The results of Lui and Meizel with a synchronous hamster sperm system (described above) can be interpreted in a similar manner. Moreover, it is certainly unlikely that all three effective synthetic trypsin inhibitors tested by Meizel and Lui blocked Ca^{2+} transport.

In addition, polypeptide inhibitors do not inhibit the acrosome reaction in the hamster studies of Meizel and Lui, and this may be due to the fact that such large inhibitors cannot enter the sperm. Wendt *et al.* (1975) have reported that polypeptide inhibitors and TLCK do not enter boar sperm, but that NPGB does. It is of interest that Zahler and Polakoski (1977) have shown that the presence of benzamidine in the washing medium prevents the activation of bovine sperm proacrosin to acrosin during sperm washing. This result suggests that at least under the conditions used by Zahler and Polakoski, benzamidine entered the sperm. Even with such strong indirect evidence, it should be pointed out that my laboratory has not yet shown directly that the synthetic acrosin inhibitors enter the acrosome in our experiments. Future experiments will attempt to obtain such direct proof.

The results of my laboratory have not completely eliminated the possibility that in addition to an internal trypsin-like enzyme, an enzyme on the sperm surface plays a role (e.g. an enzyme released from dying sperm or originating in the reproductive tract and tightly bound to sperm). A report by Wolf (1977) and a preliminary report by Talbot (1977) must be considered in regard to this possibility of exogenous trypsin-like activity being involved in the acrosome reaction. Talbot (1977, and personal communication) has observed results which suggest to her that a trypsin-like enzyme might be acting on the sperm surface during capacitation of guinea pig sperm *in vitro*. Her experiments involved a demonstration that guinea pig sperm agglutination by soybean agglutinin increases during capacitation, and that benzamidine and soybean trypsin inhibitor inhibited the agglutination while incubation of sperm with trypsin increased the agglutination. Talbot has also found that the two inhibitors also inhibit the acrosome reaction (personal communication). In Talbot's experiments, if soybean inhibitor does inhibit capacitation (as suggested by the inhibition of agglutination, a possible marker for capacitation), it is not surprising that acrosome reactions were ultimately inhibited.

Wolf (1977) has reported that soybean or lima bean trypsin inhibitors and p-aminobenzamidine inhibited the penetration of zona-free mouse ova by precapacitated sperm. If acrosome reactions had occurred in a low, but significant, percentage of the several hundred thousand sperm per ml during precapacitation, there might have been enough sperm to penetrate, and even the large polypeptide inhibitors could have inhibited acrosin in such acrosome-reacted sperm. Alternatively, acrosome reactions might have occurred in the penetrating sperm solely because of extra time and media used for incubation of the precapacitated sperm with eggs. In such a case, the presence of eggs might not have been necessary, and once again polypeptide inhibitors could inhibit acrosin. If either of these alternatives were true, the data would then support one of the two hypotheses suggested by Wolf to explain his results, namely, that acrosin has some role in sperm-egg fusion. Until those alternatives are ruled out, there is no support for the suggestion by Nicosia *et al.* (1977) that a trypsin-like enzyme from mouse cortical granules (some of which break down prior to penetration) might be responsible for the induction of the acrosome reaction in these precapacitated mouse sperm.

If we accept the evidence which suggests that an internal sperm trypsin-like enzyme is involved in the acrosome reaction, we can next discuss the question of whether that enzyme is acrosin. All the inhibitors used by Meizel and Lui to block hamster sperm acrosome reaction, also inhibit hamster acrosin (Meizel and Mukerji, 1976). Up to the present time, the only trypsin-like enzyme which has been demonstrated to be present in mammalian sperm is acrosin. Of course, one cannot yet rule out the possibility that some other trypsin-like enzyme is present within sperm in very low amounts.

In summary, there is strong evidence that a sperm trypsin-like enzyme plays some role in the *in vitro* hamster sperm acrosome reaction and that the enzyme is probably acrosin. The possibility that trypsin-like activity is also involved in some aspect of capacitation in guinea pig or hamster cannot be ruled out at this time. The possibility that some trypsin-like enzyme other than acrosin is involved in capacitation or the acrosome reaction must await demonstration of such an enzyme within sperm or on the sperm surface.

Recent freeze fracture studies of the guinea pig acrosome reaction induced by Ca^{2+} (Friend *et al.*,1977) show that certain membrane particles disappear from sites where fusion will occur. It is interesting to speculate that those results might have been due to the hydrolysis of the particle protein by acrosin.

Green (1976) has suggested that hydrolysis of the acrosomal matrix by acrosin might increase osmotic pressure causing the intake of water, which then "bursts the cell". However, this mechanism seems to be a rather severe one and unlikely to account for an organized progressive membrane fusion and vesiculation. However, water influx due to osmotic pressure is a possible explanation of the swelling and water intake observed during the guinea pig acrosome reaction by Yanagimachi and Usui (1974). If this were true, then the osmotic effects could be responsible for bringing the outer acrosomal membrane closer to the overlying plasma membrane, thus helping to promote membrane fusion.

Before we can be certain of the details of acrosin's role in the acrosome reaction, we must be certain of the intracellular site(s) of proacrosin and acrosin localization. Biochemical studies strongly suggest that all or nearly all acrosin in epididymal and ejaculated sperm is present as the zymogen proacrosin (Meizel and Mukerji, 1975, 1976; Zahler and Polakoski, 1977), and that the active acrosin is bound to the inner acrosomal membrane (Brown and Hartree, 1974; Fritz *et al.*,1975b; Hartree, 1977). However, Stambaugh *et al.*(1975) have reported light microscopic cytochemical evidence which they interpret as demonstrating that all acrosin is localized on the outer acrosomal membrane of rabbit sperm, and Stambaugh and Smith (1976) have used similar techniques and claim to have demonstrated the presence of acrosin within the sperm penetration slit in the zona pellucida. This latter result would suggest that acrosin is secreted during fertilization. Although the results of Stambaugh and his co-workers could also be interpreted to mean that at some point in time there is some acrosin present on the outer acrosomal membrane, it must be pointed out that their experiments did not include any proof of the nature of the protease activity they detected. The significance of proacrosin and acrosin localization will be pointed out when possible acrosome reaction mechanisms are discussed in section 9.

Whatever the role of acrosin in the acrosome reaction, acrosin must be produced by activation of proacrosin. What is the mechanism of proacrosin activation, and what are its controls? Meizel and Mukerji (1975, 1976) have suggested that acrosomal pH and/or ionic concentrations may play a role in the control of proacrosin autoactivation or activation by another enzyme.

Meizel and Lui (1976) suggested that the inhibition of the hamster acrosome reaction by Zn^{2+} might be related to the ability of Zn^{2+} to inhibit the autoactivation of partially purified proacrosin and/or acrosin activity (Meizel and Mukerji, 1976). Mukerji and Meizel (manuscript in preparation) have recently observed that the autoactivation of completely purified rabbit proacrosin is inhibited by Zn^{2+} and that Ca^{2+} is required to obtain maximal activity during autoactivation. In this regard, it is of interest that Zn^{2+} was particularly high in the membranes of human sperm studied by x-ray fluorescent spectrometry combined with high resolution electron microscopy (Rosada *et al.*,1974, 1977). Mann (1964) has reviewed evidence demonstrating the presence of a high concentration of Zn^{2+} in the mammalian male reproductive system. Therefore, it is possible that both internal and external Zn^{2+} may play a role in controlling the acrosome reaction.

Meizel and Deamer (1978) have shown that the pH of the hamster acrosome is pH 5 or less. Mukerji and Meizel (manuscript in preparation) have observed that there is no appreciable autoactivation of proacrosin below pH 6. Meizel and Deamer suggested that the acrosomal pH is one of the controls of proacrosin autoactivation and the acrosome reaction, and that this pH must therefore change prior to the acrosome reaction, possibly during capacitation. It is also possible that the hypothetical pH change could occur as one of the first events of the acrosome reaction.

McRorie *et al.*(1976) have reported results which suggested to them that a sperm thermolysin-like enzyme which they called acrolysin is involved in proacrosin activation within the sperm. However, this enzyme has not yet been even partially purified for characterization nor has its intracellular location been established. Thus the relationship of pH, the activity of acrolysin and proacrosin activation is not known. However,

Zn^{2+} inhibits and Ca^{2+} does stimulate thermolysin-like enzymes (McRorie *et al.*,1977).

Future studies will be necessary to establish whether proacrosin is activated by another enzyme or by autoactivation during capacitation prior to the acrosome reaction, and whether the pH of the acrosome does change during capacitation and is related to proacrosin activation.

Evidence for the involvement of a sperm phospholipase in the acrosome reaction

Allison and Hartree (1970) reported the presence of phospholipase A activity in preparations of ram sperm acrosomes. These acrosomal preparations may, however, also have included sperm head plasma membrane enzymes and full characterization of the phospholipase was not carried out. The preparation method fragmented and removed the outer acrosomal membrane and overlying plasma membrane and could solubilize enzymes from these areas as well as from the inner acrosomal membrane and acrosomal matrix. Washed hamster sperm also contain an, as yet, uncharacterized phospholipase activity (Meizel, unpublished experiments).

The question of whether phospholipase and the products of its hydrolytic activity are involved in membrane fusions has been discussed without agreement by Poste and Allison (1973), Lucy (1975) and Bach (1974). Conway and Metz (1971) reported preliminary evidence which showed that lysolecithin, a product of phospholipid hydrolysis by phospholipase, could induce the acrosome reaction in sea urchin sperm. More recently, Conway and Metz (1976) have reported on the time course of the appearance of phospholipase activity during the acrosome reaction of sea urchin sperm and have concluded that the phospholipase would probably be involved in sperm-egg fusion rather than the acrosome reaction.

The local anaesthetic dibucaine induced acrosome reactions and activation in guinea pig sperm (Yanagimachi, 1975). Although such drugs are known to induce membrane fusion in enzyme-free liposomes (Poste and Allison, 1973), it should be noted that under certain conditions such anaesthetics can also stimulate phospholipase A activity (Scherphof and Westenberg, 1975). Clegg *et al.*(1975) have reported that boar sperm do not contain a phospholipase capable of attacking choline phosphatides, but also reported that incubation of the sperm in the uterus and oviduct of the boar resulted in alterations in sperm phospholipids (probably membrane constituents).

Evidence which supports the possibility of a role for phospholipase in the mammalian sperm acrosome reaction comes from the recent experiments of Lui and Meizel (manuscript in preparation) with a nearly synchronous acrosome reaction system. The results of these experiments showed that *in vitro* hamster sperm acrosome reactions could be inhibited by two low molecular weight (< 1000), synthetic phospholipase inhibitors, p-bromophenacylbromide (Volwerk *et al.*,1974) and mepacrine (Flower and Blackwell, 1976).

For these experiments, washed hamster sperm were incubated for several hours in Tyrode's solution together with a crystalline bovine serum albumin which was a poor inducer of acrosome reactions and a protein-free ultrafiltrate of hamster adrenal. When Ca^{2+}-ionophore (A23187) was added to the sperm after several hours of incubation, as many as 50% of the sperm underwent acrosome reactions within 10-15 minutes. However, if either one of the phospholipase inhibitors was added a few minutes prior to the ionophore, acrosome reactions were greatly inhibited.

Mepacrine can also inhibit prostaglandin synthetase and certain TPCK inhibitable cathepsins. It is not known whether such enzymes are present in hamster sperm, and so it is important to note that the prostaglandin synthetase inhibitor indomethacin and TPCK did not inhibit acrosome reactions in our experiments. In addition, neither of the two phospholipase inhibitors inhibited hamster acrosin in biochemical assays. All of the inhibitor results are indirect evidence, but they at least suggest the possibility that sperm phospholipase plays a role in the acrosome reaction.

Future studies should include biochemical characterization and immuno-chemical localization of sperm phospholipase(s) and assay of the enzyme(s) during *in vitro* capacitation and acrosome reaction induction. Possible roles of phospholipase in the acrosome reaction will be discussed in section 9.

Is there a role for sperm ATPase('s) in the acrosome reaction?

There has been a great deal of speculation concerning the possibility of a role for ATPase(s) in the acrosome reaction. As discussed earlier, Yanagimachi and Usui (1974) proved that extracellular Ca^{2+} was essential to the acrosome reaction.

One of several possible mechanisms for the Ca^{2+} effect which they suggested was the inhibition of a sperm ATPase present in the plasma membrane and on the outer acrosomal membrane. The inhibition of this enzyme would reduce the energy available in the form of ATP, which is presumably used by sperm to keep water out of the acrosome. When the ATPase is inhibited by Ca^{2+}, water enters the acrosome causing swelling which would lead to a closer association of the membranes and result in fusion. In support of this hypothesis, Yanagimachi and Usui (1974) also mentioned their unpublished electron microscopic detection of a Mg^{2+}-dependent, Ca^{2+}-inhibitable ATPase on the outer acrosomal membrane and plasma membrane of guinea pig sperm.

Gordon (1973) has suggested that a plasma membrane ATPase not influenced by Ca^{2+} transports external Ca^{2+} into the area between the plasma membrane and the outer acrosomal membrane. However, in Gordon's hypothesis, this Ca^{2+} would stimulate an outer acrosomal membrane ATPase to transport Ca^{2+} into the acrosome. The Ca^{2+} influx would then stimulate acrosomal proteinases to digest areas of fused membrane, thus producing vesiculation. In her hypothesis, Gordon also suggested that fusion is induced by the Ca^{2+} mediated binding of plasma membrane phospholipids to outer acrosomal membrane phospholipids. An alternative hypothesis suggested by Gordon was that external Ca^{2+} influx might act, in part, through displacement of internal bound Ca^{2+}. This last point is of interest because Friend (1977) has suggested that the acrosome reaction is initiated at an apical site on the sperm head which contains a localized Ca^{2+} concentration.

Gordon based her theory on the ultrastructural detection of a Ca^{2+}-dependent ATPase on the guinea pig, human and rabbit sperm outer acrosomal membrane (Gordon and Barnett, 1967; Gordon, 1973) and an ATPase not requiring Ca^{2+} on the periacrosomal part of the plasma membrane of rabbit and human sperm (Gordon, 1973). The different pH's used by Yanagimachi and Usui (pH 7.4) and by Gordon (pH 9) for incubation of the substrate with sperm might explain the difference in the kind of ATPase detected in the acrosomal membrane. The concentrations of ions used for such studies could also be significant in explaining contradictory results. For example, Gonzales and Meizel (1973a,b) have carried out partial biochemical characterization of a sperm head enzyme which is apparently an ATPase and observed that 100 mM Mg^{2+}, but only 5 mM Zn^{2+} was required for stimulation. Obviously, a more complete biochemical and ultrastructural characterization of sperm head ATPases is needed in order to study their possible role.

Lindahl *et al.* (1972) have described *in vitro* and *in vivo* results which led them to hypothesize that an ATP-dependent event occurring in the sperm plasma membrane overlying the acrosome may be important in fertilization. However, as Lindahl *et al.* point out, other explanations of their results are also possible. There is, as yet, no direct evidence which demonstrates that inhibition of any sperm head ATPase will inhibit the acrosome reaction. Rogers *et al.* (1977) have reported inhibition of the guinea pig acrosome reaction by oligomycin, an inhibitor of oxidative-phosphorylation and of certain ATPases. However, these authors concluded that their other results indicated an inhibition of oxidative phosphorylation was responsible for the inhibition of acrosome reactions. Obviously, evidence for the involvement of an ATPase in the acrosome reaction would have to include inhibition by more than one ATPase inhibitor.

Gordon (1973) and Gordon and Dandekar (1977) have reported that a sperm periacrosomal plasma membrane ATPase was detectable in epididymal and capacitated, but not ejaculated, sperm. It was suggested by the latter authors that removal of a seminal plasma inhibitor of this enzyme may be part of capacitation.

Indirect support for the possible importance of an ATPase can be derived from the report of Rogers and Morton (1973) that total ATP levels decrease in the hamster sperm during *in vitro* capacitation. However, they mention that such results could also be due to activation of adenyl cyclase during capacitation.

In summary, there is much indirect, but no direct, evidence to support a role for sperm ATPase in the acrosome reaction. Future experiments should include attempts to biochemically characterize sperm head ATPase activities and to assay them at various times during *in vitro* capacitation and acrosome reaction induction.

Sperm adenyl cyclase and cyclic AMP

At the present time, there is no direct evidence to prove that adenyl cyclase and its product, cyclic AMP, play a role in the acrosome reaction. However, there are several studies which suggest that increases in these molecules might be involved in capacitation (see review by Hoskins and Casillas, 1975, for literature). Unfortunately, in those studies attempts were not made to distinguish between an effect on capacitation and one on the acrosome reaction.

Hoskins and Casillas (1975) also reviewed the convincing evidence which indicates that increased cyclic AMP is associated with increased sperm respiration and motility. It would be interesting to determine whether the activation of hamster and guinea pig sperm motility seen during *in vitro* capacitation is due to increased sperm cyclic AMP.

The suggestion that cyclic AMP could be important to the "development of the acrosome reaction" has been made by Huacuja *et al.* (1977). They based that suggestion on their observations of cyclic AMP stimulation of the phosphorylation of human sperm (membrane?) proteins. Obviously, further experiments are needed to determine whether increased sperm cyclic AMP will result directly in increased acrosome reactions. Such experiments should include the effect of exogenous cyclic AMP or its analogues and the effect of phosphodiesterase inhibitors. However, since others have suggested that cyclic AMP plays a role in capacitation, it will be particularly important to distinguish between stimulation of capacitation, and direct stimulation of the acrosome reaction.

"MOTILITY FACTOR(S)" AND THEIR IMPORTANCE IN THE HAMSTER SPERM ACROSOME REACTION

As mentioned earlier, the *in vitro* induction of the hamster sperm acrosome reaction requires the presence of "motility factors" found in a number of tissues and fluids. Bavister *et al.* (1976) have reported that hamster adrenal glands contained a particularly high concentration of motility factor(s). Cornett and Meizel (1977) have shown in a preliminary report that the motility factor is also present in bovine adrenal glands.

It must be emphasized here again that the motility factor not only seems to stimulate motility but also appears to be of importance in obtaining optimal acrosome reactions (Yanagimachi, 1970b,c; Bavister and Morton, 1974; Lui *et al.*,1977). In addition, as Bavister and Yanagimachi (1977) have pointed out, the development of fertilizing ability of hamster sperm *in vitro* also requires the combined action of the motility factor, a protein (e.g. serum albumin) and appropriate energy sources. It is not really known whether the motility factor in all these tissues and cells is identical, and we do not know how many essential components are present in the motility factor. Evidence to be discussed below suggests that more than one factor is involved at least in bovine adrenal motility factor preparations.

Bavister (1975) and Bavister and Yanagimachi (1977) have reported that the motility factor from human blood serum and the sperm of several species is heat stable and has an apparent molecular weight (as determined by Sephadex gel chromatography) of approximately 200. Lui *et al.*(1977) used Millipore PSAC ultrafiltration to show that the motility factor(s) of bovine follicular fluid is less than 1,000 molecular weight (a PSAC membrane does not allow molecules of greater than 1,000 to pass through it).

A preliminary report by Cornett and Meizel (1977) described the bovine adrenal gland motility factor(s). The protein-free PSAC membrane ultrafiltrate from bovine medulla was 2 to 3 times more effective than that of the adrenal cortex in inducing hamster sperm activation and acrosome reactions in the presence of serum albumin. This suggested that catecholamines were involved in the motility factor effect. This view was supported by the fact that two adrenergic antagonists, the alpha blocker phentolalmine and the beta blocker propranolol (Innes and Nickerson, 1975) inhibited the acrosome reaction induced by serum albumin and the bovine adrenal motility factor. In another experiment, the bovine adrenal cortex ultrafiltrate was purified further by extracting with chloroform to remove most lipids (including most steroids) and then by incubation with alumina, a procedure known to remove most catecholamines. The purified cortex preparations did support a low percentage of acrosome reactions (no higher than 20% for 4 hours) but did not support sperm activation. However, the incubation of sperm with serum albumin and purified cortex preparation plus norepinephrine or epinephrine gave much higher percentages of acrosome reactions and activation. The catecholamines did not support acrosome reactions in the absence of cortex preparation. The alpha agonist phenylephrine or the beta agonist isoproterenol (Innes and Nickerson, 1975) were able to substitute for epinephrine in incubations with cortex and serum albumin. A combination of both agonists gave even better results than incubation with one agonist.

In all these experiments, it was necessary to use high unphysiological concentrations of catecholamines (e.g. 10^{-5}M). However, it may be that in future experiments, lower physiological concentrations will prove effective if appropriate conditions are utilized. We do not yet know whether the catecholamines directly affect the hamster acrosome reaction, capacitation or both. It does appear that both alpha and beta adrenergic receptor sites are involved in this effect. The identity of the cortex factor which seems to be necessary in combination with the catecholamines is not yet known. Cornett, Bavister and Meizel (work in progress) have observed that incubation of washed sperm with catecholamines plus the cortex fraction and serum albumin increases the percentage of hamster fertilization obtained *in vitro*.

In future studies, it will be important to determine whether or not the motility factor(s) in sperm (Bavister and Yanagimachi, 1977) also includes a catecholamine and a factor similar to the cortex factor. Of course, identification of both the specific catecholamine (if present in the sperm) and the cortex factor will also be important in future experimental considerations. It would also be of interest to know the concentrations of various catecholamines in the fluid of the oviductal lumen prior to, during, and after ovulation.

Whether the catecholamine effect is on the acrosome reaction and/or capacitation, it is likely that the beta adrenergic mechanism involves stimulation of sperm adenylcyclase activity. Stimulation of cyclic AMP production is a common occurrence in somatic cell beta adrenergic effects (see review by Haber and Wrenn, 1976). Somatic cell alpha adrenergic responses are more difficult to categorize, but there is some evidence that in certain cases Ca^{2+} influx is increased during alpha adrenergic stimulation (see Assimacopoulos-Jeannet *et al.*, for literature).

THE EFFECT OF ATP PRODUCING METABOLITES ON THE ACROSOME REACTION

The ATP producing substrates present in the incubation media used for acrosome reaction studies were not discussed in section 6, but these substrates do influence the occurrence of acrosome reactions. Rogers and Yanagimachi (1975b) have shown that acrosome reactions of washed guinea pig sperm obtained with pyruvate and lactate as substrates in the incubation medium were delayed when glucose was also added. They also found that the guinea pig sperm underwent acrosome reactions much more rapidly when lactate and pyruvate were the substrates present compared to incubations in which glucose was the sole energy source.

However, Bavister and Yanagimachi (1977) showed that in the presence of serum albumin and a motility factor from sperm, the combination of lactate, pyruvate and glucose gave the highest percentage of hamster sperm activation and acrosome reactions compared to pyruvate and lactate or glucose alone. Earlier results, in which Tyrode's solution containing only glucose, supported high percentages of hamster acrosome reactions could be explained by the fact that pyruvate and lactate were undoubtedly present in the motility factors used (Bavister and Yanagimachi, 1977).

In the experiments of Rogers and Yanagimachi (1975b), the metabolizable sugars, fructose and mannose, also inhibited the acrosome reaction in a manner similar to that of glucose, but sucrose, fucose, lactose, L-glucose and galactose (sugars not metabolized to any great extent by sperm) did not inhibit the acrosome reaction.

Miyamoto and Chang (1973) reported that the addition of pyruvate in the mouse sperm incubation system increased the percentage of acrosome reactions and of fertilization. Rogers and Yanagimachi (1975b) have reviewed evidence which suggests that the concentration of lactic acid and pyruvic acid in the mammalian oviduct is high. Tsunoda and Chang (1975) demonstrated that the addition of pyruvate stimulated *in vitro* fertilization of rat eggs but that pyruvate addition was not necessary if rat eggs were still surrounded by follicular fluid cells. This suggested to Tsunoda and Chang that follicular cells surrounding the eggs were a source of pyruvate. Perhaps pyruvate was the capacitating factor which could be removed from cumulus cells and then regenerated in the experiments of Gwatkin *et al.*(1972) described in section 7.

It is not yet clear why certain metabolizable sugars such as glucose inhibit the acrosome reaction while lactate and pyruvate stimulate the reaction. Presumably, the answer to this question will involve sperm ATP production in some way.

Rogers *et al.*(1977) have shown that the respiratory inhibitors oligomycin, antimycin A and rotenone reduced the percentage of acrosome reactions occurring in hamster sperm without affecting motility in a pretreated human serum incubation system. They concluded that the inhibition by respiratory inhibitors implies a relationship between respiration, oxidative phosphorylation and capacitation in the acrosome reactions. At first, this result seems incompatible with that of Rogers and Morton (1974), which indicated that total hamster sperm levels are reduced during hamster capacitation. Perhaps this problem can be settled if future experiments can determine the ATP levels of the sperm head during capacitation (rather than that of the entire sperm). The ATP production by mitochondria may increase, but the ATP may be utilized in the head by ATPase and adenyl cyclase.

In summary, certain metabolites (pyruvate and lactate) appear to play an important role in obtaining optimal number of acrosome reactions *in vitro*. These metabolites may even be important *in vivo*. However, at the present time the question of whether the metabolites effect capacitation, the acrosome reaction directly, or both, is not answered.

DO STEROIDS INFLUENCE THE ACROSOME REACTION?

Chang and Hunter (1975) have reviewed the evidence concerned with capacitation of rabbit sperm in the uterus of estrous rabbits or of rabbits treated with estrogen. It must be remembered, however, that steroid hormones *in vivo* may not necessarily affect sperm directly, but could possibly stimulate the production of other molecules which directly influence the sperm.

Barros *et al.*(1972) have reported that the maximum *in vitro* hamster sperm acrosome reaction inducing activity of female hamster blood serum was near the time of ovulation and that estrogen treatment of hamsters increased the activity of the blood while progesterone treatment depressed the acrosome reaction activity of the blood. However, Barros *et al.*did not comment on the viability of the sperm under those conditions. Also, if the acrosome is a specialized lysosome, as suggested by Allison and Hartree (1970), one would expect that progesterone, a labilizer of lysosomal membranes (Weissman, 1969), would also labilize the acrosomal membrane. Alternatively, if the sperm were highly viable in the experiments of Barros *et al.*, blood serum estrogen might have stimulated capacitation rather than directly stimulating the acrosome reaction. Rosado *et al.* (1974) have suggested that *in vitro* capacitation of rabbit sperm by human follicular fluid might have been partly through the action of estrogen.

Austin *et al.* (1973) have suggested the possibility that steroids (particularly progesterone) produced by cumulus cells might be involved in inducing the acrosome reaction. In this regard, it is of interest that Overstreet and Bedford (1975) found that antibodies to progesterone did not inhibit fertilization by rabbit sperm of rabbit eggs. Of course, it is possible that if it were protein-bound, the progesterone might not be susceptible to antibody.

Gwatkin and Williams (1970) and Briggs (1973) reported that hamster sperm capacitation *in vitro* was inhibited by incubation of sperm with unphysiological, high concentrations of several progesterones and an estradiol. Bleau *et al.* (1975) have reported that physiological concentrations of desmosteryl sulfate (a steroid sulfate apparently present in hamster sperm membranes) inhibited *in vitro* capacitation. However, none of those reports included observations made to determine the effects of the steroid compounds on acrosome reactions.

Cornett and Meizel (unpublished) have observed that high unphysiological concentrations of estrogen and progesterone will inhibit hamster sperm acrosome reactions induced *in vitro*, but in these experiments observations of the effect on capacitation were not made.

In summary, we do not yet know whether physiological concentrations of steroids can directly influence the acrosome reaction.

ACROSOMAL ACTIN

Actin-like filaments have been detected in echinoderm sperm acrosomes (see review by Tilney, 1977) and apparently play a role in the extension of the elongated acrosomal process through the egg investments. Talbot and Kleve (1977) have reported preliminary immunofluorescence data which suggest the presence of an actin-like protein in the hamster sperm acrosome. However, the extension of an acrosomal process does not occur in the mammalian sperm acrosome reaction. Johnson (1975) had suggested that if actin-like proteins were present as microfilaments between the acrosomal and plasma membranes, the contraction of such proteins might bring the membranes closer together, but he also found that an inhibitor of microfilament contraction did not inhibit the guinea pig acrosome reaction.

THE INDUCTION OF ACROSOME REACTIONS BY DETERGENTS

Wooding (1975) has shown that incubation of bovine sperm with the detergent Hyamine, can result in acrosome reactions which appear similar to the "true acrosome reaction" at the ultrastructural level. Yanagimachi (1975) has shown that incubation of guinea pig sperm with the detergent Triton or with Hyamine can result in acrosome reactions, activation and fertilization. Yanagimachi concluded from his results that guinea pig sperm contain all of the components necessary for the occurrence of the acrosome reaction and activation by the time that they complete their epididymal maturation. However, it may be argued that the detergents stimulated some modifications in existing sperm components which also occur in the *in vivo* acrosome reaction, or that the detergents stimulated the acrosome reaction by mechanisms completely unrelated to normal events.

A WORKING HYPOTHESIS FOR THE BIOCHEMICAL EVENTS OF THE MAMMALIAN SPERM ACROSOME REACTION IN VITRO

This hypothesis is a working one and should provide a logical framework in which further logical experimental questions may be asked. Although it is hoped that more than a small fraction of the overall concept will prove correct, the results obtained in future experiments will undoubtedly modify particular aspects. Since most of my acrosome reaction studies have been with the hamster, this hypothesis is particularly designed to fit the results obtained in those experiments.

The central theme of this hypothesis involves the activation of inactive enzymes by pH and/or ionic changes within the acrosome, the involvement of the active forms of those enzymes in membrane fusion and/or vesiculation, the initiation of the process by catecholamines and the acceleration of the vesiculation by fatty acid-poor serum albumin. Several variations of the theme will be given because this hypothesis is a working one containing a number of information gaps.

SPERM COMPONENTS OF THE ACROSOME REACTION

It is suggested that the following are sperm components of the acrosome reaction: the enzyme proacrosin and its active form acrosin; a sperm phospholipase, initially present as an inactive form (either a zymogen or inhibitor-bound enzyme and activable by acrosin); one or more sperm ATPases and their product ATP; adenyl cyclase and its product cyclic AMP; acrosomal calcium, zinc and hydrogen ions.

The following have already been presented in the previous section: the fact that acrosin is present as the proenzyme (zymogen) proacrosin in sperm; evidence which strongly suggests a role for a trypsin-like enzyme, probably acrosin in the acrosome reaction; the idea that acrosomal pH and/or Ca^{2+} and Zn^{2+} might control activation of proacrosin and thus the acrosome reaction; suggestions for future experiments to elucidate the role of acrosin and the control of proacrosin activation.

The previous section also provided evidence which suggests the possibility that a sperm phospholipase is involved in the acrosome reaction and suggestions for future experiments to clarify the role of such an enzyme. Although there is no evidence, as yet, for an inactive form of sperm phospholipase, pancreatic phospholipase is known to be a zymogen activated by trypsin (Dutilh *et al.*,1975) and snake venom contains not only phospholipase but also a small polypeptide inhibitor of the enzyme (Braganca *et al.*,1970). In order to prevent general membrane destruction and random sperm head lysis, variations of this hypothesis involving phospholipase and its hydrolysis products will have to include the requirement that only a limited number of specific membrane sites are susceptible to their action.

Speculations concerning the involvement of ATPase and adenyl cyclase activities in the acrosome reaction have been presented in earlier sections, as have suggestions for future experiments to clarify their role. Other pertinent background information and suggestions for future experiments will be given throughout this section.

POSSIBLE BIOCHEMICAL EVENTS OF THE ACROSOME REACTION INVOLVING SPERM COMPONENTS (Theme and variations)

In this hypothesis, Ca^{2+} influx into the area between the plasma membrane and the outer acrosomal membrane is somehow increased. The influx of external Ca^{2+} in the area between the outer acrosomal membrane and the overlying plasma membrane might then lead to increased Ca^{2+} uptake by the acrosome due to stimulation of a Ca^{2+}-dependent ATPase (as suggested by Gordon, 1973).

One possible effect of the high calcium in the acrosome is the activation of proacrosin. This might occur in two ways: calcium ions might simply overcome the inhibition of proacrosin autoactivation or activation due to acrosomal Zn^{2+}. Calcium ions might also bring about a change in the internal pH of the acrosome by displacing hydrogen ions, thus bringing the acid pH of the acrosome to a more alkaline pH suitable for proacrosin autoactivation and acrosin activity. As mentioned earlier, the pH of the hamster sperm acrosome appears to be pH 5 or less, and Ca^{2+} influx and H^+ efflux apparently occurs during the sea urchin sperm acrosome reaction.

It is also possible that Ca^{2+} inhibits, rather than stimulates, an ATPase on the outer acrosomal membrane as suggested by Yanagimachi and Usui (1974). If that ATPase had been involved in maintaining an acid pH inside the acrosome, as has been suggested for lysosomal ATPase (Mego *et al.*,1972), then the pH gradient would run down resulting in a more alkaline acrosomal pH. Once again, it becomes obvious that future biochemical characterization and immunocytochemical localization of sperm ATPases are important.

In addition to stimulating the activation of proacrosin, Ca^{2+} would also stimulate acrosin activity (directly, as well as indirectly, by stimulating a pH change) and might also stimulate a phospholipase in the same manner. Calcium activable lysosomal phospholipases with a pH optimum as high as 7.5 do exist in some cells (Waite *et al.*,1976), and the pancreatic phospholipase produced by activation of a prophospholipase, requires Ca^{2+} and has an alkaline pH optimum of 8.7 (Dutilh *et al.*,1975).

The role of acrosin in the acrosome reaction would depend, in part, on its location in the acrosome. If acrosin plays an indirect role in the acrosome reaction, it may do so by activating another inactive enzyme so that this second enzyme would more directly affect the acrosome reaction. Based on results from my laboratory, phospholipase would be a possible choice for such an enzyme. Future experiments must establish whether or not sperm phospholipase is present as an inactive form, activable by acrosin.

If acrosin is located only on the inner acrosomal membrane, it might hydrolyze an inactive phospholipase in the soluble contents of the acrosome or an inner acrosomal membrane-bound phospholipase. In both cases, the active phospholipase would have to exert its effect on the outer acrosomal membrane. A soluble acrosomal phospholipase could hydrolyze phospholipids on the outer acrosomal membrane, and either a soluble or inner acrosomal membrane-bound phospholipase could hydrolyze an unbound phospholipid within the acrosomal contents. After any of these hypothetical enzymatic events, the lysophospholipid product of the hydrolysis might act on the outer acrosomal membrane to stimulate membrane fusion (see Lucy (1975) for a review of evidence in support of and against a role for lysophospholipids in membrane fusions).

If acrosin were present to some extent on the outer acrosomal membrane, it might still exert an indirect effect by activating a soluble acrosomal inactive phospholipase or one present on the outer acrosomal membrane. It is also possible that some acrosin might leave the inner acrosomal membrane in order to activate the hypothetical soluble or outer acrosomal membrane-bound phospholipase. If acrosin does leave the inner acrosomal membrane, it might also exert a more direct influence on the acrosome reaction by being involved in the vesiculation step (as suggested by Gordon, 1973).

Brown and Hartree (1976) have discussed the existence of an "acrosomal inhibitor" of acrosin in ram sperm which does not inhibit the membrane-bound acrosin as readily as it inhibits soluble acrosin. The presence of this inhibitor in the acrosome would make it less likely that acrosin would function in the acrosome reaction by leaving a membrane-bound site. However, the intracellular site of the inhibitor has not been conclusively established. Techniques used to remove the inhibitor would remove proteins located between the outer acrosomal membrane and plasma membrane, as well as soluble proteins of the acrosome. If the inhibitor were located in the former site, it would not interfere with acrosin released within the acrosome.

If acrosin is not involved in the activation of a phospholipase, then a soluble phospholipase or one present on the outer acrosomal membrane or plasma membrane could be activated by an increase in Ca^{2+} and/or pH or by cyclic AMP. A sperm plasma membrane phospholipase activated in such a way might conceivably be an example of an endoprotein (Rothman and Lenard, 1977) associated with the cytoplasmic side of the membrane. Thus, it would not be influenced by the environment external to the sperm, but only by internal changes in ions, pH and/or cyclic AMP.

EXTERNAL FACTORS OF THE ACROSOME REACTION

The major external factors of this hypothesis are Ca^{2+}, one or more catecholamines and serum albumin low in fatty acid content. The importance of Ca^{2+} to the acrosome reaction has already been discussed. Although endogenous Ca^{2+} may play a role, there is no doubt that, at least *in vitro*, influx of external Ca^{2+} is important. Evidence for involvement of alpha and beta-adrenergic stimulation (by catecholamines) of the *in vitro* hamster sperm acrosome reaction has been discussed in section 8. The importance of serum albumin in the hamster acrosome reaction has also been discussed in that section.

The possible involvement of the external factors in initiation of the acrosome reaction

In this hypothesis, one or more catecholamines stimulate Ca^{2+} influx by sperm. The mechanism involved is an alpha-adrenergic one. There are studies which suggest that Ca^{2+} influx mediated through alpha-adrenergic cell receptors stimulates hepatic phosphorylase (Assimacoujoulos-Jeannet *et al.*,1977) and parotid K^+ secretion (Selinger *et al.*,1974). In view of the undoubted influence of Ca^{2+} on the acrosome reaction (an influence possibly exerted through one or more of the variations suggested earlier), future studies are being planned to determine whether the alpha-adrenergic stimulation of the acrosome reaction involves increased Ca^{2+} influx.

The beta-adrenergic effect of catecholamines in this hypothesis is exerted through a stimulation of plasma membrane adenyl cyclase. This resulting increase in cyclic AMP could result in further stimulation of phospholipase activity. DeCingolani *et al.* (1972) have shown that cyclic AMP can stimulate the activity of fat cell homogenate phospholipase A. It is also possible, of course, that the alpha and beta adrenergic stimulations do not influence the acrosome reaction in a direct manner, but rather influence capacitation. Others have suggested that cyclic AMP may be involved in sperm capacitation.

Serum albumin binds many biologically active ligands, such as steroids and catecholamines (Peters, 1975). Such binding might provide a mechanism for protecting the ligands against inactivation and for presenting sperm cell receptors with localized high concentrations of the ligand(s).

Certain steroids do bind to the sperm plasma membrane *in vitro* and can influence sperm metabolism (see Hoskins and Casillas, 1975). The possibility that some serum albumin-bound ligand, such as a steroid, plays an active role in stimulating the hamster acrosome reaction, at least *in vitro* cannot be ruled out, but is still unproven. On the other hand, the possible significance to the acrosome reaction of the binding of fatty acids by albumin or other proteins is suggested by the fact that fatty acid-free and fatty acid-low serum albumin is a better inducer of acrosome reactions *in vitro* than serum albumin with a high fatty acid content, and the fact that beta-lactoglobulin, a fatty acid binding protein, will induce acrosome reactions. In a mechanism involving fatty acid binding by albumin, sperm phospholipase activity would first produce free fatty acids and lysophospholipids. One or both of these molecules might cause some membrane perturbation, which could lead to fusion of the outer acrosome membrane with the plasma membrane at specific sites. Certain fatty acids can also stimulate membrane fusions (Lucy, 1975). However, free fatty acids can inhibit phospholipase (Gatt, 1973) and it is possible that the removal of free fatty acids from a membrane may be necessary in order to promote membrane destabilization (Gul and Smith, 1972). Eventually the free fatty acids will be found outside and inside the sperm head as well as in the sperm plasma membrane. Thus, if serum albumin binds these free fatty acids, the resulting decrease of fatty acid concentration inside the sperm and/or membrane could then allow phospholipase activity to continue and/or allow membrane destabilization.

A potential interrelationship between sperm phospholipase, sperm membrane ATP levels and serum albumin is suggested by the results of Martin *et al.* (1975) and Gul and Smith (1972). Martin *et al.* reported that hydrolysis of human or guinea pig red cell membrane phospholipids by external phospholipase occurred without lysis unless red cell ATP levels were first lowered. Gul and Smith have also shown that the presence of serum albumin is necessary for the hemolysis of human red cells by phospholipase A_2. In the experiments of Gul and Smith, phospholipase hydrolyzed the membrane lipids, but there was no cell lysis in the absence of albumin, and those authors suggested that the removal of fatty acids by the albumin may have destabilized the membranes and allowed lysis. These studies of red cells involved hydrolysis by an exogenous phospholipase rather than an organized progressive membrane fusion and vesiculation. However, they do provide indirect support for the hypothesis presented in this section. It should also be remembered that ATP levels decrease during capacitation and that both ATPase activity and adenyl cyclase activity could decrease ATP.

As was mentioned earlier, this hypothesis was particularly designed to fit the results of my laboratory's studies of the hamster sperm acrosome reaction. Why are a motility factor and serum albumin essential for the hamster, but not for the guinea pig *in vitro* acrosome reaction? It is possible that the motility factor is released in high quantities from dying guinea pig sperm. This released factor might, under *in vitro* conditions, be able to add to that remaining on or in the viable sperm and thus stimulate the acrosome reaction. Bavister and Yanagimachi (1977) have reported that the guinea pig sperm have a particularly high level of motility factor compared to the hamster. Yanagimachi and Usui (1974) did include serum albumin in their guinea pig acrosome reaction studies, but proteins released from dying sperm might also be able to replace serum albumin in other guinea pig *in vitro* systems. It is also possible that some species difference in sperm membrane molecules such as fatty acids results in serum albumin being needed to destabilize hamster but not guinea pig sperm membranes after phospholipid hydrolysis.

CONCLUDING COMMENTS

In addition to the future biochemical studies of *in vitro* factors already suggested throughout this chapter, *in vivo* studies will eventually have to be carried out in order to determine whether or not the proposed hypothesis is "physiologically" valid. Such future *in vivo* studies would, of course, depend in part on results of further *in vitro* studies with the hamster and the extension of such studies to other species. In these future *in vitro* and *in vivo* experiments, the questions discussed at the beginning of section 8 must be kept in mind.

The biochemical mechanisms involved in the "physiological" *in vivo* mammalian sperm acrosome reaction will probably prove to be even more complex than those presented in the above hypothesis. In order to fully understand these molecular events, it will be necessary to continue to apply relevant knowledge obtained from the study of non-mammalian gametes and somatic cells to mammalian acrosome reaction studies and to integrate the results obtained with several techniques in future investigations of the unique aspects of mammalian gametes and mammalian fertilization. These two requirements will in turn depend on an awareness of both the similarities and differences between all three cell types and on the communication and collaboration between those of us using different research approaches.

NOTE IN PROOF

After the submission of this manuscript, the editor and Dr. D. Green were kind enough to let me see the latter's prepublication manuscript of his chapter in this book. Dr. Green's hypotheses that "acrosin inhibitors do not inhibit the acrosome reaction *per se*" but do inhibit acrosomal matrix loss and that "the loss of the matrix is the only substantial change which can unambiguously be detected by light microscopy" are certainly provocative. I look forward to seeing in published form, his detailed procedures and the detailed evidence which support his statements. Until that time and until further comparative studies are carried out, it will not be possible to resolve any conflicting views held by the two of us.

ACKNOWLEDGEMENTS

With respect to the work from my laboratory and the ideas presented in this Chapter, I wish to gratefully acknowledge the research creativity of and the intellectual stimulation by my laboratory colleagues: Dr. Sudhir Mukerji, Dr. Kevin Lui and Mr. Larry Cornett. In addition, I wish to thank the following scientists from various campuses of the University of California for valuable discussions concerning possible events of the acrosome reaction: Dr. Barry Bavister, Dr. David Deamer, Dr. Daniel Friend, Dr. James Overstreet and Dr. Prudence Talbot. However, the author bears the full responsibility for the concepts presented in this Chapter.

The author also wishes to thank the following for generously providing pre-publication information and unpublished results: Dr. Donner Babcock, Dr. George Cooper and Dr. Bennett Shapiro. The line drawing used in this Chapter was skilfully done by Ms. Celeste Warden. Finally, I wish to thank Ms. Renee Smith and Ms. Dian Francis for their secretarial skills and patience.

The work from my laboratory reported here was supported by NIH grants HD 06698 and HD 07893.

REFERENCES

ALLISON, A.C. & HARTREE, E.F. (1970) Lysosomal enzymes in the acrosome and their possible role in fertilization. J. Reprod. Fert. 21, 501-515.

AOYAGI, A. & UMEZAWA, H. (1975) Structures and activities of protease inhibitors of microbial origin. In: E. Reich, D.B. Rifkin and E. Shaw (Eds.), Proteases and Biological Control, Coldspring Harbor Laboratory, pp. 429-454.

ASSIMACOPOULOS-JEANNET, F.D., BLACKMORE, P.F. & EXTON, J.H. Studies on α-adrenergic activation of hepatic glucose output. Studies on role of calcium in α-adrenergic activation of phosphorylase. J. Biol. Chem. 252, 2662-2669.

AUSTIN, C.R. (1968) Ultrastructure of Fertilization. Holt, Rinehart and Winston, New York.

AUSTIN, C.R. (1975) Membrane fusion events in fertilization. J. Reprod. Fert. 44, 155-166.

AUSTIN, C.R. & BISHOP, M.W.H. (1958) Role of the rodent acrosome and perforatorium in fertilization. Proc. Roy. Soc. Series B. 149, 241-248.

AUSTIN, C.R., BAVISTER, B.D. & EDWARDS, R.G. (1973) Components of Capacitation. In: S.J. Segal, R. Crozier, P.A. Corfman and P.G. Condliffe (Eds.), The Regulation of Mammalian Reproduction, C.C. Thomas, Springfield, Illinois, pp. 247-254.

BACH, M.K. (1974) A molecular theory to explain the mechanisms of allergic histamine release. J. Theor. Biol. 45, 131-151.

BARROS, C. (1974) Capacitation of mammalian sperm. In: E.M. Coutinho and F. Fuchs (Eds.), Physiology and Genetics of Reproduction, Part B, Plenum Press, New York, pp. 3-24.

BARROS, C. & AUSTIN, C.R. (1967) In vitro fertilization and the sperm acrosome reaction in the hamster. J. Exptl. Zool. 166, 317-324.

BARROS, C., ARRAU, J. & HERRERA, E. (1972) Induction of the acrosome reaction of golden hamster spermatozoa with blood serum collected at different stages of the oestrous cycle. J. Reprod. Fert. 28, 67-76.

BARROS, C., BERRIOS, M. & HERRERA, E. (1973) Capacitation in vitro of guinea pig spermatozoa in a saline solution. J. Reprod. Fert. 34, 547-549.

BAVISTER, B.D. (1969) Environmental factors important for in vitro fertilization in the hamster. J. Reprod. Fert. 18, 544-545.

BAVISTER, B.D. (1973) Capacitation of golden hamster spermatozoa during incubation in culture medium. J. Reprod. Fert. 35, 161-163.

BAVISTER, B.D. (1974) The effect of variations in culture conditions on the motility of hamster spermatozoa. J. Reprod. Fert. 38, 431-440.

BAVISTER, B.D. (1975) Properties of the sperm motility stimulating component derived from human serum. J. Reprod. Fert. 43, 363-366.

BAVISTER, B.D. & MORTON, D.B. (1974) Separation of human serum components capable of inducing the acrosome reaction in hamster spermatozoa. J. Reprod. Fert. 40, 495-498.

BAVISTER, B.D. & YANAGIMACHI, R. (1977) The effects of sperm extracts and energy sources on the motility and acrosome reaction of hamster spermatozoa in vitro. Biol. Reprod. 16, 228-237.

BAVISTER, B.D., YANAGIMACHI, R. & TEICHMAN, R.J. (1976) Capacitation of hamster spermatozoa with adrenal gland extracts. Biol. Reprod. 14, 219-221.

BEDFORD, J.J. (1968) Ultrastructural changes in the sperm head during fertilization in the rabbit. Am. J. Anat. 123, 329-358.

BEDFORD, J.M. (1969) Morphological aspects of capacitation. In: G. Raspe (Ed.), Schering Symposium on Mechanisms Involved in Contraception. Adv. Biosci. Vol. 4, Pergamon Press, Oxford, pp. 35-50.

BEDFORD, J.M. (1970) Sperm capacitation and fertilization in mammals. Biol. Reprod. Suppl. 2, 128-158.

BEDFORD, J.M. (1972) An electron microscopic study of sperm penetration into the rabbit egg after natural mating. Am. J. Anat. 133, 213-254.

BEDFORD, J.M. (1974) Mechanisms involved in penetration of spermatozoa through vestments of the mammalian egg. In: E.M. Coutinho and F. Fuchs (Eds.), Physiology and Genetics of Reproduction, Part B, Plenum Press, New York, pp. 55-68.

BEDFORD, J.M. & COOPER, G.W. (1978) Membrane fusion events in the fertilization of vertebrate eggs. In: G. Poste and G.L. Nicolson (Eds.), Cell Surface Reviews, Vol. 5, Elsevier, Amsterdam. In press.

BEIER, H.M. (1974) Oviducal and uterine fluids. J. Reprod. Fert. 37, 221-237.

BLANDAU, R.J. (1961) Biology of eggs and implantation. In: W.C. Young (Ed.), Sex and Internal Secretions, Vol. II, Williams & Wilkins Co., Baltimore, pp. 797-882.

BLANDAU, R.J. (1973) Gamete transport in the female mammal. In: R.O. Greep (Ed.), Handbook of Physiology, Section 7, Part 2, Female Reproductive System, Am. Physiol. Soc., Washington, pp. 153-163.

BLANK, M., SOO, L. & BRITTEN, J.S. (1976) Absorption of albumin on rabbit sperm membranes. J. Membrane Biol. 29, 401-409.

BLEAU, G., VANDENHEUVEL, W.J.A., ANDERSEN, O.F. & GWATKIN, R.B.L. (1975) Desmosteryl sulphate of hamster spermatozoa, a potent inhibitor of capacitation in vitro. J. Reprod. Fert. 43, 175-178.

BRAGANCA, B.M., SAMBRAY, Y.M., & SAMBRAY, R.Y. (1970) Isolation of polypeptide inhibitor of phospholipase A from cobra venom. Eur. J. Biochem. 13, 410-415.

BRIGGS, M.H. (1973) Steroid hormones and the fertilizing capacity of spermatozoa. Steroids. 22, 547-553.

BROWN, C.R. & HARTREE, E.F. (1974) Distribution of a trypsin-like proteinase in the ram spermatozoon. J. Reprod. Fert. 36, 195-198.

BROWN, C.R. & HARTREE, E.F. (1976) Effects of acrosin inhibitors on the soluble and membrane bound forms of ram acrosin and a reappraisal of the role of the enzyme in fertilization. Hoppe-Seyler's. Z. Physiol. Chem. 357, 57-65.

CATYSON, C.A., VANDEZNADE, H. & GREEN, D.E. (1976) Phospholipids as ionophores. J. Biol. Chem. 251, 1326-1332.

CHANG, M.C. & HUNTER, R.H.F. (1975) Capacitation of mammalian sperm: biological and experimental aspects. In: D.W. Hamilton and R.O. Greep (Eds.), Handbook of Physiology, Section 7, Vol. 5, Male Reproductive System, Am. Physiol. Soc., 339-351.

CLEGG, E.D., MORRE, D.J. & LUNSTRA, D.D. (1975) Porcine sperm membranes: in vivo phospholipid changes, isolation and electron microscopy. In: J.G. Duckett and P.A. Racey (Eds.), The Biology of the Male Gamete, Biol. J. Linnean Soc., Vol. 7, Suppl. 1, pp. 321-335.

COLLINS, F. & EPEL, D. (1977) The role of calcium ions in the acrosome reaction of sea urchin sperm. Regulation of exocytosis. Exptl. Cell Res. 106, 211-222.

CONWAY, A.F. & METZ, C.B. (1971) A possible role of phospholipase in membrane fusion. Abstracts of the 1971 meeting. Am. Soc. Cell Biol., p.64.

CONWAY, A.F. & METZ, C.B. (1976) Phospholipase activity of sea urchin sperm: its possible involvement in membrane fusion. J. Exptl. Zool. 198, 39-48.

CORNETT, L. & MEIZEL, S. (1977) Catecholamines stimulate acrosome reactions of hamster sperm. J. Cell. Biol., Abstracts of American Soc. Cell Biol., J. Cell Biol. 75, 164a.

DAN, J.C. (1956) The acrosome reaction. Intl. Review Cytol. 5, 365-393.

DAVIS, B.K. (1976a) Influence of serum albumin on the fertilizing ability in vitro of rat spermatozoa. Proc. Soc. Exptl. Biol. Med. 151, 240-243.

DAVIS, B.K. (1976b) Inhibitory effect of synthetic phospholipid vesicles containing cholesterol on the fertilizing ability of rabbit spermatozoa. Proc. Soc. Exptl. Biol. Med. 152, 257-261.

DECINGOLANI, G.E., VAN DEN BOSCH, H. & VAN DEENEN, L.L.M. (1972) Phospholipase A and lysophospholipase activities in isolated fat cells: effect of cyclic 3', 5' -AMP. Biochim. Biophys. Acta. 260, 387-392.

DUNBAR, B.S., MUNOZ, M.G., CORDLE, C.T. & METZ, C.B. (1976) Inhibition of fertilization in vitro by treatment of rabbit spermatozoa with univalent antibodies to rabbit sperm hyaluronidase. J. Reprod. Fert. 47, 381-384.

DUTILH, C.E., VAN DOREN, P.J., VERHEUL, E.A.M. & DEHAAS, G.H. (1975) Isolation and properties of phospholipase A_2 from ox and sheep pancreas. Eur. J. Biochem. 53, 91-97.

EDWARDS, R.G. (1974) Follicular fluid. J. Reprod. Fert. 37, 189-219.

ERICCSON, R.J. (1969) Capacitation in vitro of rabbit sperm with mule eosinophils. Nature 221, 568-569.

FAWCETT, D.W. (1970) A comparative view of sperm ultrastructure. Biol. Reprod. Suppl. 2, 90-127.

FLOWER, R.J. & BLACKWELL, G.J. (1976) The importance of phospholipase A_2 in prostaglandins biosynthesis. Biochem. Pharmacol. 25, 285-291.

FRANCHI, L.L. & BAKER, T.G. (1973) Oogenesis and follicular growth. In: E.S.E. Hafez and T.M. Evans (Eds.), Human Reproduction, Harper and Row, Maryland, pp. 53-83.

FRANKLIN, L.E., BARROS, C. & RUSSELL, G.N. (1970) The acrosomal region and the acrosome reaction in sperm of the golden hamster. Biol. Reprod. 3, 180-200.

FRIEND, D.S. (1977) The organization of the spermatozoa membrane. In: M. Edidin and M.H. Johnson (Eds.). Immunobiology of Gametes. Cambridge University Press, Cambridge, pp. 5-30.

FRIEND, D.S., ORCI, L., PERRELET, A. & YANAGIMACHI, R. (1977) Membrane particle changes attending the acrosome reaction in guinea pig spermatozoa. J. Cell Biol. 74, 561-577.

FRITZ, H., SCHIESSLER, H., SCHILL, W.B., TSCHESCHE, H., HEIMBURGER, N. & WALLNER, O. (1975a) Low molecular weight proteinase (acrosin) inhibitors from human and boar seminal plasma and spermatozoa and human cervical mucus isolation, properties and biological aspects. In: E. Reich, D.B. Rifkin and E. Shaw (Eds.), Proteases and Biological Control, Coldspring Harbor Laboratory, pp. 739-766.

FRITZ, H., SCHLEUNING, D., SCHIESSLER, H., SCHILL, W.B., WENDT, V. & WINKLER, G. (1975b) Boar, bull and human sperm acrosin isolation properties and biological aspects. In: E. Reich, D.B. Rifkin and E. Shaw (Eds.), Proteases and Biological Control, Coldspring Harbor Laboratory, pp. 715-735.

GATT, S. (1973) Inhibitors of enzymes of phospholipid and sphingolipid metabolism. In: R.M. Hochster, M. Kates and J.H. Quastel (Eds.), Metabolic Inhibitors, A Comprehensive Treatise, Vol. IV, Academic Press, New York, pp. 349-387.

GONZALES, L.W. & MEIZEL, S. (1973a) Acid phosphatases of rabbit spermatozoa. 1. Electrophoretic characterization of the multiple forms of acid phosphatase in rabbit spermatozoa and other semen constituents. Biochim. Biophys. Acta. 320, 166-179.

GONZALES, L.W. & MEIZEL, S. (1973b) Acid phosphatases of rabbit spermatozoa. 2. Partial purification and biochemical characterization of the multiple forms of rabbit spermatozoan acid phosphatase. Biochim. Biophys. Acta. 320, 180-194.

GORDON, M. (1973) Localization of phosphatase activity on the membranes of the mammalian sperm head. J. Exptl. Zool. 185, 111-120.

GORDON, M. & BARNETT, R.H. (1967) Fine structural localization of phosphotases in rat and guinea pig. Exptl. Cell Res. 48, 395-412.

GORDON, M. & DANDEKAR, P.V. (1977) Fine structural localization of phosphatase activity on the plasma membrane of the rabbit sperm head. J. Reprod. Fert. 49, 155-160.

GREEN, D.P.L. (1976) The effect of trypsin inhibitors on the in vitro acrosome reaction of guinea pig spermatozoa. J. Physiol. 263, 281P (abstract).

GUL, S. & SMITH, A.D. (1972) Haemolyses of washed human red cells by the combined action of Naja Naja phospholipase A_2 and albumin. Biochim. Biophys. Acta. 299, 237-240.

GWATKIN, R.B.L. (1976) Fertilization. In: G. Poste and G.L. Nicolson (Eds.) The Cell Surface in Animal Embryogenesis and Development. Elsevier-North Holland, Amsterdam, pp. 1-54.

GWATKIN, R.B.L. & ANDERSEN, O.F. (1973) Effect of glycosidase inhibitors on the capacitation of hamster spermatozoa by cumulus cells in vitro. J. Reprod. Fert. 35, 565-567.

GWATKIN, R.B.L. & WILLIAMS, D.T. (1970) Inhibition of sperm capacitation in vitro by contraceptive steroids. Nature 227, 182-183.

GWATKIN, R.B.L., ANDERSEN, O.F. & HUTCHINSON, C.F. (1972) Capacitation of hamster spermatozoa in vitro: The role of cumulus components. J. Reprod. Fert. 30, 389-394.

GWATKIN, R.B.L., CARTER, H.W., & PATTERSON, H. (1976) Association of mammalian sperm with the cumulus cells and the zona pellucida studied by scanning electron microscopy. In: Scanning Electron Microscopy (Part VI), 1976, Proceedings of the Workshop on SEM in Reproductive Biology, ITT Research Institute, Chicago, pp. 379-384.

HARBER, E. & WRENN, S. (1976) Problems in identification of the beta adrenergic receptor. Physiological Reviews. 56, 317-338.

HARPER, M.J.K. (1970) Factors influencing sperm penetration of rabbit eggs in vivo. J. Exptl. Zool. 173, 47-62.

HARTREE, E.F. (1977) Spermatozoa, eggs, and proteinases. Biochem. Soc. Transact. 5, 375-394.

HOSKINS, D.D. & CASILLAS, E.R. (1975) Hormones, second messengers and the mammalian spermatozoa. In: J.A Thomas and R.L. Sinhal, (Eds.), Molecular Mechanisms of Gonadal Hormone Action, Vol. 1, University Press, Baltimore, pp. 293-324.

HUACUJA, L., DELGADO, N.M., MERCHANT, H., PANCARDO, R.M.A. & ROSADO, A. (1977) Cyclic AMP induced incorporation of ^{33}P into human spermatozoa membrane components. Biol. Reprod. 17, 89-96.

INNES, I.R. & NICKERSON, M. (1975) Norepinephrine, epinephrine and the sympathomimetic amines. In: L.S. Goodman and A. Gilman, (Eds.), The Pharmacological Basis of Therapeutics, 5th Edition, MacMillan, New York, p. 447-513.

IWAMATSU, T. & CHANG, M.C. (1969) In vitro fertilization of mouse eggs in the presence of bovine follicular fluid. Nature. 224, 919-920.

IWAMATSU, T. & CHANG, M.C. (1972) Capacitation of hamster spermatozoa treated with rabbit tubal fluid or rabbit and bovine cystic fluid. J. Exptl. Zool. 182, 211 219.

JOHNSON, M.H. (1975) The macromolecular organization of membranes and its bearing on events leading up to fertilization. J. Reprod. Fert. 44, 167-184.

KOEHLER, J.K. (1976) Changes in antigenic site distribution on rabbit spermatozoa after incubation in "capacitating" media. Biol. Reprod. 15, 444-456.

LINDAHL, P.E., STRINDBERG, L. & PHILIPSSON, C. (1972) Functional meaning of an ATP dependent membrane reaction in mammalian spermatozoa. Exptl. Cell Res. 72, 164-168.

LUCY, J.A. (1975) Aspects of the fusion of cells in vitro without viruses. J. Reprod. Fert. 44, 193-205.

LUI, C.W. & MEIZEL, S. (1977) Biochemical studies of the in vitro acrosome reaction inducing activity of bovine serum albumin. Differentiation 7, In press.

LUI, W., CORNETT, E. & MEIZEL, S. (1977) Identification of the bovine follicular fluid protein involved in the in vitro induction of the hamster sperm acrosome reaction. Biol. of Reprod. 17, 34-41.

MAAS, D.H.A., STOREY, B.T., & MASTROIANNI, L. Jr. (1977) pH and pCO_2 in the oviduct of the rhesus monkey during the menstrual cycle. Abstracts of 1977 Meeting Am. Fert. Soc. Fert. Steril. 28, 306.

MAHI, C.A. & YANAGIMACHI, R. (1973) The effects of temperature, osmolality and hydrogen ion concentration on the activation and acrosome reaction of golden hamster spermatozoa. J. Reprod. Fert. 35, 55-66.

MANN, T. (1964) The Biochemistry Of Semen And Of The Male Reproductive Tract. 2nd edition. Methuen and Co. Ltd., London.

MARTIN, J.K., LUTHRA, M.G., WELLS, M.A., WATTS, R.P. & HANAHAN, D.J. (1975) Phospholipase A_2 as a probe of phospholipid distribution in erythrocyte membranes. Factors influencing the apparent specificity of the reaction. Biochemistry. 14, 5400-5408.

MCRORIE, R.A., TURNER, T.B., BRADFORD, M.M. & WILLIAMS, W.L. (1976) Acrolysin the aminoproteinase catalyzing the initial conversion of proacrosin to acrosin in mammalian fertilization. Biochem. Biophys. Res. Comm. 71, 492-498.

MEGO, J.L., FARBER, R.M. & BARNES, J. (1972) An adenosine triphosphate-dependent stabilization of proteolytic activity in heterolysosomes. Biochem. J. 128, 763-769.

MEIZEL, S. & DEAMER, D.W. (1978) The pH of the hamster sperm acrosome. J. Histochem. Cytochem, In press.

MEIZEL, S., & LUI, C.W. (1976) Evidence for the role of a trypsin-like enzyme in the hamster sperm acrosome reaction. J. Exptl. Zool. 195, 137-144.

MEIZEL, S. & MUKERJI; S.K. (1975) Proacrosin from rabbit epididymal spermatozoa: Partial purification and initial biochemical characterization. Biol. Reprod. 13, 83-93.

MEIZEL, S. & MUKERJI, S.K. (1976) Biochemical studies of proacrosin and acrosin from hamster cauda epididymal spermatozoa. Biol. Reprod. 14, 444-450.

MIYAMATO, H. & CHANG, M.C. (1973) The importance of serum albumin and metabolic intermediates for capacitation of spermatozoa and fertilization of mouse eggs in vitro. J. Reprod. Fert. 32, 193-205.

MOGHISSI, K.S. (1970) Human fallopian tube fluid. 1. Protein composition. Fert. Steril. 21, 321-329.

MOORE, H.D.M. & BEDFORD, J.M. (1977) Follicular products are not needed for fertilization but do prevent polyspermy of hamster eggs in vivo. Abstracts, 1977 meeting of Society For The Study of Reproduction, pp. 55.

MORTON, D.B. (1976) Lysosomal enzymes in mammalian spermatozoa. In: J.T. Dingle and R.T. Dean (Eds.), Lysosomes in Biology and Pathology, Vol. 5, Elsevier-North Holland, Amsterdam, pp. 203-255.

MORTON, D.B. (1977) Immunoenzymic studies on acrosin and hyaluronidase in ram spermatozoa. In: M. Edidin and M.H. Johnson (Eds.) Immunobiology of Gametes, Cambridge University Press,Cambridge, pp. 115-155.

MORTON, D.B. & BAVISTER, B.D. (1974) Fractionation of hamster sperm capacitating components from human serum by gel filtration. J. Reprod. Fert. 40, 491-493.

MUKHERJEE, A.B. & LEPPES, J. (1972) Effect of human follicular and tubal fluids on human, mouse and rat spermatozoa in vitro. Can. J. Genet. Cytol. 14, 167-174.

NICOSIA, S.V., WOLF, D.P. & INOUE, M. (1977) Cortical granule distribution and cell surface characteristics in mouse eggs. Develop. Biol. 57, 56-74.

NODA, Y.D. & YANAGIMACHI, R. (1976) Electron microscopic observations of guinea pig spermatozoa penetrating eggs in vitro. Development, Growth and Differentiation. 18, 15-23.

OLIPHANT, G. (1976) Removal of sperm bound seminal plasma components as a prerequisite to initiation of the rabbit acrosome reaction. Fert. Steril. 27, 28-38.

OLIPHANT, G. & BRACKETT, B.G. (1973) Capacitation of mouse spermatozoa in media with elevated ionic strength and reversible decapacitation with epididymal extract. Fert. Steril. 24, 948-955.

OLIPHANT, G., CABOT, C.L. & SINGHAS, C.A. (1977) Nature of the rabbit acrosome reaction inducing activity of follicular fluid. J. Reprod. Fert. 50, 245-250.

O'RAND, M.G. (1977) Restriction of a sperm surface antigen's mobility during capacitation. Develop. Biol. 55, 260-270.

OVERSTREET, J.W. & BEDFORD, J.M. (1974) Comparison of the penetrability of the egg vestments in follicular oocytes, unfertilized and fertilized ova of the rabbits. Develop. Biol. 41, 185-192.

OVERSTREET, J.W. & BEDFORD, J.M. (1975) The penetrability of rabbit ova treated with enzymes or antiprogesterone antibody; a probe into the nature of a mammalian fertilizing. J. Reprod. Fert. 44, 273-284.

PETERS, T., Jr. (1975) Serum albumin. In: F. Putnam, (Ed.) The Plasma Proteins, Vol. 1, Academic Press, New York, pp. 133-181.

POSTE, G. & ALLISON, A.C. (1973) Membrane fusion. Biochim. Biophys. Acta. 300, 421-465.

ROGERS, B.J. & MORTON, B. (1973) ATP levels in hamster spermatozoa during capacitation in vitro. Biol. Reprod. 9, 361-369.

ROGERS, B.J. & YANAGIMACHI, R. (1975a) Release of hyaluronidase from guinea pig spermatozoa through an acrosome reaction initiated by calcium. J. Reprod. Fert. 44, 135-138.

ROGERS, B.J. & YANAGIMACHI, R. (1975b) Retardation of guinea pig sperm acrosome reaction by glucose: the possible importance of pyruvate and lactate metabolism in capacitation and the acrosome reaction. Biol. Reprod. 13, 568-575.

ROGERS, B.J. & YANAGIMACHI, R. (1976) Competitive effect of magnesium on the calcium dependent acrosome reaction in guinea pig spermatozoa. Biol. Reprod. 15, 614-619.

ROGERS, B.J., UENO, M. & YANAGIMACHI, R. (1977) Inhibition of hamster sperm acrosome reaction and fertilization by oligomycin, antimycin A and Rotenone. J. Exptl. Zool. 199, 129-136.

ROSADO, A., HICKS, J.J., REYES, A., & BLANCO, I. (1974) Capacitation in vitro of rabbit spermatozoa with cyclic adenosine monophosphate and human follicular fluid. Fert. Steril. 25, 821-824.

ROSADO, A., HUACUJA, L., DELGADO, N.M., MERCHANT, H., & PANCARDO, R.M. (1977) Elemental composition of subcellular structures of human spermatozoa. A study by energy dispersive analysis of X-ray. Life Science 20, 647-656.

ROTHMAN, J.E. & LENARD, J. (1977) Membrane asymmetry. Science, 195, 743-753.

SCHACKMANN, R.W., EDDY, E.M. & SHAPIRO, B.M. (1977) Ion movements during the acrosome reaction of strongylocentrotus purpuratus. Abstracts 1977 Meeting Am. Soc. Cell Biol., J. Cell Biol. In Press.

SCHERPHOF, G. & WESTENBERG, H. (1975) Stimulation and inhibition of pancreatic phospholipase A_2 by local anesthetics as a result of their interaction with the substrate. Biochim. Biophys. Acta. 398, 442-451.

SCHUEL, H., WILSON, W.L., CHEN, K., & LORAND, L. (1973) A trypsin-like proteinase localized in cortical granules isolated from unfertilized sea urchin eggs by zonal centrifugation. Role of the enzyme in fertilization. Develop. Biol. 34, 175-186.

SELINGER, Z., EIMERI, S. & SCHRAMM, M. (1974) A Calcium ionophore stimulating the action of epinephrine on the α-adrenergic receptor. Proc. Nat. Acad. Sci. USA. 71, 128-131.

SOUPART, P. (1970) Leukocytes and sperm capacitation in the rabbit uterus. Fert. Steril. 21, 724-756.

SOUPART, P. & MORGENSTERN, L.L. (1973) Human sperm capacitation and in vitro fertilization. Fert. Steril. 24, 462-478.

SOUPART, P. & STRONG, P.A. (1974) Ultrastructural observation of human oocytes fertilized in vitro. Fert. Steril. 25, 11-44.

SRIVASTAVA, P.N., MUNNELL, J.F., YANG, C.H., & FOLEY, C.W. (1974) Sequential release of acrosomal membranes and acrosomal enzymes of ram spermatozoa. J. Reprod. Fert. 36, 363-372.

STAMBAUGH, R. & SMITH, M. (1976) Sperm proteinase release during fertilization of rabbit ova. J. Exptl. Zool. 197, 121-125.

STAMBAUGH, R., BRACKETT, B.F. & MASTRIOANNI, L. (1969) Inhibition of in vitro fertilization of rabbit ova by trypsin inhibitors. Biol. Reprod. 1, 223-227.

STAMBAUGH, R., SEITZ, H.M. & MASTROIANNI, L. Jr. (1974) Acrosomal proteinase inhibitors in rhesus monkey (Macaca mulatta) oviduct fluid. Fert. Steril. 25, 352-357.

STAMBAUGH, R., SMITH, M. & FALTAS, S. (1975) An organized distribution of acrosomal proteinase in rabbit sperm acrosomes. J. Exptl. Zool. 193, 119-122.

SUMMERS, R.G., TALBOT, P., KEOUGH, E.M., HYLANDER, B.L. & FRANKLIN, L.E. Ionophore A23187 induces acrosome reactions in sea urchin and guinea pig spermatozoa. J. Exptl. Zool. 196, 381-385.

TALBOT, P. (1977) Does a trypsin-like enzyme function in sperm capacitation? Abstracts, 1977 meeting of the Society for the Study of Reproduction. pp. 55.

TALBOT, P. & FRANKLIN, L.E. (1974a) The release of hyaluronidase from guinea pig spermatozoa during the course of the normal acrosome reaction in vitro. J. Reprod. Fert. 39, 429-432.

TALBOT, P. & FRANKLIN, L.E. (1974b) Hamster sperm hyaluronidase II. Its release from sperm in vitro in relation to the degeneration and normal acrosome reaction. J. Exptl. Zool. 189, 321-332.

TALBOT, P. & FRANKLIN, L.E. (1976) Morphology and kinetics of the hamster sperm acrosome reaction. J. Exptl. Zool. 198, 163-176.

TALBOT, P. & KLEVE, M.G. (1977) The distribution of actin in hamster sperm. Abstracts of American Soc. Cell Biol. J. Cell Biol. 75, 170a

TALBOT, P., FRANKLIN, L.E. & RUSSELL, E.M. (1974) The effect of the concentration of golden hamster sperm on the acrosome reaction and egg penetration in vitro. J. Reprod. Fert. 36, 429-432.

TALBOT, P., SUMMERS, R.G., HYLANDER, B.L., KEOUGH, E.M. & FRANKLIN, L.E. (1976) The role of calcium in the acrosome reaction: an analysis using ionophore A23187. J. Exptl. Zool. 198, 383-392.

TEGNER, M.J. & EPEL, D. (1973) Sea urchin sperm-egg interactions studied in the scanning electron microscope. Science. 179, 685-688.

TILNEY, L.G. (1977) Actin: its association with membranes and the regulation of its polymerization. In: B.R. Brinkley and K.R. Porter (Eds.) International Cell Biology 1976-1977, Rockefeller University Press, New York, pp. 388-402.

TOYODA, Y. & CHANG, M.C. (1974) Fertilization of rat eggs in vitro by epididymal spermatozoa and the development of eggs following transfer. J. Reprod. Fert. 36, 9-22.

TSUNODA, Y. & CHANG, M.C. (1975) In Vitro fertilization of rat and mouse eggs by ejaculated sperm and the effect of energy sources on in vitro fertilization of rat eggs. J. Exptl. Zool. 193, 79-86.

VOLWERK, J.J., PIETERSON, W.A. & DE HAAS, G.H. (1974) Histidine at the active site of phospholipase A_2. Biochemistry 13, 1446-1454.

WAITE, M., GRIFFIN, H.D. & FRANSON, R. (1976) The phospholipase A of lysosomes. In: J.T. Dingle and R.T. Dean (Eds.), Lysosomes in Biology and Pathology. Vol. 5, Elsevier-North Holland, Amsterdam, pp. 257-305.

WALSH, K. (1970) Trypsinogens and trypsins of various species. In: G.E. Perlmann and L. Lorand (Eds.), Methods in Enzymology. Vol. 19. Proteolytic enzymes, Academic Press, New York, pp. 41-63.

WEISSMAN, G. (1969) The effects of steroids and drugs on lysosomes. In: J.T. Dingle and H.B. Fell (Eds.), Lysosomes in Biology and Pathology. Vol..1, North-Holland, Amsterdam, pp.276-295.

WENDT, V., LEIDL, W. & FRITZ, H. (1975) The influence of various proteinase inhibitors on the gelatinolytic effect of ejaculated and uterine boar spermatozoa. Hoppe Seyler's. Z. Physiol. Chem. 356, 1073-1078.

WHITAKER, J.R. & PEREZ-VILLASENOR, J. (1968) Chemical modification of papain. 1. Reaction with chloromethyl ketones of phenylalanine and lysine and with phenylomethylsulfonyl fluoride. Arch. Biochem. Biophys. 24, 70-78.

WOLF, D.P. (1977) Involvement of a trypsin-like activity in sperm penetration of zona-free mouse ova. J. Exptl. Zool. 199, 149-156.

WOODING, F.B.P. (1975) Studies on the mechanism of the hyamine-induced acrosome reaction in ejaculated bovine spermatozoa. J. Reprod. Fert. 44, 185-192.

YANAGIMACHI, R. (1969) In vitro acrosome reaction and capacitation of golden hamster spermatozoa by bovine follicular fluid and its fractions. J. Exptl. Zool. 170, 269-280.

YANAGIMACHI, R. (1970a) The movement of golden hamster spermatozoa before and after capacitation. J. Reprod. Fert. 23, 193-196.

YANAGIMACHI, R. (1970b) In vitro acrosome reaction and capacitation of golden hamster spermatozoa by bovine follicular fluid and its fractions. J. Exptl. Zool. 170, 269-280.

YANAGIMACHI, R. (1970c) In vitro capacitation of golden hamster spermatozoa by homologous and heterologous blood sera. Biol. Reprod. 3, 147-153.

YANAGIMACHI, R. (1972) Fertilization of guinea pig eggs in vitro. Anat. Rec. 174, 9-20.

YANAGIMACHI, R. (1975) Acceleration of the acrosome reaction and activation of guinea pig spermatozoa by detergents and other reagents. Biol. Reprod. 13, 519-526.

YANAGIMACHI, R. (1977) Specificity of sperm-egg interaction. In: M. Edidin and M.H. Johnson (Eds.), Immunobiology of Gametes, Cambridge University Press, pp. 225-291.

YANAGIMACHI, R. & MAHI, C.A. (1976) The sperm acrosome reaction and fertilization in the guinea pig: a study in vivo. J. Reprod. Fert. 46, 49-50.

YANAGIMACHI, R. & NODA, Y.D. (1970a) Physiological changes in the post-nuclear cap region of mammalian spermatozoa: a necessary preliminary to membrane fusion between sperm and egg cells. J. Ultrastruc. Res. 31, 486-493.

YANAGIMACHI, R. & NODA, Y.D. (1970b) Ultrastructural changes in the hamster sperm head during fertilization. J. Ultrastruc. Res. 31, 465-485.

YANAGIMACHI, R. & NODA, Y.D. (1972) Acrosome loss in fertilizing mammalian spermatozoa: a rebuttal of criticism. J. Ultrastruc. Res. 39, 217-221.

YANAGIMACHI, R. & USUI, N. (1974) Calcium dependence of the acrosome reaction and activation of guinea pig spermatozoa. Exptl. Cell Res. 89, 161-174.

YANAGIMACHI, R., YANAGIMACHI, H. & ROGERS, B.J. (1976) The use of zona-free animal ova as a test system for the assessment of the fertilizing capacity of human spermatozoa. Biol. Reprod. 15, 471-476.

ZAHLER, W.L. & POLAKOSKI, K.L. (1977) Benzamidine as an inhibitor of proacrosin activation in bull sperm. Biochim. Biophys. Acta. 480, 461-468.

ZAMBONI, L. (1970) Ultrastructure of mammalian oocytes and ova. Biol. Reprod. Suppl. 2, 44-63.

THE MECHANISM OF THE ACROSOME REACTION

D.P.L. Green

Department of Pharmacology
University of Otago Medical School,
P.O. Box 56
Dunedin
NEW ZEALAND

1. The cause of the acrosome reaction
2. The mechanism of membrane fusion
3. The mechanism of proacrosin activation
4. The physical nature of the acrosomal contents
5. The origin of acrosomal cavitation
6. The function of the acrosome reaction

The acrosome reaction is a morphological change in the sperm head. It normally occurs at, or near, the outer investment of the egg and is an essential precondition of successful fertilisation. The acrosome itself is an intracellular, membrane-bound organelle, formed during sperm development from the fusion and condensation of the proacrosomal granules (Burgos and Fawcett, 1955). In the course of the reaction, its membrane fuses with the overlying plasma membrane and becomes permanently incorporated into it with consequent exocytosis of the acrosomal contents (Colwin and Colwin, 1967; Bedford, 1968, 1970). In mammalian sperm, fusion is multiple and this leads to extensive vesiculation.

The acrosome contains a number of enzymes. One of these is the protease, acrosin, which is stored prior to the acrosome reaction as an inactive zymogen precursor, proacrosin (Meizel and Mukerji, 1976). Proacrosin is activated to acrosin during the course of the acrosome reaction (Green, 1978b). The importance of acrosin, in the context of the reaction, is that its activity has been suggested as playing an essential part (Gordon, 1973; Meizel and Lui, 1976; Meizel, this volume).

The simplest way to describe a mechanism is to identify a number of more-or-less discrete steps and arrange them, where possible, in a sequence. This I have attempted to do under a number of headings. The boundaries between each heading are not arbitrary but they imply no discontinuity. It has not been my intention to summarise all that is known about the acrosome reaction. What I have attempted to do instead is identify a number of important questions, suggest what are, in my opinion, their most likely answers and show how they form the rudiments of a mechanism for the acrosome reaction.

THE CAUSE OF THE ACROSOME REACTION

The immediate cause of the acrosome reaction is an increase in cytoplasmic free calcium. There are two arguments which can be deployed to support this contention: there is the comparative argument and the argument principally derived from the action of the divalent metal cation ionophore A23187 in inducing an acrosome reaction (Green, 1976 and 1978a; Summers *et al.*,1976; Talbot *et al.*,1976; Collins and Epel, 1977).

The comparative argument runs as follows: secretion from secretory cells is the result of calcium-mediated stimulus-secretion coupling (Douglas, 1968; Rubin, 1970; Baker, 1974) and is the direct consequence of the exocytosis of the contents of intracellular secretory granules through their fusion with the plasma membrane of the cell; the acrosome reaction is morphologically analogous to this secretory granule discharge; therefore, it itself is an example of stimulus-secretion coupling.

The concept of stimulus-secretion coupling (Douglas, 1968) rests on the truth of the following propositions: that secretion is the result of the exocytosis of the contents of intracellular secretory granules, that exocytosis is caused by fusion of the secretory granule membrane with the plasma membrane, that cytoplasmic free calcium is normally low, that it rises on the receipt by the cell of the stimulus to secretion, at least between the secretory granules and the plasma membrane, and that it is the increase in cytoplasmic free calcium which causes the fusion of the secretory granule membrane to the plasma membrane with exocytosis of the granule contents.

The comparative argument is principally, although not wholly, a morphological one. The origins and contents of the acrosome are closely similar to those of the zymogen granules of the exocrine pancreas (Burgos and Fawcett, 1955; Palade *et al.*,1962; Jamieson and Palade, 1971 a,b) and the similarities could be extended to the secretory granules of many other secretory cells. However, the zymogen granule remains, for most purposes, the best characterised.

The acrosome and the zymogen granule both contain zymogens; both are formed by the budding off of the rough endoplasmic reticulum, both undergo condensation in conjunction with the Golgi apparatus (the proacrosomal granules also fuse at this stage) and both move close to the plasma membrane (in sperm, because the whole shape of the cell is changing). Sperm are different from other secretory cells in one important respect: the Golgi apparatus and rough endoplasmic reticulum are lost early in sperm development and with them, the ability to synthesise a new acrosome. The acrosome reaction, therefore, is a much more irreversible step than is secretion in other secretory cells.

The second point, that exocytosis of the acrosomal contents takes place through the fusion of plasma and acrosomal membrane, is quickly dealt with. The morphology of the acrosome reaction is well established for a number of mammals and marine invertebrates which have either undergone an acrosome reaction in the course of fertilisation (Colwin and Colwin, 1967; Bedford, 1968, 1970; Stefanini *et al.*, 1969; Yanagimachi and Noda, 1970) or in response to one of a number of artificial stimuli (Yanagimachi and Usui, 1974; Wooding, 1975; Yanagimachi, 1975; Collins and Epel, 1977; Green, 1978a) and its principal feature is the fusion of plasma and acrosomal membranes. In the case of the mammalian acrosome reaction, fusion takes place at a number of points with the result that some vesiculation eventually occurs (Barros *et al.*, 1967).

The comparative argument unfortunately does no more than suggest a role for calcium in the acrosome reaction. It is, however, supported at this stage by the evidence that intracellular calcium plays an essential part in the reaction. It has been known for some time that the acrosome reaction in the marine invertebrates is dependent on external calcium (Dan, 1954, 1956) and the same has recently been shown for the reaction in guinea pig sperm (Yanagimachi and Usui, 1974). These results suggest that external calcium is important because it causes the increase in cytoplasmic calcium when the calcium permeability of the plasma membrane rises. This view is supported by the action of the divalent cation ionophore A23187 in inducing an acrosome reaction in the presence of external calcium (Green, 1976; Summers *et al.*, 1976; Talbot *et al.*, 1976; Collins and Epel, 1977; Green, 1978a).

A23187 is a hydrophobic antibiotic which partitions into membranes and promotes electroneutral exchange of divalent metal cations and protons (Reed and Lardy, 1972; Kafka and Holz, 1976; Babcock *et al.*,1976). The only two metal cations which need be considered, for practical purposes, are calcium and magnesium. Because exchange is tightly coupled (Kafka and Holz, 1976), net calcium movement occurs in one direction only as long as net movement of either magnesium or protons, or both, occurs in the other and it stops when the activities of the three ions collectively move into equilibrium on either side of the membrane. A23187 does not increase the calcium permeability of membranes, as is popularly but erroneously supposed (Collins and Epel, 1977), and it does not allow calcium to move into electrochemical equilibrium.

In those cells where it has been measured, cytoplasmic free calcium is low, possibly no more than 100nM (Harrison and Long, 1968; Ashley and Ridgway, 1970; Baker, 1972; Ridgway *et al.*,1977; Steinhardt *et al.*,1977). Extracellular free calcium, on the other hand, is often about 1mM, i.e. four orders of magnitude greater. There is, therefore, a steep electrochemical gradient for inward calcium movement at membrane potentials which are either negative or moderately positive. The membrane potential of guinea pig sperm is about +10mV (Rink, 1977).

The level of cytoplasmic free calcium normally prevailing in cells is the result of three fluxes: an influx due to calcium leaking through the plasma membrane and moving down its electrochemical gradient, an efflux due to either (or both) calcium pumping or coupled exchange with some other ion e.g. sodium, and calcium intake into intracellular organelles. The stimulus to secretion increases cytoplasmic free calcium by increasing the calcium permeability of the plasma membrane and allowing calcium to move down its electrochemical gradient (Douglas, 1968; Rubin, 1970; Baker, 1974) although in some cases, calcium is subsequently released from intracellular stores (Steinhardt *et al.*,1977).

Although the effects of A23187 are theoretically complicated, it is known empirically that it increases cytoplasmic free calcium (Desmedt and Hainaut, 1976; Steinhardt *et al.*,1977) as well as stimulating secretion in a wide variety of cells (Foreman *et al.*, 1973; Prince *et al.*,1973; Cochrane and Douglas, 1974; Plattner, 1974; Steinhardt and Epel, 1974; Steinhardt *et al.*,1974) for which there is evidence of secretion, either as a concomitant of increased cytoplasmic free calcium (Kanno *et al.*,1973; Ridgway *et al.*,1977; Steinhardt *et al.*, 1977) or calcium uptake (Foreman *et al.*,1973). Under certain circumstances, A23187 increases intracellular calcium in epididymal bull sperm (Babcock *et al.*,1976) and the evidence is consistent with it being an increase in cytoplasmic free calcium. The possible occurrence of an acrosome reaction was not examined.

Although the evidence is very strong that an increase in cytoplasmic free calcium is the immediate cause of the acrosome reaction, the experiments using A23187 raise two alternative possibilities. Since inward calcium movement occurs only at the expense of either magnesium or proton efflux, or possibly both, it is possible that the acrosome reaction is the consequence of intracellular magnesium depletion or a shift in intracellular pH to higher values. It is very unlikely that inward calcium movement is coupled in the natural acrosome reaction in this way and these changes would not, therefore, be expected to automatically occur. However, Collins and Epel (1977) claim that acid secretion occurs during the acrosome reaction in sea urchin sperm. Three points can be made. Firstly, the authors do not make it clear whether the acrosome reaction was induced by A23187, because if it were, acid secretion might be expected as a consequence of inward calcium movement; secondly, there is no evidence that acid secretion is not the result of the exocytosis of the acrosomal contents; and thirdly, acid secretion does not indicate the magnitude of any shift in intracellular pH, merely its direction.

If the immediate cause of the acrosome reaction is an increase in cytoplasmic free calcium, there must be some sort of stimulus for the increase. The principal, if not the sole, cause of any increase in cytoplasmic free calcium is likely to be an increase in the calcium permeability of the plasma membrane. There is no evidence for inhibition of calcium efflux or a second messenger for intracellular calcium release. The absence of any A23187-induced acrosome reaction in the absence of external calcium is, itself, fairly strong evidence that there is no pool of intracellular calcium whose release is sufficient to raise cytoplasmic free calcium above that needed to start the acrosome reaction (cf. the A23187-induced cortical reaction (Steinhardt *et al.*,1974)). A purely theoretical argument can be put forward against the development of a slow calcium leak as the means by which cytoplasmic calcium is raised. If sperm are like other cells and maintain a low cytoplasmic free calcium, it can only be by pumping calcium out, either directly or through coupled exchange with another ion which itself is pumped out. Pumping consumes energy, i.e. ATP. An increased calcium leak in the plasma membrane would, therefore, be in direct competition with motility and any other ATP-dependent process. A short-term pulse of calcium is likely to be far less deleterious. If that premise is accepted, then it follows that the increase in calcium permeability of the head plasma membrane is rapid and short-lived. For the marine invertebrates, some constituent of the jelly coat is the trigger (Dan *et al.*,1972) but in mammals, the stimulus is unknown. The obvious candidates are a constituent of the follicle cell, possibly a surface component, or the zona pellucida. The rapid decline in calcium permeability which presumably follows the initial increase could be brought about either by rapid desensitisation or the physical loss of membrane through vesiculation, although the latter does not occur in all species (Colwin and Colwin, 1967).

The simplest picture for the mechanism of the increase in cytoplasmic calcium is that there are receptors for the stimulus, broadly speaking, similar to acetylcholine receptors at the neuromuscular junction, which are associated with calcium channels in the plasma membrane of the head. The need for capacitation in mammalian sperm (Bedford, 1970) could be explained on the basis that the receptor is normally occupied by an antagonist which must first be removed before the stimulus ligand can occupy the receptor. There is no need to postulate metabolically dependent calcium uptake into the cell, i.e. inward calcium pumping, as suggested by Gordon (1973) because all the evidence suggests that inward calcium movement is down the electrochemical gradient for calcium and that, on the contrary, calcium pumping is outwards (Babcock *et al.*,1976).

THE MECHANISM OF MEMBRANE FUSION

The mechanism by which an increase in cytoplasmic free calcium causes fusion of plasma and acrosomal membranes is unknown. Yanagimachi and Usui (1974) suggested that calcium entering the cell causes the acrosome to swell and that this swelling, by pushing the acrosomal membrane against the plasma membrane, causes fusion. The mechanism proposed for the swelling is discussed below, but to anticipate the conclusions, swelling is, more strictly speaking, a cavitation in part of the acrosomal matrix (also dealt with below) and cavitation is the consequence, not the cause, of membrane fusion (Green, 1978c).

Two other, more general, arguments can, however, be put forward against the swelling mechanism of membrane fusion. Firstly, there appears to be nothing preventing the plasma membrane from touching and fusing with the acrosomal membrane in the absence of swelling and, secondly, it would be a mechanism peculiar to sperm, because only in sperm is the geometry of the secretory granule such that its swelling would increase its contact with the plasma membrane. Membrane fusion does not depend on acrosin activity or proacrosin activation, the former a suggestion of Gordon (1973), because the acrosome reaction takes place in the presence of a membrane-permeable, irreversible acrosin inhibitor (Green, 1978b).

THE MECHANISM OF PROACROSIN ACTIVATION

Zymogens generally can only be activated by proteolytic cleavage (Neurath, 1975), normally at a specific lysine residue, with loss of an N-terminal peptide. The active site of the enzyme is distorted in the zymogen but, following cleavage, it changes shape and activity is acquired. Zymogens themselves have detectable proteolytic activity (Kassell and Kay, 1973) and they react with irreversible inhibitors, although at a much slower rate than the enzymes themselves (Morgan *et al.*,1972). Proacrosin can be activated by trypsin (Meizel and Mukerji, 1976) which indicates cleavage at either an arginine or lysine residue. It can also be activated by acrosin, whose properties as a protease are closely similar to those of trypsin: it hydrolyses the same synthetic substrates and is inhibited by the same synthetic and naturally occurring inhibitors (Polakoski and McRorie, 1973). Proacrosin autoactivation (i.e. activation by acrosin) takes place in the absence of calcium (Schleuning *et al.*,1976) and is not increased, either in rate or extent, by calcium at physiological concentrations. This effectively rules out the absence of calcium as a means of preventing proacrosin autoactivation in the acrosome.

The only effective way of preventing autoactivation in the acrosome is to inhibit the acrosin. (Making proacrosin inactive as a substrate whilst leaving acrosin active is not really a solution because of the potentially damaging effects of extended proteolysis within the acrosome). How proacrosin is activated depends very much on the nature of the acrosin inhibitor. If it is small and reversible, it could rapidly dissociate from the acrosin and diffuse away as soon as the acrosomal and plasma membranes fuse, leaving acrosin behind to activate proacrosin. If, on the other hand, the inhibitor is much more irreversible, a second explanation has to be sought, for opening the acrosome would be of no use in reversing the inhibition of acrosin. In this situation, a second protease would be needed to initiate activation of proacrosin. This protease could, itself, be activated in one of two ways; by being released from inhibition by a rapidly reversible inhibitor, in much the same way as already proposed for acrosin, or by increased calcium. There is clear evidence that washed sperm contain a protein acrosin inhibitor which is, for practical purposes, irreversible under normal physiological conditions (Brown *et al.*,1975; Schleuning *et al.*,1976). If the inhibitor is a fail-safe device to inhibit any acrosin which might arise before the acrosome reaction, for example, through the proteolytic action of proacrosin, it must be present in an amount sufficient to inhibit all the acrosin normally present, but if acrosin activity is to appear at all after proacrosin activation, the amount of inhibitor must be stoichiometrically less than the total potential acrosin. Because of this, the inhibitor must be able to discriminate between proacrosin and acrosin and bind much more strongly to the acrosin.

This kind of mechanism is very similar to that already established for the exocrine pancreas, the only place where the storage and activation of zymogens has been studied in any detail (Palade *et al.*,1962; Neurath, 1975). The pancreatic acinar cells secrete the digestive enzymes trypsin, chymotrypsin, etc. as their zymogen precursors, trypsinogen, chymotrypsinogen, etc. The zymogens are stored prior to secretion in the zymogen granules together with a trypsin inhibitor (Kazal pancreatic trypsin inhibitor) (Greene *et al.*,1966). This inhibitor is secreted into the lumen with the zymogens as part of the pancreatic juice from which it can be isolated free, despite an 80-fold excess of trypsinogen (i.e. at 10^{-6}M, its approximate concentration in pancreatic juice, it does not form a stable complex with trypsinogen). Its amount in pancreatic juice is about 1% of the total potential trypsin activity. It therefore shows the two properties required of the acrosin inhibitor; it distinguishes between zymogen and active enzyme and preferentially binds to the enzyme, and it is present in sufficient quantity to inhibit all the active enzyme in the zymogen granules without preventing enzyme activity appearing on activation of the zymogen. The acrosin inhibitor in sperm is insufficient to inhibit more than a fraction of the potential acrosin (Brown *et al.*,1975) but whether it discriminates between proacrosin and acrosin is unknown, as is its location.

There is one important difference between the pancreatic zymogens and proacrosin. The pancreatic zymogens are activated, after secretion, by an external second protease, enteropeptidase (at least, trypsinogen is activated by enteropeptidase, and the trypsin formed activates the other zymogens and the trypsinogen). There is no protease external to sperm and any activating protease has to be internal.

Activation of proacrosin by a calcium-dependent protease would entail two functions for calcium in the acrosome reaction, membrane fusion and activation. This raises the question of how calcium reaches the acrosomal contents. In the normal acrosome reaction, in the absence of A23187, calcium reaches the acrosomal contents either after their exocytosis or, if the acrosomal membrane is permeable to calcium, possibly before exocytosis occurs, immediately following the rise in cytoplasmic free calcium. In many of the marine invertebrates, the acrosomal membrane is incorporated without loss into the plasma membrane of the cell, and it is, therefore, unlikely to have a substantial resting permeability to calcium. In mammalian sperm, the same must equally be true of the inner and small amount of outer acrosomal membrane which also becomes permanently incorporated, but the properties of the outer acrosomal membrane lost as a result of vesiculation may be different.

The problem of proacrosin is how, on the one hand, it is kept from activation prior to the acrosome reaction, and how, on initiation of the reaction, its activation rapidly follows. The need for an inhibitor of acrosin within the acrosome is self-evident but it is, so far, unclear whether it is reversible (that is, for example, H^+, Zn^{2+} or a small molecule) or irreversible (the protein acrosin inhibitor). Arguments against a second activating protease, on the ground of unnecessary complexity, would apply equally to the zymogens of the pancreas where they do not hold.

THE PHYSICAL NATURE OF THE ACROSOMAL CONTENTS

It is widely assumed that the acrosomal contents are soluble and that their exocytosis in the course of the reaction results in their immediate dispersal. This may be true for hyaluronidase (Talbot and Franklin, 1974) but is not the case with acrosin (Green, 1978b). If the appearance of acid-extractable acrosin, in the course of an A23187-induced acrosome reaction in guinea pig sperm, measured as a function of time, is compared with release of 'soluble' acrosin, it is found that very little 'soluble' acrosin is released at a time when proacrosin activation is complete and electron microscopy shows the acrosome reaction to have occurred in virtually the whole sperm population. The word 'soluble' has to be used carefully. For each time point in the experiments in which the appearance of 'soluble' acrosin was being measured, centrifugation at $14{,}000g_{av}$ for 15 seconds was used to separate sperm from their supernatant. This is, of course, no real test of solubility. However, the error inherent in the procedure merely reduces the time between proacrosin activation and the appearance of 'soluble' acrosin and the actual time is certain to be greater. The reason for the delay in the appearance of 'soluble' acrosin is very simple. Much of the acrosomal material is extensively aggregated (Green, 1978c) and acrosin forms part of this matrix (Green and Hockaday, 1978). The aggregate remains substantially intact after the acrosome reaction and is centrifuged down with the sperm.

When guinea pig sperm undergo an A23187-induced acrosome reaction in the presence of one of three acrosin inhibitors, either the reversible inhibitors, benzamidine or p-aminobenzamidine, or the irreversible inhibitor, phenylmethylsulphonyl fluoride (PMSF), the acrosome reaction, when assessed by Nomarski light microscopy, is apparently delayed in the sperm population as a whole (Green, 1978b). What is being assessed, however, is not, strictly speaking, the acrosome reaction at all, but the loss of the acrosomal contents, and the two are only synonymous if they occur more-or-less simultaneously.

When sperm, undergoing an A23187-induced acrosome reaction in the presence of one of the three synthetic inhibitors, are fixed at a time when, in the absence of the inhibitor, the population would have almost wholly undergone an acrosome reaction, electron microscopy shows that the reaction has proceeded normally but that detachment of the acrosomal matrix is inhibited. From this, it follows that acrosin inhibitors do not inhibit the acrosome reaction *per se* and that acrosin activity is necessary for solubilisation of the acrosomal matrix. The acrosomal matrix is optically the densest part of the whole acrosomal region and, because the vesiculation of the acrosome reaction is below the limit of resolution of the light microscope, the loss of the matrix is the only substantial change which can unambiguously be detected by light microscopy. When this loss is delayed relative to the acrosome reaction, sperm assessed for the occurrence of the reaction using light microscopy are erroneously recorded as having undergone it: in other words, loss of the acrosomal contents and their exocytosis are not always simultaneous events.

THE ORIGIN OF ACROSOMAL CAVITATION

When guinea pig sperm undergo an acrosome reaction, the acrosomal matrix cavitates (Green, 1978a). Yanagimachi and Usui (1974) regarded this cavitation as being due to swelling within an intact acrosomal membrane. Their explanation for the swelling was an osmotic one and rested on the assumption that the acrosomal membrane had a very low resting permeability to water and that, in addition, water was actively pumped out of the acrosome. On their view, a rise in intracellular calcium increases the water permeability of the acrosomal membrane and inhibits water pumping. This leads, *inter alia*, to swelling and membrane fusion. There can be little doubt that cavitation has an osmotic explanation but the problem is in identifying, firstly, the source of osmotically active particles and then the semi-permeable barrier upon which the osmotic force acts. Cavitation occurs in guinea pig sperm suspended in calcium-free medium containing both 100μM EGTA and 0.1%(w/v) of the non-ionic detergent Triton X-100 Green, 1978c). Electron micrographs also show the complete absence of all membranes. This evidence strongly suggests that cavitation occurs quite independently of calcium and that the loss of the acrosomal membrane is the cause, not the consequence, of cavitation. Other arguments can be put forward against the mechanism of Yanagimachi and Usui (1974). There is, for example, no substantiated case of water pumping, and for it to be effective, the acrosomal membrane would have to have an unusually low water permeability. In addition, the acrosome would have to be hyperosmotic inside for it to swell when the water permeability of its membrane rose.

The behaviour of guinea pig sperm in hypotonic media (Green, 1978c)suggests that the acrosomal membrane is readily permeable to water, i.e. quite normal, and it is hypo-osmotic inside the acrosome. The acrosome is apparently prevented from further contraction by the contents reaching their limit of compression. This entails that a hydrostatic pressure exists across the acrosomal membrane in an inward direction and because of the intrinsic weakness of the membrane, this can only be withstood if the membrane is supported at something close to molecular distances.

If the acrosomal membrane were to remain intact during cavitation, the membrane itself would be the obvious candidate for the semi-permeable barrier. This would leave the source of osmotically active particles to be identified.The only candidate, at present, would be those generated by the onset of proteolysis following proacrosin activation but this would require precise synchronisation between the onset of proteolysis and membrane fusion, with the two taking place in that order.

If the acrosomal membrane is not intact during cavitation, however, the only possible candidate for the semi-permeable barrier is the acrosomal matrix itself. The mechanism of generating osmotically active particles could be as before, with the additional possibility that onset of proteolysis does not occur until exocytosis of the acrosomal contents has taken place. However, it would be necessary to explain why cavitation occurs in calcium-free medium containing Triton because in that, no proacrosin activation occurs (Green, 1978b). Cavitation would occur because the detergent mimics proteolysis by internally dissolving part of the fabric of the matrix.

There is a completely different mechanism which avoids special pleading of this kind and that is to assume that the matrix contains trapped protein. (It could have been trapped during condensation of the proacrosomal granules). In the absence of the acrosomal membrane, the protein will exert a colloid osmotic pressure on the matrix tending to split it apart and produce cavitation. What prevents it from doing so inside the intact acrosome is presumably an asymmetric crystalloid distribution across the acrosomal membrane, exactly analogous to the one which exists across the plasma membrane. Cavitation, on this scheme, would be similar to the loss of cation impermeability in red blood cells which leads to swelling and haemolysis (Hofmann, 1958).

THE FUNCTION OF THE ACROSOME REACTION

It has been widely assumed that the function of the acrosome reaction is to expose on its surface a zona lysin which is needed for the formation of the characteristic penetration slit through the zona. There is really very little evidence, however, that a zona lysin is needed at all for zona penetration and the critical experiment, that of seeing whether sperm stripped of zona lysin activity actually penetrate the zona, cannot even be attempted until the potential zona lysin has been more definitively identified. It has been reported that acrosin inhibitors block fertilisation (Stambaugh *et al.*,1969; Zanefeld *et al.*,1971; Gaur *et al.*, 1972; Yang *et al.*,1976) and this is undoubtedly consistent with the view that acrosin attached to the surface of the sperm after an acrosome reaction is essential for digesting the penetration slit. However, location of the block to fertilisation has never been established and the results themselves have been contradicted (Miyamoto and Chang, 1973). The alternative mechanism of penetration is that it is purely mechanical and the function of the acrosome reaction is to alter the profile of the head, particularly at the leading edge. The two mechanisms are not mutually exclusive but the mechanical one has been given nothing like the attention it deserves.

If penetration is due to the action of a zona lysin, then the evidence supports acrosin as a major constituent because extracts of it, and trypsin, remove the zona *in vitro* (Polakoski and McRorie, 1973; Meizel and Mukerji, 1976). In addition, there is now ultrastructural evidence, using a ferritin-soybean trypsin inhibitor conjugate, that acrosin does actually remain attached to the outer surface of the inner acrosomal membrane after the acrosome reaction (Green and Hockaday, 1978).

ACKNOWLEDGEMENTS

I would like to gratefully acknowledge the invaluable criticism of C.R. Austin, P.F. Baker, R.G. Edwards, A.L. Hodgkin and T.J. Rink. The responsibility for the manuscript is wholly mine.

REFERENCES

ASHLEY, C.C. & RIDGWAY, E.B. (1970) On the relationship between membrane potential, calcium transient and tension in single barnacle muscle fibres. J. Physiol., Lond. 209, 105-130.

BABCOCK, D.F., FIRST, N.L. & LARDY, H.A. (1976) Action of ionophore A23187 at the cellular level. Separation of effects at the plasma and mitochondrial membranes. J. biol. Chem. 251, 3881-3886.

BAKER, P.F. (1972) Transport and metabolism of calcium ions in nerve. Progr. Biophys. mol. Biol. 24, 177-223.

BAKER, P.F. (1974) Excitation-secretion coupling. In: Linden, R.S. (Ed.) Recent Advances in Physiology, No.9, Churchill Livingstone, London, pp. 51-86.

BARROS, C., BEDFORD, J.M., FRANKLIN, L.E. & AUSTIN, C.R. (1967) Membrane vesiculation as a feature of the mammalian acrosome reaction. J. Cell Biol. 34, C1-C5.

BEDFORD, J.M. (1968) Ultrastructural changes in the sperm head during fertilisation in the rabbit. Am. J. Anat. 123, 329-358.

BEDFORD, J.M. (1970) Sperm capacitation and fertilisation in mammals. Biol. Reprod. Suppl. 2, 128-158.

BROWN, C.R., ANDANI, Z. & HARTREE, E.F. (1975) Studies on ram acrosin. Biochem. J. 149, 133-146.

BURGOS, M.H. & FAWCETT, D.W. (1955) Studies on the fine structure of the mammalian testis. I. Differentiation of the spermatids in the cat (Felis domestica). J. biophys. biochem. Cytol. 1, 287-300.

COCHRANE, B.E. & DOUGLAS, W.W. (1974) Calcium-induced extrusion of secretory granules (exocytosis) in mast cells exposed to 48/80 or the ionophores A23187 and X537A. Proc. natn. Acad. Sci. USA 71, 408-412.

COLLINS, F. & EPEL, D. (1977) The role of calcium ions in the acrosome reaction of sea urchin sperm. Regulation of exocytosis. Expl. Cell Res. 106, 211-222.

COLWIN, L.H. & COLWIN, A.L. (1967) Membrane fusion in relation to sperm-egg association. In: Metz, C.B. and Monroy, A. (Eds.) Fertilisation, Vol. I, pp. 295-367.

DAN, J.C. (1954) Studies on the acrosome. III. Effect of calcium deficiency. Biol. Bull. 107, 335-349.

DAN, J.C. (1956) The acrosome reaction. Int. Rev. Cytol. 5, 365-393.

DAN, J.C., KAKIZAWA, Y., KUSHIDA, H. & FUJITA, K. (1972) Acrosomal triggers. Expl. Cell Res. 72, 60-68.

DESMEDT, J.E. & HAINAUT, K. (1976) The effect of A23187 ionophore on calcium movements and contraction processes in single barnacle muscle fibres. J. Physiol. Lond. 257, 87-107.

DOUGLAS, W.W. (1968) Stimulus-secretion coupling: the concept and clues from chromaffin and other cells. Br. J. Pharmac. 34, 451-474.

FOREMAN, J.C., MONGAR, J.L. & GOMPERTS, B.D. (1973) Calcium ionophores and movement of calcium ions following the physiological stimulus to a secretory process. Nature, Lond. 245, 249-251.

GAUR, R.D., GOSWAMI, A.K. & TALWAR, G.P. (1972) Studies on sperm-ovum interaction. I. Effect of an inhibitor of proteolytic enzyme on fertilisation. Indian J. med. Res. 60, 932-937.

GORDON, M. (1973) Localisation of phosphatase activity on the membranes of the mammalian sperm head. J. exp. Zool. 185, 111-120.

GREEN, D.P.L. (1976) Induction of the acrosome reaction in guinea-pig spermatozoa *in vitro* by the Ca ionophore A23187. J. Physiol., Lond. 260, 18P-19P.

GREEN, D.P.L. (1978a) The induction of the acrosome reaction in guinea pig sperm by the divalent metal cation ionophore A23187. J. Cell Sci. (In press).

GREEN, D.P.L. (1978b) The activation of proteolysis in the acrosome reaction of guinea pig sperm. J. Cell Sci. (In press).

GREEN, D.P.L. (1978c) The osmotic properties of the acrosome in guinea pig sperm. J. Cell Sci. (In press).

GREEN, D.P.L. & HOCKADAY, A.R. (1978) The histochemical localisation of acrosin in guinea pig sperm after the acrosome reaction. J. Cell Sci. (In press).

GREENE, L.J., RIGBI, M. & FACKRE, D.S. (1966) Trypsin inhibitor from bovine pancreatic juice. J. biol. Chem. 241, 5610-5618.

HARRISON, D.G. & LONG, C. (1968) The calcium content of human erythrocytes. J. Physiol., Lond. 199, 367-381.

HOFMANN, J.F. (1958) Physiological characteristics of human red blood cell ghosts. J. gen. Physiol. 42, 9-28.

JAMIESON, J.D. & PALADE, G.E. (1971a) Condensing vacuole conversion and zymogen granule discharge in pancreatic exocrine cells: metabolic studies. J. Cell Biol. 48, 503-522.

JAMIESON, J.D. & PALADE, G.E. (1971b) Synthesis, intracellular transport and discharge of secretory proteins in stimulated pancreatic exocrine cells. J. Cell Biol. 50, 135-158.

KAFKA, M.S. & HOLTZ, R.W. (1976) Ionophores X537A and A23187. Effects on the permeability of lipid bimolecular membranes to dopamine and calcium. Biochim. Biophys. Acta 426, 31-37.

KANNO, T., COCHRANE, D.E. & DOUGLAS, W.W. (1973) Exocytosis (secretory granule extrusion) induced by injection of calcium into mast cells. Can. J. Physiol. Pharmac. 51, 1001-1004.

KASSELL, B. & KAY, J. (1973) Zymogens of proteolytic enzymes. Science N.Y. 180, 1022-1027.

MEIZEL, S. & LUI, C.W. (1976) Evidence for the role of a trypsin-enzyme in the hamster sperm acrosome reaction. J. exp. Zool. 196, 137-144.

MEIZEL, S. & MUKERJI, S.K. (1976) Biochemical studies of pro-acrosin and acrosin from hamster cauda epididymal spermatozoa. Biol. Reprod. 14, 444-450.

MIYAMOTO, H. & CHANG, M.C. (1973) Effects of protease inhibitors on the fertilising capacity of hamster sperm. Biol. Reprod. 9, 533-537.

MORGAN, P.H., ROBINSON, N.C., WALSH, K.A. & NEURATH, H. (1972) Inactivation of bovine trypsinogen by di-isopropylfluorophosphate. Proc. natn. Acad. Sci. U.S.A. 69, 3312-3316.

NEURATH, H. (1975) Limited proteolysis and zymogen activation. In: Reich, E., Rifkin, D.B. and Shaw, E. (Eds.), Cold Spring Harbor Conferences on cell proliferation. Vol. 2. Proteases and biological control. Cold Spring Harbor Laboratory, pp. 51-64.

PALADE, G.E., SIEKEVITZ, P. & CARO, L.G. (1962) Structure, chemistry and function of the pancreatic exocrine cell. In: de Reuck, A.U.S. and Cameron, M.P. (Eds.), Ciba Foundation Symposium on the exocrine pancreas. J. & A. Churchill, London, pp. 23-49.

PLATTNER, H. (1974) Intramembranous changes on cationophore-triggered exocytosis in Paramecium. Nature, Lond. 252, 722-724.

POLAKOSKI, K.L. & McRORIE, R.A. (1973) Boar acrosin. II. Classification, inhibition and specificity studies of a proteinase from sperm acrosomes. J. biol. Chem. 248, 8183-8188.

PRINCE, W.T., RASMUSSEN, H. & BERRIDGE, M.J. (1973) The role of calcium in fly salivary gland secretion analysed with the ionophore A23187. Biochim. biophys. Acta. 329, 98-107.

REED, P.W. & LARDY, H.A. (1972) A23187: a divalent cation ionophore. J. biol. Chem. 247, 6970-6977.

RIDGWAY, E.B., GILKEY, J.C. & JAFFE, L.F. (1977) Free calcium increases explosively in activating medaka eggs. Proc. natn. Acad. Sci. U.S.A. 74, 623-627.

RINK, T.J. (1977) Membrane potential of guinea-pig spermatozoa. J. Reprod. Fert. 51, 155-157.

RUBIN, R.P. (1970) The role of calcium in the release of neurotransmitter substances and hormones. Pharmac. Rev. 22, 389-428.

SCHLEUNING, W.-D., HELL, R. & FRITZ, H. (1976) Multiple forms of boar acrosin and their relationship to proenzyme activation. Hoppe-Seyler's Z. physiol. Chem. 357, 207-212.

STAMBAUGH, R., BRACKETT, B.G. & MASTROANNI, L. (1969) Inhibition of *in vitro* fertilisation of rabbit ova by trypsin inhibitors. Biol. Reprod. 1, 223-227.

STEFANINI, M., OURA, C. & ZAMBONI, L. (1969) Ultrastructure of fertilisation in the mouse. 2. Penetration of sperm into the ovum. J. submicroscop. Cytol. 1, 1-24.

STEINHARDT, R.A. & EPEL, D. (1974) Activation of sea-urchin eggs by a calcium ionophore. Proc. natn. Acad. Sci. U.S.A. 71, 1915-1919.

STEINHARDT, R.A., EPEL, D., CARROLL, E.J. & YANAGIMACHI, R. (1974) Is calcium ionophore a universal activator for unfertilised eggs? Nature, Lond. 252, 41-43.

STEINHARDT, R., ZUCKER, R. & SCHATTEN, G. (1977) Intracellular calcium release at fertilisation in the sea urchin egg. Develop. Biol. 58, 185-196.

SUMMERS, R.G., TALBOT, P., KEOUGH, E.M., HYLANDER, B.L. & FRANKLIN, L.E. (1976) Ionophore A23187 induces acrosome reactions in sea urchin and guinea pig spermatozoa. J. exp. Zool. 196, 381-386.

TALBOT, P. & FRANKLIN, L.E. (1974) The release of hyaluronidase from guinea-pig spermatozoa during the course of the normal acrosome reaction *in vitro*. J. Reprod. Fert. 39, 429-432.

TALBOT, P., SUMMERS, R.G., HYLANDER, B.L., KEOUGH, E.M. & FRANKLIN, L.E. (1976) The role of calcium in the acrosome reaction: an analysis using ionophore A23187. J. exp. Zool. 198, 383-392.

WOODING, F.B.P. (1975) Studies on the mechanism of the hyamine-induced acrosome reaction in ejaculated bovine spermatozoa. J. Reprod. Fert. 44, 185-192.

YANAGIMACHI, R. (1975) Acceleration of the acrosome reaction and activation of guinea pig spermatozoa by detergents and other reagents. Biol. Reprod. 13, 519-526.

YANAGIMACHI, R. & NODA, Y.D. (1970) Ultrastructural changes in the hamster sperm head during fertilisation. J. Ultrastruct. Res. 31, 465-485.

YANAGIMACHI, R. & USUI, N. (1974) Calcium dependence of the acrosome reaction and activation of guinea pig spermatozoa. Expl. Cell Res. 89, 161-174.

YANG, S.L., ZANEFELD, L.J.D. & SCHUMACHER, G.F.B. (1976) Effect of serum proteinase inhibitors on the fertilising capacity of rabbit spermatozoa. Fert. Steril. 27, 557-581.

ZANEFELD, L.J.D., ROBERTSON, R.T., KESSLER, M. & WILLIAMS, W.L. (1971) Inhibition of fertilisation *in vivo* by pancreatic and seminal plasma trypsin inhibitors. J. Reprod. Fert. 25, 387-392.

LIPIDS AND TRANSITIONS IN EMBRYOS

Hester P.M. Pratt

Department of Anatomy
Downing Street
Cambridge

Considerable research interest is currently being focussed on the cell membrane and its importance in the regulation of early development. It is clear that during the early differentiation of the embryo, this organelle has dual and apparently conflicting roles to perform; namely to maintain a balanced developmental sequence of cell division and metabolic activity while still retaining receptivity to changing intercellular contact and environmental stimuli. With the advent of the fluid mosaic model of membrane structure (Singer and Nicolson, 1972) and the realisation that membrane components can diffuse to greater or lesser extents within a matrix of variable fluidity, it has become easier to envisage the functioning of membranes in molecular terms. In fact the capacity of the embryo membrane to generate diverse lipid and protein interactions (Edelman, 1976) may be crucial to its regulative and permissive role in development. A picture has been built up of the morphological (Ducibella *et al.*,1975, 1977), enzymic (Borland and Tasca, 1974; Powers and Tupper, 1977) and antigenic (Johnson, 1975a) maturation of proteins and glycoproteins of the preimplantation embryo membrane. However the lipid component has been virtually ignored despite the fact that it comprises about half the membrane (Bretscher, 1972) and has been described as 'the centre, life and chemical soul of all bioplasm' (Thudichum, 1884). Consequently it has not been possible to view the embryo membrane as a whole or to examine the role that modifications in membrane structure and organisation might play in controlling or facilitating normal development.

In order to redress this balance I propose to develop the hypothesis that the structural lipids of the membrane bilayer have a regulative influence on the growth and differentiation of the pre-implantation mammalian embryo, either by affecting the plasticity of the membrane directly (Gitler, 1972), or by influencing the distribution and/or activity of pre-existing or newly synthesised components within it. To develop this argument, I shall first consider some of the physical properties of the lipid bilayer and the influence that membrane lipid composition can have on the biology of cells in general. This will then be followed by an examination of some of the membrane-mediated processes occurring during cleavage and compaction of the mouse embryo. Finally, I propose to discuss the evidence that the embryo can synthesize and modify its own lipids, to generate in association with proteins and glycoproteins, a variety of membrane complexes compatible with the spectrum of developmental changes that it has to undergo.

THE STRUCTURE OF THE LIPID BILAYER AND ITS INFLUENCE ON CELL FUNCTION

The major features of the lipid bilayer which make it such an adaptable matrix for the cell membrane are the diversity of the constituent phospholipids, glycolipids and sterols, their physical properties, and their ability to diffuse laterally and form specific associations with other components of the membrane (for reviews of membrane structure, see: Bretscher, 1972; Singer and Nicolson, 1972; Lee, 1975; Rothman and Lenard, 1977; Thompson and Nozawa, 1977). As a result of the amphiphilic nature of the majority of structural lipids the bilayer consists of two monolayers each composed of regularly packed lipid molecules with their polar head groups extending into the aqueous cytoplasmic or extracellular phase, and their hydrophobic fatty acyl chains and sterol rings extending into the non-aqueous centre of the bilayer. Both lipids and proteins are free to diffuse laterally within the plane of the membrane to extents determined by the physical state of the lipid and restrictions imposed by lipid-lipid associations (Demel *et al.*,1977) and interactions with membrane proteins including those of the cytoskeleton (Edidin, 1974; Edelman, 1976). A marked degree of lipid asymmetry also exists across the bilayer and this may be maintained by the relatively slow rate of transbilayer movement ('flip flop') of lipid (Rothman and Lenard, 1977).

The versatility of the lipid matrix also derives from the physical properties of the component phospholipid(and glycolipid) molecules and their ability to undergo transitions between two physical states; the liquid crystalline phase and the gel phase (see extensive reviews by Lee, 1975, 1977). In the more fluid and disordered state of the liquid crystal, the lipid fatty acyl chains have increased conformational freedom to rotate about C-C bonds and to flex within the hydrophobic core of the bilayer, as well as to diffuse laterally within the plane of the membrane. In the alternative conformational state, the gel phase, the fatty acyl chains are packed in ordered arrays, and molecular movement is highly anisotropic and restricted (Levine, 1972). The endothermic phase transition from the ordered gel phase to the more disordered liquid crystal can be induced experimentally by heating. This occurs at a transition temperature characteristic of the particular species of lipid. Differential scanning calorimetry, nuclear magnetic resonance spectroscopy and electron spin resonance of spin-labelled lipids have been used to follow conformational changes occurring during the phase transition (Lee, 1975, 1977; Levine, 1972), and data derived from these approaches demonstrate that the transition to the less ordered packing of fatty acyl chains in the liquid crystalline state is accompanied by an increase in volume and decrease in thickness of the bilayer. The transition temperature increases with increasing length of fatty acyl chains, decreases with increasing saturation in the chain, and depends upon the chemical nature of the polar head groups (Ladbrooke and Chapman, 1969). Consequently, the chemical nature of a particular species of lipid will determine whether it is in the liquid crystalline or gel phase at a particular temperature. The majority of phospholipids in mammalian plasma membranes have transition temperatures of 16°-18°C (Resch and Ferber, 1975) and are therefore in the fluid liquid crystalline state at 37°C. However, interactions with sterols, proteins and ions will influence the conformation of lipids and may lead to the coexistence of gel phase and liquid crystalline phase lipids within the bilayer of natural membranes as discussed below.

Cholesterol plays an important role in stabilising lipids in both natural and artificial membranes. The sterol rings are thought to align along the glycerol backbone and extend approximately ten carbon atoms into the interior of the bilayer. This interaction with sterol leads to a 'damping out' of lipid phase transitions due to the ability of cholesterol to condense those lipids in a liquid crystalline state and to disorder those lipids packed in the more regular array of the gel phase. In this manner the sterol imparts an 'intermediate fluid state' to the bilayer. The condensing effect will tend to decrease the permeability of the bilayer and increase its thickness while the introduction of disorder will act in an opposite direction to increase the volume occupied by lipid in the bilayer and consequently increase its permeability (Oldfield and Chapman, 1972; Demel and de Kruyff, 1976). These fluctuations in density within the bilayer may facilitate transport across the membrane as well as insertion of new components into it (Linden *et al.*,1973; Lee, 1975).

In addition to sterol binding, interaction with monovalent and divalent ions and variations in pH have been shown to have effects on the packing of lipid molecules in the bilayer analogous to those induced by variations in temperature (Trauble and Eibl, 1974; Lee, 1975). For example, increasing ionisation causes a decrease in transition temperature; consequently both the displacement of bound Ca^{++} and the ionisation of ionisable protons by an increase in pH can induce isothermal phase transitions in certain species of lipid (Lee, 1975). Studies of artificial bilayers of defined mixtures of lipid have demonstrated that segregated pools of lipid in the gel phase can co-exist in equilibrium with lipid in the liquid crystalline phase over a temperature and concentration range dependent upon the transition temperatures and relative amounts of the lipids making up the original mixture (Shimshick and McConnell, 1973). Evidence for similar phase separations in natural membranes at physiological temperatures has come from studies of β-galactoside transport in bacterial membranes supplemented with different fatty acids (Tsukagoshi and Fox, 1973). Discontinuities in plots of enzymatic activity, mobility of spin labelled lipids and lateral diffusion of antigens also provide evidence for heterogeneity in lipid distribution and co-existence of gel and liquid crystalline lipid within the cell membranes of higher organisms (Wisnieski *et al.*,1974; Petit and Edidin, 1974; Lee, 1975).

The most satisfactory experimental system for studying the influence of the physical state of membrane lipids on the properties of the membrane has been the fatty acid auxotrophs of *Escherichia coli* (Tsukagoshi and Fox, 1973; Overath and Trauble, 1973), since these bacteria cannot synthesise their own unsaturated fatty acids. Consequently membranes of defined lipid composition can be assembled by supplementing the medium with selected fatty acids. Using these mutants it has been shown that enrichment for unsaturated fatty acids increases the overall fluidity of both membranes and membrane lipids (Overath and Trauble, 1973) and alters the temperature at which membrane lipids begin to undergo phase separation (Linden *et al.*,1973; Tsukagoshi and Fox, 1973). Thermodynamic calculations demonstrate that lipids undergoing a phase separation into an equilibrium mixture of gel and liquid crystalline lipid have maximal lateral compressibility (Linden *et al.*,1973), which would be compatible with the optimal functioning of a transport enzyme or the insertion of a newly synthesised membrane protein. This is consistent with the observation that the temperature dependence of transport enzymes in modified membranes is altered to retain optimal activity over the temperature range of phase separation of the lipids (Linden *et al.*,1973; Tsukagoshi and Fox, 1973).

In higher organisms the enrichment of membranes with unsaturated fatty acids (Tourtellotte *et al.*,1970), and the depletion, both naturally occurring or experimentally-induced, of membrane cholesterol (Rottem *et al.*,1973; Inbar and Shinitzky, 1974a; Shinitzky and Inbar, 1974) leads to increased fluidity of membrane lipids. This decrease in order in the bilayer is also reflected in a change in the distribution of intramembranous proteins, as well as their rate of lateral mobility within the bilayer and degree of exposure above it (Edidin, 1974; Shinitzky and Inbar, 1976; Borochov and Shinitzky, 1976). The stabilising effect of cholesterol on membrane lipids surrounding a membrane enzyme (membrane ATP'ase) is well shown by the cholesterol depleted strain of *Mycoplasma* (Rottem *et al.*,1973). In the normal, unmodified cell the Arrhenius plot of ATP'ase activity against temperature is linear, however in the adapted strain a discontinuity is introduced suggesting that the lipids surrounding the enzyme undergo a thermally induced phase transition or phase separation, in the absence of the stabilising influence of cholesterol. This instability can be partially overcome by enriching the environment of the enzyme with more fluid fatty acids (Rottem *et al.*,1973). Other defined lipid environments required for optimal enzyme activity have been suggested for a number of enzymes including the Ca^{++} ATP'ase of sarcoplasmic reticulum (Warren *et al.*,1975), the Na channel of nerve (Lee, 1976) and (Na^{+}- K^{+})ATP'ase (Kimelberg and Papahadjopoulos, 1972, 1974).

Other more complicated cell functions are also crucially dependent upon the organisation of the cell membrane and seem to be related to the composition and physical state of the lipids within it. For example, the fluidity of bulk membrane lipid is known to fluctuate throughout the cell cycle (Cooper, 1977) and each stage must have a membrane of restricted phospholipid and cholesterol composition in order for the cell to progress normally through a mitotic division (Shodell, 1975; Hatten *et al.*1977; Kandutsch and Chen, 1977). In addition, the ability of a cell to receive signals and transmit information to the cell interior probably requires a degree of lipid and receptor mobility within the membrane (Bennett *et al.*,1975; Houslay *et al.*,1976). In fact, the interaction of differentiative stimuli (mitogens) with responsive cells (lymphocytes) is known to induce changes in lipid fluidity (Barnett *et al.*,1974) and stimulate the synthesis and turnover of membrane phospholipid (Resch *et al.*,1976) and cholesterol (Pratt *et al.*,1977). These modulations in membrane lipid composition occur very soon after interaction of the mitogen with the cell and may be associated with rearrangements of the membrane necessary for transmission of the signal to the genome since differentiation does not occur if these early lipid changes are inhibited (Alderson and Green, 1975; Pratt *et al.*,1977). Other membrane-mediated functions, also presumably dependent on intramembranous mobility, have been shown to have defined lipid requirements, for example, adhesion of cells (Curtis *et al.*,1975a,b,c), aggregation of platelets (Cooper, 1977), mitogen induced agglutination (Rittenhouse *et al.*,1974; Horwitz *et al.*,1974),phagocytosis (Cooper, 1977) and pinocytosis (Heiniger *et al.*,1976).

Another series of observations consistent with the notion that membrane lipid composition has to be maintained within certain limits in order for a cell to function normally are the well documented examples of human pathological conditions with cells of abnormal lipid composition (Cooper, 1977). For example, plasma membrane phospholipid and cholesterol are generally present in a 1:1 molar ratio (Bretscher, 1972): however deficiencies in synthesis or exchange (Cooper, 1977) may lead to cholesterol accumulation as in atherosclerotic aortae (Papahadjopoulos, 1974) and the red blood cells and platelets of spur cell anaemia (Cooper, 1977), or depletion of cholesterol as in leukaemic lymphocytes (Inbar and Shinitzky, 1974b; Pratt *et al.*,1978). Both forms of modification are associated with defects in membrane deformability and permeability to ions and macromolecules (Papahadjopoulos, 1974). In addition to cholesterol, the choline-containing phospholipids phosphatidylcholine and sphingomyelin have an important role to play in controlling membrane fluidity. Sphingomyelin decreases the fluidity of artificial bilayers as compared with phosphatidylcholine (Shinitzky and Barenholz, 1974) and the two lipids generally vary reciprocally within natural membranes. However, in red blood cells from patients with abetalipoproteinaemia an increase in sphingomyelin is associated with a decreased fluidity of membrane lipid, and similar changes are also observed in the aortic media with ageing or atherosclerosis. In contrast, the ratio of sphingomyelin:phosphatidylcholine of human pulmonary surfactant decreases during the final days of gestation inducing an increase in surfactant fluidity necessary for normal respiration (Cooper, 1977).

The available evidence therefore points to a limited range of lipid composition and organisation being compatible with normal cell function. Modifications in the physical state of membrane lipids which influence the organisation or activity of membrane proteins can presumably be introduced by local ionic or pH effects (Trauble and Eibl, 1974; Lee, 1975), or by the synthesis, turnover and exchange of the major lipids phospholipid and cholesterol. The enzymes for phospholipid synthesis are located almost exclusively on the endoplasmic reticulum, however other membranes can be assembled and modified by the transfer of phospholipids between membranes, a process which is catalysed by specific exchange proteins, and by the local action of deacylating and reacylating enzymes (for review, see Dawson, 1973). Membrane cholesterol is acquired by a combination of *de novo* synthesis, uptake from sterol containing lipoproteins and exchange between cell membranes (for review, see Brown and Goldstein, 1976). Thus there are a variety of lipid modifying mechanisms which can be invoked to modulate the properties of the cell membrane and adapt the cell to any change in function.

MEMBRANE LIPIDS AND THE EARLY DEVELOPMENT OF THE MAMMALIAN EMBRYO

In this section I shall discuss evidence to support the hypothesis that the lipids of the preimplantation mouse embryo are important for maintaining membrane properties compatible with normal cleavage, mitosis, compaction and subsequent differentiation. This will include the results of some experiments suggesting that the mammalian embryo is capable of synthesising and possibly modifying its own membrane lipids from a very early stage. Similar lipid synthesising and modifying mechanisms could probably be invoked to explain changes in membrane behaviour during earlier processes of oogenesis, spermatogenesis and fertilisation. In a recent review, Johnson (1975b) has discussed changes in the organisation of sperm membranes during maturation, capacitation and the acrosome reaction which are compatible with transitions between gel and liquid crystalline states in the lipid moiety of the membrane. Membrane changes occurring during the acrosome reaction and fertilisation are also discussed by Meizel in this volume.

CLEAVAGE AND CELL DIVISION

The bulk of plasma membrane components are thought to be synthesised during interphase (Graham *et al.*,1973) and it is likely that proteins and lipids are inserted coordinately into the membrane (Overath *et al.*,1971; Tsukagoshi and Fox, 1973; Rothman and Lodish, 1977). This is accompanied by some movement of lipid across the bilayer to maintain asymmetry (Rothman and Kennedy, 1977) and lateral diffusion of both lipid and protein to restore the configuration of the cell surface. During the G_1 and S phases of the cell cycle the membrane becomes highly villated, possibly to accommodate the additional synthesised membrane. The transition from the G_2 to the M phase is then envisaged as a physical stretching process culminating in the division of the cell with the aid of microfilaments (Graham *et al.*,1973). These processes evidently impose certain constraints on the properties of the membranes of dividing cells. For example, though the exact mechanism of insertion or fusion of newly synthesised lipoprotein into existing membrane is unclear (Dawson, 1973), the process itself implies a degree of fluidity within the pre-existing lipid to permit a balanced distribution of macromolecules between daughter cells and to allow the development of energetically and enzymically favourable associations between the lipids and proteins (Kimelberg and

Papahadjopoulos, 1972, 1974; Warren *et al.*,1975, Demel *et al.*,1977). The lipid composition of the cell membrane is apparently kept within narrow limits to achieve the necessary plasticity and fluidity to undergo membrane synthesis, reorganisation and division. Thus, cells enriched with (Alderson and Green, 1975) or depleted of (Pratt *et al.*,1977; Kandutsch and Chen, 1977) cholesterol are unable to progress from G_1 to S phase and phospholipids in a liquid crystalline state are essential for the passage of cells through the G_1 and G_2 phases (Shodell, 1975; Hatten *et al.*,1977).

However, the cleavage of the mammalian embryo may not be analogous to cell division in other cells since the cell cycle changes during early development (Graham, 1973), and cleavage does not represent a period of growth but rather a partitioning of the original egg cytoplasm. Using the simplest model which is to consider the individual blastomeres as perfect spheres, three cleavage divisions of the fertilized egg (diameter 100 μm) result in a two fold increase in surface area by the pre-compacted 8-cell stage (Izquierdo, 1977). This estimate is clearly complicated by the presence of microvilli on all embryonic cells including the fertilized egg (but excluding the smooth region surrounding the polar body (Eager *et al.*,1976)). If it is assumed that the micro-villi are uniformly distributed over the cell surface (Ducibella *et al.*,1977) then they probably increase the surface area by approximately one-third at all stages (Knutton *et al.*,1976). However, the necessity for assembly of new plasma membrane by the 8-cell stage is still evident, as well as the presumed need for replication of other membrane bound organelles including nuclei, mitochondria and smooth and rough endoplasmic reticulum. The structural lipids required (principally phospholipid and choles-terol) could either be synthesised *de novo*, or derived from lipid stores in the egg cytoplasm or taken up from genital tract fluids by pinocytosis or receptor mediated processes (Dawson, 1973; Brown and Goldstein, 1976). Yolky eggs like those of the sea urchin contain relatively large stores of lipid (approximately 28 μg per 1000 unfertilised eggs) of which one-third is phospholipid (Mohri, 1964), in contrast to the small amount in the mouse egg (approxi-mately 3 μg per 1000 unfertilized) eggs of which about one half is cholesterol and its esters (Loewenstein and Cohen, 1964; Pratt, unpublished observations). On the basis of previous estimates of the phospholipid and cholesterol content of resting lymphocytes (Allan and Crumpton, 1970; Pratt *et al.*,1978), the lipid stored in the unfertilized mouse egg is sufficient to assemble the internal and plasma membranes of >1000 preimplantation embryos. However, these lipids may not be used exclusively for membrane assembly since mammalian embryos are known to utilise lipids as an energy source (Kane, 1976). Another potential source of lipid or lipid precursors during early development is the oviducal or uterine fluid surrounding the embryo. Sterols are present in uterine fluids (Fowler *et al.*,1977; Pratt, unpublished observations) and could be taken up into embryonic cells via a receptor-mediated or pinocytotic process (Brown and Goldstein, 1976) or via an exchange process (Cooper, 1977) partitioning between oviducal or uterine lipoproteins and embryo membranes. Phospholipid exchange between cells and their medium is also well documented (Peterson and Rubin, 1969; Dawson, 1973). However, in view of the fact that embryos can be cultured *in vitro* in the absence of exogenous lipid provided that an energy source is supplied (Brinster, 1965), it is unlikely that genital tract fluids provide an important source of lipid for the preimplantation embryo. Nevertheless, microheterogeneity could be introduced into membranes by local lipid exchange or other modifications and these will be discussed more fully in association with later development of the embryo.

It still remains to be ascertained, therefore, whether the preimplantation mouse embryo is metabolically equipped to synthesise the major structural lipids of its cell membranes, phospholipid and cholesterol. Studies of the sea urchin have demonstrated *de novo* synthesis of phospholipid during cleavage (Mohri, 1964; Pasternak, 1973) and some incorporation of acetate into cholesterol (Mohri, 1964). Since unesterified cholesterol is predominantly membrane associated and is deficient in intracellular membranes (Bretscher, 1972), it can probably be assumed that this synthesis of cholesterol is associated with assembly of new plasma membrane. Phospholipids, on the other hand, are distributed generally throughout cellular membranes (Kuff *et al.*,1966) and consequently their synthesis is only indicative of membrane assembly in general. There is, however, very little information about the ability of the mammalian embryo to synthesise its own membrane lipids and it is clearly necessary to resolve this question before elaborating any hypotheses concerning the role of membrane lipids in modulating membrane properties and influencing embryonic differentiation.

Phospholipid synthesis

Phosphatidylcholine (lecithin) and phosphatidylethanolamine (cephalin) together comprise the majority of membrane phospholipid and hence are major contributors to the bulk properties of the lipid bilayer (Resch and Ferber, 1975; Bretscher, 1972). Synthesis of the choline-containing lipids, phosphatidylcholine, lysolecithin (lysophosphatidylcholine) and sphingomyelin, was studied using methyl-^{3}H choline as a specific precursor (Bergeron *et al.*,1970). In view of the anticipated large and variable size of the cold choline pool in these mouse embryos, external levels of labelled choline were chosen which saturated choline incorporation into lipid irrespective of the amount of unincorporated choline transported into the cells (Daentl and Epstein, 1971). Under these conditions, the amount of choline incorporated into individual phospholipids could be used as an index of phospholipid, and hence membrane, synthesis (Bergeron *et al.*,1970). It was possible to saturate the lipid precursor choline pool with methyl-^{3}H choline at all embryonic stages except the blastocyst. Phospholipids and their biosynthetic intermediates (Kennedy, 1957) were analysed and identified using conventional lipid extraction and thin layer chromatography techniques (Figure 1) (Dawson, 1960; Plagemann, 1968; Pasternak, 1973). Uptake and incorporation of methyl-^{3}H choline was linear with time and Table 1 illustrates the extent of choline incorporation into major biosynthetic intermediates of the Kennedy pathway for phosphatidylcholine synthesis (Kennedy, 1957; Plagemann, 1968) by 2-cell embryos, compacting morulae and blastocysts. Methyl-^{3}H choline transported into the cells is phosphorylated to form phosphorylcholine inferring an active choline kinase at all stages, and this has been subsequently confirmed using an *in vitro* enzyme assay (Schneider, 1969). However, only about 10% of the methyl-^{3}H choline taken up by embryos is phosphorylated, in accordance with results obtained from cleaving sea urchin eggs (Pasternak, 1973) but in direct contrast to the 100% phosphorylation of transported choline observed in rapidly metabolising hepatoma cells (Plagemann, 1968; Plagemann and Roth, 1969). This suggests that choline kinase activity may be a limiting factor in the synthesis of phospholipid in immature cells, such as those of the preimplantation mouse embryo and the cleaving sea urchin. Following the formation of phosphorylcholine, the Kennedy pathway involves the generation of a cytidine diphosphocholine intermediate

TABLE 1: UPTAKE OF METHYL-^{3}H CHOLINE AND ITS INCORPORATION INTO PHOSPHOLIPID BY MOUSE EMBRYOS* (Pratt, unpublished)

	Choline (cpm x 10^{-3})	Phosphorylcholine (cpm x 10^{-3})	Lysolecithin (cpm x 10^{-3})	Phosphatidylcholine (cpm x 10^{-3}	(pmol)
Embryonic stage					
2-cell	0.590	0.090	0.017	0.310	0.3
8-cell morula	63.673	8.143	0.145	0.845	0.8
Blastocyst	108.355	11.763	0.409	2.200	2.0

* Embryos were incubated for 4 hours in (methyl-^{3}H) choline chloride (8.4 Ci/mmol), 0.25mM, washed in medium containing 1 mM choline, lysed and the lipids extracted as described by Pasternak (1973) and analysed on thin layer plates according to Plagemann (1968) and Plagemann and Roth (1969) (see legend to Figure 1). Results of a typical experiment are shown, counts (or pmol radioactive choline) incorporated are the mean of duplicates and expressed per 10 embryos.

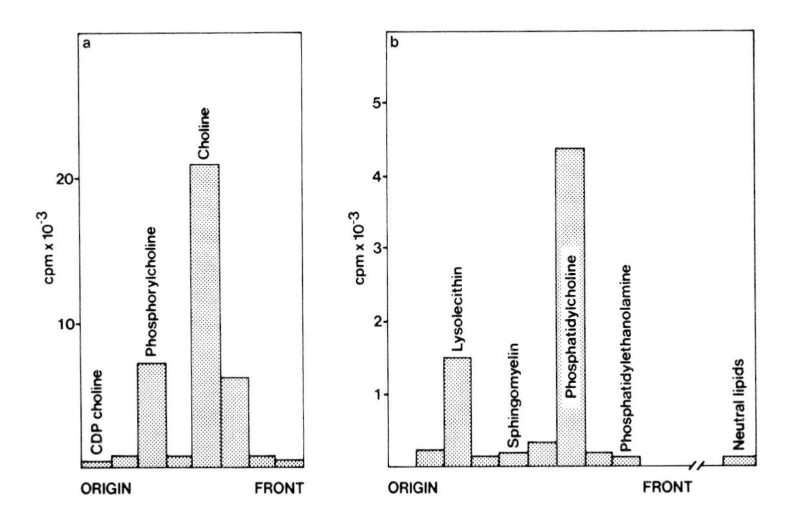

Figure 1: Representative chromatographic separations of methyl-^{3}H choline labelled lipids and lipid precursors extracted from mouse embryos. Embryos were incubated in culture medium containing 4mg/ml bovine serum albumin, 0.25mM methyl-^{3}H choline chloride (8.4Ci/mmol), for 4 hours, washed in non-radioactive medium containing 1mM choline and the lipid and non-lipid material separated by acid precipitation and centrifugation as described by Pasternak (1973). (1a) Acid soluble material was applied to a cellulose thin layer plate and developed with the solvent n-butanol: ethanol:acetic acid: H_2O (8:2:1:3 vol/vol). Reference standards were chromatographed on every plate and identified by exposure to iodine vapour or elution and spectrophotometric assay (as described by Plagemann, 1968; Plagemann and Roth, 1969). Experimental samples were analysed by cutting the cellulose into strips, eluting the material into 0.5ml H_2O and counting the eluate in a Triton-toluene scintillation cocktail. Recovery of applied radioactivity was 75-80%. (1b) The acid insoluble precipitate containing the lipid was washed twice with 5% trichloroacetic acid and then dissolved in 0.2ml chloroform:methanol (2:1 vol/vol). This lipid extract was washed twice with 0.1ml 0.58% NaC1 (Pasternak, 1973) dried down, dissolved in 5-10µl of chloroform:methanol (2:1 vol/vol) and applied to a thin layer plate of silica gel in addition to samples of standard known lipids. The plates were developed with the solvent:chloroform:methanol: NH_4OH (65:25:5 vol/vol) (Resch *et al.*,1976), and standard lipids identified by exposure to iodine vapour. Experimental samples were scraped from the plate and counted directly in Triton-toluene scintillant. Recovery of applied counts was 85-95%.

(CDP-choline) which then combines with diacylglycerol to form phosphatidylcholine (Kennedy, 1957). The CDP-choline intermediate was not detected in extracts of mouse embryos; however, this is not surprising since cells are known to have a very small pool of this short-lived metabolite (Kennedy, 1957;

Plagemann, 1968). Choline phosphotransferase activity has not yet been measured in preimplantation embryos but has been shown to be present in the embryo, placenta and yolk sac of the 14 day old rat (Chepenik *et al.*,1977).

The only lipids to incorporate methyl-^{3}H choline were phosphatidylcholine and lysolecithin. The absence of labelled sphingomyelin, even after long labelling periods, may be interesting since it is also deficient in the membranes of cleaving sea urchins (Pasternak, 1973) and a deficiency of sphingomyelin has been correlated with immaturity in other cells (Gottfried, 1967, 1971). Sphingomyelin is also known to decrease the fluidity (increase the microviscosity) of artificial lipid bilayers (Shinitzky and Barenholz, 1974) and in natural membranes marked increases in membrane sphingomyelin are found in association with pathogenic reductions in the fluidity of membrane lipids (Cooper, 1977). Sphingomyelin is also a more stable membrane lipid than phosphatidylcholine, with a much lower rate of turnover (Pasternak and Friedrichs, 1970). It appears therefore, that during this period of developmental lability from the cleavage to blastocyst stage, embryos incorporate lipids into their cell membranes which impart fluidity to the bilayer and facilitate reorganisation of pre-existing components as well as permitting possible local modulations in composition through lipid turnover. The predominance of lysolecithin among the methyl-^{3}H labelled lipids may be of interest in this context, since lysolecithin is known to have a disordering effect on the lipid bilayer which facilitates protein insertion (Eytan *et al.*, 1975). Its presence also indicates that embryos have phospholipase A type activity, which is responsible for cleavage of fatty acyl chains from phospholipids and hence can contribute to fatty acid turnover (Dawson, 1973). Since lysolecithin itself, phospholipase A and fatty acid turnover have been discussed in relation to the adhesion and fusion of cell membranes (Poste and Allison, 1973; Curtis *et al.*,1975c), further consideration of the relevance of these molecules will be deferred to the section on the compaction of the morula.

Incorporation of labelled choline into phospholipid can only be taken as indicative of phospholipid synthesis if the possibility of extensive turnover is eliminated. When morulae and blastocysts were prelabelled for 24 hours, and then washed and cultured in non-radioactive medium, the labelled phosphatidylcholine declined with a half life of approximately 24 hours, consistent with the rate of turnover of this lipid in other cells (Pasternak and Friedrichs, 1970). However, the labelled phospholipid remained stable for short periods and hence incorporation of methyl-^{3}H choline into lipid during the 4 hour labelling period can probably be equated with phospholipid and consequently membrane synthesis (Bergeron *et al.*,1970). Table 1 gives values for the picomoles of radioactive choline (diluted to an unknown extent by the internal choline pool) incorporated into lipid at different embryonic stages. In the case of the 2-cell and 8-cell embryos, it was possible to dissociate incorporation of choline into lipid from uptake of choline into the embryo (Daentl and Epstein, 1971), and the results indicate that synthesis of phospholipid increases approximately 3-fold during these two cleavage divisions. It therefore seems likely that the demand for new membranes resulting from a doubling in surface area and associated proliferation of membrane bound organelles is met by *de novo* synthesis of phospholipid in addition to any utilisation of pre-existing stores. The rate of synthesis increases up to the blastocyst, however it was not possible to saturate the lipid precursor pool at this stage and the value obtained is probably an underestimate of the extent to which blastocysts are synthesising phosphatidylcholine.

In conclusion, the mammalian embryo appears to possess the necessary endoplasmic reticulum enzymes and transport proteins (Plagemann and Roth, 1969; Dawson, 1973) to synthesise phospholipids via the Kennedy pathway (Kennedy, 1957) from the 2-cell stage onwards. In addition, from the limited information available, it appears that the phospholipid composition of embryo membranes is designed to allow maximum lability and receptivity during this period of incipient differentiation. However, nothing as yet is known about the regulation of this membrane lipid synthesis, for example whether it is restricted to one stage in the cell cycle and coupled with DNA synthesis, as appears to be the case in the cleaving sea urchin (Pasternak, 1973), and whether synthesis is transcriptionally controlled or dependent upon messenger RNA templates stored in the cytoplasm.

Cholesterol synthesis

The second important structural lipid of the cell membrane to be considered is cholesterol. As discussed previously, this sterol in combination with phospholipid is a primary modulator of the fluid and enzymic properties of the mammalian cell membrane

(Demel and de Kruyff, 1976). In contrast to phospholipids which are distributed throughout cellular membranes, cholesterol is located primarily in the plasma membrane, in 1:1 molar ratio with phospholipid (Bretscher, 1972; Rothman and Lenard, 1977), and is often used as a marker molecule for this organelle. The synthesis of cholesterol from acetate takes place in the microsomes and involves the condensation of three molecules of acetyl-Co A to form mevalonic acid, catalysed by hydroxymethylglutaryl Co A reductase (HMG Co A reductase) the rate limiting enzyme in cholesterol biosynthesis (Brown and Goldstein, 1974). The mevalonate thus formed is converted via a series of reactions into squalene, a dihydrotriterpene, which undergoes cyclisation to form the first sterol in the sequence, lanosterol, and subsequently cholesterol (Popjak and Cornforth, 1966). However, *de novo* synthesis is not the only way in which cells obtain cholesterol for membrane assembly. In some cells (including fibroblasts, lymphocytes and hepatocytes) synthesis of sterols is suppressed by the action of cholesterol, bound to low density lipoproteins, on the rate limiting enzyme HMG Co A reductase (Brown and Goldstein, 1976). In the face of this physiological suppression, one way in which the cells can acquire membrane cholesterol is via a low density lipoprotein receptor-mediated pathway (Brown *et al.*,1975; Brown and Goldstein, 1976), with surplus sterol being stored in the form of cholesterol esters. Alternatively, cholesterol can be taken up from lipoproteins by exchange processes (Bailey and Butler, 1973) or as a consequence of pinocytosis (Goldstein and Brown, 1974).

The only recent studies of sterol synthesis in early embryos are those of Mohri and Huff and Eik-Ness who demonstrated incorporation of acetate into sterols of early sea urchin embryos (Mohri, 1964) and 6 day old rabbit blastocysts (Huff and Eik-Ness, 1966). The evidence for sterol synthesis in mouse embryos that I propose to discuss has been obtained from experiments utilising radioactive precursors of cholesterol and a series of studies analysing the effects of treating embryos with specific inhibitors of cholesterol synthesis.

Tritium labelled mevalonic acid was used as a specific, high activity precursor for embryo sterols (Kandutsch and Saucier, 1969) and embryos were incubated with this precursor for varying lengths of time. Mouse lymphocytes and P815 mastocytoma cells were used in comparison as examples of rapidly metabolising cell types known to synthesise sterols *in vitro* (Pratt *et al.*,1977, 1978). The results of preliminary experiments of this type are presented in Table 2. The lymphocytes and mastocytoma cells readily transported mevalonic acid, of which approximately 7% was incorporated into cholesterol as identified by its Rf value on thin layer plates (Figure 2) and precipitation with digitonin (Pratt

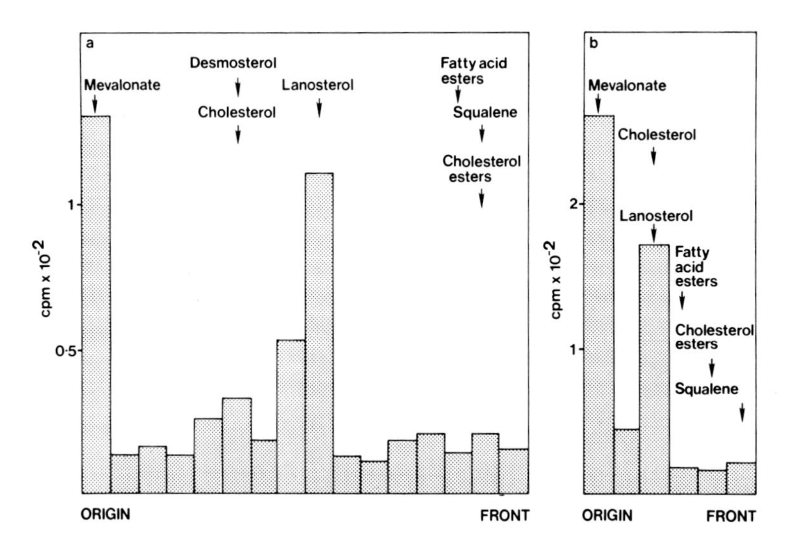

Figure 2: Representative chromatographic separations of lipids synthesised from ^{3}H mevalonic acid by mouse embryos. Embryos were cultured in medium containing 4mg/ml bovine serum albumin, 0.08mM mevalonic acid. DL - (mevalonic acid-5-^{3}H(N)) DBED salt, 5Ci/mmol New England Nuclear Enterprises. The dibenzylethylene diamine was removed with alkali and extraction with ether as described by the suppliers. After culture, embryos were washed in non-radioactive medium containing 1mM mevalonic acid, lysed by freeze-thawing twice in 20 µl of H_2O and the lipids extracted with 0.2ml chloroform:methanol (2:1 vol/vol) containing 50 µg of cholesterol as carrier, as described by Kandutsch and Chen (1973). (2a) Chromatography in one dimension. Experimental samples and standard lipids were dissolved in 5-10 µl of chloroform: methanol (2:1 vol/vol), applied to a thin layer plate of silica gel and developed with the solvent petroleum ether:ether:acetic acid, (40:60:0.1 (vol/vol) (Bowyer and Davies, 1976). Standard lipids were visualised by exposure to iodine vapour and experimental samples were scraped from the plate and counted directly in Triton-toluene scintillation fluid. (2b) Chromatography in two dimensions. The sample of ^{3}H mevalonic acid labelled lipid was combined with standard lipids and separated in one dimension using the solvent system described in Fig.2a. The plate was then dried, rotated through 90° and the lipids separated in the second dimension using the solvent petroleum ether:acetic acid:ethyl acetate (95:2:2) (vol/vol) (Bowyer and Davies, 1976). The plate was then dried, the lipids were visualised with iodine vapour and analysed by scraping samples from the plates and counting them directly in Triton-toluene scintillation fluid.

TABLE 2: ASSAYS FOR THE SYNTHESIS OF STEROLS (Pratt, unpublished)

Precursor		^{3}H mevalonic acid*			^{14}C-HMG Co A**		
Product:		^{3}H Sterol (pmol)	^{3}H Sterol (cpm x 10^{-3})	Total uptake (cpm x 10^{-3})			^{14}C mevalonate (cpm x 10^{-3})
Cells or embryos	No.				Cells or embryos	No.	
P815	10^{6}	10.0	10.0	100.0	P815	10^{6}	50.0
Mouse lymphocytes	10^{6}	6.0	6.0	76.0			
Morulae	10	0.180	0.180	0.200	Morulae	10	0.100
Blastocysts	10	0.250	0.250	0.350	Blastocysts (grown from morulae *in vitro*)	10	0.350
					Blastocysts	10	0.400

*P815 mastocytoma cells, mouse lymphocytes and embryos were incubated for 6 hours in 0.08 mM ^{3}H mevalonic acid and the lipid and non-lipid material extracted as described by Brown and Goldstein (1974). Lipids were chromatographed in one dimension and analysed as described in the legend to Figure 2a. The ^{3}H mevalonic acid incorporated into mastocytoma cells and lymphocytes chromatographed with cholesterol whereas ^{3}H mevalonic acid incorporated into morulae or blastocysts chromatographed with lanosterol. 75-90% of incorporated radioactivity was precipitated with digitonin (Pratt *et al.*,1977).

**P815 mastocytoma cells, lymphocytes or embryos were assayed for HMG Co A reductase activity following lysis in the reaction buffer (Shapiro *et al.*,1969), the addition of 2 μCi of Hydroxymethyl glutaryl-3^{14}C coenzyme A (49.5 mCi/mmol) to each tube and incubation at 37°C for 1 hour. ^{14}C mevalonate was extracted and analysed on thin layer plates as described by Shapiro *et al.*,(1969) with reference to labelled standards. Results given are the mean of duplicates. Reaction buffer alone gave no incorporation. Similar levels of incorporation of ^{14}C-HMG Co A into mevalonate were obtained when intact cells and embryos were incubated with ^{14}C-HMG Co A.

et al.,1977). The uptake of mevalonate by embryos, when calculated on a per cell basis, was equivalent to that observed with the mastocytoma cells and lymphocytes. However, the majority of the precursor transported into the cells (60-80%) was incorporated into a lipid, which on analysis in a 2 dimensional chromatographic system migrated with lanosterol (Figure 2) and was also precipitable with digitonin, sustaining its sterol nature (Pratt *et al.*,1977). This metabolite has been found in both morulae and blastocysts but not in 2-cell embryos which convert transported ^{3}H mevalonic acid into a molecule behaving like squalene after lipid extraction and thin layer chromatography. The implication that preimplantation mouse embryos are capable of synthesising membrane sterols is further strengthened by evidence for the activity of the rate limiting enzyme, HMG Co A reductase, at these early stages. Embryos, either intact or osmotically lysed in the enzyme assay buffer, were analysed for their capacity to catalyse the conversion of ^{14}C-HMG Co A to mevalonate (Shapiro *et al.*,1969; Goldfarb and Pitot, 1971) in comparison with mastocytoma cells or medium alone. Intact embryos and mastocytoma cells were both permeable to ^{14}C-HMG Co A and converted the label to a faster migrating molecule with an Rf value equivalent to that of mevalonate. A similar conversion was observed in the *in vitro* enzyme assay (Shapiro *et al.*, 1969; Goldfarb and Pitot, 1971) when lysates of morulae or blastocysts were used. When calculated on a per cell basis, the extent of conversion of ^{14}C-HMG Co A to mevalonate was similar for mastocytoma and embryo cells, though the blastocysts were more active than the morulae (Table 2). The possibility that embryonic sterol synthesis was being suppressed by sterols and steroids in the genital tract (Fowler *et al.*,1977) was examined by removing morulae from the oviduct, allowing them to develop to blastocysts *in vitro* (in a medium devoid of cholesterol and containing bovine serum albumin as the only macromolecule) and comparing them with blastocysts which had developed *in vivo*. Under these conditions lymphocytes or hepatocytes would be anticipated to increase the synthesis of sterols 3-6 fold (Pratt *et al.*,1977; Edwards *et al.*,1976). However, there was no difference in sterol synthesis between the two groups of embryo when either ^{3}H mevalonic acid or ^{14}C-HMG Co A (Table 2) were used as precursor, and consequently no evidence for environmental regulation of sterol synthesis in embryos.

Though these results need to be confirmed and extended, the available evidence suggests that mouse embryos from 8-16 cell morula

TABLE 3: GROWTH OF 2 CELL MOUSE EMBRYOS SUPPLEMENTED WITH STEROLS*

Sterol	μg/ml	% of 2 cell embryos undergoing cleavage	(number of embryos assayed)
None	-	87	(23)
Cholesterol	100	86	(14)
	50	100	(6)
25-hydroxycholesterol	100	83	(6)
	25	100	(6)
	10	88	(8)
6-ketocholestanol	25	0	(10)
	10	17	(24)
	1	100	(6)
7-ketocholesterol	25	0	(10)
	10	11	(18)
	1	10	(6)
6-ketocholestanol	10	8	(12)
+ 1 mM mevalonate	10	0	(6)
+ 10 mM mevalonate	10	0	(6)
+ 100 μg/ml cholesterol	10	67	(12)
+ 50 μg/ml cholesterol	10	38	(8)
7-ketocholesterol	10	4	(26)
+ 1 mM mevalonate	10	0	(6)
+ 10 mM mevalonate	10	0	(14)
+ 100 μg/ml cholesterol	10	30	(20)
+ 50 μg/ml cholesterol	10	38	(8)
10 mM mevalonate	-	67	(6)

* Embryos were cultured in 0.1 ml of medium in the wells of microtitre plates, in an atmosphere of 95% air, 5% CO_2 at 37°C. The plates were covered with non toxic adhesive covers which prevent evaporation but permit gas exchange. Paraffin oil was not used as it would solubilise the sterols. Stock solutions of sterols were made up in benzene and stored under N_2 at 4°C. These stock solutions were diluted into a 5% solution of bovine serum albumin (BSA) as described by Kandutsch and Chen (1973) and then added at the appropriate concentrations to the culture medium (Pratt and Keith, unpublished).

stage onward possess the enzymes required to synthesise sterols, though the earlier cleavage stages may not have the later enzymes in the sequence. However, the fact that the only sterol(s) found did not appear to be cholesterol and was tentatively identified as lanosterol, a cholesterol precursor generally found in only small amounts in mammalian membranes (Demel and de Kruyff, 1976), suggests that either the sterol composition of the embryonic cell membrane is somewhat unusual, or else that the newly synthesised sterol is stored and only subsequently converted to cholesterol for membrane assembly. As discussed previously, the unfertilised egg contains 1-2 ng of cholesterol and cholesterol esters, which on the basis of 40 pmol cholesterol per 10^{10} lymphocytes (diameter 10 μm) (Pratt *et al.*,1978) exceeds the cholesterol required for the plasma and internal membranes of all the cells in the blastocyst. Consequently there is no apparent immediate demand for these newly synthesised sterols. Further work will be needed to assess the relative importance of *de novo* synthesis as opposed to utilisation of pre-existing stores or uptake from the environment in contributing to the cholesterol composition of embryo membranes.

Another approach to the study of embryonic sterol synthesis and the relationship between membrane composition and the ability of cells to undergo DNA synthesis and cytokinesis has been the use of certain oxygenated derivatives of cholesterol which specifically depress hydroxymethylglutaryl Co A reductase activity in human and mouse cells (Brown and Goldstein, 1974) and deplete cellular membranes of cholesterol (Kandutsch and Chen, 1977). When 2-cell mouse embryos were cultured in a number of these inhibitory sterols (Table 3), cleavage was inhibited in a dose dependent manner with the severity of inhibition varying according to the effectiveness of the particular sterol in suppressing HMG Co A reductase activity (Brown and Goldstein, 1974). In comparison, 2-cell mouse embryos cleave and develop normally to the blastocyst stage in media containing up to 100 μg/ml cholesterol. Cells of inhibited embryos swell within 12 hours of addition of the inhibitor, and ultimately lyse. This behaviour would be anticipated of cells with membranes depleted of cholesterol since, as discussed previously, cholesterol has a stabilising effect on membranes condensing the phospholipids, increasing bilayer thickness and decreasing membrane fluidity (Demel and de Kruyff, 1976). Removal of cholesterol destabilises the membrane, increases the motional freedom of the phospholipid fatty acyl chains and results in increased permeability and fragility of

TABLE 4: GROWTH OF PREIMPLANTATION MOUSE EMBRYOS SUPPLEMENTED WITH STEROLS*
(Keith and Pratt, unpublished)

Sterol	μg/ml	% of morulae undergoing compaction	(number of embryos assayed)
None	-	100	(19)
Cholesterol	100	82	(11)
	50	100	(13)
25-hydroxycholesterol	100	100	(5)
	25	100	(13)
	10	86	(16)
6-ketocholestanol	25	0	(6)
	10	6	(17)
	1	80	(5)
7-ketocholesterol	25	0	(8)
	10	5	(19)
	1	80	(5)
6-ketocholestanol	10	6	(17)
+ 1 mM mevalonate	10	67	(6)
+ 10 mM mevalonate	10	64	(14)
+ 100 μg/ml cholesterol	10	75	(20)
+ 50 μg/ml cholesterol	10	100	(8)
7-ketocholesterol	10	0	(24)
+ 1 mM mevalonate	10	0	(14)
+ 10 mM mevalonate	10	29	(14)
+ 100 μg/ml cholesterol	10	21	(19)
+ 50 μg/ml cholesterol	10	0	(8)
10 mM mevalonate	-	83	(6)

* For experimental details, see Table 3.

the membrane, events which are evidently incompatible with normal membrane assembly, DNA synthesis and cell division. Though cleavage of 2-cell embryos was clearly affected within 12 hours of inhibition, treated morulae continued through one or two more cycles of cell division before lysing, as noted for mouse foetal liver cells (Kandutsch and Chen, 1977). This increased susceptibility of early cleavage stages may indicate that there are more constraints on the sterol composition of cell membranes during cleavage than during mitosis of later stage embryos. However, the membranes of inhibited morulae though capable of cytokinesis cannot form the close intercellular appositions necessary for compaction. This presumed consequence of membrane cholesterol depletion will be discussed more fully in a later section.

Though it has been assumed that these inhibitory sterols are acting on an embryonic HMG Co A reductase enzyme in an analogous manner to that described for human and murine cells (Kandutsch and Chen, 1973, 1977; Brown and Goldstein, 1974; Pratt *et al.*,1977), it is possible that 7-ketocholesterol and other derivatives of cholesterol are incorporated into the membrane and exert a direct effect on membrane properties by modifying phospholipid packing and disrupting specific phospholipid-protein interactions (Demel *et al.*, 1972a,b). However, the fact that products of the HMG Co A to mevalonate conversion, mevalonate and cholesterol, can overcome inhibition of mitosis and lead to normal development of morulae (Table 4) argues that 7-ketocholesterol and 6-ketocholestanol are acting, at least in part, via specific repression of the HMG Co A reductase rather than by direct incorporation into the membrane. These two alternatives can be distinguished experimentally by measuring HMG Co A reductase activity directly in inhibited embryos. Inhibition of 2-cell embryos, on the other hand, can only be overcome by cholesterol supplementation - mevalonate is ineffective (Table 3). This may indicate that the inhibitory sterols are being incorporated directly into the membrane, possibly by an exchange process (Bailey and Butler, 1973) and that a membrane composition compatible with normal cleavage is only restored by further exchange with supplemented cholesterol.

These studies of lipid metabolism strongly suggest that cleavage of early embryos and mitosis of morulae and blastocysts are associated with synthesis of phospholipid and sterols, and hence are assumed to involve the assembly of new internal and plasma membranes. This conclusion is supported bv the appearance of new surface antigens during this phase of development (Muggleton-Harris and Johnson, 1976; Krco and Goldberg, 1976). Phospholipid synthesis is detectable at the 2-cell stage (the earliest stage so far studied) and increases throughout development to the blastocyst (Table 1), turnover of the choline moiety being apparent from the precompacted 8-cell stage onwards. This phospholipid turnover coupled with the absence of a stabilising lipid like sphingomyelin, suggests that embryo membranes retain a fluid, adaptable lipid matrix during this period which is capable of modifying the lipid and protein interactions within it and hence adapting to developmental changes. The available evidence suggests that the synthesis of sterols is probably occurring in morulae and later stages, though at a level appreciably lower than the synthesis of phospholipid (compare Tables 1 and 2), which may not be surprising, considering the more restricted membrane distribution of cholesterol (Bretscher, 1972) and its much greater stability (Kandutsch and Saucier, 1969). The results of these experiments highlight the role that membrane lipids play in the normal cleavage and mitosis of the preimplantation embryo and specifically emphasise the importance of maintaining the relative levels of phosphatidylcholine and sphingomyelin, and phospholipid and cholesterol within a limited range compatible with the membrane reorganisations necessary for normal cytokinesis (Cooper, 1977).

COMPACTION OF THE MORULA AND SUBSEQUENT DIFFERENTIATION OF THE EMBRYO

Having discussed the ability of the mouse embryo to synthesise its own membrane lipids, I now propose to consider mechanisms whereby the embryo could modify these lipids and alter the properties of the membrane in ways conducive to further development and differentiation.

I have chosen to use the compacting morula as a model for this discussion since it represents a crucial turning point in development during which 'inside' cells are isolated from 'outside' cells and their respective differentiation into inner cell mass and trophectoderm is initiated (Johnson *et al.*,1977). This isolation is accomplished by the formation of tight intercellular junctions (Ducibella and Anderson, 1975; Ducibella *et al.*,1975, 1977), which may be associated with the transmission of developmental signals and is accompanied by a general increase in membrane transport activities and the accumulation of fluid to form the beginning of the blastocoelic cavity (Daentl and Epstein, 1971; Epstein and Smith, 1973; Borland and Tasca, 1974). The development of intercellular adhesions, together with the putative capacity to emit and respond to developmental stimuli and the changes in permeability to ions and metabolites, are all indicative of modifications in the dynamic properties of embryo membranes at this stage. I propose to develop the argument that changes in lipid composition could form the basis of, or be critically associated with, these membrane changes.

As discussed earlier, the physical state of the lipid bilayer is now known to have an important influence on the organisation and activity of membrane proteins (Lee, 1975) as well as affecting the overall plasticity of the membrane (Gitler, 1972). The introduction of 'order' or 'disorder' into the lipid bilayer could occur over a wide range possibly involving the whole cell membrane, while short range effects could be restricted to regions of lipid surrounding transport enzymes (Warren *et al.*,1975) or cytoskeletal proteins (Berlin, 1975; Edelman, 1976). It seems unlikely that modulations in specific aspects of cell behaviour will be induced by changes in lipid organisation throughout the cell membrane, since these long range changes might not be compatible with other normal functions of the cell. One possible exception to this is the membrane of the mammalian egg, where lateral mobility of both protein and lipid appears to be restricted after fertilisation, possibly due to incorporation of cortical granule membranes throughout the egg surface (Johnson and Edidin, 1978). Short-range changes in lipid organisation could be introduced into the bilayer by localised increases in turnover of phospholipid head groups and fatty acyl chains or the induction of isothermal phase transitions by mono or divalent cations and pH as described in studies of artificial bilayers (Trauble and Eibl, 1974; Lee, 1975). Addition or depletion of cholesterol will influence the thickness of the bilayer through its condensing effect on phospholipid fatty acyl chains and preferred interactions with specific phospholipids (Demel *et al.*,1977). These lipid-lipid interactions may also influence the conformation of membrane proteins which could, in turn, favour certain lipid-protein associations and lead to regions of immobilised 'boundary lipid' (Warren *et al.*,1975). Any, or all of these local modifications in lipid composition could form the basis of changes in membrane properties observed during the compaction and subsequent differentiation of mouse embryos and, as discussed previously, could derive from the embryo's ability to synthesise its own membrane lipids.

The formation of intercellular junctions

The first process which merits consideration in these terms is the means by which the blastomeres of the 8-16 cell embryo become tightly apposed to one another, subsequently developing focal tight junctions between the outermost blastomeres and gap junctions between the developing inner cells (Ducibella *et al.*, 1975). The partial fusion of membranes that occurs during the formation of intercellular junctions is likely to involve mechanisms analogous to those underlying the complete fusion of two cell membranes, as occurs for example in the acrosome reaction of the sperm and fusion of egg and sperm at fertilization (Johnson, 1975b; Meizel, this volume), and is probably dependent upon similar reorganisation of membrane proteins within a specialised and permissive lipid environment (for review of membrane fusion, see Poste and Allison, 1973).

The first stage in establishing cell contact is the close approximation of the two cell membranes, which, though favoured by attractive Van der Waals forces, will require energy to overcome the opposing electrostatic forces derived from charged groups on the surface of the membrane and the associated electrical double layer (Poste and Allison, 1973; Lee, 1975). Poste (1972) has calculated that molecular approximation between membranes can only be achieved if the approach and contact are made by microvilli with a small enough radius of curvature (<0.1 μm) to encounter minimal electrostatic repulsion and thus overcome the potential energy barrier. Microvilli of this size are a consistent finding in cell fusion induced by viral and non-viral agents and the ability of different cells to fuse corresponds directly to their ability to form microvilli (Poste, 1970). The distribution of microvilli on the mammalian embryo is interesting since they are only excluded from areas of the membrane which do not seem to be required to undergo fusion type reactions; for example, the smooth region of membrane overlying the second metaphase spindle in the unfertilised egg (Johnson *et al.*,1975; Eager *et al.*,1976), membrane not associated with the cleavage furrow during cytokinesis of blastomeres and areas of spreading cell contact between adjacent cells once compaction has been initiated (Ducibella *et al.*,1977). The microfilament assisted retraction of these microvilli is assumed to bring the membranes into closer contact, overcoming the electrostatic barrier and eliminating the microvilli (Ducibella *et al.*, 1977). The enhanced fusibility of the cells at this stage (Mintz, 1971) suggests that the cells have acquired a predisposition to fusion either as a consequence of changed configurations in the microvilli membranes, or following maturation of the cortical microtubule and microfilament retractile apparatus (Ducibella *et al.*, 1977). The activation or maturation of a cortical developmental programme has been implicated in studies of many developing organisms (Whittaker, 1973; Izquierdo, 1977; Ducibella *et al.*,1977) but its nature remains a matter of speculation. One possibility is the progressive reduction with every cell division of a membrane component (protein or lipid) which reaches a limiting concentration at about the 8-cell stage and changes the membrane to a fusigenic state.

The initial stage of cell fusion or cell interaction is generally agreed to involve the introduction of disorder into the apposed membranes, and in view of the well-known membrane stabilising properties of Ca^{++} and ATP (Poste and Allison, 1973) it has been suggested that localised susceptibility to membrane fusion is induced by displacement of membrane-associated Ca^{++} and ATP (Woodin, 1968). However, the stimulus initiating these changes is still a matter of debate. Lysolecithin has been suggested as a potential initiator of membrane fusion due to its ability to produce substantial disorder when inserted into a regular array of lipid molecules (Haydon and Taylor, 1963; Eytan *et al.*,1975), and the fact that it can cause cells to fuse *in vitro* (Lucy, 1970). Also the redistribution of Ca^{++} associated with membrane fusion (Poste and Allison, 1973) could be interpreted as facilitating the activation of phospholipases which generate lysolecithin (Ferber, 1971). However, in view of the fact that a wide variety of lipophilic and lipolytic agents have been shown to induce cell fusion (Poste and Allison, 1973) and that there is little evidence for the involvement of this molecule during natural fusion of cells (Elsbach *et al.*,1969) lysolecithin is unlikely to be the initiator of fusion though it may be generated during the associated lipid turnover to be considered later.

A more likely candidate for the stimulus causing membrane destabilisation and thus initiating cell fusion is the close apposition of cell membranes *per se*. As discussed by Poste and Allison (1973) when cells come into contact the approach and interaction of their electrical double layers will alter the electrostatic potential within the membrane and cause radical changes in membrane properties due to reorientation of charged and dipolar groups within it. Gingell (1967, 1968) has provided theoretical treatments to support the assumptions that this altered electrostatic potential will lead to conformational changes in membrane macromolecules binding Ca^{++}, as well as displacement of Ca^{++} from the membrane which will favour further structural alterations, increase membrane permeability to cations, and could lead to activation of membrane ATP'ases. This sequence of putative changes developing from close approximation of membranes will also result in an asymmetric distribution of fixed charges and counterions across each membrane (Poste and Allison, 1973), a condition known to induce instability in artificial phospholipid membranes leading to the inversion of molecules across the bilayer and significantly increasing their permeability (Papahadjopoulos and Ohki, 1969). Alterations in distribution of fixed membrane charges such as described above, have been shown to influence the packing of lipid molecules in a bilayer. In general, increasing ionisation of any phospholipid will tend to decrease its transition temperature introducing more mobility and more fluidity into the bilayer (Trauble and Eibl, 1974; Lee, 1975). Consequently displacement of the counterion, Ca^{++}, from its binding sites on negatively charged phospholipids, e.g. phosphatidic acid and phosphatidylserine (Seimiya and Ohki, 1973) will tend to increase the ionisation of the head groups and lead to a decreased packing density and increased mobility of fatty acyl chains coupled with an increase in permeability of the bilayer (Papahadjopoulus, 1968). In addition to its effects on the fluidity of phospholipids, Ca^{++} is found to influence the distribution of lipids within bilayers by inducing phase separations and forming segregated domains of lipid in a gel or liquid crystalline state (Papahadjopoulos *et al.*,1974). Thus the displacement of Ca^{++} thought to be associated with membrane apposition and the early stages of membrane fusion could induce an overall increase in disorder of membrane lipid coupled with the separation of membrane lipid into regions of more or less fluidity. It is known from studies of artificial bilayers that negatively charged fluid lipids provide the most fusible membranes (Papahadjopoulos *et al.*,1974) and localised regions of these relatively unstable and fusible lipids could occur within regions of apposed cell membrane permitting molecular rearrangements necessary for the establishment of stable intercellular molecular linkages, as for example, in the tight and gap junctions of the compacting morula.

If the compaction of the morula proceeds by mechanisms analogous to cell fusion, then the early stages might be expected to involve localised changes in the fluidity of membrane lipids. Close approximation of blastomere membranes could lead to displacement of Ca^{++} from charged membrane lipids, which would induce local isothermal phase transitions (Trauble and Eibl, 1974) and phase separations within the lipids (Papahadjopoulos *et al.*,1974). Lipid fluidity could also be modified by increased turnover of phospholipid to enrich for either short, unsaturated (fluid) fatty acids or longer, saturated (more rigid) fatty acids, and the localised enrichment or depletion of cholesterol in areas of the bilayer where its associations with liquid crystalline or gel phase lipids tend to regulate the motional freedom of the fatty acyl chains (Phillips, 1972).

Studies of cellular adhesion have emphasised the turnover of phospholipid fatty acids and consequent modifications of bulk membrane properties in determining the degree to which cells can adhere to one another and to the substratum (Curtis *et al.*,1975a,b, c). It was suggested that endogenous phospholipase A influences the adhesion potential of cells by producing lysolecithin and other lysolipids in the plasma membrane (Curtis *et al.*,1975a). Conditions which would be expected to stimulate reacylation of lysolipids to the diacyl state, i.e. the inclusion of Co A, ATP and a fatty acid in the medium led to maintenance or recovery of adhesion while those increasing the cellular content of lysolipids, e.g. addition of phospholipase A or incubation of cells in a balanced salt solution with no other substrates, diminished adhesion (Curtis *et al.*,1975b). The plasma membrane was found to contain the enzymes necessary to effect the turnover of fatty acids, namely phospholipase A, a Co A ligase and acyl transferase. This turnover led to the replacement of approximately 20% of plasma membrane fatty acids in 30 minutes, and enrichment for saturated long chain fatty acids with a resulting marked change in the bulk properties of the cell membrane favouring adhesion. A similar increase in fatty acid turnover is stimulated by the interactions of mitogens with lymphocytes (Ferber and Resch, 1973). In fact, it seems that the modification of localised or bulk membrane properties by deacylation and reacylation of phospholipids and lysolipids may be an important mechanism for regulating cell interaction and cell function (Dawson, 1973).

Similar mechanisms could be operating during development. For example, as discussed by Meizel (this volume), phospholipases are present in sperm, and the acrosome reaction could be associated with the deacylation of phospholipids in the membrane. The observation that the acrosome reaction is potentiated in media containing delipidated as opposed to fatty acid enriched bovine serum albumin (Lui *et al.*,1977) could be explained by assuming that the delipidated bovine serum albumin is a more efficient acceptor of fatty acids displaced by the phospholipase than the bovine serum albumin with fatty acids already bound to it. It is possible that an analogous deacylation-reacylation reaction is operating in the membranes of compacting morulae, since the levels of labelled lysolecithin observed (Table 1), whether physiological or representing enzymic deacylation of phospholipid during extraction, indicate the existence of phospholipase A type activity in both morulae and blastocysts. This enzyme has previously been demonstrated in 14 day old rat placentae (East *et al.*,1975). Phospholipase action is known to modify cell function in a number of different ways, for example by unmasking antigens (Hanaumi *et al.*,1976) or hormone receptors (Cuatrecasas, 1971), by preventing intercellular recognition and normal fusion of myoblasts (Nameroff *et al.*,1973) and by stimulating active Na transport (Yorio and Bentley, 1976). The phospholipase activity observed in morulae could have a similar membrane modifying function to perform, independently or in association with reacylation of lysolipids. The turnover of choline-containing phospholipid in morulae mentioned earlier and the observation (Pratt, unpublished) that morulae tend to decompact in media devoid of bovine serum albumin or an energy source, conditions which inhibit the reacylation of lysolipids in parallel with decreased adesion (Curtis *et al.*,1975b), are compatible with the idea that localised modulation of fatty acid composition could promote the formation of the junctional complexes of compacting morulae. In order to substantiate this scheme, it will be necessary to demonstrate turnover in the fatty acyl, as well as the choline moiety, of morula phospholipids and to show that morulae incubated in a 'decompacting medium', e.g. in the absence of macromolecules or an energy source, or in the presence of lysolecithin, can be induced to recompact if the components required for reacylation of lysolipids (Co A, ATP and a fatty acid) are added back. It should then be possible to demonstrate other components of the deacylation-reacylation system in addition to phospholipases, notably Co A ligase and an acyl transferase (Curtis *et al.*,1975b). If this fatty acid turnover system exists in morulae (and other embryonic stages) it may be feasible to enrich membrane phospholipids for fatty acids of differing chain lengths and degrees of saturation, and analyse in detail the developmental consequences of modifying the fluid characteristics of the membrane lipid bilayer.

Local modifications of cholesterol composition could also be a means of generating the specialised lipid environment necessary for the formation of intercellular contacts, due to its stabilising effect on phospholipid bilayers and ability to inhibit fusion (Papahadjopoulos *et al.*,1974; Lee, 1975). The developmental consequences of incubating 8-cell morulae with cholesterol derivatives which block sterol synthesis (Table 4), support this contention. As mentioned previously, precompaction 8-cell morulae grown in 6-ketocholestanol or 7-ketocholesterol progress through one or two cycles of cell division, however they do not compact to form intercellular junctions. This inability to compact can be overcome by the inclusion of mevalonate or cholesterol in the medium (Table 4). In contrast, embryos which have already undergone compaction continue to develop normally to the blastocyst stage. Cholesterol itself does not inhibit development at any stage (Tables 3 and 4), in fact embryos treated with cholesterol tend to divide faster, develop more microvilli and compact more rapidly than untreated controls (Chakraborty and Pratt, unpublished results). However, the anticipated mode of action of these inhibitory sterols (Brown and Goldstein, 1974) has not yet been confirmed for mouse embryos, and the details of cholesterol uptake by these embryos still remain to be investigated. Consequently, it is not yet possible to conclude that the membranes of inhibited morulae are depleted of cholesterol as compared to membranes of morulae incubated with cholesterol which are anticipated to be enriched for this sterol (Brown and Goldstein, 1976). Nevertheless, there are marked differences in the membrane properties of the two types of embryo. The ability of inhibitory sterols to arrest compaction but not cell division in precompaction embryos while having no effect on post compaction embryos, suggests that the sterol composition of locally apposed membrane required for junction formation is more stringent and specialised than that needed to maintain cell viability and normal cytokinesis. A similar effect has been observed in lymphocytes induced to agglutinate by non-mitogenic agents, where the enforced cell contact is found to stimulate the synthesis of cholesterol (Pratt *et al.*,1977).

In normal plasma membranes, cholesterol tends to 'damp out' phase transitions (Phillips, 1972). However, in a membrane depleted of cholesterol (possibly including membranes of inhibitor-treated morulae) this lipid stabilising influence will no longer exist, and some lipid previously bound with cholesterol in a lipid crystalline state might be expected to undergo an isothermal phase transition to the gel phase, with an opposite transition to a more fluid state occurring in regions of lipid previously stabilised in the gel phase. This 'disordering' effect would also be accompanied by the disruption of specific cholesterol-phospholipid (Demel and de Kruyff, 1976) and cholesterol-protein associations (Warren *et al.*, 1975). It is probable that these new lipid-protein interactions would no longer be compatible with the development of intercellular apposition and junction formation.

I therefore suggest that the acquisition of localised and specialised microenvironments within the lipid bilayer, either derived from Ca^{++}-induced isothermal phase transitions or phase separations or other modifications of phospholipid or cholesterol content, may be necessary at the sites of membrane contact to promote junction formation. It is possible that this region of 'conditioned lipid' then acts as a permissive matrix allowing the aggregation of membrane glycoproteins as in virally-induced fusion (Bächi and Howe, 1972) or mucocyst secretion in protozoa (Satir *et al.*,1973), as well as determining the extent to which these proteins are exposed or hidden in the bilayer (Borochov and Shinitzky, 1976). If phase transitions or phase separations are induced within the lipid bilayer at the sites of membrane apposition, then the associated increase in lateral compressibility might also facilitate the insertion of newly synthesised junction proteins at these sites (Linden *et al.*,1973). Membrane reorganisation within this 'conditioned lipid' could then lead to the interdigitation of protein and lipid across the two membranes and the establishment of direct molecular contact between them (Poste and Allison, 1973; Papahadjopoulos *et al.*,1974). Intercellular apposition would then be followed by the formation of tight or gap junctions from the proteins assembled in this region of the membrane and the complexes of cytoskeletal microfilaments and microtubules underlying it (Curtis, 1973; Ducibella *et al.*,1975). Once established, the integrity of these junctions would presumably require a defined lipid environment in addition to specific junction proteins and both of these components would be dependent upon the presence of Ca^{++} to maintain their stability (Ducibella and Anderson, 1975).

Changes in membrane permeability

The establishment of intercellular junctions in the morula is also associated with changes in the permeability of the embryonic membranes, notably an increase in the transport of sugars (Wales and Biggers, 1968), nucleosides (Daentl and Epstein, 1971) and amino acids (Epstein and Smith, 1973; Borland and Tasca, 1974). The number (or mobility) of carrier sites for leucine and methionine increases in association with the activation (or development) of a Na^{+} dependent amino acid transport system (Borland and Tasca, 1974), presumably indicative of qualitative and quantitative changes in the membranes of the morula as compared to earlier stages. These modulations in membrane transport may occur as a result of synthesis of new carrier molecules or the post-translational modification of pre-existing ones. However, some of these changes in transport activity could be explained on the basis of lipid phase separations or transitions and other modifications of the bilayer introduced by ions, or phospholipid and cholesterol turnover or exchange.

Increases in permeability will be due to the activation of pre-existing carriers or the synthesis and insertion of new ones. In the former case, the reorganisation of local domains of lipid might lead to the association of an inactive enzyme with a lipid microenvironment favourable for its activity; for example, within a phase separation where the equilibrium mixture of liquid crystalline and gel phase lipid is optimal for the action of a transport protein (Linden *et al.*,1973). Other specific examples of lipid modifications which would be expected to activate or increase the activity of the carriers involved include: the removal of cholesterol from the internal cation site of the Na pump (Giraud *et al.*,1976), and an increase in the fluidity of lipids surrounding (Ca^{++}-Mg^{++}) ATP'ase (Warren *et al.*,1975) or (Na^{+}-K^{+}) ATP'ase (Kimelberg and Papahadjopoulos, 1972, 1974). This type of activity modulation could explain the increased transport of many metabolites by morulae, including the development of Na dependent amino acid transport (Borland and Tasca, 1974), and might form the basis of fundamental changes in permeability to K^{+} and Na^{+} during pre-implantation development (Powers and Tupper, 1977). Alternatively, if carrier proteins consist of more than one subunit, modifications of local lipid composition might permit previously restricted molecules to diffuse laterally and interact to form a stable active transport complex, as for example, postulated in the mobile receptor hypothesis of Cuatrecasas. In this theory, the hormone receptor and adenyl cyclase are assumed to exist independently within the bilayer until the interaction of the hormone with the receptor leads to the formation of an active enzyme-receptor complex and to the stimulation of the cell (Cuatrecasas, 1974). Lipid modifications favourable to transport activity could also be brought about either by localised phospholipase activity unmasking pre-existing macromolecules (Cuatrecasas, 1971; Yorio and Bentley, 1976) or the local exchange of phospholipid (Dawson, 1973) or cholesterol (Brown and Goldstein, 1976) between apposed membranes.

As discussed earlier, the activities of several membrane proteins exhibit breaks at characteristic temperatures that are dependent on the lipid microenvironments of the proteins (Linden *et al.*,1973; Wisnieski *et al.*,1974). Most transport proteins require a fluid environment for optimal activity (Lee, 1975), consequently decreasing the temperature and inducing a phase transition to the more regular gel phase will decrease the compressibility of the bilayer and reduce the transport across it. However, an anomalous increase in enzyme or carrier activity may occur at a characteristic temperature close to the onset of the phase transition of the lipids in the immediate vicinity of the protein. The increased lateral compressibility found in these equilibrium mixtures of gel and liquid phase lipid is thought to be the explanation for this anomalous increase in activity (Linden *et al.*,1973; Petit and Edidin, 1974). The interpretation of thermal effects on biological membranes is complex (see Lee, 1975 for a discussion of the parameters involved). Nevertheless, it should be possible to investigate whether any changes in membrane permeability that occur during development are lipid-mediated by measuring the temperature dependent activity of specific transport enzymes at different embryonic stages. For example, the increased temperature sensitivity of methionine transport in blastocysts, as compared to early morulae (Borland and Tasca, 1974), could be associated with changes in the physical properties of lipids surrounding these carrier proteins.

Alternatively, some of the enhanced transport activity observed at the morula stage could be due to *de novo* synthesis of enzymes and carrier molecules inserted into regions of pre-existing lipid facilitating this insertion, or synthesised co-ordinately with membrane lipid (Tsukagoshi and Fox, 1973; Rothman and Lodish, 1977).

Finally, in addition to the establishment of lipid and protein complexes for the transport of metabolites, active ion transport and fluid accumulation responsible for blastocyst expansion begin at the morula stage (Cross and Brinster, 1970; Cross, 1973). This process starts by the accumulation of fluid into cytoplasmic vesicles and its release into the intercellular spaces inside the embryo (Calarco and Brown, 1969). Both processes, the activation of ion and water transport, and the fusion of cytoplasmic vesicles with intercellular spaces will require specialised areas of lipid in the membrane (Poste and Allison, 1973; Heiniger *et al.*,1976). It is therefore interesting that the morulae inhibited from compacting by treatment with 7-ketocholesterol and assumed to have membranes depleted of cholesterol, though able to divide, are incapable of active fluid transport. Further work will be necessary to establish whether this incapacity is due to inactive transport, the lack of tight junctions or the inability of cytoplasmic fluid vesicles to fuse with intercellular spaces, all of which could be consequences of depletion of membrane cholesterol (Poste and Allison, 1973; Cooper, 1977; Heiniger *et al.*,1976).

Membranes and developmental signals

There remains one final aspect of morula membranes to be discussed and that is their role as transducers of developmental information. The 'inside-outside' theory of development as proposed by Tarkowski and Wroblewska (1967) (see also discussion by Johnson *et al.*,1977) suggests that inside cells develop in an inner cell mass direction due to their position inside the morula. Inside cells are tightly apposed to one another and maintained within a specialised environment of ions and possibly other informational molecules, due to the formation of focal tight junctions between the outer cells of the morula (Ducibella *et al.*,1975). The divergent differentiation of inner cell mass and trophoblast might then be initiated by local changes in the concentrations of ions and small molecules (McMahon, 1974), which could alter the distribution of fixed charges on the phospholipids or interact with specific receptors in the cell membrane. Alternatively, the important factor could be the degree to which the cells are in contact and hence able to exchange or modify lipid (Dawson, 1973) or protein components (Roth, 1973) of the apposing membrane. Whatever the nature of the differentiation stimulus, it is clear that the plasma membrane will need to transmit this information to the interior of the cell for interpretation and subsequent transcription by the genome or activation of a pre-existing cytoplasmic developmental programme. Membrane lipids have been shown to play a role in the transmission of environmental signals and similar mechanisms may operate in the morula. For example, the 'internal' position of the inner cells could be translated into ion induced phase transitions or phase separations in the lipid bilayer (Papahadjopoulos *et al.*,1974), or result in increased turnover and modification of phospholipid and cholesterol, of the type that is associated with mitogen stimulation (Resch and Ferber, 1975; Pratt *et al.*,1977) or induced cell contact (Curtis *et al.*,1975c; Pratt *et al.*,1977). Lipid changes could then lead to local reorganisation of the membrane, possibly associated with the formation of active receptor-enzyme type complexes which initiate a sequence of catalytic reactions analogous to those induced by hormones (Cuatrecasas, 1974). This catalytic sequence, depending upon its nature and the inducing stimulus, could then stimulate the transcription of genes for inner cell mass or trophoblast specific proteins (Johnson *et al.*,1977), or initiate their production by activating the necessary translational complexes stored in the cytoplasm. The subsequent differentiation of the embryo is then likely to involve repetition of this sequence of information reception and transmission by the membrane and its interpretation and manipulation by the cell. A more complete understanding of this process will only be achieved by further investigation of the interrelated molecular events occurring in the plasma membrane, nucleus and cytoplasm during early development.

CONCLUSION

The hypothesis for regulation of embryonic differentiation described here rests on the assumption that the maturation of the embryonic plasma membrane is associated with changes in the physical state or composition of its constituent phospholipid and cholesterol molecules.

A number of simple predictions can be made concerning the behaviour of membrane lipids during preimplantation differentiation and these will require confirmation before the hypothesis merits further investigation. Firstly, membrane lipid may be modified metabolically. Consequently it will be necessary to demonstrate that embryos have the capacity to regulate the turnover of phospholipid and the uptake or synthesis of cholesterol, in response to demands for changes in membrane activity. Secondly, these modifications, in combination with ionic or pH effects or the insertion of protein into the bilayer, could result in local phase transitions or separations within the lipid. If such transitions occur in association with certain developmental stages, then spin labelled lipids (Lee, 1975) non-penetrating lipid labelling agents (Rothman and Lenard, 1977) the laser bleach-regeneration system (Zagyansky and Edidin, 1976; Johnson and Edidin, 1978) and studies of the temperature dependence of enzyme activity (Wisnieski *et al.*,1974), could all be used to identify alterations in the organisation and lateral mobility of membrane lipids.

The ability to inhibit compaction reversibly by manipulating the Ca^{++} composition of the medium (Ducibella *et al.*,1975) or exposing embryos to 7-ketocholesterol or cholesterol (Table 4) provides a useful controlled system for monitoring lipid rearrangements necessary for membrane apposition, intercellular junction formation, and the consequent changes in permeability and subsequent differentiation of the embryo. Another approach to the analysis of the relation between lipid composition and membrane function during development is the study of the effects of introducing molecules into the bilayer which would tend to 'disorder' the regular array of phospholipids, e.g. phospholipases, lysolipids, anaesthetics, unsaturated fatty acids and agents depleting the membrane of cholesterol (Lee, 1975). In addition, the cultivation of embryos with differing membrane lipid compositions (viz. enriched or depleted for specific fatty acids or cholesterol) should permit an evaluation of the role of the membrane in regulating the transcriptional or translational control of differentiation as determined by the synthesis of tissue and stage specific proteins (Johnson *et al.*,1977; Handyside and Johnson, 1978).

In conclusion, the paucity of information on lipid synthesis by early mammalian embryos has, of necessity, rendered this review somewhat speculative. However, there are sufficient precedents derived from studies of other cell types to indicate that the organisation and composition of membrane lipids is an important factor in determining the viability and normal differentiation of cells, and the limited studies of mouse embryo lipids described here support this view. The plasma membrane is an extraordinarily adaptable organelle, maintaining as it does the internal milieu and normal functions of a viable cell, while remaining susceptible to any environmental perturbations which can be translated into a variety of physiological responses. The range of potential activities of a developing embryo may be even more diverse than that of a terminally differentiated cell. However, the spectrum of 'balanced instability' created by the lipid-protein interactions within the plasma membrane of the embryo is evidently sufficient to generate a variable response to environmental stimuli while still contributing to the orderly developmental sequence leading to the differentiation and implantation of the mouse blastocyst. The physical and metabolic transitions occurring in the lipid moiety of the membrane may be a vital, yet hitherto unexplored, element in maintaining the order and instability necessary for regulating these complex processes.

ACKNOWLEDGEMENTS

The work described in this article was supported by a grant from the Ford Foundation to Professor C.R. Austin. I am grateful to Dr. M.H. Johnson and Dr. P. Braude for helpful discussions, to Dr. K. Bighouse for a gift of P815 mastocytoma cells and to Miss J. Keith for her collaboration in the sterol synthesis experiments.

REFERENCES

ALLAN, D. & CRUMPTON, M.J. (1970) Preparation and characterisation of the plasma membrane of pig lymphocytes. Biochem. J. 120, 133-143.

ALDERSON, J.C.E. & GREEN, C. (1975) Enrichment of lymphocytes with cholesterol and its effect on lymphocyte activation. FEBS letters 52, 208-211.

BACHI, T. & HOWE, C. (1972) Fusion of erythrocytes by Sendai Virus studied by electronmicroscopy. Proc. Soc. Exp. Biol. Med. 141, 141-149.

BAILEY, J.M. & BUTLER, J. (1973) Cholesterol uptake from doubly-labelled α-lipoproteins by cells in tissue culture. Archiv. Biochem. Biophys. 159, 580-581.

BARNETT, R.E., SCOTT, R.E., FURCHT, L.T. & KERSEY, J.H. (1974) Evidence that mitogenic lectins induce changes in lymphocyte membrane fluidity. Nature 249, 465-466.

BENNETT, V., O'KEEFE, E. & CUATRECASAS, P. (1975) Mechanism of action of cholera toxin and the mobile receptor theory of hormone receptor-adenylate cyclase interactions. Proc. Nat. Acad. Sci. 72, 33-37.

BERGERON, J.J.M., WARMSLEY, A.M.H. & PASTERNAK, C.A. (1970) Phospholipid synthesis and degradation during the life cycle of P815Y mast cells synchronised with excess of thymidine. Biochem. J. 119, 489-492.

BERLIN, R.D. (1975) Microtubule membrane interactions: Fluorescence techniques. In: M. Borgers and M. de Brabander (Eds.), Microtubules and microtubule inhibitors. North Holland Pub. Co., Amsterdam, pp. 327-339.

BORLAND, R.M. & TASCA, R.J. (1974) Activation of a Na^+-dependent amino acid transport system in preimplantation mouse embryos. Dev. Biol. 30, 169-182.

BOROCHOV, H. & SHINITZKY, M. (1976) Vertical displacement of membrane proteins produced by changes in microviscosity. Proc. Nat. Acad. Sci. 73, 4526-4530.

BOWYER, D.E. & DAVIES, P.F. (1976) Effect of EPL on the metabolism of lipids in the arterial wall. In: H. Peeters (Ed.), Phosphatidylcholine.Biochemical and clinical aspects of essential phospholipids. Springer-Verlag, Berlin, Heidelberg, pp. 160-186.

BRETSCHER, M.S. (1972) Membrane structure: some general principles. Science, 181, 622-629.

BRINSTER, R.L. (1965) Studies on the development of mouse embryos *in vitro*. II. The effect of energy source. J. Exp. Zool. 158, 59-68.

BROWN, M.S., FAUST, J.R. & GOLDSTEIN, J.L. (1975) Role of the low density lipoprotein receptor in regulating the content of free and esterified cholesterol in human fibroblasts. J. Clin. Invest. 55, 783-793.

BROWN, M.S & GOLDSTEIN, J.L. (1974) Suppression of 3-Hydroxy-3-methylglutaryl coenzyme A reductase activity and inhibition of growth of human fibroblasts by 7-ketocholesterol. J. Biol. Chem. 249, 7306-7314.

BROWN, M.S. & GOLDSTEIN, J.L. (1976) Receptor-mediated control of cholesterol metabolism. Science 191, 150-154.

CALARCO, P.G. & BROWN, E.H. (1969) An ultrastructural and cytological study of preimplantation development of the mouse. J. Exp. Zool. 171, 253-283.

CHEPENIK, K.P., BORST, D.E. & WAITE, B.M. (1977) Cholinephosphotransferase activities in early rat embryos and their associated placentas. Dev. Biol. 60, 463-473.

COOPER, R.A. (1977) Abnormalities of cell membrane fluidity in the pathogenesis of disease. New Eng. J. Med. 297, 371-377.

CROSS, M. (1973) Active sodium and chloride transport across the rabbit blastocoele wall. Biol. Reprod. 8, 556-575.

CROSS, M. & BRINSTER, R.L. (1970) Influence of ions, inhibitors and anoxia on transtrophoblast potential of rabbit blastocyst. Exp. Cell Res. 62, 303-309.

CUATRECASAS, P. (1971) Unmasking of insulin receptors in fat cells and fat cell membranes. J. Biol. Chem. 246, 6532-6542.

CUATRECASAS, P. (1974) Membrane receptors. Ann. Rev. Biochem. 43, 169-214.

CURTIS, A.S.G. (1973) Cell adhesion. Progr. Biophys. Molec. Biol. 27, 315-386.

CURTIS, A.S.C., CAMPBELL, J. & SHAW, F.M. (1975a) Cell surface lipids and adhesion. I. The effects of lysophosphatidyl compounds, phospholipase A_2 and aggregation-inhibiting protein. J. Cell Sci. 18, 347-356.

CURTIS, A.S.G., SHAW, F.M. & SPIRES, V.M.C. (1975b) Cell surface lipids and adhesion. II. The turnover of lipid components of the plasmalemma in relation to cell adhesion. J. Cell Sci. 18, 357-373.

CURTIS, A.S.G., CHANDLER, C. & PICTON, N. (1975c) Cell surface lipids and adhesion. III. The effects on cell adhesion of changes in plasmalemmal lipids. J. Cell Sci. 18, 375-384.

DAENTL, D.L. & EPSTEIN, C.J. (1971) Developmental interrelationships of uridine uptake, nucleotide formation and incorporation into RNA by early mammalian embryos. Dev. Biol. 24, 428-442.

DAWSON, R.M.C. (1960) A hydrolytic procedure for the identification and estimation of individual phospholipids in biological samples. Biochem. J. 75, 45-53.

DAWSON, R.M.C. (1973) The exchange of phospholipids between cell membranes. Subcell. Biochem. 2, 69-89.

DEMEL, R.A., BRUCKDORFER, R. & VAN DEENEN, L.L.M. (1972a) Structural requirements of sterols for the interaction with lecithin at the air-water interface. Biochim. Biophys. Acta 255, 311-320.

DEMEL, R.A., BRUCKDORFER, R. & VAN DEENEN, L.L.M. (1972b) The effect of sterol structure on the permeability of liposomes to glucose, glycerol and Rb^+. Biochim. Biophys. Acta 255, 321-330.

DEMEL, R.A. & DE KRUYFF, B. (1976) The function of sterols in membranes. Biochim. Biophys. Acta 457, 109-132.

DEMEL, R.A., JANSEN, J.W.C.M., VAN DIJCK, P.W.M. & VAN DEENEN, L.L.M. (1977) The preferential interaction of cholesterol with different classes of phospholipid. Biochim. Biophys. Acta 465, 1-10.

DUCIBELLA, T., ALBERTINI, D.F., ANDERSON, E. & BIGGERS, J.D. (1975) The preimplantation mammalian embryo: characterisation of intercellular junctions and their appearance during development. Dev. Biol. 45, 231-250.

DUCIBELLA, T. & ANDERSON, E. (1975) Cell shape and membrane changes in the eight-cell mouse embryo: prerequisites for morphogenesis of the blastocyst. Dev. Biol. 47, 45-58.

DUCIBELLA, T., UKENA, T., KARNOVSKY, M. & ANDERSON, E. (1977) Changes in cell shape and cortical cytoplasmic organisation during early embryogenesis in the preimplantation mouse embryo. J. Cell Biol. 74, 153-167.

EAGER, D.D., JOHNSON, M.H. & THURLEY, K.W. (1976) Ultrastructural studies on the surface membrane of the mouse egg. J. Cell Sci. 22, 345-353.

EAST, J.M., CHEPENIK, K.P. & WAITE, B.M. (1975) Phospholipase(s) A activities in rat placentas of 14 days' gestation. Biochim. Biophys. Acta 388, 106-112.

EDELMAN, G.M. (1976) Surface modulation in cell recognition and cell growth. Science 192, 218-226.

EDIDIN, M. (1974) Rotational and translational diffusion in membranes. Ann. Rev. Biophys. Bioeng. 3, 179-201.

EDWARDS, P.A., FOGELMAN, A.M. & POPJAK, G. (1976) A direct relationship between the amount of sterol lost from rat hepatocytes and the increase in activity of HMG CoA reductase. Biochem. Biophys. Res. Commun. 68, 64-69.

ELSBACH, P., HOLMES, K.V. & CHOPPIN, P.W. (1969) Metabolism of lecithin and virus-induced cell fusion. Proc. Soc. Exp. Biol. Med. 130, 903-908.

EPSTEIN, C.J. & SMITH, S.A. (1973) Amino acid uptake and protein synthesis in preimplantation mouse embryos. Dev. Biol. 33, 171-184.

EYTAN, G., MATHESON, M.J. & RACKER, E. (1975) Incorporation of biologically active proteins into liposomes. FEBS letters 57, 121-125.

FERBER, E. (1971) In: Wallach, D.F.H. and Fischer, H. (Eds.) The dynamic structure of cell membranes. Springer-Verlag, Berlin, pp. 125-149.

FERBER, E. & RESCH, K. (1973) Phospholipid metabolism of stimulated lymphocytes: activation of acyl-CoA:lysolecithin acyl transferase in microsomal membranes. Biochim. Biophys. Acta 296, 335-349.

FOWLER, R.E., JOHNSON, M.H., WALTERS, D.E. & EAGER, D.D. (1977) The progesterone content of rabbit uterine flushings. J. Reprod. Fert. 50, 301-308.

GINGELL, D. (1967) Membrane surface potential in relation to a possible mechanism for intercellular interactions and cellular responses: a physical basis. J. Theor. Biol. 17, 451-482.

GINGELL, D. (1968) Computed surface potential changes with membrane interaction. J. Theor. Biol. 19, 340-344.

GIRAUD, F., CLARET, M. & GARAY, R. (1976) Interactions of cholesterol with the Na pump in red blood cells. Nature 264, 646-648.

GITLER, C. (1972) Plasticity of biological membranes. Ann. Rev. Biophys. Bioeng. 1, 51-92.

GOLDFARB, S. & PITOT, H.C. (1971) Improved assay of 3-hydroxy-3-methylglutarylcoenzyme A reductase. J. Lipid Res. 12, 512-515.

GOLDSTEIN, J.L. & BROWN, M.S. (1974) Binding and degradation of low density lipoproteins by cultured human fibroblasts. J. Biol. Chem. 249, 5153-5162.

GOTTFRIED, E.L. (1967) Lipids of human leukocytes: relation to cell type. J. Lipid Res. 8, 321-327.

GOTTFRIED, E.L. (1971) Lipid patterns in human leukocytes maintained in long-term culture. J. Lipid Res. 12, 531-537.

GRAHAM, C.F. (1973) The cell cycle during mammalian development. In: Balls, M.S. and Billett F.S. (Eds.), 'The cell cycle in development and differentiation'. Cambridge University Press, pp. 293-310.

GRAHAM, J.M., SUMNER, M.C.B., CURTIS, D.H. & PASTERNAK, C.A. (1973) Sequence of events in plasma membrane assembly during the cell cycle. Nature 246, 291-295.

HANAUMI, K., ABO, T. & KUMAGI, D. (1976) Exposure by phospholipase A of receptors for sheep erythrocytes on human B cells. Nature 259, 142-126.

HANDYSIDE, A.H. & JOHNSON, M.H. (1978) Temporal and spatial patterns of the synthesis of tissue-specific polypeptides in the preimplantation embryo. J. Emb. Exp. Morph. (In press).

HATTEN, M.E., HORWITZ, A.F. & BURGER, M.M. (1977) The influence of membrane lipids on proliferation of transformed and untransformed cell lines. Exp. Cell Res. 107, 31-34.

HAYDON, D.A. & TAYLOR, J. (1963) The stability and properties of bimolecular lipid leaflets in aqueous solutions. J. Theor. Biol. 4, 281-296.

HEINIGER, H.J., KANDUTSCH, A.A. & CHEN, H.W. (1976) Depletion of L-cell sterol depresses endocytosis. Nature 263, 515-517.

HORWITZ, A.F., HATTEN, M.E. & BURGER, M.M. (1974) Membrane fatty acid replacements and their effect on growth and lectin-induced agglutinability . Proc. Nat. Acad. Sci. 71, 3115-3119.

HOUSLAY, M.D., HESKETH, T.R., SMITH, G.A., WARREN, G.B. & METCALFE, J.E. (1976) The lipid environment of the glucagon receptor regulates adenylate cyclase activity. Biochim. Biophys. Acta 436, 495-504.

HUFF, R.L. & EIK-NESS, K.B. (1966) Metabolism *in vitro* of acetate and certain steroids by six-day-old rabbit blastocysts. J. Reprod. Fert. 11, 57-63.

INBAR, M. & SHINITZKY, M. (1974a) Increase of cholesterol in the surface membrane of lymphoma cells and its inhibitory effect on ascites tumour development. Proc. Nat. Acad. Sci. 71, 2128-2130.

INBAR, M. & SHINITZKY, M. (1974b) Cholesterol as a bioregulator in the development and inhibition of leukaemia. Proc. Nat. Acad. Sci. 71, 4229-4231.

IZQUIERDO, L. (1977) Cleavage and differentiation. In: M.H. Johnson (Ed.), Development in Mammals, Vol. 2, North Holland Pub. Co. Amsterdam, New York and London, pp. 99-118.

JOHNSON, M.H. (1975a) Antigens of the peri-implantation trophoblast. In: R.G. Edwards, C.W.S. Howe and M.H. Johnson (Eds.), Immunobiology of Trophoblast. Cambridge University Press, pp. 87-112.

JOHNSON, M.H. (1975b) The macromolecular organisation of membranes and its bearing on events leading up to fertilisation. J. Reprod. Fert. 44, 167-184.

JOHNSON, M.H., EAGER, D., MUGGLETON-HARRIS, A. & GRAVE, H.M. (1975) Mosaicism in organisation of concanavalin A receptors on surface membrane of mouse egg. Nature, 257, 321-322.

JOHNSON, M.H., HANDYSIDE, A.H. & BRAUDE, P. (1977) Control mechanisms in early mammalian development. In: M.H. Johnson (Ed.), Development in Mammals. Vol. 2, North Holland Pub. Co. Amsterdam, New York and London, pp. 67-97.

JOHNSON, M.H. & EDIDIN, M. (1978) Lateral diffusion in the plasma membrane of the mouse egg is restricted after fertilisation. Nature. (In press).

KANDUTSCH, A.A. & CHEN, H.W. (1973) Inhibition of sterol synthesis in cultured mouse cells by 7-αhydroxycholesterol, 7β-hydroxycholesterol and 7-ketocholesterol. J. Biol. Chem. 248, 8408-8417.

KANDUTSCH, A.A. & CHEN, H.W. (1977) Consequences of blocked sterol synthesis in cultured cells. J. Biol. Chem. 252, 409-415.

KANDUTSCH, A.A. & SAUCIER, S.E. (1969) Regulation of sterol synthesis in developing brains of normal and jimpy mice. Archiv. Biochem. Biophys. 135, 201-208.

KANE, M.T. (1976) Growth of fertilised one-cell rabbit ova to viable morulae in the presence of pyruvate or fatty acids. J. Physiol. 263, 235-236P.

KENNEDY, E.P. (1957) Metabolism of lipids. Ann. Rev. Biochem. 26, 119-148.

KIMELBERG, H.K. & PAPAHADJOPOULOS, D. (1972) Phospholipid requirements for ($Na^+ + K^+$) ATP'ase activity; head group specificity and fatty acid fluidity. Biochim. Biophys. Acta. 282, 277-292.

KIMELBERG, H.K. & PAPAHADJOPOULOS, D. (1974) Effects of phospholipid acyl chain fluidity, phase transitions, and cholesterol on ($Na^+ + K^+$) stimulated adenosine triphosphatase. J. Biol. Chem. 249, 1071-1080.

KNUTTON, S., JACKSON, D., GRAHAM, J.M., MICKLEM, K.J. & PASTERNAK, C.A. (1976) Microvilli and cell swelling. Nature 262, 52-53.

KUFF, E.L., HYMER, W.C., SHELTON, E. & ROBERTS, N.E. (1966) The *in vivo* protein synthetic activities of free versus membrane-bound ribonucleoprotein in a plasma-cell tumour of the mouse. J. Cell Biol. 29, 63-75.

KRCO, C.J. & GOLDBERG, E.H. (1976) H-Y (Male) antigen: detection on eight-cell mouse embryos. Science 193, 1134-1135.

LADBROOKE, B.D. & CHAPMAN, D. (1969) Thermal analysis of lipids, proteins and biological membranes. A review and summary of recent studies. Chem. Phys. Lipids, 3, 304-319.

LEE, A.G. (1975) Functional properties of biological membranes: a physical-chemical approach. Progr. Biophys. Molec. Biol. 29, 3-56.

LEE, A.G. (1976) Model for action of local anaesthetics. Nature 262, 545-548.

LEE, A.G. (1977) Lipid phase transitions and phase diagrams. I. Lipid phase diagrams. Biochim. Biophys. Acta 472, 237-281.

LEVINE, Y.K. (1972) Physical studies of membrane structure. Progr. Biophys. Molec. Biol. 24, 1-74.

LINDEN, C.D., WRIGHT, K.L., McCONNELL, H.M. & FOX, C.F. (1973) Lateral phase separations in membrane lipids and the mechanism of sugar transport in *Escherichia coli*. Proc. Nat. Acad. Sci. 70, 2271-2275.

LOEWENSTEIN, J.E. & COHEN, A.I. (1964) Dry mass, lipid content and protein content of the intact and zona-free mouse ovum. J. Emb. Exp. Morph. 12, 113-121.

LUCY, J.A. (1970) The fusion of biological membranes. Nature, 227, 815-817.

LUI, C.W., CORNETT, L.E. & MEIZEL, S. (1977) Identification of the bovine follicular fluid protein involved in the *in vitro* induction of the hamster sperm acrosome reaction. Biol. Reprod. 17, 34-41.

McMAHON, D. (1974) Chemical messengers in development: a hypothesis. Science, 185, 1012-1021.

MINTZ, B. (1971) Allophenic mice of multi-embryo origin. In: J.C. Daniel (Ed.), Methods in Mammalian Embryology. W.H. Freeman & Co., San Francisco, pp. 186-214.

MOHRI, H. (1964) Utilisation of ^{14}C-labelled acetate and glycerol for lipid synthesis during the early development of sea urchin embryos. Biol. Bulletin 126, 440-455.

MUGGLETON-HARRIS, A. & JOHNSON, M.H. (1976) The nature and distribution of serologically detectable alloantigens on the preimplantation mouse embryo. J. Emb. Exp. Morph. 35, 59-72.

NAMEROFF, M., TROTTER, J.A., KELLER, J.M. & MUNAR, E. (1973) Inhibition of cellular differentiation by phospholipase C. J. Cell Biol. 58, 107-118.

OLDFIELD, E. & CHAPMAN, D. (1972) Dynamics of lipids in membranes: heterogeneity and the role of cholesterol. FEBS letters 23, 285-297.

OVERATH, P., HILL, F.F. & LAMNEK-HIRSCH, I. (1971) Biogenesis of *E. coli* membrane: evidence for randomisation of lipid phase. Nature, 234, 264-267.

OVERATH, P. & TRAUBLE, H. (1973) Phase transitions in cells, membranes, and lipids of *Escherichia coli*. Detection by fluorescent probes, light scattering and dilatometry. Biochemistry 12, 2625-2634.

PAPAHADJOPOULOS, D. (1968) Surface properties of acidic phospholipids: interaction of monolayers and hydrated liquid crystals with uni- and bi-valent metal ions. Biochim. Biophys. Acta. 163, 240-254.

PAPAHADJOPOULOS, D. (1974) Cholesterol and cell membrane function: a hypothesis concerning the etiology of atherosclerosis. J. Theor. Biol. 43, 329-337.

PAPAHADJOPOULOS, D. & OHKI, S. (1969) Stability of asymmetric phospholipid membranes. Science, 164, 1075-1077.

PAPAHADJOPOULOS, D., POSTE, G., SCHAEFFER, B.E. & VAIL, W.J. (1974) Membrane fusion and molecular segregation in phospholipid vesicles. Biochim. Biophys. Acta. 352, 10-28.

PASTERNAK, C.A. (1973) Phospholipid synthesis in cleaving sea urchin eggs: model for specific membrane assembly. Dev. Biol. 30, 403-410.

PASTERNAK, C.A. & FRIEDRICHS, B. (1970) Turnover of mammalian phospholipids. Biochem. J. 119, 481-488.

PETERSON, J.A. & RUBIN, H. (1969) The exchange of phospholipids between fibroblasts and their growth medium. Exp. Cell Res. 58, 365-378.

PETIT, V.A. & EDIDIN, M. (1974) Lateral phase separation of lipids in plasma membranes: effect of temperature on the mobility of membrane antigens. Science 184, 1183-1185.

PHILLIPS, M.C. (1972) The physical state of phospholipid and cholesterol in monolayers, bilayers and membranes. In: J.F. Danielli, M.D. Rosenberg and D.A. Cadenhead (Eds.), Progress in Surface and Membrane Science. Vol. 5, Academic Press, New York.

PLAGEMANN, P.G.W. (1968) Choline metabolism and membrane formation in rat hepatoma cells grown in suspension culture. Archiv. Biochem. Biophys. 128, 70-87.

PLAGEMANN, P.G.W. & ROTH, M.F. (1969) Permeation as the rate limiting step in the phosphorylation of uridine and choline and their incorporation into macromolecules by Novikoff hepatoma cells. Competitive inhibition by phenethyl alcohol, persantin and adenosine. Biochemistry 8, 4782-4789.

POPJAK, G. & CORNFORTH, J.W. (1966) Substrate stereochemistry in squalene biosynthesis. Biochem. J. 101, 553-568.

POSTE, G. (1970) Virus-induced polykaryocytosis and the mechanism of cell fusion. Adv. Virus Res. 16, 303-356.

POSTE, G. (1972) Mechanism of virus-induced cell fusion. Int. Rev. Cytol. 33, 157-252.

POSTE, G. & ALLISON, A.C. (1973) Membrane fusion. Biochim. Biophys. Acta. 300, 421-465.

POWERS, R.D. & TUPPER, J.T. (1977) Developmental changes in membrane transport and permeability in the early mouse embryo. Dev. Biol. 56, 306-315.

PRATT, H.P.M., FITZGERALD, P.A. & SAXON, A. (1977) Synthesis of sterol and phospholipid induced by the interaction of phytohaemagglutinin and other mitogens with human lymphocytes and their relation to blastogenesis and DNA synthesis. Cellular Immunol. 32, 160-170.

PRATT, H.P.M., SAXON, A. & GRAHAM, M.L. (1978) Membrane lipid changes associated with malignant transformation and normal maturation of human lymphocytes. Leukaemia Research. (In press).

RESCH, K. & FERBER, E. (1975) The role of phospholipids in lymphocyte activation. In: Proceedings of the Ninth Leucocyte Conference. Academic Press, New York and London, pp. 281-312.

RESCH,K., FERBER, E., PRESTER, M. & GELFAND, E.W. (1976) Mitogen-induced membrane changes and cell proliferation in T lymphocyte subpopulations. Eur. J. Immunol. 6, 168-173.

RITTENHOUSE, H.G., WILLIAMS, R.E. & FOX, C.F. (1974) Effect of membrane lipid composition and microtubule structure on lectin interactions of mouse LM cells. J. Supramol. Struct. 2, 629-645.

ROTH, S. (1973) A molecular model for cell interactions. Quart. Rev. Biol. 48, 541-563.

ROTHMAN, J.E. & LENARD, J. (1977) Membrane asymmetry. Science 195, 743-753.

ROTHMAN, J.E. & LODISH, H.F. (1977) Synchronised transmembrane insertion and glycosylation of a nascent membrane protein. Nature 269, 775-780.

ROTHMAN, J.E. & KENNEDY, E.P. (1977) Rapid transmembrane movement of newly synthesised phospholipids during membrane assembly. Proc. Nat. Acad. Sci. 74, 1821-1825.

ROTTEM, S., CIRILLO, V.P., DE KRUYFF, B., SHINITZKY, M. & RAZIN, S. (1973) Cholesterol in *Mycoplasma* membranes: correlation of enzymic and transport activities with physical state of lipids in membranes of *Mycoplasma mycoides* var. *capri* adapted to grow with low cholesterol concentrations. Biochim. Biophys. Acta. 323, 509-519.

SATIR, B., SCHOOLEY, C. & SATIR, P. (1973) Membrane fusion in a model system. Mucocyst secretion in *Tetrahymena*. J. Cell Biol. 56, 153-176.

SEIMIYA, T. & OHKI, S. (1973) Ionic structure of phospholipid membranes and binding of calcium ions. Biochim. Biophys. Acta 298, 546-561.

SCHNEIDER, W.C. (1969) Enzymatic preparation of labelled phosphorylcholine, phosphorylethanolamine, cytidine diphosphate choline, cytidine diphosphate ethanolamine and deoxycytidine diphosphate ethanolamine. In: Methods in Enzymology 14, 684-690.

SHAPIRO, D.J., IMBLUM, R.I. & RODWELL, V.W. (1969) Thin-layer chromatographic assay for HMG CoA reductase and mevalonic acid. Anal. Biochem. 31, 383-390.

SHIMSHICK, E.J. & McCONNELL, H.M. (1973) Lateral phase separation in phospholipid membranes. Biochemistry 12, 2351-2360.

SHINITZKY, M. & BARENHOLZ, Y. (1974) Dynamics of the hydrocarbon layer in liposomes of lecithin and sphingomyelin containing dicetylphosphate. J. Biol. Chem. 249, 2652-2657.

SHINITZKY, M. & INBAR, M. (1974) Difference in microviscosity induced by different cholesterol levels in the surface membrane lipid layer of normal lymphocytes and malignant lymphoma cells. J. Mol. Biol. 85, 603-615.

SHINITZKY, M. & INBAR, M. (1976) Microviscosity parameters and protein mobility in biological membranes. Biochim. Biophys. Acta 433, 133-149.

SHODELL, B. (1975) Reversible arrest of 3T6 cells in G_2 phase of growth by manipulation of a membrane-mediated G_2 function. Nature 256, 578-580.

SINGER, S.J. & NICOLSON, G.L. (1972) The fluid mosaic model of the structure of cell membranes. Science 175, 720-731.

TARKOWSKI, A.K. & WROBLEWSKA, J. (1967) Development of blastomeres of mouse eggs isolated at the 4 and 8-cell stage. J. Emb. Exp. Morph. 18, 155-180.

THOMPSON, G.A. & NOZAWA, Y. (1977) Tetrahymena: a model system for studying dynamic membrane alterations within the eukaryotic cell. Biochim. Biophys. Acta 472, 55-92.

THUDICHUM, J.L.W. (1884) A treatise on the chemical constitution of the brain. Bailliere, Tindall & Cox, London.

TOURTELLOTTE, M.E., BRANTON, D. & KEITH, A. (1970) Membrane structure: spin labelling and freeze etching of *Mycoplasma laidlawii*. Proc. Nat. Acad. Sci. 66, 909-916.

TRAUBLE, H. & EIBL, H. (1974) Electrostatic effects on lipid phase transitions: membrane structure and ionic environment. Proc. Nat. Acad. Sci. 71, 214-218.

TSUKAGOSHI, N. & FOX, C.F. (1973) Transport system assembly and the mobility of membrane lipids in *Escherichia coli*. Biochemistry 12, 2822-2829.

WALES, R.G. & BIGGERS, J.D. (1968) The permeability of two and eight cell mouse embryos to L-malic acid. J. Reprod. Fert. 15, 103-111.

WARREN, G.B., HOUSLAY, M.D., METCALFE, J.C. & BIRDSALL, N.J.M. (1975) Cholesterol is excluded from the phospholipid annulus surrounding an active calcium transport protein. Nature 255, 684-687.

WHITTAKER, J.R. (1973) Segregation during ascidian embryogenesis of egg cytoplasmic information for tissue specific enzyme development. Proc. Nat. Acad. Sci. 70, 2096-2100.

WISNIESKI, B.J., PARKES, J.G., HUANG, Y.O. & FOX, C.F. (1974) Physical and physiological evidence for two phase transitions in cytoplasmic membranes of animal cells. Proc. Nat. Acad. Sci. 71, 4381-4385.

WOODIN, A.M. (1968) In: E.E. Bittar and N. Bittar (Eds.), The Biological Basis of Medicine, Vol. 2. Academic Press, New York, pp. 373-396.

YORIO, T. & BENTLEY, P.J. (1976) Stimulation of active Na transport by phospholipase C. Nature 261, 722-723.

ZAGYANSKY, Y. & EDIDIN, M. (1976) Lateral diffusion of concanavalin A receptors in the plasma membrane of mouse fibroblasts. Biochim. Biophys. Acta 433, 209-214.

VIRUS-LIKE PARTICLES IN EMBRYOS AND FEMALE REPRODUCTIVE TRACT

Joseph C. Daniel Jr.* and Beverly S. Chilton**

**Department of Zoology*
University of Tennessee
Knoxville, Tennessee 37916

*** Department of Obstetrics and Gynecology*
University of Pennsylvania
Philadelphia, Pennsylvania 19174

1. Virus-like particles in embryos
2. Uterine VLPs
   - *RDDP inhibition by Blastokinin*
   - *Crystalline inclusion bodies*
3. VLPs in Placenta
4. Relationships to other RNA viruses
5. VLPs in milk and mammary glands
6. Vertical Transmission
7. Biological Function
8. Hypothetical Model

Tiny, particulate, roughly spherical organelles, best described as "virus-like" are regular components of the cellular ultrastructure of the genital tract, gametes and early embryos from a variety of mammals. These virus-like particles (VLP) resemble the RNA tumor viruses in that they are about 50-110 nm in diameter, have electron-dense shells and may exhibit other viral-like characteristics in the composition of their proteins and nucleic acid. Dalton (1972) describes the three main types of RNA tumor viruses, namely A, B and C types as follows:

> A-type particles average 70 nm in diameter and consist of two concentric spherical electron-dense shells with electron-lucent centers. Two types exist, both are intracellular and are distinguished by their localization within the cell. The first is present in the cytoplasmic matrix, accordingly named intracytoplasmic A; the second

type, intracisternal A (IAP), buds into the cisternae of the endoplasmic reticulum (ER).

B-types are particles of approximately 110 nm diameter and composed of two moderately electron-dense shells surrounding a central nucleoid with an electron-lucent center. An outer unit membrane, covered with spikes and radiating spokes, connects the inner border of this envelope with the outermost shell. The nucleoid may be an acquired intracytoplasmic A particle or specifically formed under the plasma membrane of a developing bud and commonly is in an accentric position.

C-types are generally extracellular particles which differ from B-type in that the inner of the two shells is much more electron-dense than the outer layer. The nucleoid is of uniform electron density and centrally located. Surface projections are sometimes seen from the loosely fitting envelope but are not consistently found. C-type particles average 100 nm in diameter. (Speculation exists that intracisternal A particles may mature into C-type morphology (e.g. Maugh, 1974a; Luftig *et al.*,1974; deHarven, 1974).

In the C-type RNA tumor viruses, the nucleoid typically contains a 60S-70S RNA similar to cellular messenger RNA. The 3' of the virion RNA and messenger RNA are both highly polyadenylate (poly A) and the 5' end of the minus strand is polyuridylate (poly U) (Baltimore, 1976).

The structural and enzymatic protein components of RNA viruses are located in either the envelope or the core of the virion. Three of these, which are especially characteristic of C-type viruses, are of special interest to the reproductive biologist, namely: group specific antigen (gs), transforming protein and RNA-dependent DNA polymerase (RDDP). Group specific antigens may be of several kinds and are determinants shared within a single species. The major C-type gs antigens are about 30,000 daltons in molecular weight and are found associated with the viral core. Transforming protein is the agent effective in transforming cells following virus infection, and each type of virus presumably makes a specific type of transforming protein which probably acts at the cell surface, though Baltimore (1976) indicates this remains to be proven. The detailed analysis by SDS-PAGE (sodium dodecyl sulfate-polyacrylamide gel electrophoresis) of virus induced changes in membrane glycoproteins (Smart and Hogg, 1976) supports this probability. From his review of the literature on leukemia, Maugh (1974b) concludes that cell transformation by RNA viruses is unequivocally associated with RNA-dependent DNA polymerase. This enzyme catalyses the transcription of the viral RNA into a DNA copy in the host cell (obviously in the opposite direction from the more common transcription of DNA into RNA) and is appropriately named "reverse transcriptase". This enzyme associates with the core of all RNA tumor viruses and in murine leukemia virus (MuLV) is typically about 70,000 molecular weight sedimenting at 4-4.5S on a glycerol gradient. The absence of RDDP activity renders a virus non-infectious (Maugh, 1974b). Major envelope glycoproteins are also about 70,000 daltons in M.W.

RNA viruses replicate in mammalian cells by reverse transcribing an intermediate DNA (provirus) which becomes integrated into cellular DNA where it can be regularly expressed. Alternatively the integrated genetic material may remain inactive while being replicated during successive cell divisions. The reader is referred to Gross (1970) for background and early history; Tooze (1973) and Baltimore (1976) for general details of oncorna virus biology; Gillespie and Gallo (1975) for RNA virus origin and evolution; to deHarven (1974) for ultrastructure and Schafer and Bolognesi (1977) for immunological aspects of viral components.

VLPs IN EMBRYOS

Most studies of virus-like particles in early embryos have been done on rodents. Particles have been reported for the 8-cell and blastocyst guinea pig (Enders and Schlafke, 1965), implantation stage hamster (Chapman *et al.*,1974) and 2-cell to egg cylinder mouse (Biczysko *et al.*,1973; Chase and Piko, 1973). Chase and Piko (1973) note that for the mouse, there were " four morphologically distinct virus-like particles whose expression was markedly dependent on the developmental state of the embryos". One was a dense-cored vesicle, two others were "A-type" particles (small and large respectively), and a fourth was a "C-type" particle (Figure 1). The small A-type was common in early stages but absent in blastocysts, while C-type only occurred in blastocysts. The other two were identifiable at all stages, most prominently in blastocysts (see Table 1). The sequence was the same in a dozen different strains of mice but the numbers of particles were slightly increased in highly leukemogenic AKR mice (Piko, 1975). Chase and Piko (1973) report no particles in one-cell embryos. Calarco and Szollosi (1973) also found two types of A particles in oocytes and 8-cell mouse embryos, but no transition stage was identified. One type was intracytoplasmic and resembled the nucleoid of a B-type particle, the other was intracisternal

TABLE 1: DISTRIBUTION OF VLPs IN PREIMPLANTATION MOUSE EMBRYOS

Type of Particle	Size nm	Stage of Embryo Development: 1 Cell	2-4	8	Morula	Blastocyst
Dense-Cored	50	0	±	±	±	++
Small A	75-85	0	++++	+++	++	+
Large A	85-100	0	0	±	+	++
C	90-110	0	0	0	0	+

Reprinted from Chase and Piko (1973)

(Figure 2). They found the latter in the dictyate stage but not in metaphase II oocytes. Fertilized eggs of four different strains of mice studied had no particles but budding particles were detected by the late 2-cell stage and even in the first polar body. The number of IAPs peaks at the 4 to 8 cell stage but they are rarely seen in the morula (16 cells) or early blastocyst (32 cells) stages. These authors note the coincidence between the onset of ribosomal RNA synthesis by nucleoli and the appearance of type A particles in the 2 to 4 cell stage suggesting that the appearance of particles may depend on the cells' RNA synthetic processes. The studies of Biczysko *et al.*(1973) also failed to reveal the presence of the particles in the mouse ovum either just before or after fertilization. In other studies, Calarco (1975) found A particles in the 4-cell mouse with some "incomplete" structures recognizable at the 2-cell stage. Cultured zygotes also had early A formation in the 2-cell stage.

Similar particles appear in primordial germ cells of guinea pigs (Figure 3) and Black (1974) concluded that they were immature C particles as they had characteristics of both A and C types. They were 85-110 nm in diameter with two electron dense shells

Figure 1: Thin sections of early mouse embryos showing various kinds of virus-like particles or budding particles; a and d are dense-cored vesicles; b and e are small A-type; c and f are large A-type; g, h, and i are C-type. (X 88,000). C = crystalloid. Reprinted, with permission, from Chase and Piko (1973). (In a personal communication from Dr. David Chase, he expressed concern that the "vesicle" shown in (a) may instead be a fibrous-cored tubule).

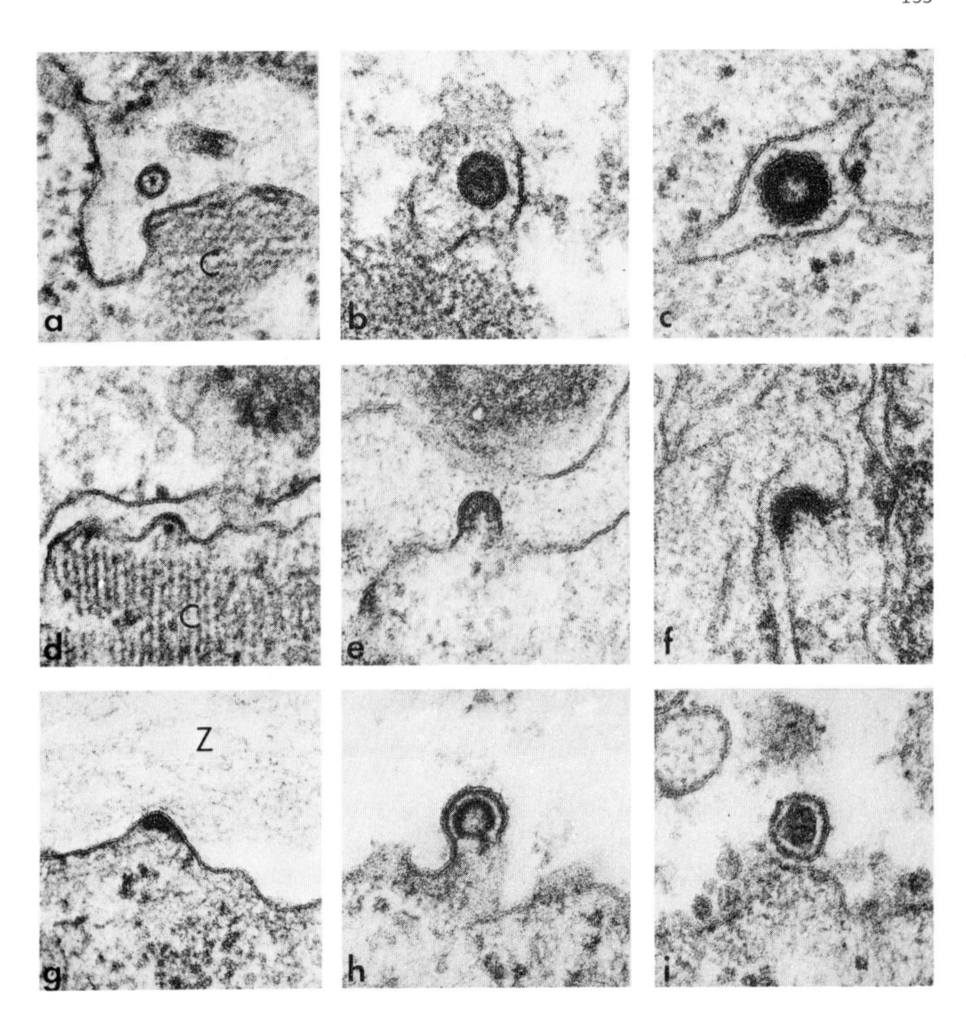
a
C
b
c
d
C
e
f
Z
g
h
i

surrounding a less dense core. "They acquired the outer shell by budding into cisternae of ER and had a definite layer separating this membranous envelope from the inner shell. Filamentous projections were associated with the outer envelope". The particles were seen in migrating germ cells and in cells of developing testes and ovaries from 19-65 day old fetuses (most commonly in gonocytes and oogonia at 27-35 days) and occasionally in 1-13 day old neonates. They were not found in intercellular spaces or somatic cells except for the parietal endoderm of the yolk sac placenta near the site of origin of the germ cells.

Both A and C types are found in rabbit blastocysts (Manes, 1974) but we are not aware of any definitive reports of VLPs in earlier stages. Pedersen and Seidel (1972) described spherical, membrane-bound, electron-dense "granules", 30-70 nm in diameter, abundant in the zona pellucida of the rabbit oocyte; and similar surface protrusions or micropapillae, that might represent budding granules, associated with the inner leaflet of the plasmalemma of oocytes and adjacent follicular cell processes. The granules appeared when the zona was formed and were present through fertilization. They were not perceptible in primordial follicles and never associated with other membranes in the interior of the cell. These authors noted several earlier papers where similar granules appeared in the published electron-micrographs but were not specifically mentioned, or were thought to be glycogen granules. Pedersen and Seidel (1972), however, claim the granules have a "superficial resemblance to viruses but do not closely correspond in their fine structure to any viruses known to us".

Baboon embryos of 4-8 days of age (presumably morulae and blastocysts) have C-type particles (Kalter *et al.*,1974) especially in the zona pellucida and perivitelline space. They were not found in coronal cells. Follicular oocytes and tubal ova, isolated on the day of ovulation and one day thereafter, also had C-type particles but no A-type (Kalter *et al.*,1975d). The particles were adjacent to the plasma membrane and in the perivitelline space along the inner margin of the zona pellucida. These investigators were unable to find C-type particles in Baboon sperm.

Figure 2: Cytoplasmic region from an eight-cell mouse embryo showing intracisternal A particles within agranular ER cisternae and crystalloid aggregates. (X 56,250). Reprinted, with permission from Calarco and Szollosi (1973).

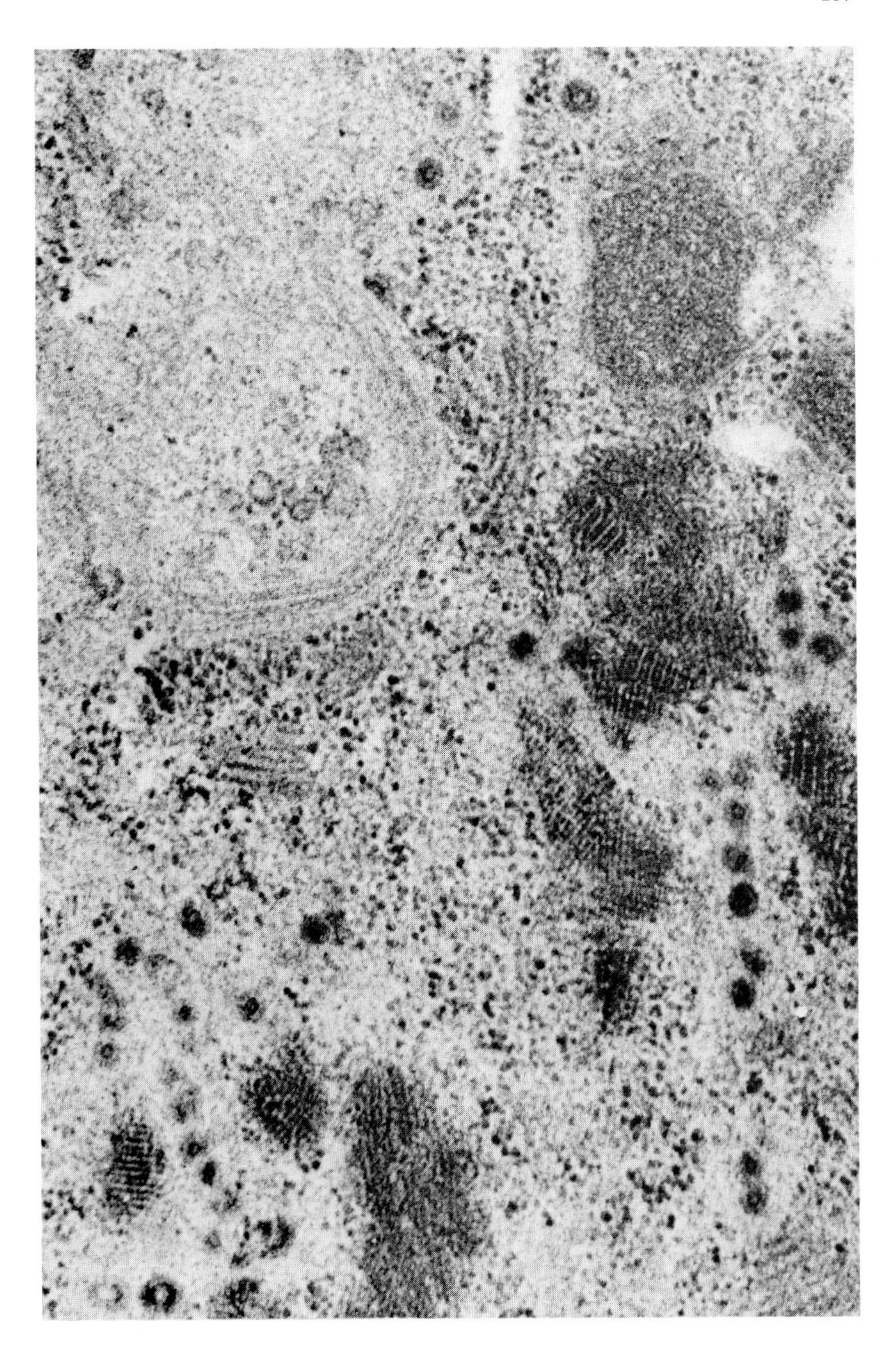

Calarco and McLaren (1976) were unable to find VLP's in preimplantation embryos of sheep. Also to the best of our knowledge, and confirmed in a personal communication by Dr. Allen Enders, VLPs have not been reported from dormant embryos of mammals having a delayed implantation.

Fertilization is not required for VLP production as indicated by their presence in oocytes and oogonia. It is also shown by studies of parthenogenotes: (1) Biczysko *et al.* (1974) reported the initial appearance of type-A particles in 3-4 celled mouse ova after parthenogenetic stimulation and subsequent cleavage *in vitro*. The particles were evident after culture through at least the 8-cell stage. These investigators also noted the existence of "round granular structures of high electron optical density found in the nucleoplasm and between the duplicated inner leaflets of nuclear membrane which might represent the most immature stage of virus development". Following this the "next stage would be represented by particles budding from the outer leaflet of the nuclear membrane". "Finally the mature virus particles are found in protrusions of the nuclear membrane into the cytoplasm, and later in the cytoplasm in the cisternae of the endoplasmic reticulum". (2) VanBlerkom and Runner (1976) also found virus-like particles resembling A-type in parthenogenetic mouse embryos. These particles appeared rather suddenly at the two-cell stage and more gradually reduced in number through the morula stage. Particles tend to occur in clusters between the inner and outer membrane or within elongated and distended cisternae of the smooth endoplasmic reticulum. By the morula stage some were free in the cytoplasm. No structures resembling C-type viruses were encountered in blastocyst-stage parthenogenotes.

Figure 3: Ovary of 39-day guinea pig fetus. Virus particles (p) occur in the ER of fetal guinea pig primordial germ cells, but not of adjacent somatic cells (lower right). Particles occur in dilated cisternae with few ribosomes (r) on their membranes. They can be seen within such cisternae, near the Golgi complex (G), but are rare within the flattened sacs of the Golgi elements of the associated small vesicles. Some particles have electron-lucent, others electron-dense centers. Particles are attached to the reticulum membranes (single arrows). Others seem free within the cisternae. Most occur singly; some are joined by narrow bridges into pairs (double arrows). (X 18,200).
Inset: Occasionally, such bridges link several particles in series, as shown in this micrograph from a germ cell of a sexually undifferentiated gonad. (X 28,200). Reprinted, with permission, from Black (1974) and legend quoted from that paper.

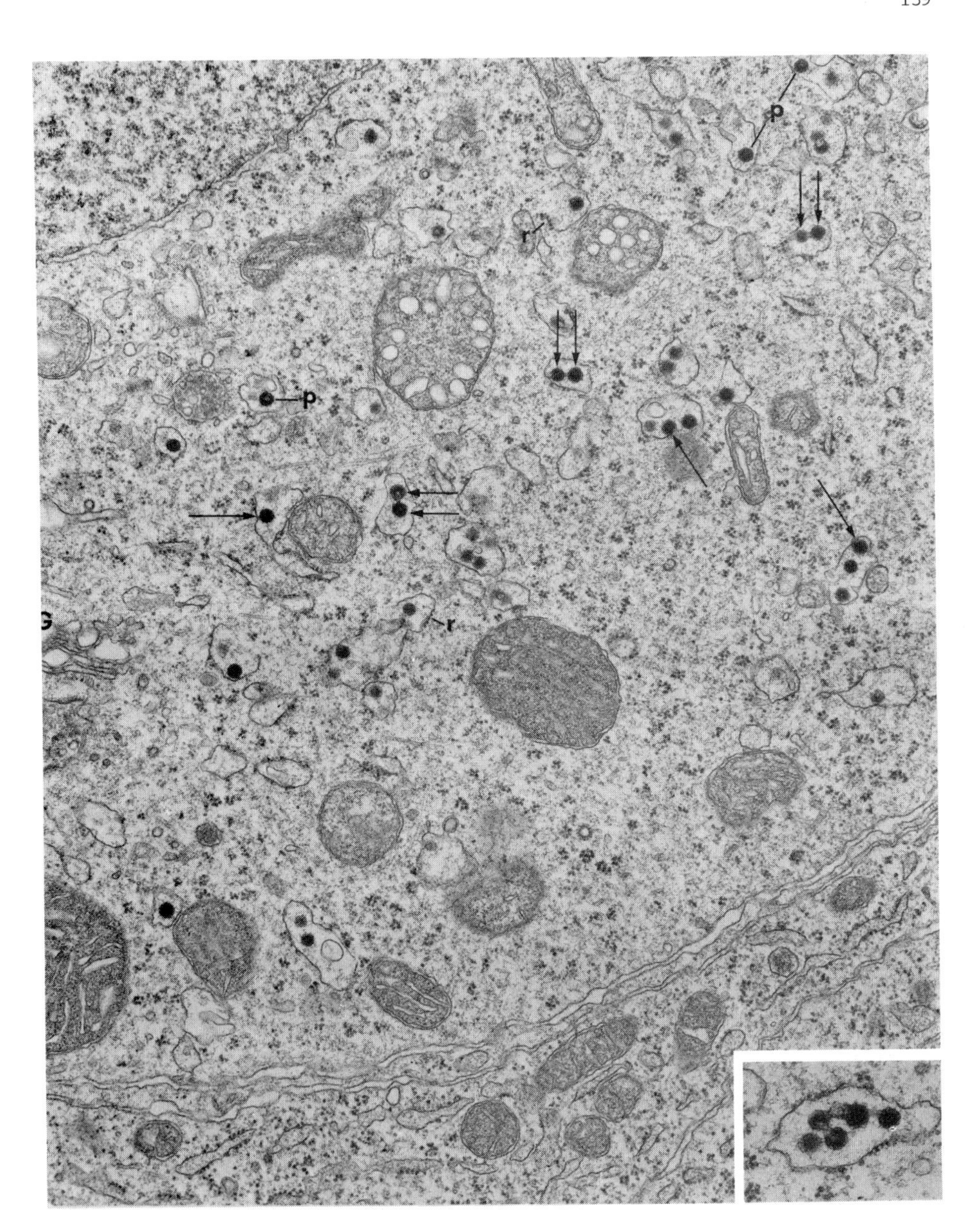
p
r
p
r
G

Some tissues taken from fetal stage embryos also have VLPs. Chandra *et al.*(1970) reported C-type particles in human embryonic muscle, liver and possibly thymus, but were unable to demonstrate any VLPs in kidney, spleen, pancreas, lung or testicle. Conversely, Schidlovsky and Ahmed (1973) found C-type (but not A-type) particles in erythroblasts and erythrocytes of fetal spleen and kidney tissues as well as smooth muscle of the umbilical cord of 67-105 day old Rhesus monkey embryos. They found no particles in capillaries, endothelial cells, striated muscle, or chorionic and amniotic membranes. These latter authors suggest that steroid and gonadotropin target tissues favour virus replication. C-particles were found in liver, spleen and thymus but not muscle of 85% of embryonic and newborn mice tested from four genetic strains (and a few substrains) and a wild type (Vernon *et al.*,1973) but the incidence varied with the strain and was progressively reduced with advancing embryonic age (Table 2). Chang *et al.*(1974) isolated type C-RNA viruses from spleen extracts of apparently normal mice at one to twelve months of age but not at eighteen months. Particles were detected in 60% of the four month old animals.

Cultured embryonic cell lines may also exhibit VLPs. C-type particles have been recovered or identified in normal rat (Ebert and Pearson, 1974; Rasheed *et al.*,1976), mouse (Gardner *et al.*, 1973; Stephenson *et al.*,1974; Simons *et al.*,1972; Levy, 1973), cat (Todaro *et al.*,1973) and human (Panem *et al.*,1975, 1977) cells. Replication of VLPs is also potentiated in cultured cells by exposure to certain chemicals, particularly the nucleoside analogs 5-bromo-2'-deoxyuridine (5-BrdU) and 5-iododeoxyuridine (IdU) (Green and Barron, 1975). Stephenson and Aaronson (1972) reported the inducibility of C-type RNA viruses from mouse embryo cell cultures by the use of IdU, and Lieber *et al.*(1973) demonstrated the superinduction of endogenous type-C virus from transformed mouse clones by 5-BrdU. Verwoerd and Sarma (1973) detected particles after treatment of rat embryo fibroblast cultures with 5-BrdU. Similarly, using secondary cell cultures from 16 day old rat embryos, Schwartz *et al.*(1974) demonstrated the release of type-C particles following exposure of the cells to 5-BrdU for 36 hours. Maximal release was observed between 8 and 10 days and ceased 16 to 19 days after removal of the analog. Van Blerkom found C-type particles in cultured rabbit blastocysts after exposure to 5-BrdU

TABLE 2: PREVALENCE OF C-TYPE PARTICLES IN VISCERAL TISSUES OF EMBRYONIC AND NEWBORN MICE

Strain	Days of Gestation	Number of particles/100 squares examined in: Liver	Spleen	Thymus
C3Hf/Mai	10-13	35	-	-
	18	1	13	70
BALB/cCr	12-14	28	26	0
	17-18	6	3	0
	Newborn	0	1	8
C57BL/6Cum	11-14	28	-	-
	16-18	1	8	16
NIH Swiss	11-14	35	16	0
	15-18	14	0	21
	Newborn	9	7	20
Wild	10-14	17	103	14
	17-18	18	0	10
	Newborn	1	9	-

Reprinted from Vernon *et al.* (1973)

(as reported by Manes, 1974) but similar particles were also found in untreated controls.

Attempts have been made to characterize biochemically the VLPs from early embryos or fetal tissues but the results are not always in agreement. Piko (1975) claims that the intracisternal A particles from tumors and embryonic tissues are distinct from oncogenic RNA viruses in structure and antigenic properties and have no demonstrated infectivity. Very recently Piko (1977) reported the presence of a "germinal vesicle antigen" in the mouse which cross reacts with antiserum made against p30 core protein of MuLV and which can be detected during oogenesis and cleavage up to the morula stage. It is absent in blastocysts and in a variety of tissues from juvenile mice. The antiserum can be absorbed with some mouse viruses but not others, indicating the antigen is type specific. The author hypothesizes that this germinal vesicle antigen is a cellular gene product which has a normal function in early zygote development. Huebner *et al.* (1970) showed that gs antigens of the C-type RNA tumor viruses were detectable in whole extracts made from 10, 12 and 14 day old mouse embryos and in the spleen, thymus and liver specifically from 18 day embryos and neonates. Lerner *et al.* (1976) using RIAs and immunofluorescence also

TABLE 3: ONCORNAVIRUS PROTEINS IN TISSUES OF MICE

(ng/mg of total protein)

Protein	Strain	Whole Embryo Extract at Age (Day P.C.): 10	12	14	16	18	18th Day Fetal Organs: Brain	Thymus	Spleen	Kidney	Liver	Neonate Extract	Adult Organs: Liver	Kidney	Spleen
Internal Core Protein (p 30)	BALB/c	<1	<1	<1	2	2	–	–	–	–	–	9	<1	<1	1
	C_3H	1	1	4	–	4	–	–	–	–	–	2	1	10	9
	AKR	–	<1	2	9	5	5	20	20	5	5	3	1	14	100
Envelope Glyco-protein (gp 69/71)	BALB/c	<1	<1	<1	<1	8	–	–	–	–	–	18	<1	<1	<1
	C_3H	33	33	33	–	40	–	–	–	–	–	20	<1	4	35
	AKR	–	<1	<1	<1	<1	5	25	25	5	<1	18	<1	18	60

Reprinted from Strand *et al.* (1977)

found expression of both p30 antigen and gp 70, the major envelope glycoprotein from MuLV, in 10, 14 and 18 day old mouse embryos. These were also detectable in adult tissues but were restricted to certain anatomical sites and cell types prominent among which were lymphoid and epithelial cells. The search of Hilgers *et al.* (1974) for MuLV gs antigen was consistently negative for embryos of 10-18 days of gestation. They did, however, find it in a variety of organs from neonatal and postnatal mice.

C-type particles also share common group specific antigens, and structural genes for C-type viruses appear to be present in all mouse cells. Todaro *et al.* (1973) found viral antigens similar to the RD-114 virus in normal cat embryo cells but Reznikoff *et al.* (1973) were unable to demonstrate viral antigens in a cloned line of mouse embryo cells. According to Gooding *et al.* (1975) murine ova and morulae have three cell surface antigens which are immunologically similar to those found on the surface of murine tumor cells induced by viruses. Calarco (1975) further underscored the similarities between intracisternal A particles and mouse oncoviruses by their sensitivity to actinomycin D. She identified regions in 2-cell cultured mouse embryos as presumptive stages of early IAP formation in embryos cultured in α-amanitin but not in those cultured in actinomycin D.

RNA complementary to DNA transcripts synthesized from an endogenous type-C virus of BALB/c 3T3 was detected in 14, 16 and 18 day old mouse embryos as well as many tissues from newborn and adult animals (Mukherjee and Mobry, 1975). Because the complementarity was highest in tissues with greater proliferative activity, the authors suggest a correlation between the endogenous C-type virus genome transcription pattern and cell proliferation.

Strand *et al.* (1977) analyzed embryos, neonates, and adult organs from three strains of mice for the concentrations of the major internal (30,000 daltons) protein and the major envelope (70,000 daltons) glycoprotein of endogenous C-type oncornaviruses, by use of a competitive radioimmunoassay. Each strain showed a different pattern of protein expression at any given stage of development compared to other stages and adult tissue. Their results are summarized in Table 3. No distinct, consistent pattern emerges which could be attributed to a developmental process, with the possible exceptions of an increase in envelope protein at birth and higher levels of both proteins in fetal and adult spleens.

TABLE 4: REPORTED OCCURRENCE IN MAMMALIAN GAMETES AND EMBRYOS OF VIRUS-LIKE

		1 Mouse	2 Mouse Parthenote	3 Rat	4 Hamster	5 Guinea Pig	6 Rabbit	7 Sheep
	Primordial Germ Cell					VL		
Gametes	Sperm	P,gs**					VL,(P)	
Gametes	Oocyte	A,gs		gs		VL	VL	
Gametes	Ovum	(A)(C)gs						(VL)
	2-Cells	A,VL,gs	A,(C)				(P)	
	4-Cells	A,VL,gs	A,(C)					
	8-Cells	A,VL,gs	A,(C)			VL	(P)	(VL)
	16-Cells	A,gs	A,(C)					(VL)
	Morula	A,VL,gs	A,(C)				(P)	(VL)
	Blastocyst	A,C,VL,gs	(A)(C)			VL	A,C,gs (P)	(VL)
Developmental Stages	Egg Cylinder	A,C			C		(P)	
Developmental Stages	Early Fetus	C,gs,P						
Developmental Stages	Mid Fetus	gs		C,gs,R,P		VL		
Developmental Stages	Late Fetus	C,gs		C,gs,R,P		VL		
Developmental Stages	Neonate	C,gs				VL		
Developmental Stages	Placenta	C,A		C		A,C	(C)	
Embryonic Tissue	Skeletal muscle	(C)						
Embryonic Tissue	Smooth muscle							
Embryonic Tissue	Liver	C,gs						
Embryonic Tissue	Thymus	C,gs						
Embryonic Tissue	Spleen	C,gs						
Embryonic Tissue	Kidney							
Embryonic Tissue	Pancreas							
Embryonic Tissue	Lung							
Embryonic Tissue	Testis							

* For references relating to each species, see Table 4a.

PARTICLES OR OTHER CHARACTERISTICS ASSOCIATED WITH RNA TUMOR VIRUSES*

8 Human	9 Baboon	10 Rhesus Monkey	11 Woolly Monkey	12 Patas Monkey	13 Cynonolgus Monkey	14 Cebus Monkey	15 Chimpanzee	16 Marmoset
P	(C)							
	(A),C							
	(A),C							
	C							
	C							
C,VL,(C) (P)	(A),C,gs P	(A),C,VL, P	C,P	C	C	(C)	C	C
C								
		(A),C						
C								
C(?)								
(A),(C),C		(A),C						
(A),(C),C		(A),C						
(C)								
(C)								
(C)								

** KEY:
A - A-type particles
C - C-type particles
VL - Virus-like particles
gs - Group specific antigen
R - 60S - 70S RNA
P - RNA-dependent - DNA polymerase activity
( ) - Searched for but not found

TABLE 4A: REFERENCES TO EACH OF THE OBSERVATIONS

1.	Mouse	Kalter *et al.*(1975b); Kajima and Pollard (1968); Vernon *et al.*(1973); Todaro and Huebner (1972); Chase and Piko (1973); Biczysko *et al.*(1970); Calarco and Szollosi (1973); Calarco (1975); Huebner *et al.*(1970); Sherman and Kang (1973); Witkin *et al.*(1975); DelVillano and Lerner (1976); Piko (1977); Piko (1975); Van Blerkom and Manes (1977); Hilgers *et al.*(1974); Lerner *et al.*(1976); Smith *et al.*(1975).
2.	Mouse Parthenote	Biozysko *et al.*(1974); Van Blerkom and Runner (1976).
3.	Rat	Rasheed *et al.*(1976); Cross *et al.*(1975); Piko (1977).
4.	Hamster	Chapman *et al.* (1974).
5.	Guinea Pig	Kalter *et al.* (1975b); Hsuing *et al.*(1974); Black (1974); Anderson and Jeppesen (1972); Enders and Schlafke (1965).
6.	Rabbit	Manes (1974); Chilton and Daniel (1978); Van Blerkom and Manes (1977); Kalter *et al.*(1975b); Pedersen and Seidel (1972).
7.	Sheep	Calarco and McLaren (1976).

RECORDED IN TABLE 4.

8.	Human	Chandra *et al.*(1970); Vernon *et al.*(1974); Dalton *et al.*(1974); Kalter *et al.*(1975b); Schidlovsky and Ahmed (1973); Witkin *et al.*(1975); Bargmann and Knoop (1959).
9.	Baboon	Kalter *et al.*(1975b); Kalter *et al.* (1974); Kalter *et al.*(1973); Dalton *et al.*(1974); Sherr and Todaro (1974); Hellman *et al.*(1974); Strickland *et al.*(1973).
10.	Rhesus Monkey	Schidlovsky and Ahmed (1973); Feldman (1975); Mayer *et al.*(1974); Kalter *et al.* (1975b).
11.	Woolly Monkey	Sherr *et al.*(1974).
12.	Patas Monkey	Kalter *et al.*(1975b).
13.	Cynonolgus Monkey	Kalter *et al.*(1975b).
14.	Cebus Monkey	Kalter *et al.*(1975b).
15.	Chimpanzee	Kalter *et al.*(1975b).
16.	Marmoset	Seman *et al.*(1975); Kalter *et al.*(1975b).

TABLE 5: BIOCHEMICAL CRITERIA FOR CHARACTERIZATION OF RDDP ACTIVITY IN VARIOUS RNA TUMOR VIRUSES COMPARED TO THAT IN MAMMALIAN REPRODUCTIVE TISSUE

Viral Property and References		Cellular Property and References
Located in particulate cytoplasmic fraction (Sarngadharan *et al.*,1976)	+	(Strickland *et al.*,1973; Mayer *et al.*,1974; Yang *et al.*,1976; Chilton and Daniel, 1978)
Solubilization with NP-40 and KC1 (Wu and Gallo, 1975; Sarngadharan *et al.*,1976)		(Strickland *et al.*,1973; Mayer *et al.*,1974; Yang *et al.*,1976; Chilton and Daniel, 1978)
Approximately 4.5s in size (Wu and Gallo, 1975)	4-4.5s 5-6s	(Sherman and Kang, 1973) (Yang *et al.*,1976)
Molecular weight approximately 70,000 (Wu and Gallo, 1975)	62,000	(Mayer *et al.*,1974)
α and β subunits (Sarngadharan *et al.*,1976)	NR	
Assay requirements		
Tris buffer, pH 7.5-8.3 (Wu and Gallo, 1975; Sarngadharan *et al.*,1976)	+	(Fowler *et al.*,1972; Strickland *et al.*,1973; Mayer *et al.*,1974; Witkin *et al.*,1975; Yang 1976; Chilton and Daniel, 1978)
Template preference (Wu and Gallo, 1975; Sarngadharan *et al.*,1976)		
$(rA)_n \cdot (dT)_{9-18}$	+	(Fowler *et al.*,1972; Sherman and Kang, 1973; Strickland *et al.*,1973; Mayer *et al.*,1974; Witkin *et al.*,1975; Yang *et al.*,1976; Chilton and Daniel , 1978)
$(rC)_n \cdot (dG)_{6-18}$	+	(Mayer *et al.*,1974; Yang *et al.*,1976)
70s RNA	-	(Yang *et al.*,1976)
Cation requirements (Wu and Gallo, 1975)		
1.0 - 10.0 mM $MnCl_2$	0.2 - 0.8 mM	(Mayer *et al.*,1974; Witkin *et al.*,1975;
0.5 - 1.0 mM $MgCl_2$	0.1 -10.0 mM	Yang *et al.*,1976; Chilton and Daniel, 1978)

Optimum incubation temperature of 37°C (Wu and Gallo, 1975)	35-37°C	(Fowler *et al.* 1972; Sherman and Kang, 1973; Strickland *et al.* 1973; Mayer *et al.*,1074; Witkin *et al.*, 975; Yang *et al.*,1976; Chilton and Daniel, 1978).
RNA-DNA hybrid intermediate product (Green and Gerard, 1974; Wu and Gallo, 1975; Sarngadharan *et al.*,1976)	NR	
RNase sensitive final product (Wu and Gallo, 1975; Sarngadharan *et al.*,1976)	+	(Witkin *et al.* 1975)
Association of polymerase with Ribonuclease H enzyme (Wu and Gallo, 1975; Sarngadharan *et al.*,1976)	NR	
Inhibition of enzyme reaction		
Rifamycin SV (Wu and Gallo, 1975)	NR	
Antipolymerase sera (Wu and Gallo, 1975; Sarngadharan *et al.*,1976)	±	(Fowler *et al.*,1972; Strickland *et al.*,1973; Mayer *et al.*,1974)
Stability of stored enzyme (-20 to -70°C; 50% glycerol) (Sarngadharan *et al.*,1976)	±	(Yang, personal communication; Witkin *et al.*,1975)
RDDP activity in tissue known to have C-type particles (Wu and Gallo, 1975)	+	(Strickland *et al.*,1973; Yang *et al.*,1976; Chilton and Daniel, 1978)
gs antigen (Huebner *et al.*,1970)	+	(Fowler *et al.*,1972; Strickland *et al.*,1973; Del Villano and Lerner, 1976).

\+ indicates that the same is true, - is not true, or NR not reported for the cellular RDDP activity.

Verwoerd and Sarma (1973) showed reverse transcriptase activity and gs antigen in cultured rat embryo fibroblasts from six normal strains of domestic and wild rats after IdU treatment.

Rasheed *et al.*(1976) isolated type-C viruses containing 60-70S RNA, RDDP, and a 30,000 MW virus-specific protein, from cell cultures derived from 14-20 day old embryos of two different strains of rats. Sherman and Kang (1973) also found RDDP activity in cells from 11 day mouse embryos. The specific activity of this enzyme was highest in rapidly dividing embryonic tissue, decreased approximately 50% in the cells of the invasive trophoblast where DNA replication continues in the absence of cell division; and further decreased to a low but detectable level in association with the non-dividing decidua. We (Chilton, 1976; Chilton and Daniel, 1978) were unable to find evidence of RNA-dependent DNA-polymerase activity in rabbit embryos recovered from the genital tract on any of days 1-9 *post coitum*.

Studies of the presence of virus-like particles, or their identifying qualities, in the male have produced inconsistent results. Witkin *et al.*(1975) report an RDDP associated with human sperm and seminal fluid and similar complexes from murine sperm. However, preliminary data from our laboratory indicates an absence of RDDP activity in rabbit sperm and sperm-free seminal fluid (unpublished observation). Croker *et al.*(1976) found C-type virions in vas deferens and epididymis of the mouse, but not in the testis. A protein, related immunologically to gp70, the major envelope glycoprotein of the murine leukaemia viruses, is present in the testis (Lerner *et al.*,1976) and in the secretion of the epididymis and ductus deferens of the mouse (Del Villano and Lerner, 1976) and is also associated with the sperm surface, possibly being related to capacitation.

Table 4 summarizes the reported occurrence in mammalian gametes and embryos of virus-like particles or other characteristics commonly associated with the presence of oncorna viruses.

UTERINE VLPs

VLPs, or markers indicating their expression, also exist in the uterus. The first evidence of this came from the work of Hellman and Fowler (1971) and Fowler *et al.*(1972). They showed the presence of group specific antigen p30, and RDDP in the uteri of ovariectomized NIH Swiss mice 2-4 days after treatment with estradiol-17β, synthetic estrogen compounds (both steroidal and non-steroidal), or by X-irradiation. Both of these viral markers from the uterus were of larger molecular weight than those from Rauscher leukemia virus (Strickland *et al.*,1974) leading to the suggestion that the uterine enzyme might be a precursor of the DNA polymerase associated with mature viral particles. This estrogen-induced polymerase activity was inhibited by an antibody to murine virus RDDP. The level of viral proteins depends on the time interval following treatment and the "potency" of the estrogen (Fowler *et al.*,1977a) and is strain-dependent (Fowler *et al.*,1973). The uteri of ovariectomized, aged mice (2-2.5 years old) respond to estrogen treatment much the same as young animals, relevant to p30 elevation, but RDDP activity remains low or absent (Strickland *et al.*,1976), perhaps implying "...that the production of these viral proteins is under separate control mechanisms".

The presence of endogenous VLPs (C-type) in the uteri of adult NIH Swiss mice was confirmed by electron microscopy (Hellman *et al.* 1976; Fowler *et al.*, 1977a). Particles were found in the endometrium budding from, or adjacent to, epithelial cells, mainly near the basal lamina. Intracisternal A-particles were also present in some specimens but no particles were ever observed in the lumen (Figure 4). When uterine specimens were taken throughout gestation, the number of particles seen was much greater between days 4 to 9 than either earlier or later (an association with implantation was implied). These investigators used a double antibody radioimmune precipitation assay to determine the level, and location in the uterus, of the major internal core C-type virus group specific protein, p30 (as obtained from Rauscher leukemia virus). They found that uterine levels of p30 were low and did not differ significantly during the estrous cycle, but did increase and change predictably during pregnancy. Within the first four days *post coitum* the level doubled over precoital levels. It remained elevated through day 12 after which a second acceleration began which culminated in a 10-fold increase by about day 17. This high level was retained through day 1 *post-partum* after which it dropped precipitously. Levels of p30 were noted to correlate with plasma estrogen concentrations but not progesterone. Figure 5 shows this coincidence by diagramming the p30 levels with circulating estradiol-17β and progesterone levels as reported by McCormick and Greenwald (1974). The strongest fluorescence used for localization of p30 in the mouse uterus was reported for the

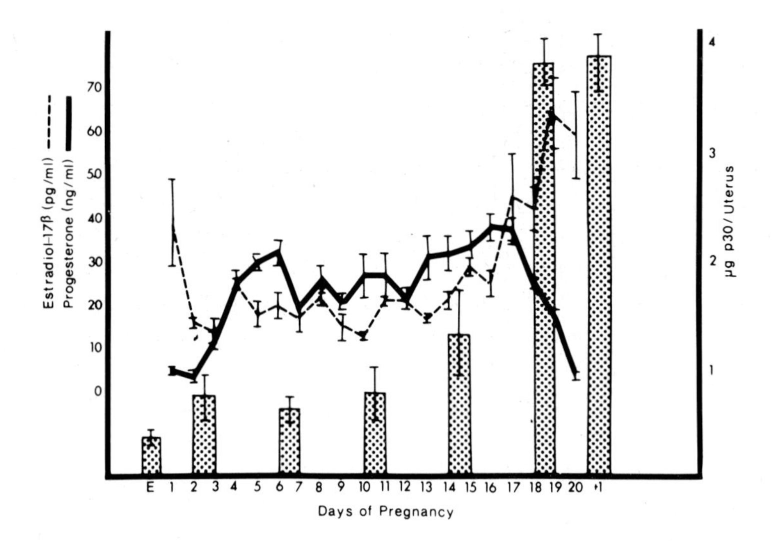

Figure 5: Comparison of p30 in the mouse uterus with circulating levels of ovarian steroid hormones. Derived from the data of Fowler *et al.*(1977c) and of McCormack and Greenwald (1974).

luminal and glandular epithelia; it was never seen in stromal cells. The intensity was greatest in the epithelia of pregnant uteri and very weak in uteri from ovariectomized mice (Figure 6). Lerner *et al.*(1976) also found p30 and gp70 from MuLV in the mouse uterus.

Direct innoculation of mink, cat and human target cell cultures with cell-free uterine extracts of normal adult NIH Swiss mice resulted in the isolation of xenotropic type-C particles from about half of the cultures (Allen *et al.*,1977). Further characterization of the virus isolated from the mink cell cultures (designated M55) showed it to be infective in cells of human, dog, mink, cat and bat origin, but not in mouse cells. It had RNA which sedimented at 50-60S and exhibited both major and minor polypeptides common to other mammalian C-type viruses but some of

Figure 4: Gravid mouse uterine endometrium. Budding type-C particles (a, c) from epithelial cells, mature type-C particle (b) in the intercellular space, and an intracisternal A particle (d) in an epithelial cell. Reprinted, with permission, from Fowler *et al.*(1977b), and legend quoted from that paper.

Figure 6: Immunofluorescence to Rauscher leukemia virus p30 in the uterine secretory epithelium of NIH Swiss mice during the estrous cycle. Reprinted with permission, from Fowler *et al.*(1977b), and legend quoted from that paper.

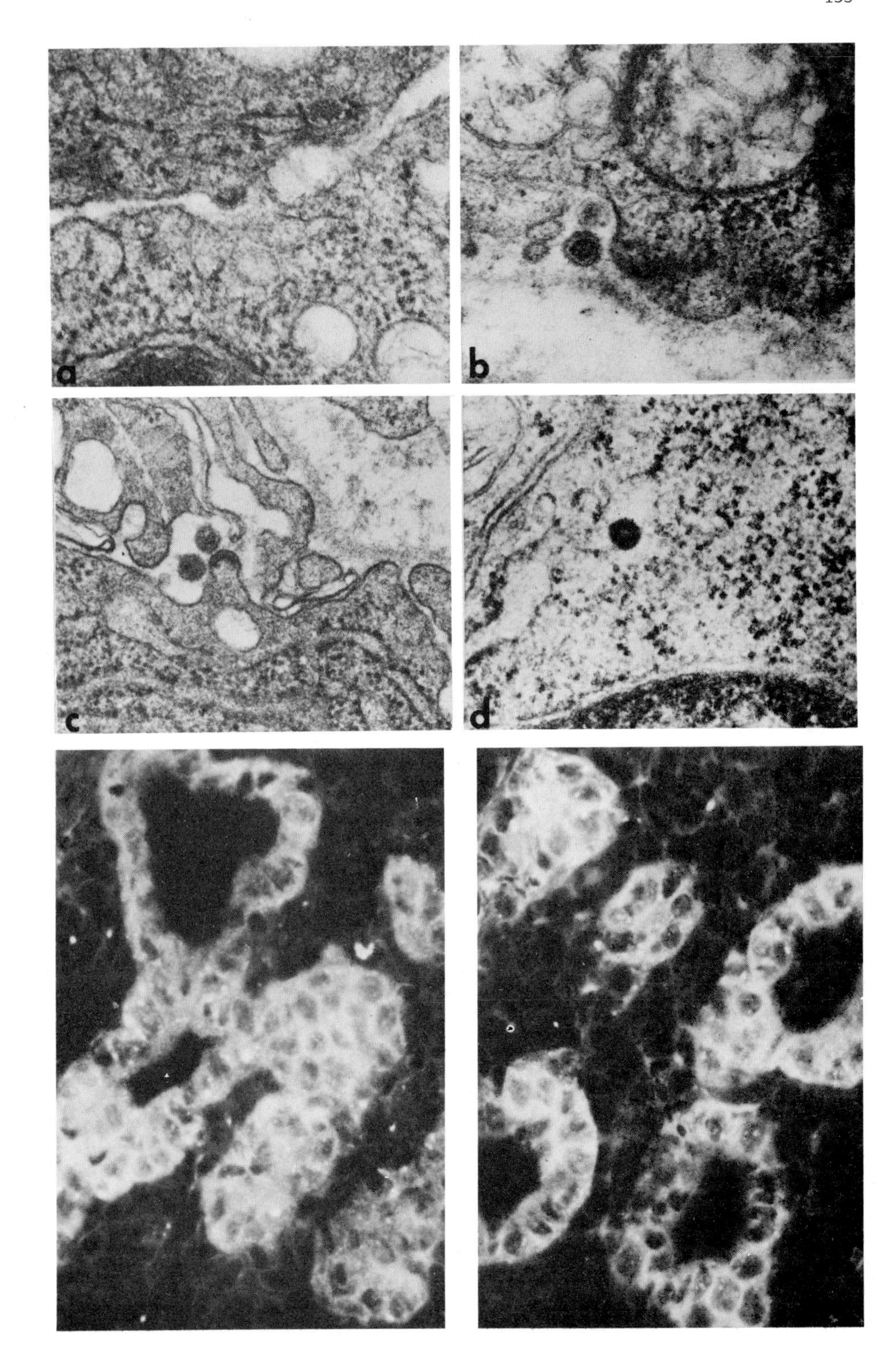
a
b
c
d

these differed slightly in staining intensity and electrophoretic mobility indicating at least a qualitative difference from MuLV polypeptides.

Particles resembling mouse hepatitis virus have also been reported in the uteri of nude mice (Smith *et al.*,1976).

Contrary to the situation in the mouse, VLPs were first found in the uterine lumina of rabbits, specifically in pellets of flushings taken on days 4-6 *post coitum*. In unpublished work, Richard Tyndall of Oak Ridge National Laboratory and Joseph Daniel found cross-reactivity between RD-114, Rauscher leukemia virus, and some mouse and human tumor extracts with an antiserum to uterine flushings from a 5 day pregnant rabbit (Johnson *et al.*,1972). The search for the common antigen led to the discovery (with electron microscopy) of VLPs in uterine flushings. The presence of particles was erratic from one animal to another but particles were never found in flushings taken from estrous rabbits or on days 1 and 2 *post coitum*. Subsequent studies revealed the presence of VLPs in sections of progestational uterine luminal epithelium and in mitochondrial and microsomal pellets from endometrial extracts (Figure 7) for the same days of early pregnancy but were absent in estrous stage uteri. These findings were reported in 1973 at the Symposium on the Uterus, given before the Houston meeting of the American Society of Zoologists, but were never published. However, Yang *et al*,(1976) were unable to find VLPs in uterine flushings and could neither release nor rescue viruses after coculture of endometrial cells with human embryonic rhabdomyosarcoma cells (TE_{32} cell line) and S^+L^- human amnion cells (F_{160} cell line). These studies did, however, demonstrate the presence of RDDP in rabbit uterine endometrium, activity varying predictably with the state of early pregnancy and differing in various ways, including molecular size,from the reverse transcriptase of RNA tumor viruses. Unpublished work from this laboratory shows another difference in that uterine RDDP activity is rapidly lost after freezing whereas Rauscher leukemia virus standards remain active after repeated freezer-thaw cycles.

Figure 7: VLPs in rabbit uterus. The primary plate is of a negatively stained (2% aqueous KPTA at pH 7) pellet from the mitochondrial fraction obtained after differential centrifugation of an endometrial homogenate collected on day 5 *post coitum*. (This was an especially rich preparation; we have not consistently seen such particles in all samples). Insert: Uranyl acetate-lead citrate stained particles isolated from uterine flushings. (X 40,000).

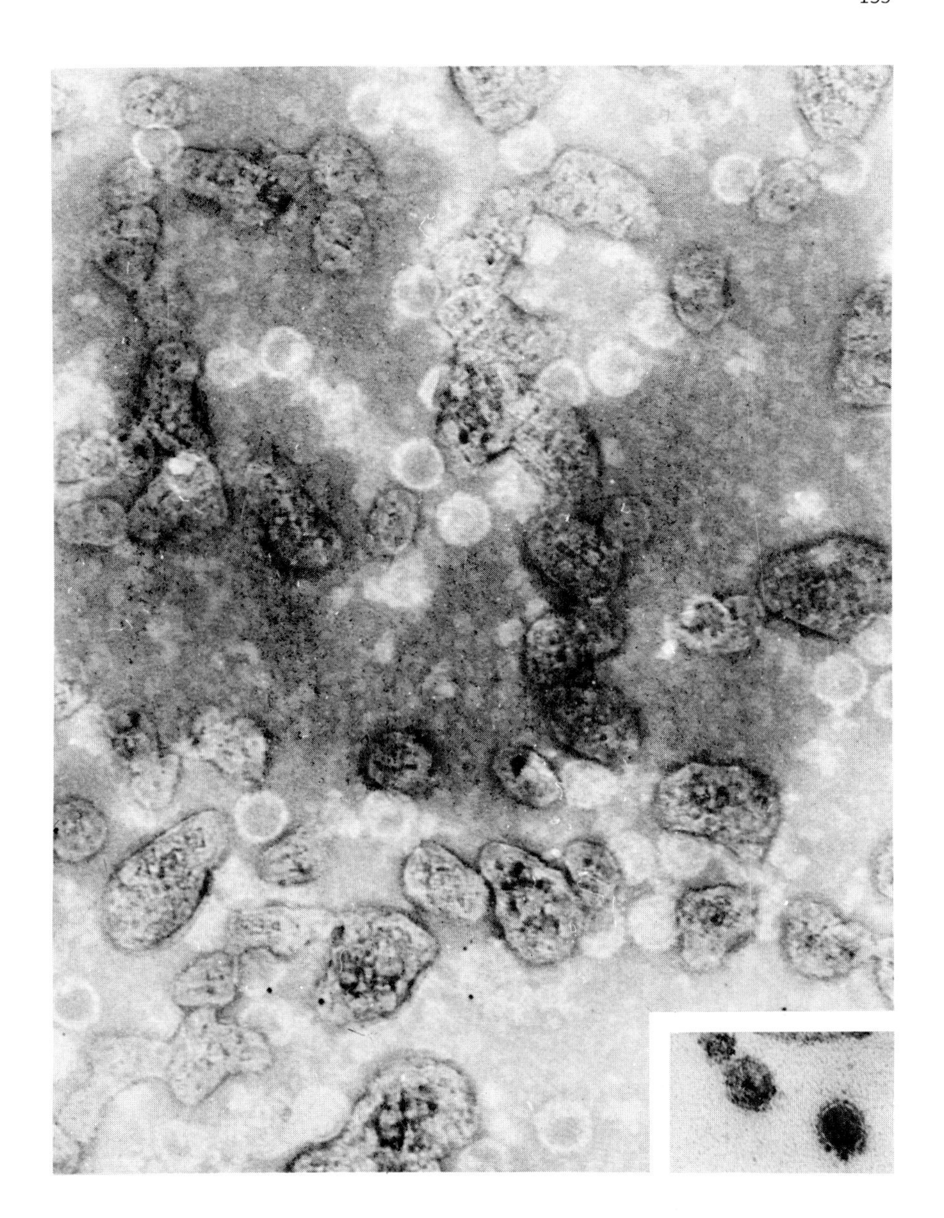

RDDP was independently described by Temin and Mizutani (1970) and Baltimore (1970) as an RNA tumor virus-specific enzyme. Thus detection of this enzyme is a sensitive means of assaying for type-C RNA tumor viruses in transformed cells. It remains to be seen if the RNA-dependent DNA polymerase enzyme associated with mammalian reproduction is the same as the viral reverse transcriptase (see Table 5).

The RDDP activity which Yang *et al.*(1976) identified appears in subcellular fractions of rabbit endometrium 24 hours after coitus and reaches a maximum on days 3-4 of pregnancy concurrent with the appearance of virus-like particles in the endometrium. This enzyme was extracted from tissues using high salt and non-ionic detergent, sedimented at 5-6S and showed a preference for viral-specific template-primers [$(rA)_n \cdot (dT)_9$ and $(rC)_n \cdot (dG)_6$]. Further studies by Chilton and Daniel (1978) have also demonstrated that this uterine RDDP activity reaches a peak on day 3-4 *post coitum*, the time when embryos first encounter the uterine milieu and remains high through the time of implantation (days 7-9 of pregnancy) as shown in Figure 8. Polymerase activity levels in control tissues, namely, homogenates of lung, liver and spleen from the same animals remain low to undetectable for the same time period. As seen in Figure 9, enzyme activity levels in the endometrium of pseudopregnant does rise initially but drop by day 8 in the absence of viable embryos and the initiation of implantation.

Correlative with the RDDP activity during pseudopregnancy is the maintenance of plasma progesterone levels not significantly different from those measured during the first 8 days of pregnancy (Holt *et al.*,1976). Therefore, in order to determine the effects of the ovarian steroid hormones in RDDP activity, virgin, estrous does were bilaterally ovariectomized, maintained untreated for 3 weeks, and then given supplementary injections of progesterone, estradiol 17-β, both in sequence of 6 days of estradiol followed by two days of progesterone, or corn oil carrier alone (Chilton and Daniel, 1978). As seen in Figure 8, the histogram representing induced polymerase activity superimposed on the curve established for normal pregnancy shows that the synergistic effects of estrogen followed by progesterone are necessary to induce polymerase activity levels comparable to those normally seen on day 3-4 *post coitum*.

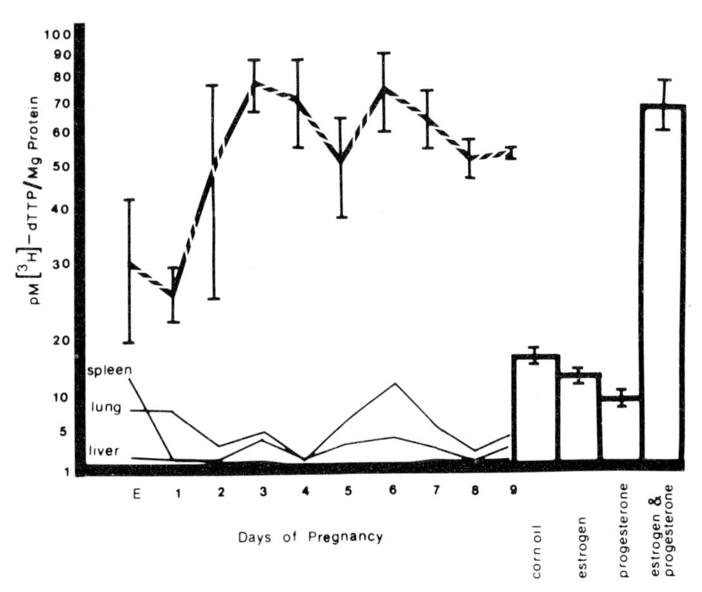

Figure 8: RDDP activity levels in the uteri of hormonally primed ovariectomized rabbits are presented by the bar graphs together with the line graph illustrating RDDP activity during pregnancy. Corn oil injected, control animals produced detectable levels of polymerase activity comparable to that measured in estrous does. Injections of either estrogen or progesterone resulted in RDDP levels equivalent to the control animals; while the dual effects of both hormones resulted in RDDP activity equivalent to days 3-7 of pregnancy. Spleen, lung and liver activities were consistently low or absent. Compiled from the data of Chilton and Daniel (1978).

RDDP INHIBITION BY BLASTOKININ

Closer examination of the graph representing normal RDDP activity (Figure 8) reveals a decrease in enzyme activity on day 5 of pregnancy, concomitant with the peak in availability of the uterine protein blastokinin (BKN) (Krishnan and Daniel, 1967; Daniel and Booher, 1977). Because BKN binds certain divalent cations, including manganese (Daniel, 1976) and since the endometrial RDDP has some cellular and some viral characteristics (Yang *et al.*,1976) including the need for Mn^{2+} cofactor, it was thought that a chelating effect might account for the change in activity. We measured the inhibitory effects of the BKN on the uterine RDDP activity and also the RDDP activity of the cat endogenous virus RD-114 by the substitution of BKN for BSA in the standard polymerase assay mixture. Selection of RD-114 for this study was based on the fact that it possesses an RDDP enzyme; it has nucleic acid homology with

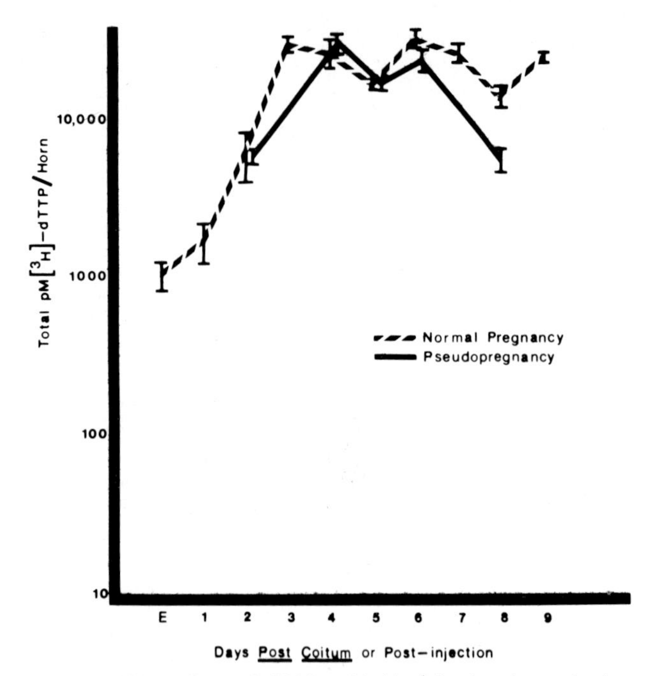

Figure 9: The profile of total RDDP activity/uterine horn during early pregnancy and pseudopregnancy in the rabbit. The Y-axis is a logarithmic scale and the vertical lines represent the standard error term. Compiled from the data of Chilton and Daniel (1978).

the cellular DNA of the domestic cat, several primate species and the rabbit; and is suspect for germ line transmission (Benveniste *et al.*,1975).

The reaction mixture for each polymerase assay contained BSA at a final concentration of 0.17 μg/μl. This protein was added in excess to protect the enzyme from denaturation. In order to test the effects of blastokinin on polymerase activity, BKN was substituted for BSA in increasing increments of 20, 40, 60, 80 and 100%. The protein concentrations represented by the percentages used ranged from .06 to .32 μg BKN/μl.

In the first series of experiments, the effect of these BKN concentrations on the RDDP activity in the cat endogenous virus RD-114 was tested and the results are shown in Figure 10. The ratio of 0.06 μg/μl BKN : 0.14 μg/μl BSA gives a 45% reduction in the RDDP activity found in the absence of BKN. A ratio of 0.13 μg/μl BKN : 0.10 μg/μl BSA results in 75% polymerase inhibition and when BKN is at a concentration of 0.19μg/μl, almost three times the BSA concentration, a 94% loss

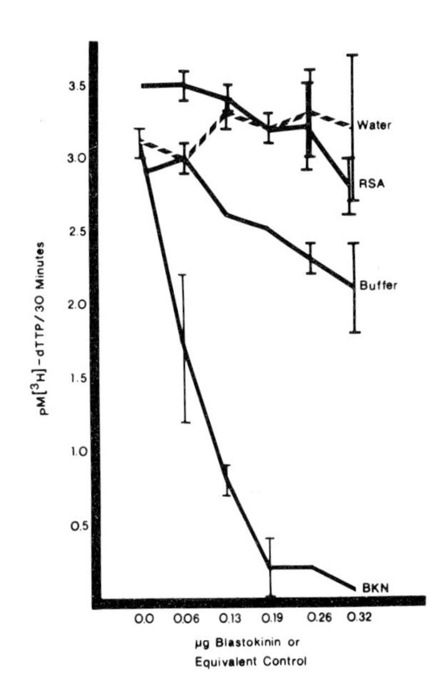

Figure 10: Analysis of inhibition of cat endogenous virus RD-114 RNA-directed DNA polymerase activity by the uterine protein blastokinin (BKN). Included is the graphic representation of water blank, buffer and rabbit serum albumin (RSA) controls. Vertical lines represent the standard error (S.E.). See text for details.

of activity is obtained. RDDP activity was completely lost by the total replacement of BSA with BKN.

Experimental controls were designed to demonstrate that the reduction in activity was not caused simply by a decrease in the amount of BSA in the reaction mixture, possible traces of sodium azide in the elution buffer, or some other factor introduced by the procedures used to purify the BKN (Murray *et al.*,1972). Therefore, aliquots of distilled water or 0.02 M citrate buffer (pH 7.4) with 0.02% sodium azide were substituted for the comparable volume of BKN originally tested. Rabbit serum albumin (RSA) was purified by the same procedures used for BKN through elution from the Sephadex G-200 column; subsequently the appropriate fractions were pooled, concentrated with Aquacide (26.9 μg/μl) and diluted to a final protein concentration of 3.2 μg/μl with distilled water. This RSA was then substituted for the same concentrations of BKN in the final assay mixture. None of these variations inhibited

RDDP activity to the same degree as blastokinin, although the buffer modification had a slight effect (Figure 10).

To further demonstrate that the loss of RDDP activity is due to BKN and not just the reduction in BSA, the assay was performed using 0.19 μg/μl BKN while maintaining the BSA concentration at the standard 0.17 μg/μl. The RDDP activity was reduced to 0.3 ± 0.1 pM (^{3}H)-dTTP/30 minutes, thus demonstrating that BKN is responsible for the loss of viral RDDP activity.

In a preliminary experiment, an inhibitory effect of BKN on Rauscher leukemia virus RDDP activity was partially reversed by adding back manganese to the reaction mixture (W. Yang, personal communication). Because of this result and because BKN seems to bind certain divalent cations, including manganese (Daniel, 1976), we challenged the BKN inhibition of RD-114 RDDP activity with higher manganese concentrations. Increasing concentrations of $MnCl_2$ were added to the assay mixture in an attempt to restore normal levels of activity. In increments from 0.2 - 1.5 mM, no restoration of normal polymerase activity was obtained. The higher concentrations of $MnCl_2$ were beyond the concentration (1.0 mM) at which the salt itself becomes inhibitory to the reaction (Walters and Yang, 1974).

The second series of experiments tested the effect of BKN on the uterine RDDP activity expressed on day 4 of pregnancy. At 0.06 μg/μl BKN gives approximately 22% inhibition of RDDP activity. Increasing the concentration of BKN three-fold (0.19 μg/μl) only serves to maintain this level of reduced polymerase activity.

These experiments present evidence for an interaction between a uterine protein, a hormonally regulated RDDP enzyme or a viral RDDP enzyme which has nucleic acid homology with rabbit cellular DNA and the potential for vertical transmission (Benveniste *et al*.. 1975). How BKN inhibits RDDP remains unknown but, by stimulating protein synthesis, BKN could be having the opposite effect of that noted by Aaronson and Dunn (1974) where inhibitors of protein synthesis induced C-type viruses with high frequency.

Bedigian *et al*.(1976) also found RDDP activity in the rabbit uterus. Polymerase activity in uterine homogenates collected on the tenth day of pregnancy was associated with particles that banded at a density characteristic of type-C RNA viruses when separated by sucrose gradient centrifugation. Activity, at a lower level, was also detectable in uterine fluids accumulated for one month after ligation of the uterine horns. Uterine tissues collected on the 25th day of gestation were essentially devoid of RDDP activity, thus indicating the enzyme's association with earlier phases of gestation.

CRYSTALLINE INCLUSION BODIES

VLPs of embryonic cells may be associated with another structure located in the same cells, whose presence seems to be dependent on uterine provision. Calarco and Szollosi (1973) and Chase and Piko (1973) found VLPs in preimplantation stage mouse embryos to be related to crystalline inclusion bodies (Figures 1 and 2). Daniel and Kennedy (1978) in reviewing the literature, note that similar bodies have been reported in a variety of pre-implantation embryos and placental tissue and are especially prominent in both blastocyst and endometrial cells of the rabbit coincidentally with the presence of VLPs. However, embryos confined to the oviduct (Daniel and Kennedy, 1978) or grown *in vitro* to the blastocyst stage (Van Blerkom *et al.*,1973) do not develop crystalline bodies. When crystalloid-free embryos are transplanted to a synchronized progestational uterus, they develop the bodies within 36 hours. Presumably the uterus provides some of the substance of the crystals or the conditions appropriate to their formation. According to Van Blerkom and Runner (1976) crystalline bodies are rarely found in mouse parthenotes.

VLPs IN PLACENTA

Considering the presence of VLPs in both embryos and uterus, their occurrence in placental tissues is not surprising. C-type particles were first demonstrated in baboon placenta (Kalter *et al.*, 1973; Dalton *et al.*,1974; Kalter *et al.*, 1975a) shown in Figure 11, and have also been reported in placentae of a number of other primates, notably chimpanzee, rhesus monkey, woolly monkey, Patas monkey, Cynomolgus monkey, and two species of marmosets but not the Cebus monkey (Kalter *et al.*, 1975b; Kalter *et al.*, 1975c; Feldman, 1974; Schidlovsky and Ahmed, 1973; Mayer *et al.*,1974; Sherr *et al.*, 1974; Seman *et al.*,1975). They also occur in placental tissue from the human (Kalter *et al.*,1973; Dalton *et al.*,1974; Vernon *et al.*,1974), cat (Gardner, 1971), rat (Gross *et al.*,1975), mouse and guinea pig (Kalter *et al.*,1975b), but have not been detected in the rabbit. Dalton *et al.* (1974) made ultrastructural comparisons between the particles associated with baboon and human placenta, and a variety of other type-C particles. These authors concluded that the particles displayed the typical morphology of type-C particles but that the nucleocapsid was closely applied directly to the envelope. Thus, the possibility exists that these two kinds of type-C particles represent two different viral types. Dirksen and Levy (1977) also found particles in term human placentas which may have been C-type but were distinguishable by the absence of a space between the envelope and the nucleocapsid. First trimester placentae had no particles.

A-type particles occur in the mouse (Smith *et al.*,1975) and guinea pig (Hsuing *et al.*,1974) placenta, but apparently not in that of the baboon (Kalter *et al.*,1973) or Rhesus monkey (Schidlovsky and Ahmed, 1973). Smith *et al.*(1977) identified intracytoplasmic type-A immature virus particles and mature particles similar in morphology to type-D particles in the squirrel monkey placenta. (They described "D" particles as having eccentric nucleoids, and sometimes displaying rod-shaped, poorly demonstrable envelope projections. Protrusions may exist on intermediate layers and short surface projections are occasionally seen).

VLPs occur in both the embryonic and maternal placental tissues but reports of their locations differ somewhat depending on the species studied. Kalter *et al.*(1973) demonstrated endogenous C-type particles in the placental villi of the baboon between the 27th to the 170th day of pregnancy. Both budding and mature forms were observed in the syncytiotrophoblast. In the chimpanzee, particles were found in the spaces between the syncytiotrophoblast and the basement membranes (Kalter *et al.*,1975c). They were not seen budding from the microvillous brush border of the syncytiotrophoblast or from cytotrophoblast and were rare in placental stroma. Feldman (1974) found C-type particles budding from syncytial trophoblast, pericytes, Hofbauer cells and mesenchyme in

Figure 11: VLPs in baboon placenta. (a) Electron micrograph showing complete particle and double budding structure in the syncytiotrophoblast cell. Virus particles were consistently seen at sites opposite the microvilli, usually in areas of convoluted plasma membrane (X 10,000). (b) Budding structure at convoluted plasma membrane (X 60,000). (c) Complete particle among podocytic processes of plasma membrane (X 60,000). (d) Single budding structure and doublet particle at plasma membrane (X 60,000). (e) Single particle budding into an intracytoplasmic cisterna (X 20,000). (f) Several C-type particles and various pleomorphic structures within an intracytoplasmic vesicle (X 60,000). All preparations were fixed in glutaraldehyde, and then in osmium tetroxide; sections were stained with uranyl acetate and lead citrate. Reprinted with permission, from Kalter *et al.*(1973) and legend quoted from that paper.

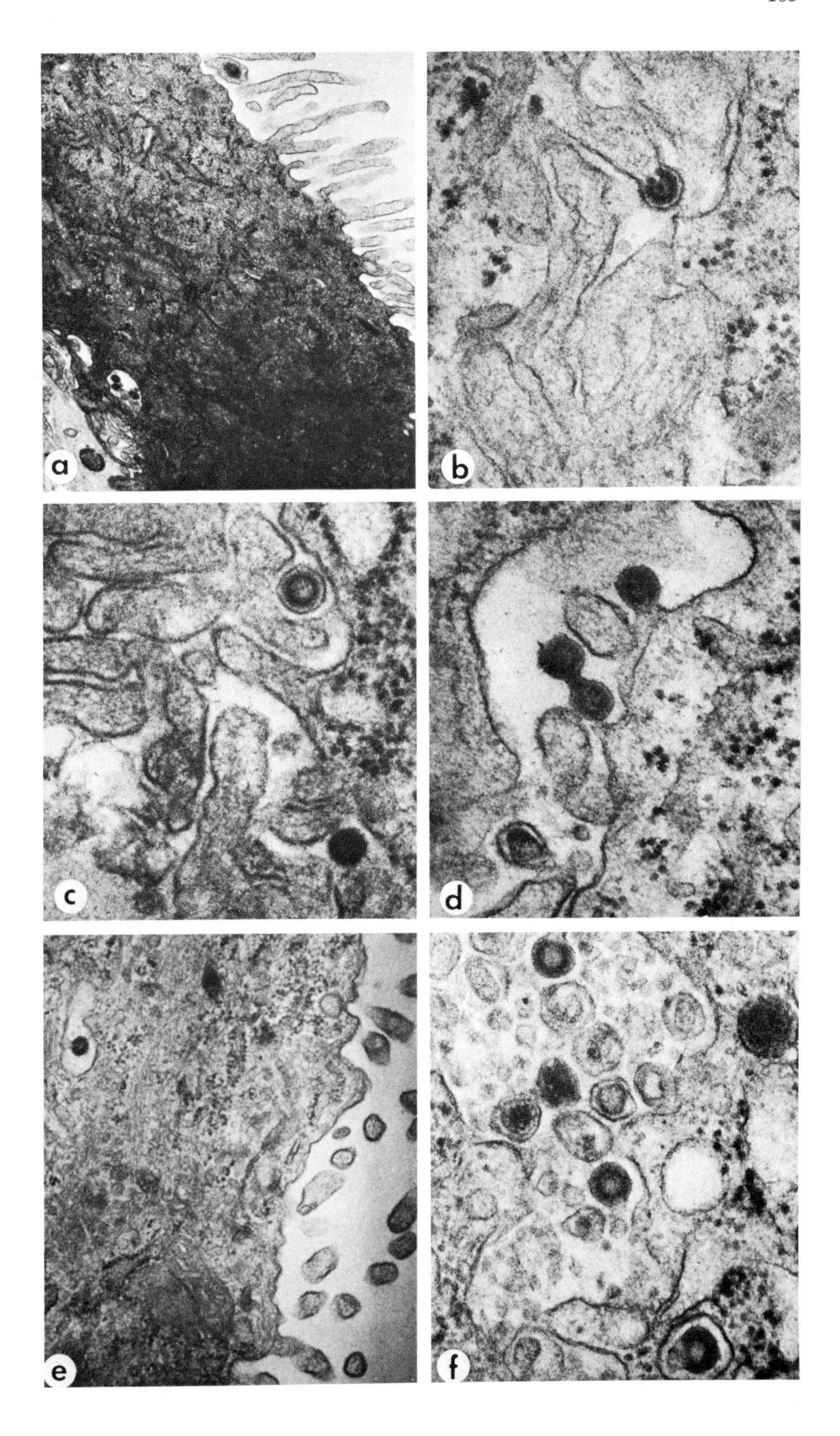
a
b
c
d
e
f

chorionic villi from Rhesus monkey placenta. She states that "Other particles having a dense central core surrounded by a narrow electron lucid zone and enclosed by an outer unit membrane collectively of about 30 nm in diameter were associated with the maternal components of the placenta but absent from the chorionic villi". These additional particles were found budding from cells of the cytotrophoblastic cell column and decidual basalis. In the cottontop marmoset, both budding and mature particles were associated with the basal trophoblast; they were in the plasma membrane of syncytiotrophoblast with fewer in extracellular spaces (Seman *et al.*,1975). They were also in the syncytiotrophoblasts of squirrel monkey (Smith *et al.*,1977) and human (Dirksen and Levy, 1977). In the human they were never found within the cytoplasm or near the microvillous borders or in the endothelial cells of the fetal blood capillaries within the villi. Gross *et al.*(1975) reported C-type particles in the junctional region where fetal and maternal cell layers meet in the placenta of Sprague-Dawley rats. In mice, Smith *et al.*,(1975) identified intracisternal A-type particles in the cytotrophoblastic layer of the placental labyrinth but they were absent in the trophoblastic layers adjacent to the fetal endothelium.

RELATIONSHIPS TO OTHER RNA VIRUSES

Placental C-type particles of primates are morphologically similar (Kalter *et al.* 1975c) and also have other common characteristics, which may or may not be shared with certain RNA viruses. The xenotropic endogenous type-C virus from the baboon placenta fails to replicate in the baboon outside the reproductive tract, but it will replicate in the canine brain or thymus (Sherr *et al.*,1974) and some other cultured cells. Baboons of either sex possess naturally occurring virus-binding antibodies against envelope but not core proteins (Weislow *et al.*,1976). Benveniste *et al.*(1974b) isolated this virus, identified as M-7, after co-cultivation of first trimester baboon placenta with various mammalian cell lines (fetal canine, rhesus monkey, bat lung and human rhabdomyosarcoma) and demonstrated by nucleic acid hybridization studies that M-7 is genetically different from other mammalian C-type viruses. It resembles other C-type viruses in that its reverse transcriptase activity elutes at a position corresponding to a molecular weight of 70,000 and has a preference for Mn^{2+} over Mg^{2+}. In addition to the placenta, liver DNA also had the genetic information for this virus but no overt particles were identified. Type A particles were seen after co-cultivation of squirrel monkey placenta with mink lung cells (Smith *et al.*,1977). Isolates from these cultures had an RDDP with a preference for Mg^{2+} over Mn^{2+}. Hellman *et al.*(1974) showed M-7 to have a group specific antigen (gs-1) immunologically identical with RD-114 and that anti-RD-114 DNA polymerase IgG inhibits M-7 polymerase by 57%. Strickland *et al.*(1973) report an RDDP enzyme in normal baboon placenta, where C-type particles had been previously located, and interspecies specific antigens of the C-type RNA tumor virus had been demonstrated. Sedimenting like the RDDP isolated from estrogen-activated mouse uterine tissue (Strickland *et al.*,1974), this polymerase activity is not inhibited by specific antisera to other purified primate viruses. Reverse transcriptase has been isolated from normal rhesus monkey placenta (Mayer *et al.*, 1974) known to harbor C-type particles (Schidlovsky and Ahmed, 1973). This enzyme has the same extraction requirements, template-primer preference and divalent cation dependency as the type-C viral reverse transcriptase, but can be distinguished immunologically from the RDDP activity of known oncornaviruses. Thus, it can be suggested that the RDDP enzyme associated with reproduction is endogenous and has a normal physiological function.

Benveniste *et al.*(1974a) detected nucleic acid sequences homologous to a single stranded DNA transcript prepared from a baboon C-type virus in the DNA of several old world monkeys including the baboon, patas, African green monkey and two species of macaques, namely Rhesus and stumptail, but similar sequences were not detected in human or new world monkey tissues. These results demonstrate that certain primates other than the baboon also contain endogenous type-C viral genes and these authors feel that these homologies suggest the possibility of horizontal transmission between the progenitors of these animals at some point in evolution. No homologies could be demonstrated between the DNA transcripts of tumor tissue of a woolly monkey and a gibbon ape with any other primate tissue examined.

Bedigian *et al.*(1976) found the same RDDP activity described for rabbit uterine tissue to be present in placental extracts from 10 day pregnant animals but absent in samples taken on day 25. They characterized this placental polymerase by column chromatographic characteristics, template primer preference (for oligo (dT)·poly(A) and oligo (dG)·poly(C) and also utilization of viral 70S RNA), molecular weight (of about 70,000 daltons), and an absolute requirement for Mn^{2+} as the divalent cation and concluded that it was distinct from other known cellular DNA polymerases and similar to the RDDP of mammalian type-C RNA viruses.

MuLV antigens p30 and gs70 are detectable in mouse placentae (Hilgers *et al.*,1974; Lerner *et al.*,1976) and appear more prominent in mid to late gestation.

VLPs IN MILK AND MAMMARY GLANDS

Virus-like particles also appear in milk and mammary tissues. This subject is outside the scope of this chapter but is called to the reader's attention because of the reproduction-support function of the secretion of these tissues and the obvious potential for transmission of VLPs to neonates. Most of the research has been done with mice and humans. In both species: 1) the predominant particle is the B-type but C-type is also common and cytoplasmic A-type is sometimes seen; 2) the particles have been implicated in breast cancer but the correlation is far from absolute. An average of about 14% of human milk samples contain particles; 3) RNA viral-like characteristics of RDDP activity (inhibited by RNAse), gs antigens, 60S - 70S RNAs with about 1.5% poly A regions and particle densities of 1.16 - 1.19 g/ml are reported.

VLPs IN OTHER TISSUES

As noted, the finding of VLPs in reproductive tissues of some species (mice particularly) does not, in itself, generate special attention from the developmental biologist because viruses of many kinds are common to the other tissues of these species. But in the case of rabbits and some primates, systemic endogenous RNA viruses are either unreported or extremely rare. Neubauer *et al.* (1974) after repeated examination of normal and neoplastic tissue from four non-human primates (Rhesus monkey, crab-eating macaque, African green monkey and owl monkey) were unable to find evidence for the presence of type-C viruses. In a more detailed investigation of owl monkey tissues, they could find no positive evidence from reverse transcriptase assay, group specific antigen radioimmunoassay, or molecular hybridization studies and concluded "....if type-C viral expression is common, it is not easily detectable..."

Most attempts to find RNA type viruses in human tissues have been inconclusive and conflicting. Stephenson and Aaronson (1976) could not find virus in over 200 human tissue cultures examined. Even covert expressions of the viruses are not easily demonstrated: Gallo (1972) for example, after extensive review of the previous literature on reverse transcriptase, concluded that with one possible preliminary exception "...there are no published data to indicate the existence in normal cells of RNA dependent DNA polymerase capable of using natural templates". Stephenson and Aaronson (1976) were unsuccessful in their attempts to detect woolly monkey viral antigens in human tissues or antibodies to them in human sera, but Aoki *et al.*(1976) claimed that 76% of the human serum samples they examined contained naturally occurring antibodies against gs antigen on the envelope of type-C viruses isolated from four primates, including the woolly monkey, simian sarcoma virus and the baboon endogenous type-C virus. They suggest that type-C related RNA viruses may be widespread in humans. Strand and August (1974), in summarizing many attempts, noted "Evidence for a type-C virus in human cells has been sought in many laboratories by a variety of techniques. An infectious particle generally recognized as being of human origin, has not been isolated".

In 1975, Panem *et al.* published the first documented report of a type-C virion isolated from a normal diploid human cell strain. The cells in this case had been cultured for 6 months after derivation from the lungs of a spontaneously aborted 8 week old embryo. The cells did, however, have the interspecies p30 antigen of simian sarcoma virus and similar reverse transcriptase (also see Panem *et al.*, 1977). Other cultured human cells have also yielded viruses (e.g. Gallagher and Gallo, 1975; Prochownik and Kirsten, 1976).

After analysis of human tissues by competitive radioimmunoassay for two principal structural components of type-C RNA viruses, the major core proteins (p27 and p30) and the major envelope glycopeptides (gp 67/71), Strand and August (1974) concluded that many, if not all, humans harbor at least part of the genome of one or more type-C viruses. Donehower *et al.*(1977) offer evidence that one human RNA tumor virus may have been transmitted in nature from the baboon and Wong-Staal *et al.*(1976) suggest that the baboon endogenous type-C RNA virus may also be horizontally transmitted among humans because its RNA hybridizes to DNA from tissues of leukaemic patients. The findings of naturally occurring antibodies to structural proteins of baboon endogenous virus in 100% of the normal healthy humans examined by Kurth *et al.*(1977) would seem to support this possibility. Conversely, Stephenson and Aaronson (1977) found by immunoassay, evidence for baboon virus structural protein p30 in many species of old world monkeys but not in a large number of human tissues suggesting that productive infection with this virus "is likely to be infrequent if it occurs at all in humans".

In a series of papers, Benveniste and Todaro (1974a, 1974b, 1976) have explored the evolutionary implications of endogenous C-type virus relationships, particularly using evidence from DNA hybridization studies. They feel that the existence of a variety of mammalian cells containing multiple divergent copies of endogenous C-type virogenes suggests that C-type genes have been selectively preserved during evolution and that their expression has some normal physiological function. In one study (1976) they hypothesize on an Asian origin for man because his viral gene sequences are more closely related to those of Asian (gibbon and orang-utan) rather than African primates (gorilla, chimpanzee and baboon). Endogenous viruses from primates may also have infected and even become part of the germ line of evolutionarily distinct groups - for example, the cat family, since the cellular DNA of anthropoid primates contains genes related to the nucleic acids of the domestic cat virus RD-114. The work of Stephenson *et al.*(1976) supports this belief by showing that the immunological properties of two structural polypeptides of type-C RNA viruses endogenous to old world monkeys (including the baboon placenta) and to the domestic cat showed a very close relationship, even though the viruses may have undergone different rates of change in the course of evolution within their respective host cell genome. Smith *et al.*(1976), however, were unable to find evidence of infection in nude mice after experimental innoculation with baboon type-C virus.

Except in reproductive tissues, overt C-type RNA viruses have not been detected in the domestic rabbit (Bedigian *et al.*,1976) but may exist in covert form as demonstrated by Meier and Fox (1973) using antisera to group specific antigen p30.

VERTICAL TRANSMISSION

Extensive speculation has been offered concerning the significance of virus-like particles associated with mammalian reproduction, as reviewed here. VLPs in maternal reproductive tissues, gametes, early embryos and fetal tissues but not in other mature tissues to any extent, supports the hypothesis of a vertical mode of transmission (i.e. passing from parent to offspring), as discussed by numerous authors (e.g. Huebner *et al.*,1970; Calarco and Szollosi, 1973; Kalter *et al.*,1975d; Schidlovsky and Ahmed, 1973; Jaenisch and Mintz, 1974; Chase and Piko, 1973; Benveniste *et al.*, 1974b; Patillo, 1973; Gross *et al.*,1975). Essentially this would require passage of the viral genome to the conceptus along with, and as an inherent part of, the total genetic contribution of the parents, via their gametes. Alternatively, VLPs might also be transmitted across the placenta from mother to fetal tissues or to the primordial germ cells of the fetus. Schidlovsky and Ahmed (1973) note: "The presence of viruses on both sides of the trophoblastic basement membrane in the mesodermal core of the placental villi and the syncytial trophoblast suggests that C-type virus particles are transported across the placental barrier into the fetus. An alternative explanation to the presence of viruses in the fetus may be activation of the viral genome in utero". Jaenisch and Mintz (1974) have demonstrated the possibility of vertical transmission of a viral genome and its integration into embryonic host genome by injecting SV40 viral DNA into mouse blastocysts and later showing, by molecular hybridization, the presence of SV40-specific DNA sequences in DNA extracted from tissues of the healthy adults that developed from these blastocysts. Jaenisch (1976) further showed that infection of preimplantation embryos can lead to integration of a virus into the germ line. He infected preimplantation stage mouse embryos (4 to 8 cell stage) with Moloney leukemia viruses and demonstrated that viremic males of the first backcross generation transmitted the virus to 50% of their offspring when mated with uninfected females. Likewise a 50% transmission was observed from viremic N-2 and N-3 generations of males to the next generation. Molecular hybridization experiments revealed that viremic N-1 and N-2 animals carried one copy of Moloney leukemia virus per diploid mouse genome equivalent in all non-target organs tested. These experiments indicated that the exogenous Moloney virus can be converted to an endogenous virus after infection of preimplantation embryos.

This concept of vertical transmission has become commonly accepted in relation to the RNA viruses associated with tumorigenesis in vertebrates (e.g. Meier and Huebner, 1971; Stephenson and Aaronson, 1973). Overt particles may not be detectable in tumors but the viral genomes can frequently be demonstrated by the presence of group specific antigens (e.g. Ioachim *et al.*,1972; Gilden and Oroszlan, 1972). These covert viruses can provide information serving as an "endogenous source of oncogenic activity" (Meier and Heubner, 1971).

BIOLOGICAL FUNCTION

Even though the exact nature of their function(s) remains obscure, it is generally agreed that VLPs play roles in both carcinogenesis and reproduction. Huebner *et al.*(1970) for example, "suggest that the genes for RNA tumor virus, which later in life act as determinants of cancer, may be important also as gene determinants in the developing embryo", and Chase and Piko (1973) speculated that "Expression of these endogenous viral genomes may be linked to the developmental program of the embryos". Mayer *et al.*(1974) feel that particles from Rhesus monkey and baboon placentae represent endogenous viruses that may function in the transfer of genetic information during embryogenesis. Indeed, Loeb and Tartof (1976) have expressed objection to some of the experiments with endogenous baboon type-C viruses because of their possible potential relationship with human tumor viruses.

Baltimore (1976) in his Nobel prize lecture, reviewed studies of the role of viruses and their polymerases in the etiology of cancer. He concluded that: "...not only will such work help us to understand carcinogenesis, it may also be important to the study of developmental biology because of the intimate relationship between the differentiated state of cells and the type of virus able to transform them". Virus related genes are an aspect of the normal genetic complement of the animal and are virus related because they are the progenitors of retroviruses. He feels that these genes play some important role in the life of the animal and so are not dispensable. This explanation is basically the protovirus hypothesis put forth by Temin (1971).

Two related hypotheses have arisen from attempts to describe the role of viruses in cancer and embryogenesis. The viral oncogene concept (Huebner and Todaro, 1969) says that all vertebrate cells contain genetic information (virogenes) for producing RNA viruses and that some of that information (oncogenes) transforms a normal cell into a tumor cell. The information is commonly transmitted vertically in a covert form, and normally maintained unexpressed by repressors in those cells; inhibition or loss of the repressor leading to tumorigenesis. The protovirus hypothesis (Temin, 1971) noted above differs mainly in that cells of the germ line do not contain the cancer virus but rather the potential for genetic evolution of the information in somatic cells. Specific repressors are therefore unnecessary. Thus a protovirus can direct an RNA transcript which is packaged for transport to a neighbouring cell where it can be reverse transcribed into DNA and integrated into the host cell genome. Temin believes that the protovirus hypothesis could also account for the orderly conversion of embryonic tissues to a new differentiated state such as that seen in cases of embryonic induction. Todaro and Huebner (1972) note that "...information for various expressions of the RNA tumor virus genome in the embryo on the one hand and cancer on the other may have played some essential roles in the development of species and subsequently in the embryonic development of the individual organism". If the virus-like particles that are associated with some kinds of cancer, are common and natural to the normal tissues of the organism, and by some unexplained mechanism are activated to induce or permit development of tumors in these tissues, then it is entirely possible that these same particles also constitute a normal mode for information transmission during the period of embryogenesis and their presence in the adult is a carryover from the developmental period, or evidence of the need for continuation of that process in some mature tissues.

Hellman *et al.*(1976) express disbelief in the possibility that the uterine-embryonic VLPs have no beneficial significance and discuss a variety of possible functions, most notably as messengers for cell communication between uterus and mother during development. [They generously credit this concept, utilizing an unknown "uterine factor", to Daniel and Krishnan (1969)]. They further emphasize that to perform such a function the messenger would have to fulfil certain criteria, such as species specificity; ability to induce genetic and/or metabolic activity; presence in the reproductive tract; ready transmissibility from mother to fetus; and inducibility at the proper time and under specific physiological controls. They recount, as circumstantial evidence of this possibility, the work of Ikawa *et al.*(1974) who showed that cells which did not synthesize haemoglobin began to do so after receiving type-C viruses which were carrying messenger RNA for haemoglobin from cells which did synthesize it. As a variation on this theme, Lieberman and Sachs (1977) suggest that these viruses may influence differentiation by modifying the competence of cells induced to differentiate by a normal regulator. They present data to show that "mice myeloid leukaemic cells that can be induced to differentiate to mature macrophages and granulocytes by the normal inducer protein MGI, (MGI = macrophage and granulocyte inducer = colony simulating factor or activity) produced a higher amount of type-C RNA virus than clones of myeloid leukaemic cells that can be only partially or not at all induced to differentiate by this normal regulator".

Hellman and colleagues (1976) felt that VLPs might be involved in maternal acceptance of the fetus as an allograft. One of us (in the discussion following Daniel, 1976) has also suggested that the cells of the uterine lining and those of the implantation stage embryo have a number of common qualities, including the presence of VLPs, which might serve to minimize the differences between them. Such convergence might eliminate, or reduce, the tendency for an immunological reaction. Moroni and Schumann (1977) also believe that endogenous C-type viruses may be involved in the immune system. They note that, "Mammalian endogenous viruses in general are restricted for growth in their autologous species but replicate in heterologous species and are termed xenotropic. In mice and some other species, a further class of endogenous viruses has evolved which is restricted in heterologous cells but replicates in homologous cells; they are termed ecotropic. Endogenous viral genes are inherited in the germ line and have co-evolved in general with nonviral host genes as indicated by evolutionary data". "Whereas B cells can be induced by mitogens to release virus, we have evidence that T cells are refractory to induction when using T cell mitogens as well as 5-bromo-2-deoxyuridine. These results suggest that the expression of endogenous xenotropic C-type viral genes may be physiologically required for B cells to participate in the immune response. As a test of this hypothesis, we examined the effect of antiserum directed against xenotropic endogenous C-type virus on the humoral immune response of mice. We show here that such sera are immuno-suppressive". Dirksen and Levy (1977) believe that the presence of VLPs in human placentas might represent the result of an immunological reaction because, of six women whose placentas contained VLPs, five had had more than one pregnancy. The one exception was a patient with autoimmune disease.

In a very recent series of studies, Fowler and colleagues (Fowler *et al.*,1977a, 1977b, 1977d; Twardzik *et al.*,1977) have advanced the concept of an immuno-suppressive role for oncornaviruses. They have shown that:

(1) The major envelope glycoprotein (gp71) of Rauscher-MuLV specifically binds to murine lymphoid cells.

(2) A protein (probably of about 13,000 daltons) isolated from R-MuLV suppresses PHA and ConA induced lymphocyte transformation.

(3) Certain isolated thymocyte surface proteins bind R-MuLV envelope proteins and may be products of the major histocompatibility locus.

They suggest that "...a virion envelope component interferes with cellular mediated immunity by altering cell recognition sites".

Van Blerkom and Manes (1977) note that A-type particle formation in mouse oocytes and embryos is correlated with nucleolar activity, perhaps in paraprocessing "...in association with extraordinarily rapid transport of genetic transcripts out of the nucleus". They also support the suggestion of Clegg and Piko (1975) of a possible correlation with a polyadenylated 55s RNA in the preimplantation mouse embryo.

There is not, of course, unanimity of opinion about a developmental role for VLPs. Strand *et al.*(1977), for example, after determining that the embryos of different strains of mice showed different patterns of viral protein expression (see Table 3) concluded that the state of differentiation of the cell may influence this pattern but that their results "do not support the concept that oncornavirus gene expression plays an important role in mammalian embryogenesis". (Also, see Jaenisch and Berns, 1977).

In recognition of the capacity of certain viruses to facilitate cell fusion, we feel moved to speculate that the presence of VLPs at the time of sperm-egg interaction, syncytial trophoblast formation and placentation, may be evidence for their involvement in those cell fusion phenomena; one of these producing a zygote or genetically hybrid cell and the others yielding multinucleate heterokaryons.

HYPOTHETICAL MODEL

From the minimal and sometimes conflicting data available for presenting strong evidence of a definitive role for the virus-like particles associated with reproductive tissues of the mammal, one would be well advised to exercise restricted speculation. Currently, the weight of evidence causes us to favour an immunological role but considering that authors of the chapters in this volume were encouraged to "...be as provocative as possible", we feel that another model could also prove to be stimulating. Also we believe that the scientist, like the artist, to advance his understanding must sometimes claim freer licence in exercising the interpretation of his data than might seem justifiable by the existing state of knowledge. Recognizing the limited state of our current information, we will nonetheless attempt to model a reproductive role for VLPs.

It is generally accepted that RNA viruses represent packaged genetic messages originating in one cell and being carried to another where, after incorporation into the genome of that host cell, by reverse transcription of the RNA core, they could result in the repeated forward transcription of viral components. Some of these are necessary for the perpetuation of the particles while others, by influencing the host cell surface, can cause permanent differentiative changes. In reviewing the work leading to formulation of his Provirus Hypothesis, Temin (1976) emphasizes that: (1) cell transformation results from the action of viral genes and that these genes mutate at a high rate; (2) a normal replicative cell cycle is needed for initiation of virus production; and (3) the recombination between viruses and cells is not random but is primarily with specific cellular genes.

The VLP-related sequence of events in embryogenesis might begin with C-type maturation, at the blastocyst stage, from the intracellularly-bounded A-type particles already present in oocytes from primordial germ cell stages. It may be that conversion of A to C type requires fertilization of the ovum and embryonic development to an active blastocyst stage because C-types have not been found in parthenogenotes or in diapausing embryos during delayed implantation. Alternatively, A-type particles might also represent vehicles for amplification of repeated sequences of the embryonic genome, or become incorporated from the maternal system prior to ovulation as "masked messengers". In these cases, C-types would arise independently thereafter.

An active blastocyst being supported in a normally synchronized, progestational uterus would bud off C-type particles which would enter uterine endometrial cells, informing them of the embryonic presence and stimulating appropriate responses, the decidual reaction being one possibility. These "informed" uterine cells would, in turn, synthesize a VLP message which could influence other uterine cells or return to the blastocyst and assist in initiating differentiation. Because it always develops after the zygote enters the uterus, and because it is the first stage in embryogenesis where a visible differentiative event occurs, namely in the distinction between inner cell mass and trophoblast, the blastocyst is the logical receptor of a uterine "message". Once received by the blastocyst, the uterine VLP message would "trigger" a specific initial differentiation step, perhaps that of defining the primary germ layers. Then through production of still newer intraembryonic VLPs, the next generation of tissue differentiation would be induced. This progression would continue thereafter in a cascading series of differentiative events punctuated by the synthesis of new, presumably slightly different, VLPs carrying the messages of induction from one embryonic tissue to another until embryogenesis was complete.

Strong objections can be made against the need of the female reproductive tract to promote embryonic cell differentiation early in pregnancy (e.g. see Herbert and Graham, 1974; Van Blerkom and Manes, 1977). These are based mainly on the facts that mammalian embryos can develop in extrauterine sites and *in vitro* where, for short periods of time, their structure, growth, metabolism and synthesizing activities do not differ significantly from embryos grown *in vivo*. Although they are not entirely convincing, we concede some merit to these arguments and offer the alternative of endogenous C-type particles in the blastocyst as the induction initiating "trigger". Subsequently, the same "cascading VLP induction" hypothesis would suffice.

In support of this hypothesis is the work of Todaro *et al.* (1976) showing that certain tumor viruses may cause transformation by means of a viral gene product similar to epidermal growth factor, the work of Roby *et al.* (1976) showing that the synthesis of specific products of differentiated cells is suppressed after transformation by temperature sensitive viruses, and the work of

Rosenberg and Baltimore (1976) showing specificity in the types of mouse cells that can be infected by a particular leukemia virus. These and other related studies have been reviewed by Kolata (1977).

The recent exciting studies of Elder *et al.* (1977) also especially support this possibility. We note them by quoting the abstract from the paper: "Structural comparison of the major envelope glycoprotein (gp 70) from 35 different mouse type C viruses and free gp 70 expressed at various anatomical sites in the mouse showed that the gp 70s are polymorphic products of a large multi gene family encoding viral and differentiation antigens. Different proviruses are expressed in cells following distinct pathways of differentiation. When the various gp 70s are grouped according to primary structure they fall naturally into viral host range classes, confirming the suspicion that C-type viral tropism is largely determined by the nature of the gp 70 product expressed".

This concept of particulate-type messengers involved in cell-to-cell communication during embryogenesis is not new, except possibly in relation to the preimplantation stages of development. Slavkin *et al.* (1972) presented some morphological evidence for the sequential formation of "matrix vesicles" during incisor tooth development in embryonic rabbits, particularly in relation to the role of preodontoblast mesenchyme in initiating calcification during dentinogenesis. The authors did not define them as being virus-like but in some of their electronphotomicrographs the vesicles resembled C-type viruses. Furthermore, they were described as being 50-100 nm in diameter with variable shape and electron density of their contents and limited by a unit trilaminar membrane often coated with a filamentous mat. The vesicles contained ribosome-like granules that had an affinity for indium which was ribonuclease labile and believed to be an RNA-protein complex. Slavkin *et al.* (1972) proposed that the "matrix vesicles may be involved in the transmission of embryonic induction".

Some of the special products of specific stage embryonic tissue genome expression may become identifiable as precursor substances, isozymes, fetal antigens, etc. and a garbled message or one generated at the wrong time would result in a malformed or malfunctioning product. We wonder if the recent report of Pottathil and Meier (1977) showing that the injection into 12 and 14 day pregnant mice of specific RNA(s) isolated from 10-12 day old embryos, results in embryotoxicity, may be related to this concept?

Carcinogenic agents may act to initiate a reactivation of production of VLPs, or their effective products, in mature, fully differentiated tissues, thus causing the aberrant growths we classify as cancers. Friend *et al.*(1974) reviewed the evidence "...to support the hypothesis that some, if not all, malignant diseases may be disorders of cell differentiation", but that hypothesis remains for the present in the same realm of speculation as the one offered here on the role of VLPs in regulating embryogenesis.

When the manuscript for this chapter was in final draft, our attention was called to the newly published paper by Jaenisch and Berns (1977) reviewing tumor virus expression during mammalian embryogenesis. It contains a thought-provoking appraisal of the interaction of both exogenous and endogenous tumor viruses with developing mouse embryos. After posing a number of challenging questions which emphasize the difficulty of demonstrating a functional relationship between embryogenesis and C-type virus expression, Jaenisch and Berns decided that a "safe conclusion" is not possible. They do, however, offer an explanation of this relationship which could easily be integrated with the one we offer here and is based on the "...assumption that in each instance a virus gene is integrated at a different chromosomal region, which is active at specific stages of development or in certain tissues but not at other stages of development or in other tissues".

We feel that this "cascading VLP induction" model satisfies much of the current information about VLPs in the mammalian reproductive tract and many of the conditions associated with embryonic tissue induction (see review by Wolff, 1968). It is expected to encourage new investigations into the physiological role of these virus-like particles and their expression as common characteristics in both tumorigenesis and embryogenesis.

ACKNOWLEDGEMENT

Much of the authors' research which is reported here was supported by NIH grant R01 HD06226. We appreciate the helpful suggestions of Drs. Richard Tyndall, Wen Yang, Kerry Foresman, and technical assistance of Patsy Boyce and Phyllis Bice.

REFERENCES

AARONSON, S.A. & DUNN, C.Y. (1974) High frequency C-type virus induction by inhibitors of protein synthesis. Science 183, 422-424.

ALLEN, P.T., MULLINS, J.A., SAVIOLAKIS, G.A., STRICKLAND, J.E., FOWLER, A.K. & HELLMAN, A. (1977) Direct isolation of xenotropic retroviruses from the NIH Swiss mouse uterus. Virology 79, 239-243.

ANDERSON, H.K. & JEPPESEN, T. (1972) Virus-like particles in guinea pig oogonia and oocytes. J. Natl. Cancer Inst. 49, 1403-1410.

AOKI, T., WALLING, M.J., BUSHER, G.S., LIU, M. & HSU, K.C. (1976) Natural antibodies in sera from healthy humans to antigens on surfaces of type C RNA viruses and cells from primates. Proc. Nat. Acad. Sci. USA 73, 2491-2495.

BALTIMORE, D. (1970) RNA-dependent DNA polymerase in virions of RNA tumor viruses. Nature (London) 226, 1209-1211.

BALTIMORE, D. (1976) Viruses, polymerases and cancer. Science 192, 632-636.

BARGMANN, W. & KNOOP, A. (1959) Elektronenmikroskopische Untersuchungen an Plazentarzotten des Menschen.Zeitschrift fur Zellforsch. 50, 472-493.

BEDIGIAN, H.G., FOX, R.R. & MEIER, H. (1976) Evidence for a particle-associated RNA-directed DNA polymerase in rabbit placental and uterine tissues. Cancer Research 36, 4687-4692.

BENVENISTE, R.E. & TODARO, G.J. (1974a) Multiple divergent copies of endogenous C-type virogenes in mammalian cells. Nature 252, 170-172.

BENVENISTE, R.E. & TODARO, G.J. (1974b) Evolution of C-type viral genes: inheritance of exogenously acquired viral genes. Nature 252, 456-458.

BENVENISTE, R.E. & TODARO, G.J. (1976) Evolution of type C viral genes: evidence for an Asian origin of man. Nature 261, 101-108.

BENVENISTE, R.E., HEINEMANN, R., WILSON, G.L., CALLAHAN, R. & TODARO, G.J. (1974a) Detection of baboon type C viral sequences in various primate tissues by molecular hybridization. J. Virol. 14, 56-57.

BENVENISTE, R.E., LIEBER, M.M., LIVINGSTON, D.M., SHERR, C.J., TODARO, G.J. & KALTER, S.S. (1974b) Infectious C-type virus isolated from a baboon placenta. Nature 248, 17-20.

BENVENISTE, R.E., SHERR, C.J. & TODARO, G.J. (1975) Evolution of type C viral genes: origin of feline leukemia virus. Science 190, 886-888.

BICZYSKO, W., PIENKOWSKI, M., SOLTER, D. & KOPROWSKI, H. (1973) Virus particles in early mouse embryos. J. Natl. Cancer Inst. 51, 1041-1050.

BICZYSKO, W., SOLTER, D., GRAHAM, C. & KOPROWSKI, H. (1974) Synthesis of endogenous type-A virus particles in parthenogenetically stimulated mouse eggs. J. of Nat'l Cancer Inst. 52, 483-489.

BLACK, V.H. (1974) Virus particles in primordial germ cells of fetal guinea pigs. J. Nat'l Cancer Inst. 52, 545-551.

CALARCO, P.G. (1975) Intracisternal A particle formation and inhibition in preimplantation mouse embryos. Biol. of Reproduction 12, 448-454.

CALARCO, P.G. & McLAREN, A. (1976) Ultrastructural observations of preimplantation stages of the sheep. J. Embryol. Exper. Morphol. 36, 609-622.

CALCARO, P.G. & SZOLLOSI, D. (1973) Intracisternal A particles in ova and preimplantation stages of the mouse. Nature (New Biol.) 243, 91-93.

CHANDRA, S., LISZCZAK, T., KOROL, W. & JENSEN, E.M. (1970) Type-C particles in human tissue I. Electron microscopic study of embryonic tissues *in vivo* and *in vitro*. Int. J. Cancer 6, 40-45.

CHANG, K.S., LAW, L.W. & AOKI, T. (1974) Type-C RNA virus isolated from SJL-J mice. J. Nat'l Cancer Inst. 52, 777-784.

CHAPMAN, A.L., WEITLAUF, H.M., & BOPP, W. (1974) Effect of feline leukemia virus on transferred hamster fetuses. J. Nat'l Cancer Inst. 52, 583-586.

CHASE, D.G. & PIKO, L. (1973) Expression of A- and C-type particles in early mouse embryos. J. Nat'l Cancer Inst. 51, 1971-1975.

CHILTON, B.S. (1976) Rabbit endometrial RNA- and DNA-dependent DNA polymerase activity during the first trimester of pregnancy. PhD. Thesis, Univ. of Tennessee.

CHILTON, B.S. & DANIEL, J.C. (1978) Rabbit endometrial RNA- and DNA-dependent DNA polymerase activity. Biol. Reprod. (In press).

CLEGG, K.B. & PIKO, L. (1975) Patterns of RNA synthesis in early mouse embryos. J. Cell Biol. 67 (Suppl.), 72a.

CROKER, B., McCONAHEY, P. & DIXON, F. (1976) Quantitation of oncorna virus expression in normal, lymphomatous and immunopathologic mice. Fed. Proc. 35, (Abstract # 1025).

DALTON, A.J. (1972) RNA tumor viruses. Terminology and ultrastructural aspects of virion morphology and replication. J. Nat'l. Cancer Inst. 49, 323-327.

DALTON, A.J., HELLMAN, A., KALTER, S.S. & HELMKE, R.J. (1974) Ultrastructural comparison of placental virus with several type-C oncogenic viruses. J. Nat'l. Cancer Inst. 52, 1379-1381.

DANIEL, J.C. (1976) Blastokinin and analogous proteins. J. Reprod. Fertl. 25, 71-83.

DANIEL, J.C. & BOOHER, C.B. (1977) Synthesis of the rabbit uterine protein, blastokinin: Changes in rate prior to implantation. J. Tenn. Acad. Sci. 52, 35-37.

DANIEL, J.C. & KENNEDY, J.R. (1978) Crystalline inclusion bodies in rabbit embryos. J. Embryol. Exp. Morph. (In press).

DANIEL, J.C. & KRISHNAN, R.S. (1969) Studies on the relationship between uterine fluid components and the diapausing state of blastocysts from mammals having delayed implantation. J. Exp. Zool. 172, 267-282.

DeHARVEN, E. (1974) Remarks on the ultrastructure of type A, B, and C virus particles. Advances in Virus Research 19, 221-264.

DEL VILLANO, B.C. & LERNER, R.A. (1976) Relationship between the oncornavirus gene product gp70 and a major protein secretion of the mouse genital tract. Nature 259, 497-499.

DIRKSEN, E.R. & LEVY, J.A. (1977) Virus-like particles in placentas from normal individuals and patients with systemic lupus erythrematosus. J. Nat'l. Cancer Inst. 59, 1187-1192.

DONEHOWER, L., WONG-STAAL, F. & GILLESPIE, D. (1977) Divergence of baboon endogenous type C virogenes in primates: Genomic viral RNA in molecular hybridization experiments. J. Virology 21, 932-941.

EBERT, P.S. & PEARSON, G.R. (1974) Correlation of C-type virus (WF-1) production and heme synthesis in a rat fibroblastic cell line. Proc. Soc. Exp. Biol. Med. 145, 298-301.

ELDER, J.H., JENSEN, F.C., BRYANT, M.L. & LERNER, R.A. (1977) Polymorphism of the major envelope glycoprotein (gp70) of murine C-type viruses: virion associated and differentiation antigens encoded by a multi-gene family. Nature 267, 23-28.

ENDERS, A.C. & SCHLAFKE, S.J. (1965) The fine structure of the blastocyst: some comparative studies. In: G.E.W. Wolstenholme and M. O'Connor (Eds.), Preimplantation Stages of Pregnancy, Little, Brown: Boston.

FELDMAN, D. (1974) An electron microscopic study of virus particles in rhesus monkey placenta. Proc. Nat. Acad. Sci. USA 72, 118-121.

FOWLER, A.K., McCOHANEY, P.J. & HELLMAN, A. (1973) Strain dependency of hormonally activated C-type RNA tumor virus markers in mice. J. Nat'l Cancer Inst. 50, 1057-1059.

FOWLER, A.K., REED, C.D., TODARO, R.J. & HELLMAN, A. (1972) Activation of C-type RNA virus markers in mouse uterine tissue. Proc. Nat'l. Acad. Sci. USA 69, 2254-2257.

FOWLER, A.K., STRICKLAND, J.E., KOUTTAB, N.M. & HELLMAN, H. (1977a) RNA tumor virus expression in mouse uterine tissue during pregnancy. Biol. of Reprod. 16, 344-348.

FOWLER, A.K., TWARDZIK, D.R., REED, C.D., WEISLOW, O.S. & HELLMAN, A. (1977b) Binding characteristics of Rauscher leukemia virus envelope glycoprotein, gp71, to murine lymphoid cells. J. Virology 24, 729-735.

FOWLER, A.K., TWARDZIK, D.R., REED, C.D., WEISLOW, O.S. & HELLMAN, A. (1977c) Inhibition of lymphocyte transformation by disrupted murine oncornavirus. Cancer Research 37, 4529-4531.

FOWLER, A.K., TWARDZIK, D.R., WEISLOW, O.S., REED, C.D. & HELLMAN, A. (1977d) Modification of *in vitro* cell mediated immune response by protein components of a leukemogenic virus. Presented at: VIII Internal Symposium on Comparative Leukemia Research. Amsterdam August 1977.

FRIEND, C., PREISLER, H.D. & SCHER, W. (1974) Studies on the control of differentiation of murine virus-induced erythroleukemic cells. In: A.A. Moscona and A. Monroy (Eds.), Current topics in Developmental Biology 8, 81-101. Academic Press, New York and London.

GALLAGHER, R.E. & GALLO, R.C. (1975) Type C RNA tumor virus isolated from cultured human acute myelogenous leukemia cells. Science 187, 350-353.

GALLO, R.C. (1972) RNA-dependent DNA polymerase in viruses and cells: views on the current state. Blood 39, 117-137.

GARDNER, M.B. (1971) Current information on feline and canine cancers and relationship or lack of relationship to human cancer. J. Nat'l. Cancer Inst. 46, 281-290.

GARDNER, M.B., OFFICER, J.E., RONGEY, R.W., CHARMAN, H.P., HARTLEY, J.W., ESTES, J.D. & HUEBNER, R.J. (1973) C-type RNA tumor virus in wild house mice (*Mus musculus*). Biblio. Haemato. 39, 335-344.

GILDEN, R.V. & OROSZLAN, S. (1972) Group-specific antigens of RNA tumor viruses as markers for subinfectious expression of the RNA virus genome. Proc. Nat. Acad. Sci. USA 69, 1021-1025.

GILLESPIE, G. & GALLO, R.G. (1975) RNA processing and RNA tumor virus origin and evolution. Science 188, 802-811.

GOODING, L.R., HSU, Y.C. & EDIDIN, M. (1975) Expression of teratoma-associated antigens on murine ova and early embryos. Develop. Biol. 49, 479-486.

GREEN, J.A. & BARRON, S. (1975) 5-Iododeoxyuridine potentiation of the replication *in vitro* of several unrelated RNA and DNA viruses. Science 190, 1099-1101.

GREEN, M. & GERARD, G.F. (1974) RNA-directed DNA polymerase - Properties and functions in oncogenic RNA viruses and cells. Progr. Nucl. Acid Res. Mol. Biol. 14, 187-334.

GROSS, L. (1970) Oncogenic Viruses. 2nd Ed. Pergamon Press, Oxford.

GROSS, L., SCHIDLOVSKY, G., FELDMAN, D., DREYFUSS, Y. & MOORE, L.A. (1975) C-type virus particles in placenta of normal healthy Sprague-Dawley rats. Proc. Nat. Acad. Sci. USA 72, 3240-3244.

HELLMAN, A. & FOWLER, A.K. (1971) Hormone activated expression of the C-type RNA tumor virus genome. Nature (New Biol.) London 233, 142-144.

HELLMAN, A., FOWLER, A.K., STRICKLAND, J.E. & KOUTTAB, N.M. (1976) Type C virus modulation by estrogens: Its possible biological function. In: E. Deutsch, K. Moses, H. Rainer and A. Stacher (Eds.), Molecular Base of Malignancy. Georg Thieme Publishers, Stuttgart, Germany.

HELLMAN, A., PEBBLES, P.T., STRICKLAND, J.E., FOWLER, A.K., KALTER, S.S., OROSZLAN, K.S. & GILDEN, R.V. (1974) Baboon virus isolate M-7 with properties similar to feline virus RD-114. J. Virol. 14, 133-138.

HERBERT, M.C. & GRAHAM, C.F. (1974) Cell determination and biochemical differentiation of the early mammalin embryo. Current Topics in Develop. Biol. 8, 151-178. A.A. Moscona and A. Monry (Eds.), Academic Press, New York and London.

HILGERS, J., DECLEVE, A., GALESLOOT, J. & KAPLAN, H.S. (1974) Murine leukemia virus group-specific antigen expression in AKR mice. Cancer Res. 34, 2553-2561.

HOLT, J.A., HEISE, W.F., WILSON, S.M. & KEYES, P.L. (1976) Lack of gonadotropic activity in the rabbit blastocyst prior to implantation. Endocrinology 98, 904-909.

HSUING, G.D., FONG, C.K.Y. & EVANS, C.H. (1974) Prevalence of endogenous oncornavirus in guinea pigs. Intervirology 3, 319-331.

HUEBNER, R.J. (1970) Identification of leukemogenic viruses; specifications for vertically transmitted, mostly "switched off" RNA tumor viruses as determinants of the generality of cancer. Biblio. Haemato. 36, 22-44.

HUEBNER, R.J. & TODARO, G.J. (1969) Oncogenes of RNA tumor viruses as determinants of cancer. Microbiology 64, 1087-1094.

HUEBNER, R.J., KELLOFF, G.J., SARMA, P.S., LANE, W.T., TURNER, H.C., GILDEN, R.V., OROSZLAN, S., MEIER, H., MYERS, D.D. & PETERS, R.L. (1970) Group-specific antigen expression during embryogenesis of the genome of the C-type RNA tumor virus: Implications for ontogenesis and oncogenesis. Proc. Natl. Acad. Sci. USA 67, 366-376.

IKAWA, Y., ROSS, J. & LEDER, P. (1974) An association between globin messenger RNA and 60S RNA derived from Friend leukemia virus. Proc. Natl. Acad. Sci. USA 71, 1154-1158.

IOACHIM, H.E., DORSETT, B., SABBATH, M. & KELLER, S. (1972) Loss and recovery of phenotypic expression of gross leukemia virus. Nature New Biology 237, 215-218.

JAENISCH, R. (1976) Germ line integration and Mendelian transmission of the exogenous Moloney leukemia virus. Proc. Nat. Acad. Sci. USA 73, 1260-1264.

JAENISCH, R. & BERNS, A. (1977) Tumor virus expression during mammalian embryogenesis. In: M. I. Sherman (Ed.) Concepts in Early Mammalian Development, M.I.T. Press.

JAENISCH, R. & MINTZ, B. (1974) Simian virus 40 DNA sequences in DNA of healthy adult mice derived from preimplantation blastocysts injected with viral DNA. Proc. Natl. Acad. Sci. USA 71, 1250-1254.

JOHNSON, M.H., COWAN, B.D. & DANIEL, J.C. (1972) An immunological assay for blastokinin. Fertil. and Steril. 23, 93-100.

KAJIMA, M. & POLLAND, M. (1968) Wide distribution of leukemia virus in strains of laboratory mice. Nature 218, 188-189.

KALTER, S.S., HEBERLING, R.L., HELLMAN, A., TODARO, G.J. & PANIGEL, M. (1975a) C-type particles in baboon placenta. Proc. R. Soc. Med. 68, 135-140.

KALTER, S.S., HEBERLING, R.L., HELMKE, R.J., PANIGEL, M., SMITH, G.C., KRAEMER, D.C., HELLMAN, A., FOWLER, A.K., & STRICKLAND, J.E. (1975b) A comparative study on the presence of C-type viral particles in placentas from primates and other animals. Biblio. Haemato. 40, 391-401.

KALTER, S.S., HEBERLING, R.L., SMITH, G.C., & HELMKE, R.J. (1975c) C-type viruses in chimpanzee (*Pan sp.*) placentas. J. Natl. Cancer Inst. 55, 735-736.

KALTER, S.S., HEBERLING, R.L., SMITH, G.C., PANIGEL, M., KRAEMER, D.C., HELMKE, R.J. & HELLMAN, A. (1975d) Vertical transmission of C-type viruses: their presence in baboon follicular oocytes and tubal ova. J. Natl. Cancer Inst. 54, 1173-1176.

KALTER, S.S., HELMKE, R.J., PANIGEL, M., HEBERLING, R.L., FELSBURG, P.J. & AXELROD, L.R. (1973) Observations of apparent C-type particles in baboon (*Papio cynocephalus*) placentas. Science 179, 1332-1333.

KALTER, S., PANIGEL, M., KRAEMER, D.C., HEBERLING, R.L., HELMKE, R.J., SMITH, G.C. & HELLMAN, A. (1974) C-type particles in baboon (*Papio cynocephalus*) preimplantation embryos. J. Natl. Cancer Inst. 52, 1927-1928.

KOLATA, G.B. (1977) Animal viruses: Probes of cell function. Science 196, 417-418.

KRISHNAN, R.S. & DANIEL, J.C. (1967) "Blastokinin": Inducer and regulator of blastocyst development in the rabbit uterus. Science 158, 490-492.

KURTH, R., TEICH, N.M., WEISS, R. & OLIVER, R.T.D. (1977) Natural human antibodies reactive with primate type-C viral antigens. Proc. Nat'l. Acad. Sci. USA 74, 1237-1241.

LERNER, R.A., WILSON, C.B., DELVILLANO, B.C., McCONAHEY, P.J. & DIXON, F.J. (1976) Endogenous oncornaviral gene expression in adult and fetal mice: quantitative, histologic, and physiologic studies of the major viral glycoprotein, gp70. J. Exp. Med. 143, 151-166.

LEVY, J.A. (1973) Xenotropic viruses: Murine leukemia viruses associated with NIH Swiss, NZB and other mouse strains. Science 182, 1151-1153.

LIEBER, M.M., LIVINGSTON, D.M. & TODARO, G.J. (1973) Super-induction of endogenous Type C virus by 5-Bromodeoxyuridine from transformed mouse clones. Science 182, 56-59.

LIEBERMANN, D. & SACHS, L. (1977) Type C RNA virus production and cell competence for normal differentiation in myeloid leukaemic cells. Nature 269, 173-175.

LOEB, L.A. & TARTOF, K.D. (1976) Construction of human tumor viruses. Science 193, 272.

LUFTIG, R.D., McMILLAN, P.N. & BOLOGNESI, D.P. (1974) An ultra-structural study of C-type virion assembly in mouse cells. Cancer Res. 34, 3303-3310.

MANES, C. (1974) Phasing of gene products during development. Cancer Res. 34, 2044-2052.

MAUGH, T.H. (1974a) RNA viruses: The age of innocence ends. Science 183, 1181-1185.

MAUGH, T.H. (1974b) Leukemia: Much is known, but the picture is still confused. Science 185, 48-57.

MAYER, R.J., SMITH, R.G., & GALLO, R.C. (1974) Reverse trans-criptase in normal rhesus monkey placenta. Science 185, 864-867.

McCORMACK, J.T. & GREENWALD, G.S. (1974) Progesterone and oestradiol-17β concentrations in the peripheral plasma during pregnancy in the mouse. J. Endocr. 62, 101-107.

MEIER, H. & FOX, R.R. (1973) Hereditary lymphosarcoma in WH rabbits and hereditary hemolytic anemia associated with thymoma in strain X rabbits. Biblio. Haematol. 39, 79-92.

MEIER, H. & HUEBNER, R.J. (1971) Host-gene control of C-type tumor virus-expression and tumorigenesis: Relevance of studies in inbred mice to cancer in man and other species. Proc. Nat. Acad. Sci. USA 68, 2664-2668.

MORONI, C. & SCHUMANN, G. (1977) Are endogenous C-type viruses involved in the immune system? Nature 269, 600-601.

MUKHERJEE, B.B. & MOBRY, P.M. (1975) Variations in hybridization of RNA from different mouse tissues and embryos to endogenous C-type virus DNA transcripts. J. Gen. Virol. 28, 129-135.

MURRAY, F.A., McGAUGHEY, R.W. & YARUS, M.J. (1972) Blastokinin: its size and shape, and an indication of the existence of subunits. Fertil. and Steril. 23, 69-77.

NEUBAUER, R.H., WALLEN, W.C., PARKS, W.P., RABIN, H. & CICMANEC, J.L. (1974) Attempts to demonstrate type-C virus in normal and neoplastic tissues of nonhuman primate origin. Lab. Anim. Sci. 24, 235-240.

PANEM, S., PROCHOWNIK, E.V., REALE, F.R. & KIRSTEN, W.H. (1975) Isolation of type C virions from a normal human fibroblast strain. Science 189, 297-298.

PANEM, S., PROCHOWNIK, E.V., KNISH, W.M. & KIRSTEN, W.H. (1977) Cell generation and type C virus expression in the human embryonic cell strain HEL-12. J. Gen. Virol. 35, 487-495.

PATTILLO, R.A. (1973) Trophoblastic cancers: Chorionic gonadotropin hormone production, antigenic expression, and trophoblast redifferentiation in multiple forms of malignancy. Pathobiology Annual 3, 269-390.

PEDERSEN, H. & SEIDEL, G. (1972) Micropapillae: A local modification of the cell surface observed in rabbit oocytes and adjacent follicular cells. J. Ultrastructure Res. 38, 540-548.

PIKO, L. (1975) Expression of mitochondrial and nuclear genes during early development. In: M. Balls and A.E. Wild (Eds.) The Early Development of Mammals, Cambridge University Press.

PIKO, L. (1977) Immunocytochemical detection of a murine leukemia virus-related nuclear antigen in mouse oocytes and early embryos. Cell 12, 697-707.

POTTATHIL, R. & MEIER, H. (1977) Antitumor effects of RNA isolated from murine tumors and embryos. Cancer Res. 37, 3280-3286.

PROCHOWNIK, E.V. & KIRSTEN, W.H. (1976) Inhibition of reverse transcriptases of primate type C viruses by 7S immunoglobulin from patients with leukemia. Nature 260, 64-67.

RASHEED, S., BRUSZEWSKI, J., RONGEY, R.W., ROY-BURMAN, P., CHARMAN, H.P. & GARDNER, M.B. (1976) Spontaneous release of endogenous ecotropic type C virus from rat embryo cultures. J. Virology 18, 799-803.

REZNIKOFF, C.A., BRANKOW, D.W. & HEIDELBERGER, C. (1973) Establishment and characterization of a cloned line of C3H mouse embryo cells sensitive to postconfluence inhibition of division. Cancer Res. 33, 3231-3238.

ROBY, K., BOETTIGER, D., PACIFICI, M. & HOLTZER, H. (1976) Effects of Rous sarcoma virus on the synthetic programs of chondroblasts and retinal melanoblasts. Am. J. Anat. 147, 401-405.

ROSENBERG, N. & BALTIMORE, D. (1976) A quantitative assay for information of bone marrow cells by Abelson murine leukemia virus. J. Exp. Med. 143, 1453-1463.

SARNGADHARAN, M.G., ALLAUDEEN, H.S. & GALLO, R.C. (1976) Reverse transcriptase of RNA tumor viruses and animal cells. In: H. Busch (Ed.) Methods in Cancer Research, Vol. XII, Academic Press.

SCHAFER, W. & BOLOGNESI, D.P. (1977) Mammalian C-type oncornaviruses: Relationship between viral structure and cell-surface antigens and their possible significance in immunological defense mechanisms. In: M.G. Hanna Jr., and F. Rapp (Eds.) Contemporary Topics in Immunobiology 6, 127-167, Plenum Press, New York and London.

SCHIDLOVSKY, G. & AHMED, M. (1973) C-type virus particles in placentas and fetal tissues of Rhesus monkeys. J. Natl. Cancer Inst. 51, 225-233.

SCHWARTZ, S.A., PANEM, S., STEFANSKI, E. & KIRSTEN, W.H. (1974) Endogenous type C particles from rat embryo cells treated with 5-bromodeoxyuridine. Cancer Res. 34, 2255-2259.

SEMAN, G., LEVEY, B.M., PANIGEL, M. & DMOCHOWSKI, L. (1975) Type-C virus particles in placenta of the cottontop marmoset (*Saguinus oedipus*). J. Natl. Cancer Inst. 54, 251-252.

SHERMAN, I.I. & KANG, H.S. (1973) DNA polymerases in midgestation mouse embryo, trophoblast, and decidua. Dev. Biol. 34, 200-210.

SHERR, C.J., LIEBER, M.M., BENVENISTE, R.E. & TODARO, G.J. (1974) Endogenous baboon type C virus (M7): biochemical and immunologic characterization. Virology 58, 492-503.

SIMONS, P.J., RANKIN, B.J. & PAPADIMITRIOU, J.M. (1972) Isolation and properties of cells releasing C-type particles from BALB-c embryo cell cultures. J. Natl. Cancer Inst. 49, 1539-1552.

SLAVKIN, H.C., BRINGAS, P., CROISSANT, R. & BAVETTA, L.A. (1972) Epithelial-mesenchymal interactions during odontogenesis. II. Intercellular matrix vesicles. Mech. Age. Dev. 1, 139-161.

SMART, J.E. & HOGG, N. (1976) Alteration in membrane glycoproteins after type-C virus infection of murine fibroblasts. Nature 261, 314-316.

SMITH, G.C., HEBERLING, R.L., HELMKE, R.J., BARKER, S.T. & KALTER, S.S. (1977) Oncornavirus-like particles in squirrel monkey (*Saimiri sciureus*) placenta and placenta culture. J. Natl. Cancer Inst. 59, 975.

SMITH, G.C., KALTER, S.S., HELMKE, R.J., HEBERLING, R.L., PANIGEL, M. & KRAEMER, D.C. (1975) A-type particles in placentas of four mouse strains. Proc. Soc. Exp. Biol. Med. 148, 1212-1216.

SMITH, G.C., KALTER, S.S., HEBERLING, R.L. & HELMKE, R.J. (1976) Particles morphologically resembling mouse hepatitis virus in nude mouse uterus. Intervirology 6, 90-97.

STEPHENSON, J.R. & AARONSON, S.A. (1972) Genetic factors influencing C-type RNA virus induction. J. Exp. Med. 136, 175-184.

STEPHENSON, J.R. & AARONSON, S.A. (1973) Segregation of loci for C-type virus induction in strains of mice and high and low incidence of leukemia. Science 180, 865-866.

STEPHENSON, J.R. & AARONSON, S.A. (1976) Search for antigens and antibodies crossreactive with type C viruses of the wooly monkey and gibbon ape in animal models and in humans. Proc. Natl. Acad. Sci. USA 73, 1725-1729.

STEPHENSON, J.R. & AARONSON, S.A. (1977) Endogènous C-type viral expression in primates. Nature 266, 469-472.

STEPHENSON, J.R., REYNOLDS, R.K. & AARONSON, S.A. (1976) Comparisons of the immunological properties of two structural polypeptides of type C RNA viruses endogenous to old world monkeys. J. Virology 17, 374-384.

STEPHENSON, J.R., TRONICK, S.R. & AARONSON, S.A. (1974) Analysis of type specific antigenic determinants of two structural polypeptides of mouse RNA C-type viruses. Virology 58, 1-8.

STRAND, M. & AUGUST, J.T. (1974) Type-C RNA virus gene expression in human tissue. J. Virology 14, 1584-1596.

STRAND, M., AUGUST, J.T. & JAENISCH, R. (1977) Oncornavirus gene expression during embryonal development of the mouse. Virology 76, 886-890.

STRICKLAND, J.E., FOWLER, A.K., KIND, P.P., HELLMAN, A., KALTER, S.S., HEBERLING, R.L. & HELMKE, R.J. (1973) Group specific antigen and RNA-directed DNA polymerase activity in normal baboon placenta. Proc. Soc. Exp. Biol. Med. 144, 256-258.

STRICKLAND, J.E., KIND, P.D., FOWLER, A.K. & HELLMAN, A. (1974) Comparison of viral marker proteins in murine leukemia virus and mouse uterus. J. Natl. Cancer Inst. 52, 1161-1165.

STRICKLAND, J.E., SAVIOLAKIS, G.A., FOWLER, A.K., KOUTTAB, N.M. & HELLMAN, A. (1976) Impaired estrogen-mediated production of type C viral DNA polymerase in aged NIH Swiss mouse uteri (39481) Proc. Soc. Exp. Biol. Med. 153, 63-69.

TEMIN, H.M. (1971) The protovirus hypothesis: Speculations on the significance of RNA-directed DNA synthesis for normal development and for carcinogenesis. J. Natl. Cancer Institute 46, 3-6.

TEMIN, H.M. (1976) The DNA provirus hypothesis. Science 192, 1075-1080.

TEMIN, H.M. & MIZUTANI, S. (1970) RNA-dependent DNA polymerase in virions of Rous sarcoma virus. Nature (London) 226, 1211-1213.

TODARO, G.J., BENVENISTE, R.E., LIEBER, M.M., & LIVINGSTON, D.M. (1973) Infectious type C viruses released by normal cat embryo cells. Virology 55, 506-515.

TODARO, G.J., DeLARCO, J.E. & COHEN, S. (1976) Transformation by murine and feline sarcoma viruses specifically blocks binding of epidermal growth factor to cells. Nature (London) 264, 26-31.

TODARO, G.J. & HUEBNER, R.J. (1972) New evidence as the basis for increased efforts in cancer research. Proc. Nat. Acad. Sci. USA 69, 1009-1015.

TOOZE, J. (1973) The molecular biology of tumour viruses. Cold Spring Harbor Laboratory Publ., New York, 743 pg.

TWARDZIK, D.R., FOWLER, A., WEISLOW, O. & HELLMAN, A. (1977) Interaction of oncornavirus proteins with cell surface products of the major histocompatibility complex. IRCS Medical Science: Microbiology, Parasitology and Infectious Diseases 5, 451.

VANBLERKOM, J. & MANES, C. (1977) The molecular biology of the preimplantation embryo. In: M.I. Sherman (Ed.) Concepts in Early Mammalian Development, M.I.T. Press.

VANBLERKOM, J., MANES, C. & DANIEL, J.C. (1973) Development of preimplantation rabbit embryos *in vivo* and *in vitro*. I. An ultrastructural comparison. Dev. Biol. 35, 262-282.

VANBLERKOM, J. & RUNNER, M.N. (1976) The fine structural development of preimplantation mouse parthenotes. J. Exp. Zool. 196, 113-124.

VERNON, M.L., LANE, W.T. & HUEBNER, R.J. (1973) Prevalence of type-C particles in visceral tissues of embryonic and newborn mice. J. Natl. Cancer Inst. 51, 1171-1175.

VERNON, M.L., McMAHON, J.M. & HACKETT, J.J. (1974) Additional evidence of type-C particles in human placentas. J. Natl. Cancer Inst. 52, 987-989.

VERWOERD, D.W. & SARMA, P.S. (1973) Induction of type C virus-related functions in normal rat embryo fibroblasts by treatment with 5-iododeoxyuridine. Int. J. Cancer 12, 551-562.

WATERS, L.C. & YANG, W.K. (1974) Comparative biochemical properties of RNA-directed DNA polymerases from Rauscher murine leukemia virus and avian myeloblastosis virus. Cancer Res. 34, 2585-2593.

WEISLOW, O.S., SCHNEIDER, S., HEBERLING, R.L., KALTER, S.S. & HELLMAN, A. (1976) Autogenous humoral immunity to baboon xenotropic endogenous type C virus. J. Natl. Cancer Inst. 57, 561-566.

WITKIN, S.S., KORNGOLD, G.C. & BENDICH, A. (1975) Ribonuclease-sensitive DNA-synthesizing complex in human sperm heads and seminal fluid. Proc. Nat. Acad. Sci. USA 72, 3295-3299.

WOLFF, E. (1968) Specific interactions between tissues during organogenesis. Current Topics in Developmental Biology 3, 65-94.

WONG-STAAL, F., GILLESPIE, D. & GALLO, R.C. (1976) Proviral sequences of baboon endogenous type C RNA virus in DNA of human leukaemic tissues. Nature 262, 190-195.

WU, A.M. & GALLO, R.C. (1975) Reverse transcriptase. Critical Rev. in Biochem. 3, 289-347.

YANG, W.K., TYNDALL, R.L. & DANIEL, J.C. (1976) DNA- and RNA-dependent DNA polymerases: Progressive changes in rabbit endometrium during preimplantation stage of pregnancy. Biol. Reprod. 15, 604-613.

BIOCHEMICAL STUDIES ON MAMMALIAN X-CHROMOSOME ACTIVITY

Marilyn Monk

MRC Mammalian Development Unit
Wolfson House, University College London
4 Stephenson Way, London, NW1 2HE

1. Biochemical Tools
   - *Linkage studies*
   - *Variant enzyme forms*
   - *Gene dosage effects*
2. The XO female mouse
3. X-chromosome activity in germ cells
4. X-chromosome activity in teratocarcinoma cells
5. X-chromosome activity in early embryos
6. Preferential expression of the maternal X-chromosome
7. X-chromosome differentiation : an interpretation

During early development of female eutherian mammals, one or other of the X-chromosomes becomes inactive. The hypothesis of a "single-active-X" was originally proposed by Lyon (1961), Russell (1961) and Beutler *et al.* (1962) to explain the different cytological properties of the two chromosomes and the genetic mosaicism observed in female mammals heterozygous for X-linked genes or X-autosome translocations. Since the choice of which X-chromosome will be inactive is random, half the cells of the embryo at the time this event occurs express the maternally derived X-chromosome and the other half express the paternally derived X-chromosome. The differentiation of the X-chromosomes in each cell is irreversible so that the same X-chromosome is active in all the progeny of a particular cell (see West, this volume). The presence of a single active X-chromosome in all female somatic cells ensures that these cells have the same effective X-chromosome dosage as male cells, although why this is necessary is not clear. Cytogenetic evidence for a single active X and studies on mosaicism in females heterozygous for X-linked characters will be briefly presented here. For more comprehensive information, see reviews by Ohno (1969), Lyon (1968; 1972; 1974), Eicher (1970) and Gartler and Andina (1976).

Associated with the inactivation of an X-chromosome is its transformation into heterochromatin which is believed to be genetically inert. The inactive X-heterochromatin remains condensed throughout the cell cycle and may be seen as a discrete mass during interphase, the sex chromatin or Barr body. It has been established that the DNA of the inactive X-chromosome is replicated late during the S phase of the cell cycle in many female mammals, e.g. in humans (Morishima *et al.*,1962), in mules (the late replicating X-chromosome is randomly of horse or donkey origin, Mukherjee and Sinha, 1964) and in cows (Gartler and Burt, 1964). In the mouse it has been difficult to identify unequivocally a late labelling sex chromosome (Galton and Holt, 1965; McLaren, 1972), even with the use of a cytologically distinct X-autosome translocation (Evans *et al.*,1965). Difficulties are increased by the fact that many autosomes also show late replicating regions. Better resolution was obtained by Nesbitt and Gartler (1970) by exploiting the fact that both the mouse Y-chromosome and the late replicating X-chromosome could be more readily identified as unlabelled chromosomes when the cultures were labelled early in the S phase. Subsequently Takagi and Oshimura (1973) have used special staining techniques for more efficient visualisation of the heterochromatic X-chromosome in metaphase spreads prepared from early post-implantation mouse embryos.

The other major line of evidence for a single active X-chromosome comes from observations on female mice heterozygous for sex-linked mutant genes with some measurable effect, for example, coat colour. Such females show a "mottled" or "dappled" phenotype due to inactivation of one or other of the X-chromosomes early in development (Lyon, 1961). A similar mosaicism occurs when translocated autosomal genes are carried by one of the X-chromosomes; in this situation inactivation appears to spread into the autosomal segment (see Eicher, 1970). Similarly, biochemical studies of a population segregating two alleles of an X-linked gene which determine variant forms of an enzyme, show that males, with one X-chromosome, express one or the other variant enzyme form, whereas females can express one or the other, or both forms. The latter, heterozygous, females are mosaics with respect to X-chromosome-linked enzyme activity: one or other allele being active in different somatic cells. This can be demonstrated if somatic cells from the heterozygous female are cloned *in vitro*; Davidson *et al.* (1963) showed two distinct types of clones with respect to the X-linked G6PD (glucose-6-phosphate dehydrogenase) locus in heterozygous females, thus providing strong evidence that X-chromosome dosage compensation involves inactivation of one or the other X-chromosome. Further, clones from individuals heterozygous for two X-linked markers show *cis* expression of the enzymes coded by one or the other X-chromosome (Gartler *et al.*,1972a). The relationship between X-chromosome inactivity and late replication was established in clones from a female mule where the G6PD enzyme form expressed correlates with the cytologically identifiable active X-chromosome (Ratazzi and Cohen, 1972; Ray *et al.*,1972).

There has been considerable interest in the timing of X-chromosome differentiation in early development (see Lyon, 1972; 1974). It has been established that a late replicating X-chromosome, or a heterochromatic X-chromosome, is not present until the late blastocyst stage in many mammals. In embryos of the monkey and human, sex chromatin is first found in nuclei at the late blastocyst stage (Park, 1957). Late replication of an X-chromosome can be detected as early as the morula stage in the rabbit (though sex chromatin appears later, Issa *et al.*,1969) and by the late blastocyst stage in the mouse (Mukherjee, 1976). Similarly, Takagi (1974) showed that whereas all chromosomes appeared euchromatic during cleavage and at the morula stage, one X-chromosome could be shown to be heterochromatic and brightly fluorescent after quinacrine mustard staining by the blastocyst stage; however, this author did not observe predominant late replication until the ninth day of gestation.

The available cytogenetic evidence thus indicates that X-chromosome differentiation occurs around the time of implantation and perhaps as early as the blastocyst stage. Nesbitt (1971) has calculated a similar time for X-chromosome differentiation from the degree of mosaicism in tissues of adult females heterozygous for an X-linked marker (see also Nesbitt and Gartler, 1971). Gardner and Lyon (1971) have obtained evidence that both X-chromosomes are potentially active in cells of the inner cell mass of the blastocyst; injection of a single inner cell mass cell heterozygous for X-linked coat colour markers into a recipient blastocyst can result in the appearance of both colours in the coat of a resultant chimaeric mouse. Lyon (1972) has suggested that X-chromosome differentiation occurs during the fifth day of pregnancy (late blastocyst stage). In general, these approaches have assumed that X-inactivation occurs in all tissues of the embryo at the same time. A direct biochemical assay of X-chromosome-linked gene activity is required to test this assumption.

This chapter will review biochemical studies on X-chromosome activity and will be predominantly concerned with observations in the mouse (see also Chapman *et al.*,1977). The following sections will deal with the biochemical tools available, the XO female mouse, X-chromosome activity in germ cells and teratocarcinoma cells, and recent results from our laboratory on X-chromosome activity in individual pre-implantation embryos. Recent work showing the predominance of an active maternally derived X-chromosome in some tissues of the conceptus will be briefly reviewed. Finally, the nature of the events which initiate and maintain the inactivity of one X-chromosome in female somatic cells will be considered. The following interpretation is presented: X-chromosome differentiation occurs as a sequential inactivation event in cells as they depart from the stem cell line, showing (1) a preferential recall that the paternal X-chromosome was previously inactive; (2) a later loss of this recall in embryo precursor cells; and (3) the maintenance of two active X-chromosomes through early development and throughout the germ line.

BIOCHEMICAL TOOLS

The most direct analysis of X-chromosome function involves studies on enzymes known to be coded for by genes on the X-chromosome. The most studied are:

- G6PD (glucose-6-phosphate dehydrogenase, EC 1.1.1.49),
- HPRT (hypoxanthine phosphoribosyl transferase, EC 2.4.2.8),
- PGK (phosphoglycerate kinase, EC 2.7.2.3) and
- α-gal (α-galactosidase, EC 3.2.1.22).

LINKAGE STUDIES

Location of G6PD, HPRT and PGK on the mouse X-chromosome was demonstrated by Chapman and Shows (1976) by a genetic analysis of somatic cell hybrids from an interspecific hybrid foetus. The two parent mice, *Mus musculus* and *Mus caroli*, have different electrophoretic forms of a number of enzymes. The different forms of the three enzymes G6PD, HPRT and PGK were present in female hybrid embryos, whereas male embryos expressed the forms characteristic of one or the other parent. Individual clones from cultured female cells also expressed forms characteristic of one or the other parent due to X-inactivation having occurred, thus providing further evidence of X-linkage. Linkage of α-gal to the mouse X-chromosome was shown by chromosome segregation studies in somatic cell hybrids of Chinese hamster and mouse cells by Kozak (1975). Subsequently, Lusis and West (1976) showed that a thermolabile variant form of α-gal was inherited as an X-linked gene product in *Mus musculus molossinus*. A recently discovered electrophoretic variant form of PGK (Nielsen and Chapman, 1978) has provided further confirmation of X-linkage of this enzyme.

Further evidence for X-linkage in the mouse comes from X-chromosome dosage effects. Since both X-chromosomes are active in female germ cells (see sections 3 and 5), oocytes of XX mothers have been shown to have approximately twice the activities of G6PD (Epstein, 1969), HPRT (Epstein, 1972; Monk and Kathuria, 1977), and PGK (Kozak *et al.*,1974), compared with oocytes of XO mothers. The activities of control autosomal enzymes (lactic dehydrogenase, guanine deaminase and adenine phosphoribosyl transferase) were approximately the same in the two groups of eggs (Epstein, 1969; 1972; Monk and Kathuria, 1977).

Linkage of the above enzymes to the X-chromosome has been demonstrated in other mammals: for example, for G6PD in man (Marks and Gross, 1959; Davidson *et al.*,1963; Ricciuti and Ruddle, 1973; Shows and Brown, 1975), HPRT in man (Seegmiller *et al.*,1967); Ricciuti and Ruddle, 1973; Shows and Brown, 1975), PGK in man (Valentine *et al.*,1968; Chen *et al.*,1971; Ricciuti and Ruddle, 1973; Shows and Brown, 1975) and kangaroo (Cooper *et al.*,1971), and α-gal in man (Grzeschik *et al.*,1972). Westerveld *et al.*,(1972) have shown that G6PD, HPRT and PGK are on the same chromosome in the Chinese hamster. In man, enzyme deficiencies of G6PD, HPRT, PGK and α-gal have been associated, respectively, with drug-induced haemolytic anaemia (Marks and Gross, 1959), Lesch-Nyhan disease (Seegmiller *et al.*,1967), chronic haemolytic anaemia (Valentine *et al.*,1968; Chen *et al.*,1971) and Fabry's disease (Kint, 1970). In general, genes found to be X-linked in one mammalian species are also X-linked in other mammalian species. This was predicted by Ohno (1969) on the expectation that the X-chromosome remained genetically isolated after evolution of placental mammals.

VARIANT ENZYME FORMS

Investigations of X-linked gene expression would be facilitated by the availability of suitable biochemical variants of the enzymes under study. Such variants are at present rare. Embryos of the interspecific cross between *Mus musculus* and *Mus caroli*, which were used to assign G6PD, HPRT and PGK to the X-chromosome (see above), can be obtained by artificial insemination but they do not survive beyond about 16 days of gestation (Chapman and Shows, 1976). Such hybrid embryos could prove invaluable in the future for studies of X-chromosome expression in early development. The PGK variant (Nielsen and Chapman, 1978) has provided interesting results (see section 6) but its use at earlier stages of development is hampered by the high levels of PGK activity in the maternal cytoplasm of the egg. Two other enzymes with variant forms have also been shown to be inherited as X-linked products in the mouse. These are phosphorylase *b* kinase (Lyon *et al.*, 1967) whose expression is limited to skeletal muscle, and ornithine carbamoyltransferase (De Mars *et al.*,1976). A variant form of this latter enzyme is associated with the X-linked sparse-fur (spf) mutation in the mouse and was identified on the basis of urinary bladder stones in affected male mice. These latter variants, along with the thermolabile variant of α-gal (Lusis and West, 1976), have not yet been utilised in critical analyses of the "single-active-X" hypothesis and will not be further discussed in this chapter.

GENE DOSAGE EFFECTS

In the absence of enzyme variant forms, one can study X-chromosome activity by an analysis of the effects of X-chromosome dosage, particularly in comparisons of XX and XO eggs (section 3), XX and XY preimplantation embryos (section 5), and early embryos derived from XX and XO females (section 5). In such investigations relying on X-chromosome dosage, the implicit assumption is that enzyme activity is directly proportional to the number of gene copies. This has been found to be true for many loci in man and other mammals. Westerveld *et al.*(1972) assayed PGK in diploid, hyperdiploid, tetraploid and hybrid cells, originating from a Chinese hamster cell line, and showed a relationship between gene multiplicity and enzyme activity. Some other examples may be given for man, viz. trisomy for chromosome 16 causes a proportional increase in adenine phosphoribosyl transferase (APRT, EC 2.4.2.7) coded by a gene on this chromosome (Marimo and Gianelli, 1975), and a similar situation was found by Magenis *et al.* (1975) for trisomy involving an erythrocyte acid phosphatase. Other examples of proportionality between gene dosage and enzyme activity may be found in the literature.

It should be emphasised, however, that there are special cases which do not show the correlation between gene dosage and the amount of gene product. RNA polymerase activity in both bacterial cells (Scaife, 1976) and in eukaryotic cells (Guialis *et al.*, 1977) can be regulated independently of gene dosage. In the case of bacteria, it is known that such regulation is autogenous, i.e. it is mediated by the gene products themselves. For other examples of autogenous regulation, see Backman *et al.* (1976), Gold *et al.* (1976), Reed *et al.* (1976), Russel *et al.* (1976) and Goldberger (1974).

Also it must be remembered that in addition to gene dosage effects, activity of individual genes may be affected by a variety of factors such as developmental stage, sex, specific tissue, genetic background and various manipulations, such as *in vitro* culture. Steele and Owens (1973) and Steele and Migeon (1973) have demonstrated sex differences in activity of G6PD in erythrocytes from newborn infants and in cultured fibroblasts. Sex differences were not found for another X-linked enzyme, HPRT. Since X-inactivation has already occurred, X-chromosome dosage cannot be the cause of the differences observed. Rather some other regulatory mechanism dependent on sex must be involved.

THE XO FEMALE MOUSE

Many of the experiments reported below (sections 3 and 5) depend on comparisons of X-chromosome dosage in the female germ line of XO and XX female mice and in embryos arising from these two types of female. In this section some of the properties of these exceptional females which have proved so useful in studies of X-chromosome activity will be briefly described.

In mice, as in man, it is the presence or absence of the Y-chromosome which determines sex (Ohno, 1976). However, in terms of viability the X-chromosome is more important. XO female mice appear quite normal (Cattanach, 1962; Morris, 1968), though they exhibit a shorter reproductive life span (Lyon and Hawker, 1973). The near normality of the XO female indeed tells us that only one X-chromosome is necessary for development and later survival and is taken as support for the "single-active-X" hypothesis (Lyon, 1961; Russell, 1961); Embryos which are OY, on the other hand, are inviable and die by the 8-cell stage (Burgoyne and Biggers, 1976) suggesting that some X-linked function is required by this stage. The litters of XO females consist of XX, XO, and XY offspring. The relative proportions of these three types would be expected to be equal. However, the number of XO females among the offspring (Cattanach, 1962; Morris, 1968; Burgoyne and Biggers, 1976) is approximately only one-third the number expected (13.2% XO, 43.2% XX, 43.6% XY; Burgoyne and Biggers, 1976).

Earlier work by Cattanach (1962) had established that the shortage of XO daughters was not due to post-natal inviability and he suggested that there may be a preferential loss of chromosome sets lacking an X-chromosome into the polar bodies during meiotic divisions. This possibility was confirmed by Kaufman (1972) who showed that only approximately 30% of ovulated eggs in the XO female lacked an X-chromosome. Similar results were obtained by Luthardt (1976).

Because both X-chromosomes are active during oogenesis, the embryos of the XO female will be deficient in X-linked products compared with XX-derived embryos. This deficiency may be the reason for the small litter sizes of XO mothers and the slower development of XO-derived preimplantation embryos *in vitro* (Burgoyne and Biggers, 1976). Morris (1968) found a higher level of preimplantation loss of embryos in XO females than in XX females which he interpreted as being due to the lethality of the YO embryos. There was also some suggestion of decreased viability of XO-derived embryos *in utero*. The embryos lost were assumed to be XO embryos although they were not identified as such with respect to sex chromosome constitution. In the light of Burgoyne and Biggers' observations (1976) that all the XO-derived preimplantation embryos show delayed preimplantation development, the embryos lost, in addition to the lethal YO, may be randomly XO, XX or XY.

In summary, in the XO female, (a) there is non-random segregation resulting in preferential loss of chromosome sets lacking an X-chromosome to the polar bodies; (b) early death of YO embryos; and (c) some general embryonic loss presumably due to the fact that XO-derived embryos begin development with half the normal complement of X-linked gene products. The following section will include studies on X-chromosome activity in the germ line of XX and XO females.

X-CHROMOSOME ACTIVITY IN GERM CELLS

In contrast to the situation in somatic cells of adult female mammals, female germ cells do not contain sex chromatin (Ohno *et al.*,1961; 1962) suggesting that both chromosomes are active. In contrast, follicle cells, like other female somatic cells, show formation of sex chromatin by a single heteropycnotic X-chromosome. Cytogenetic evidence for the activity of both X-chromosomes has been obtained for the mitotic (oogonial) phase of germ cell differentiation, as well as for later meiotic stages in the foetal rat (Ohno *et al.*,1961) and human (Ohno *et al.*,1962).

In addition to the cytogenetic data there is biochemical evidence for two active X-chromosomes in the germ line of human females heterozygous for variant electrophoretic forms of the X-linked enzyme G6PD (Gartler *et al.*,1972b). Somatic tissues of such heterozygous females consist of a mixture of cells expressing either the G6PD A or G6PD B forms as demonstrated by cell cloning in culture (Davidson *et al.*,1963). Mature oocytes at the dictyotene stage produce the A and B bands, and in addition the AB hybrid band. The presence of the AB hybrid band in the oocytes from adult females (Gartler *et al.*,1972b) implies that both X-chromosomes are active in the individual cells. Follicle cells gave only the A and B bands.

The important question is whether both X-chromosomes are active in earlier germ line stages, i.e. are the original primordial germ cells determined and partitioned off before X-inactivation has occurred, hence "escaping" X-inactivation: or have they been through a cycle of inactivation and re-activation? Gartler *et al.*(1973) have demonstrated full expression of both X-chromosomes in human foetal oocytes at 16 weeks of gestation. In a subsequent work, Gartler *et al.*(1975) have claimed that, in the earlier mitotic period (in a 12 week foetus), germ cells heterozygous for G6PD electrophoretic variant forms show banding patterns of pure A and B forms of enzyme, a result which is more consistent with the interpretation that only one X-chromosome is active in each oogonium as is the case of X-inactivated somatic tissue. This would argue for a cycle of inactivation during early embryonic development and reactivation in the germ line at meiosis. In contrast to these observations, Migeon and Jelalian (1977) have observed the AB hybrid band in an 8 week old female foetus at a time when the germ cell population is pre-meiotic. These authors claim that two

X-chromosomes are active in the germ line before meiosis. Their results are more consistent with the cytological observations (Ohno *et al.*,1961; 1962) showing that pre-meiotic germ cells do not possess a heteropycnotic X-chromosome. On the other hand, Ohno (1964) has claimed that earlier primordial germ cells in the yolk sac do have an inactive X-chromosome.

In the mouse a suitable electrophoretic X-linked variant enzyme form has not been available for studies on X-chromosome activity in the developing female germ line. However, there is good evidence, based on gene dosage effects, that both X-chromosomes are active in the production of mature mouse oocytes. Ovulated oocytes of XX mice have approximately twice the activity of G6PD, HPRT and PGK as do ovulated oocytes of XO mice (Epstein, 1969; 1972; Monk and Kathuria, 1977; Kozak *et al.*,1974), whereas activities of autosomally coded enzymes, lactic dehydrogenase, guanine deaminase and adenine phosphoribosyl transferase, are approximately the same in the two groups of eggs. Assays of X-coded enzymes at earlier stages of germ line differentiation may tell us whether there exists an early population of primordial germ cells with a single active X-chromosome. If this were the case, X-chromosome dosage of XX and XO germ cells would be equivalent, and a doubling in X-coded enzyme activities might be expected in the XX germ line, though not in the XO germ line, at some later stage prior to meiosis.

In contrast to the situation in the female, in the male germ line the X- and Y-chromosomes are inactive at later stages of spermatogenesis. The X- and Y-chromosomes become condensed (Ohno *et al.*,1961) and late replicating (Odartchenko and Pavillard, 1970) before or during meiosis, and there is little, if any, detectable G6PD activity (Vanha-Perttula *et al.*,1970). It has been proposed that inactivation of the single X-chromosome in spermatocytes is required for normal spermatogenesis (Lifschytz and Lindsley, 1972; see also O, this volume).

X-CHROMOSOME ACTIVITY IN TERATOCARCINOMA CELLS

Teratocarcinomas are transplantable tumours composed of embryonic stem cells and a range of differentiated derivative cells. They result from disorganised growth following grafting of early embryos, or parts of them, into ectopic sites in syngeneic hosts, or, spontaneously from germ cells in the testes or ovaries of certain strains of mice (Stevens, 1970; Damjanov and Solter, 1974;

Stevens and Varnum, 1974). Provided they still contain the undifferentiated embryonic stem cells, they are immortal, i.e. transplantable and potentially adaptable to *in vitro* culture (see also Martin and Heath, this volume).

The stem cells resemble early embryonic inner cell mass cells (epiblast cells) or primordial germ cells in appearance (Martin, 1975; Pierce and Beals, 1964). They characteristically form piled-up colonies of small rounded cells in culture which in long term culture may differentiate into a variety of cell types. Under free floating conditions (in suspension or in the peritoneal cavity) clumps of stem cells become surrounded by endoderm and resemble embryonic egg cylinders in appearance. These embryoid bodies on attachment and outgrowth *in vitro* may also produce a broad range of differentiated cell types (Martin and Evans, 1975). The differentiated cells are subject to ageing and are no longer transplantable.

Linder (1969) found that the differentiated cells from human female teratocarcinomas contain a single active X-chromosome. However, the morphological resemblance of the stem cells to germ cells suggests that both X-chromosomes might be active in female stem cells. If such were the case, teratocarcinomas would provide a valuable system to study X-chromosome inactivation *in vitro*. The initial attempts to investigate this possibility suggested that only one X-chromosome is active in the stem cells. McBurney and Adamson (1976) examined a number of different stem cell lines derived from teratocarcinomas produced by embryo grafts. They determined the specific activities of the X-linked enzymes G6PD, HPRT, PGK and α-gal in five lines (two XX and three XO in sex chromosome constitution). Two autosomally-coded enzymes 6PGD (phosphogluconate dehydrogenase, EC 1.1.1.44) and APRT (adenine phosphoribosyl transferase, EC 2.4.2.7) were also analysed as controls. Similar enzyme activities in the five cell lines were found for all enzymes whether X-linked or autosomal, except for one cell line (XX in sex chromosome constitution) which had approximately two times the activity of α-gal. Since this result is not supported by increases in activity of the other three X-linked enzymes, it is difficult to interpret. These authors also looked for evidence of a late replicating X-chromosome, and found it in one of the XX cell lines, but not in the other XX line which had shown the high α-gal activity. Could this be evidence for partial inactivation of one X-chromosome so that it failed to become late replicating with the α-gal gene remaining active while the genes for G6PD, HPRT and PGK became inactivated?

Recently Martin *et al.* (1978) have obtained clear evidence for two active X-chromosomes in cells of a spontaneously derived teratocarcinoma stem line from the ovary of an LT female mouse. They determined the specific activities of three X-linked enzymes (G6PD, HPRT, and α-gal), and seven autosomally coded enzymes, in the LT-XX cell line, and in several XO cell lines derived from a testicular teratocarcinoma and from two embryo-derived tumours. The ratios of specific activities for each enzyme in the LT-XX versus the XO cells were approximately two for the X-linked enzymes and approximately one for the autosomal enzymes. When the cells were allowed to differentiate, the ratios of the X-linked enzymes in the XX versus XO cells approached the value of one. These elegant experiments provide the first demonstration of a process of X-inactivation in differentiating cells *in vitro* and a valuable model system for a further analysis of this process. The discrepancy between the results of Martin *et al.* (1978) and McBurney and Adamson (1976) may be related to the parthenogenetic origin of the XX cells used by the former workers, compared with the embryograft origin of the cells used by McBurney and Adamson.

X-CHROMOSOME ACTIVITY IN EARLY EMBRYOS

As discussed earlier, cytogenetic evidence has placed the timing of X-chromosome differentiation in female embryos at around the time of implantation. In this section, the biochemical evidence bearing on this question will be discussed. Kozak and Quinn (1975) analysed batches of preimplantation embryos from XX mothers (XX and XY) and from XO mothers (XX, XY, XO). In both cases the PGK activities remain constant until the 8-cell stage and then decrease. However, the embryos from the XX mothers express twice the level of PGK activity compared with those from XO mothers throughout preimplantation development, thus reflecting the X-chromosome dosage effect during oogenesis (Kozak *et al.*, 1974). It can be inferred that the activity observed before implantation is maternal in origin. Immediately after implantation, PGK activities increase about 100-fold and now are approximately the same in single embryos from either XX or XO mothers, i.e. activity is now embryo-coded and is the same irrespective of X-chromosome dosage. In summary, this represents biochemical evidence for a single active

X-chromosome in female , early postimplantation embryos. What is the situation in preimplantation embryos?

To ask whether both chromosomes are active in preimplantation embryos, ideally we require embryos heterozygous for X-linked variant forms of an enzyme which is absent, or virtually absent, in the maternal cytoplasm, and which is synthesised *de novo* from embryonic genes in sufficient amount to enable assays on single embryos. We could then ask whether both enzyme forms exist in half the embryos assayed (presumptive females) whereas only the maternally-inherited form is expressed in presumptive males. Unfortunately such a set of ideal conditions do not exist at the present time. However, what we can do is assay individual embryos (XX or XY) for dosage effects on activities of an X-linked enzyme. Again we require the enzyme to be low in amount in the egg, to show an embryo-coded increase in preimplantation development, and to be detectable in single embryos. Two enzymes possibly satisfy these criteria: HPRT and α-gal.

Epstein's studies on HPRT expression in early development have been introduced in section 3. In measurements on batches of embryos deriving from XX and XO mothers he established that whereas at the 2-cell stage the two groups of embryos showed a two-fold difference, reflecting the number of X-chromosomes active during oogenesis, the markedly increased HPRT activities in batches of embryos from XX and XO mothers on the fourth day of pregnancy were approximately the same (Epstein, 1972). Hence we can conclude that HPRT is embryo-coded at the blastocyst stage. Theoretically, if both X-chromosomes had been active in generating the activity observed at this stage, then the XO-derived embryos should show a predictable fraction (0.89 or higher) of the HPRT activity of the XX-derived embryos. This fraction is derived from the expectation of four active X-chromosomes in three embryos (XX, XO, XY) compared with three active X-chromosomes in two embryos (XX, XY). The value would be greater than 0.89 if the proportions of XO embryos from the XO mothers is less than one-third, which we have seen to be the case (section 2). Epstein could not conclude from his data whether both X-chromosomes were active in preimplantation development.

Recently, Adler *et al.* (1977) have demonstrated a 300-fold increase in α-gal activity during preimplantation development, and have used a highly sensitive assay technique to analyse α-gal activities in single preimplantation embryos. On the expectation that half the embryos are XX and half are XY the distribution of activities in a number of single embryos would be bimodal if both X-chromosomes were active in female embryos. Although they showed a possible suggestion of bimodality at the 8-cell and morula stage in some experiments the result was not always reproducible and the spread in the single embryo assays was too great for any definite conclusion to be drawn.

We have recently approached the question of X-chromosome activity in preimplantation embryos using a highly sensitive assay for HPRT activity in single embryos. By measuring the ratio of HPRT activity to the activity of a functionally related autosomal enzyme, APRT (adenine phosphoribosyl transferase, EC 2.4.2.7. on chromosome 8, Kozak *et al.*,1975), in the same reaction mix, we have succeeded in reducing the variance or spread in the individual embryo assays produced by differences in recovery of single embryo extracts.

The greater accuracy achieved has allowed us to ask the following questions:

(a) Are both X-chromosomes active in female embryos during preimplantationdevelopment?

(b) If so, when does X-chromosome inactivation occur?

(c) Does inactivation occur at the same time in all tissues of the embryo?

Our recent work has answered the first two questions. Both X-chromosomes probably are active in preimplantation female embryos at least at the morula stage, and inactivation occurs in most, if not all, of the cells by the blastocyst stage. Work is in progress to attempt to answer the third question. Our results are reported in more detail below.

The simultaneous assay of the X-linked enzyme (HPRT) and the autosomally-coded enzyme (APRT) is based on the conversion of ^{3}H-guanine and ^{14}C-adenine to their respective monophosphates (McBurney and Adamson, 1976). The labelled phosphates are precipitated by lanthanum chloride, collected on filters, and the amounts of the products produced by HPRT and APRT in the embryo extract calculated, as well as the ratio of HPRT:APRT activities (Monk and Kathuria, 1977). This assay could be criticised because the products are not directly identified and there may be factors interfering with the assay. Such factors could include breakdown of the substrates, variation in the pool sizes of guanine and adenine, and nucleotidases which might remove the products. However, it is reasonable to assume that such factors will be constant for a particular stage of development and will not themselves vary between male and female embryos (i.e. they are not X-linked). Confidence in our assay is substantiated by the following controls:

(a) Omission of the substrate, phosphoribosyl pyrophosphate, required for both enzyme reactions, gives no incorporation of ^{3}H or ^{14}C into precipitable products.

(b) Assay of an HPRT negative cell line drastically reduces the incorporation of ^{3}H into precipitable product while still showing high levels of APRT.

(c) The levels of HPRT and APRT activities we observe for particular stages of development are very similar to those obtained by Epstein (1970) who used a different assay procedure on batches of preimplantation embryos.

(d) Assays of batches of 5 to 10 eggs from XX and XO females show the approximately two-fold difference in HPRT expected. Figure 1 shows HPRT and APRT activities and their ratios for a number of separate batches of eggs from XX and XO females.

The distribution of activities of HPRT, or HPRT:APRT ratios, in individual preimplantation mouse embryos might be expected to reflect the relative gene doses in female (XX) and male (XY) embryos and hence to show bimodality. We have determined these values in extracts prepared from a series of individual 8-cell embryos, 9- to 16-cell morulae and blastocysts. The results are shown in Figures 2, 3 and 4. The distributions for the 8-cell embryos (Figure 2) and blastocysts (Figure 4) are clearly unimodal with over 90% of the HPRT:APRT ratios falling within a two-fold range (Monk and Kathuria, 1977). Even the HPRT activities alone in these experiments show a low variance which is not consistent with a bimodal distribution. In contrast, the morulae (Figure 3) show a bimodal distribution of ratios (Monk, unpublished) with a two-fold separation of the modes. The results for the morulae are consistent with the hypothesis that there are two populations of embryos with respect to HPRT dosage. We assume at this stage that the two populations represent male and female morulae. Work is in progress to verify this point.

At first sight the transition, unimodal to bimodal to unimodal, for the three stages of preimplantation development tested might seem somewhat unexpected. However, further work (Monk and Harper, 1978) has suggested that the initial 20-fold increase in HPRT

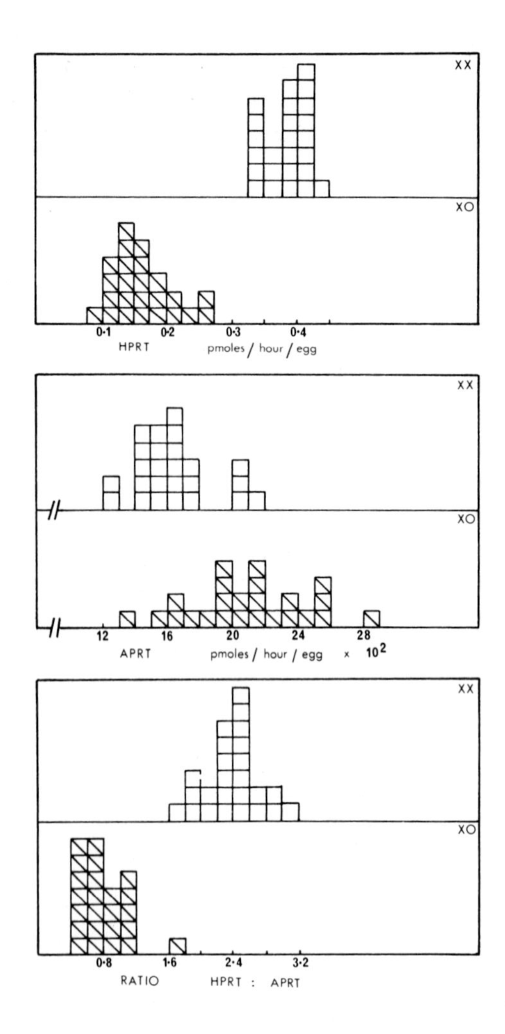

Figure 1: Activities of HPRT, APRT and ratios of activities HPRT:APRT in batches of eggs (5 to 10 eggs per batch) from XX and XO (hatched squares) females (genotypes Ta/+ and +/O). Each square represents a separate batch of eggs. The mice were superovulated and unfertilised eggs collected at approximately 21 hours following HCG injection. Mean activities (pMoles per hour per egg ± S.E.), XX eggs: HPRT 0.38 ± 0.01, APRT 0.16 ± 0.005, HPRT/APRT 2.4 ± 0.07; XO eggs: HPRT 0.17 ± 0.01, APRT 0.21 ± 0.01, HPRT/APRT 0.8 ± 0.06. (Data from Monk and Harper, 1978). Note, throughout figures, the absolute values for both enzyme activities, and hence their ratio, for a particular stage of development, may vary from one figure to another. However, the data in each figure are internally consistent and the unavoidable day to day variation does not affect the interpretation of the results.

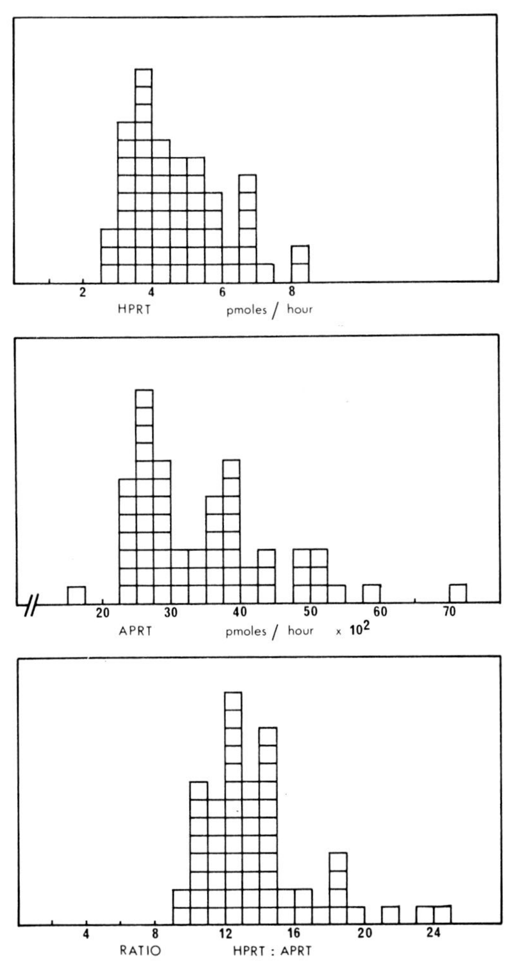

Figure 2: Activities of HPRT, APRT and their ratios in single 8-cell embryos isolated from MF1 females early on the third day of pregnancy. Each square represents a value obtained for a separate embryo. The results from two experiments are pooled. Mean activities (pMoles per hour per embryo ± S.E.), HPRT 4.70 ± 0.17, APRT 0.35 ± 0.01, HPRT/APRT 13.88 ± 0.40 (Data from Monk and Kathuria, 1977).

during cleavage to the 8-cell stage is synthesised on active stable maternal messenger RNA. Monk and Harper (1978) made a comparison of HPRT and APRT activities and HPRT:APRT ratios in embryos deriving from XX and XO mothers throughout pre-implantation development. It was argued that for as long as there remained the two-fold difference in X-linked HPRT activity shown by the eggs from these two types of female (Figure 1: Epstein, 1972), then it could be concluded that the

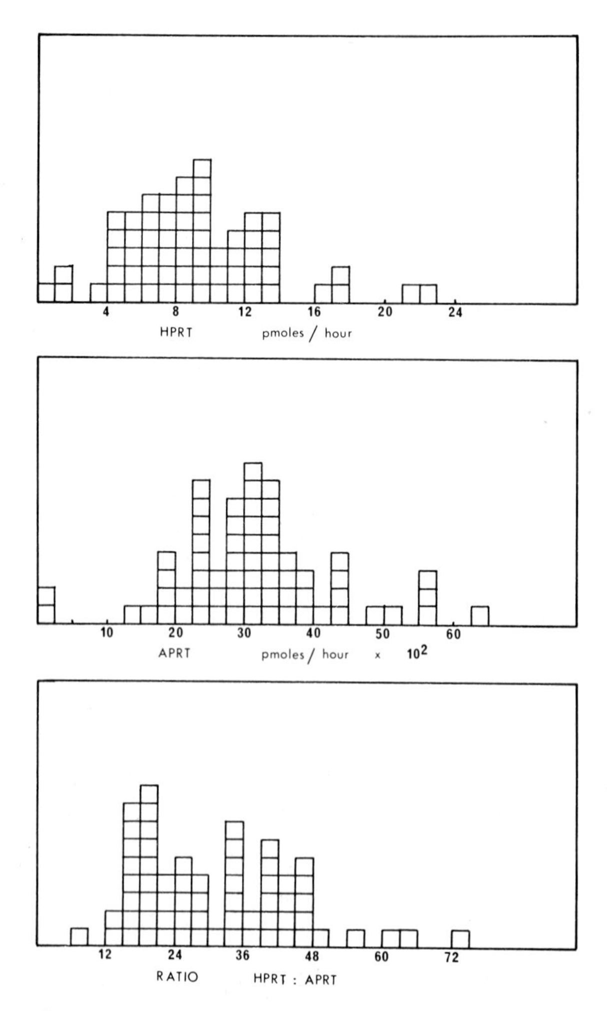

Figure 3: Activities of HPRT, APRT and their ratios in single 9- to 16-cell morulae isolated late on the third day of pregnancy (72 hours post HCG injection). The results from two experiments are pooled. Mean activities (pMoles per hour per embryo ± S.E.), HPRT 9.14 ± 0.54, APRT 0.32 ± 0.02, HPRT/APRT 30.87 ± 1.70 (Monk, unpublished).

activity observed was dependent on the maternal cytoplasm of the egg. Figure 5 compares 8-cell embryos from XX and XO (hatched squares) mothers showing that HPRT activity in XO-derived embryos remains half that in XX-derived embryos. Since there is an overall 20-fold increase in HPRT by the 8-cell stage we suggest that activity in 8-cell embryos is due to stable active maternal messenger RNA. We were able to show that the differences observed were not due to the general

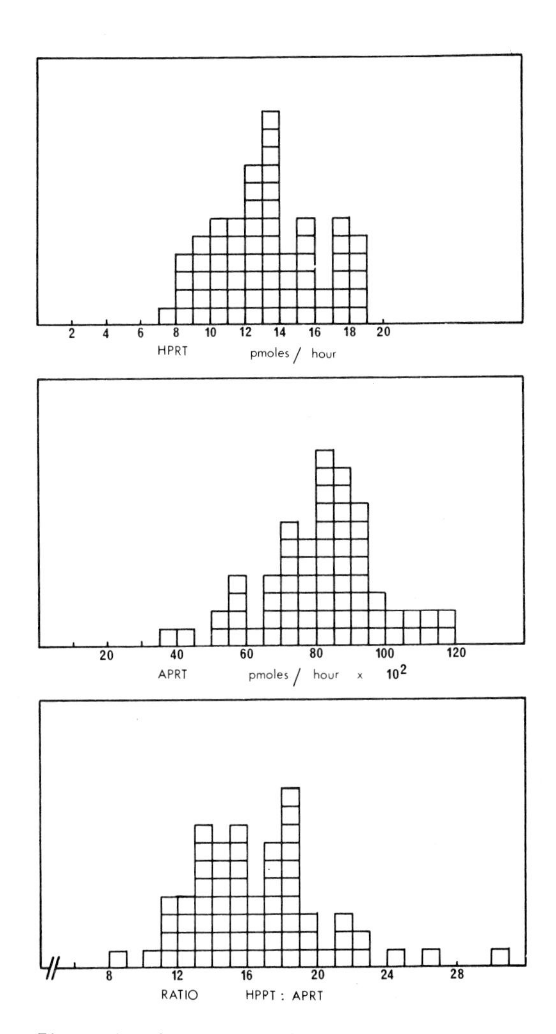

Figure 4: Activities of HPRT, APRT and their ratios in single blastocysts isolated on the fourth day of pregnancy. The results from two experiments are pooled. Mean activities (pMoles per hour per embryo ± S.E.), HPRT 13.27 ± 0.37, APRT 0.82 ± 0.02, HPRT/APRT 16.44 ± 0.48 (Data from Monk and Kathuria, 1977).

developmental lag shown by the XO-derived embryos (Burg ne and Biggers, 1976; Monk and Harper, 1978).

By the 9- to 16-cell stage, however, XX- and XO-derived embryos have equivalent HPRT, APRT and HPRT:APRT ratios showing that embryo-coded activity was predominant by the morula stage on the third day of pregnancy (Figure 6). We have thus demonstrated a transition from maternal- to embryo-coded

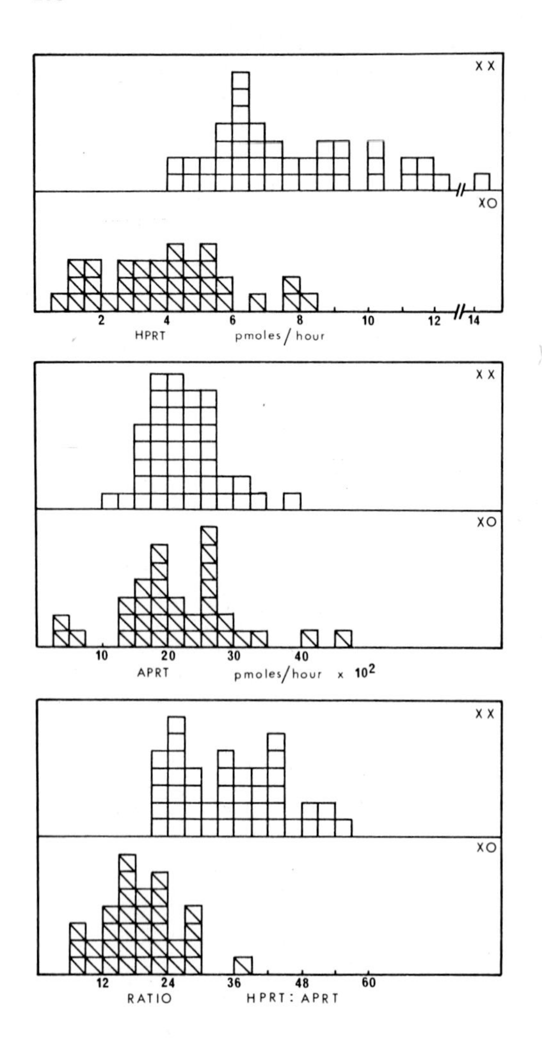

Figure 5: Activities of HPRT, APRT and their ratios in single 8-cell embryos isolated from XX and XO (hatched squares) females on the third day of pregnancy at 64 and 68 hours respectively, after HCG injection. The results from two experiments are pooled. Mean activities (pMoles per hour per embryo ± S.E.), XX embryos: HPRT 7.5 ± 0.4, APRT 0.22 ± 0.01, HPRT/APRT 35.4 ± 1.4; XO embryos: HPRT 4.0 ± 0.3, APRT 0.22 ± 0.01, HPRT/APRT 18.9 ± 1.2 (Data from Monk and Harper, 1978).

activity for an X-linked enzyme between the 8-cell stage and 9- to 16-cell stage of preimplantation development. Many other changes occur at this stage where the equivalence and multipotency of the separate blastomeres is lost as the onset of morphogenesis leads to embryo compaction and formation of

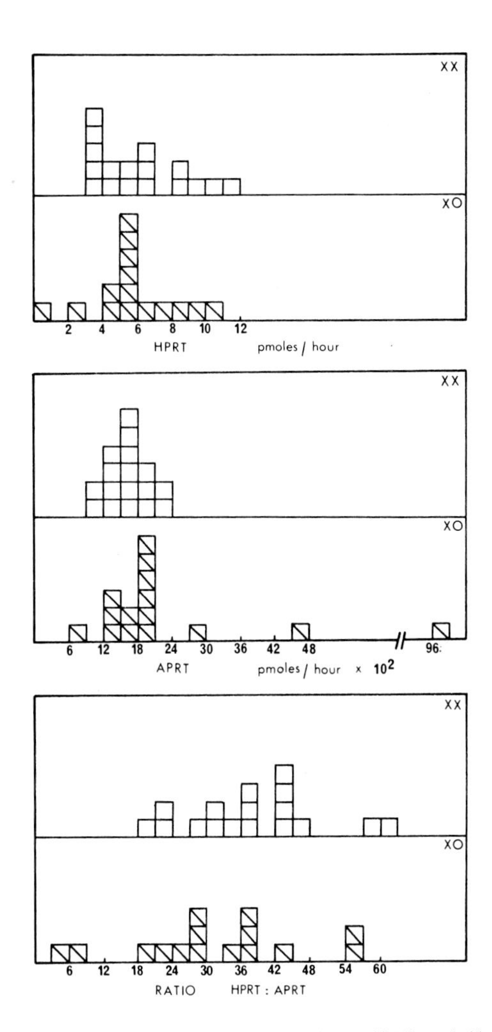

Figure 6: Activities of HPRT, APRT and their ratios in single 9- to 16-cell morulae isolated from XX and XO (hatched squares) females at 70 and 76 hours, respectively, after HCG injection. Mean activities (pMoles per hour per embryo ± S.E.), XX embryos: HPRT 6.2 ± 0.6, APRT 0.16 ± 0.01, HPRT/APRT 37.8 ± 2.9; XO embryos: HPRT 7.1 ± 1.3, APRT 0.25 ± 0.06, HPRT/APRT 32.3 ± 3.3 (Data from Monk and Harper, 1978).

intercellular junctions of the trophoblast (Ducibella, 1977). These transitions may reflect a general shift from maternally-influenced development to embryonic control.

With respect to X-chromosome dosage compensation between XX and XY embryos at the 8-cell stage, we can now say that this is due to HPRT activity which is maternal in origin; hence we can say nothing about the activity status of embryonic X-chromosomes at this stage. In morulae and blastocysts where HPRT activity is embryo-coded, our evidence suggests that both X-chromosomes are active in female morulae and X-inactivation occurs in most, if not all, of the cells at the blastocyst stage.

Gardner and Lyon (1971) have presented evidence that both X-chromosomes are potentially active in single cells of the inner cell mass. In the blastocyst on the fourth day of pregnancy, approximately one quarter of the cells are inner cell mass cells (Horner and McLaren, 1974). Our results could be consistent with the hypothesis that X-inactivation may have occurred in the developmentally determined trophectoderm, though not in the developmentally labile inner cell mass.

PREFERENTIAL EXPRESSION OF THE MATERNAL X-CHROMOSOME

It is now well established in marsupials that the inactive X-chromosome is paternally-derived. A cytologically identifiable paternal X-chromosome was shown to be late replicating (Sharman, 1971). Cooper *et al.*(1971; 1975) analysed variant electrophoretic forms of G6PD and PGK in blood cells of heterozygous females; only the form specified by the maternally inherited X-chromosome was expressed. However, in tissues other than blood, and in cultured fibroblasts, partial expression of the paternal PGK allele was observed. Also in fibroblast cultures some expression of the paternal G6PD, and of the hybrid AB band, was detectable, suggesting that both alleles might be active in the same cell. These findings contrast with human G6PD expression in cultured fibroblasts (Davidson *et al.*, 1963). Could there be partial reactivation of the paternal X-chromosome in some tissues of the kangaroo?

Although the paternal X-chromosome is preferentially inactive in somatic cells of the female kangaroo, it may be transmitted by her to her offspring in active form; inactivation either does not occur in the germ line, or reactivation must occur during oogenesis. However, no evidence for two active X-chromosomes in oocytes could be obtained by Johnston *et al.*(1976). It remains possible that the G6PD patterns analysed by these authors reflect enzyme forms present before reactivation occurred. This could account for the absence of the heteropolymer AB band.

Recently, evidence has been presented that the paternal X-chromosome is preferentially inactive in certain extra-embryonic tissues of the developing mouse conceptus. The evidence is both cytogenetic and biochemical. Takagi and Sasaki (1975) analysed various tissues of female mouse embryos heterozygous for an identifiable X-autosome translocation (Cattanach's translocation) to see which X-chromosome was inactive. In both the reciprocal crosses (either female or male contributing the translocated X-chromosome) they found evidence that the inactive X-chromosome was predominantly paternally derived. This preference was already discernible in cells of the 6.5 day old embryo. In 8.5 day old embryos the inactive paternal X-chromosome could be localised and was shown to be present in the majority of cells of extra-embryonic tissues, the yolk sac (>89 per cent) and chorion (> 95 per cent), whereas amongst embryonic cells approximately equal numbers of cells showed inactive paternal or maternal X-chromosomes. Subsequently, Wake *et al.*(1976) were able to demonstrate a similar situation in the rat. The possibility that the preferential expression of the maternal X-chromosome in extra-embryonic cells resulted from a "switch" of activity from the paternal to the maternal X-chromosome was excluded by an elegant double labelling technique (Takagi, 1976).

Biochemical evidence for non-random inactivation in extra-embryonic tissues has recently been obtained by West *et al.* (1978). Various tissues from female mouse embryos heterozygous for the variant form of PGK were examined electrophoretically and scored for the relative amounts of enzyme specified by maternal and paternal X-chromosomes. In agreement with the cytological data of Takagi and Sasaki (1975) they found that the maternal X-chromosome was preferentially expressed in the mouse yolk sac whereas embryonic tissue showed a more equivalent expression of the enzymes specified by the two X-chromosomes. Further, West *et al.*(1978) were able to demonstrate that the preferential expression of the maternal X-chromosome in the yolk sac was predominantly confined to the endoderm component, where possibly the paternal X-chromosome is not expressed at all; the mesodermal component was more similar to embryonic regions (although still showing some preponderance of the active maternal enzyme form).

In conclusion, in marsupials, and in mouse and rat extra-embryonic tissues, the two X-chromosomes are distinguished at the time of X-chromosome differentiation in such a way that the maternal X-chromosome is preferentially active. In addition, Ropers, Wolff and Hitzeroth (personal communication) have reported preliminary results suggesting preferential expression of the maternal X-chromosome in placenta membranes of newborn girls heterozygous for G6PD variant enzyme forms. Perhaps connected with these observations is evidence for a "vulnerable" state of the paternal X-chromosome in early development; in mice (Russell, 1961), and possibly also in humans (Race, 1965), it appears that the origin of XO females in the majority of cases is due to loss of the paternal X-chromosome in early cleavage.

The following alternative explanations for the preferential expression of the maternal X-chromosome may be considered:
(a) Inactivation is random but some selection process favours cells with the maternal X-chromosome active. Such a selection process cannot be ruled out but is considered to be unlikely for various reasons (Wake *et al.*,1976; West *et al.*,1978). (b) The paternal X-chromosome is preferentially inactivated. (c) The paternal X-chromosome is inactive in development until implantation where its potential activation is restricted to embryo-precursor cells.

These possibilities will be discussed together with the observations on X-chromosome activity in preimplantation embryos in the final section.

X-CHROMOSOME DIFFERENTIATION : AN INTERPRETATION

In a consideration of mechanisms involved in the regulation of X-chromosome activity in early mammalian development, we can delineate two aspects: the event of X-chromosome differentiation itself, and its maintenance. Consider the first event. Is it as favoured by Ohno (1969) and Riggs (1975), random inactivation, or a "switch" mechanism? When does it occur in different tissues? How is the X-chromosome which will be inactive chosen? The following findings must be taken into account:

1. The bimodality shown for 9- to 16-cell morulae for HPRT activity in single mouse embryos (section 5) strongly suggests that both X-chromosomes are active in preimplantation development of female embryos. This assumes that the activity of the HPRT gene reflects the activity status of the X-chromosome. We cannot say whether both X-chromosomes are active at earlier stages since the increase in HPRT activity up to the 8-cell stage is dependent on the maternal cytoplasm of the egg (section 5).
2. The unimodal distribution of HPRT activities at the 3.5 day old blastocyst stage suggests that X-chromosome inactivation has occurred in most, if not all, cells at this stage (section 5). The cytogenetic data shows the earliest signs of inactivation at the late blastocyst stage in the mouse.

3. The suggestion that X-inactivation has occurred by the blastocyst stage appears at first sight to be in conflict with the results of Gardner and Lyon (1971). Their experiments show two potentially active X-chromosomes in individual cells of the inner cell mass of the 3.5 day old blastocyst and suggest inactivation occurs later than this (Lyon, 1972).
4. The choice of which X-chromosome will be inactive is not random in marsupials and in extra-embryonic tissues of the developing mouse and rat; the paternal X-chromosome is preferentially inactive.

The following hypothesis is proposed to account for the above facts. It is suggested that during early development of female mammals, genes on both X-chromosomes may become active. This could be true also of marsupials but has not yet been tested. It is further proposed that the X-chromosome differentiation is an inactivation event, as favoured by West *et al.*,1978, and is linked in some way to a cellular differentiation step; hence it occurs in sequence in tissues as they "depart" from a multipotent stem cell line, first in trophectoderm, then in primary endoderm (the precursor cells of yolk sac endoderm), and finally, in the differentiating definitive germ layers at gastrulation. It is notable that Park, as early as 1957, observed that sex chromatin could be detected earlier in trophoblast than in embryo precursor cells in human and monkey embryos. Superimposed upon this sequential inactivation occurring in differentiating tissues may be some memory mechanism in the X-chromosome recalling its origin, maternal or paternal. Eicher (1970) has reviewed earlier cytological evidence for such a memory mechanism, and models involving "imprinting" of the X-chromosome in the male or female germ line have been proposed (Cooper, 1971; Brown and Chandra, 1973) to link paternal X-inactivation in kangaroos to random inactivation in eutherian mammals (see also Chandra and Brown, 1975). More recent evidence is given in section 6. The simplest memory mechanism may be some structural difference in the paternally inherited X-chromosome, which is retained in early development, so that this X-chromosome is preferentially inactivated in those tissues differentiating early, i.e. the trophectoderm, the primary endoderm, and their derivatives. Then some change, which may be gradual, may occur in the paternal X-chromosome by the time of gastrulation so that it is no longer distinguishable from the maternal X-chromosome and inactivation becomes random. On this hypothesis, one could argue that in kangaroos the memory mechanism inbuilt in the paternal X-chromosome persists until gastrulation so that the paternal X-chromosome is the one chosen to be inactive throughout development. Cooper (1971) has proposed a different model suggesting that random inactivation in eutherian mammals has evolved from paternal X-inactivation in kangaroos. He has suggested that a controlling element introduced into the paternal X-chromosome may, in early development of eutherian mammals, be excised and reinserted at random so as to transfer activity from the maternal to the paternal X-chromosome in approximately 50% of the cells. Such a model is not consistent with the bimodality of the distribution of HPRT activities in single morulae.

Given that inactivation is non-random in certain tissues of the developing mouse, it is not surprising that it is the paternal X-chromosome which is chosen to be inactive. Both maternal X-chromosomes are active during oogenesis whereas the X- and Y-chromosomes are inactive in spermatogenesis (section 3). Although some genes, such as that for HPRT, are expressed on the paternal, as well as the maternal, X-chromosome early in development, it is possible that the paternal X-chromosome does not become fully active (structurally indistinguishable) until the postulated time of inactivation in the embryo precursor cells. It should be borne in mind, however, that the paternal X-chromosome is not cytologically distinguishable in early development and that XO female embryos do quite well with only a paternal X-chromosome.

On the hypothesis that definitive X-chromosome inactivation may not occur in embryo precursor cells until gastrulation, it is suggested that X-inactivation does not occur in the female germ line at all. This would seem attractive from the point of view of continuity in the female germ line, i.e. the maintenance of two active X-chromosomes in the germ line, through early development, to the germ line again. However, it has been claimed (Ohno, 1964; see section 3) that cytological evidence exists for a cycle of inactivation and reactivation, and one could equally well argue for a requirement for X-chromosome dosage equivalence in the germ cells *en route* to the genital ridges.

In this discussion, we have been concerned with the choice of which X-chromosome is to be inactive and when the choice is made. What is the mechanism of inactivation and how is it maintained? These questions have been considered at length (e.g. Ohno, 1969;

Eicher, 1970; Lyon, 1971; 1972; 1974; Cattanach, 1975; Holliday and Pugh, 1975; Riggs, 1975; Gartler and Andina, 1976) and will not be reviewed fully here. There are two aspects, however, that it may be profitable to consider further at this stage - the relationship between the number of active X-chromosomes to autosome sets and the number of potential sites (one or more?) along the X-chromosome regulating its activity. Brown and Chandra (1973) have suggested that a "quantum" of activating information, provided by the set of maternal autosomes, randomly activates the maternal or paternal X-chromosome at some later period of development. However, this model fails to account for cases of individuals with multiple X-chromosomes where the number of active X-chromosomes is not always correlated with their origin. Nor does it account for the inactive X-chromosome in diploid parthenogenote female teratoma cells (Linder, 1969) and in diploid parthenogenote embryos (Kaufman *et al.*,1978). Given a lack of consistent evidence for autosomal control of X-chromosome activity perhaps the simplest model is one where the X-chromosome itself monitors its numbers in the cell as well as cell ploidy. We will return to this point later.

There is little information concerning the number of sites on the X-chromosome controlling its activity. Russell (1963), from her studies of variegation produced by X-autosomal translocations, interpreted the "single active-X" hypothesis in terms of inactivation spreading from a site, the inactivation centre, in such a way that inactivation could spread from the inactive X-chromosome to variable extents into attached autosomal material in a polarised fashion (see Cattanach, 1970; 1974; 1975). Eicher (1970) suggested that inactivity of the X-chromosome is controlled at a number of sites along its length. It seems that both situations may be operative - an inactivation centre may be initially involved in setting up regulation of inactivity at a number of sites on the X-chromosome. Certainly a single inactivation centre by itself is a very loose control mechanism; the inactivation process would presumably be "open" to mutation along the entire length of the X-chromosome. In bacteriophage (Hershey, 1971), regulation of the developmental pathway, or of the maintenance of the quiescent phase of the virus, is dependent on "cascade" effects which rigidly ensure the irreversibility of determination of the chosen state. Regulatory molecules act at a number of sites on the viral DNA to turn on, or turn off, blocks of genes which themselves have regulatory functions. The elegance of such a cascade system is the tight control operative once a decision is taken.

For some time the irreversibility of X-chromosome inactivation was unassailable. Comings (1966) failed to reactivate the inactive X-chromosome with a wide range of treatments. Migeon (1972) also failed to reactivate the inactive X-chromosome using selective media on interspecific mouse-human cell hybrids: in more than 10^7 cells plated to select for the HPRT allele on the inactive X-chromosome not one resistant colony was found. Kahan and De Mars (1975) using a similar approach to that of Migeon have succeeded in selecting rare (10^{-6}) localised de-repression of an X-chromosomal gene in mouse-human cell hybrids. The human female cells possessed genetically and morphologically distinctive active and inactive X-chromosomes. Selection was made for hybrid cells expressing an allele ($HPRT^+$) present only on the inactive human X-chromosome (the active X-chromosome was $HPRT^-$). The recovery of $HPRT^+$ cells was associated with a loss of the previously active X-chromosome. The neighbouring G6PD and PGK genes on either side of the derepressed HPRT gene on the inactive X-chromosome were not expressed. Derepression was limited therefore to a block of genes including the HPRT gene, and Kahan and De Mars concluded that there were at least three repressible "blocks" of genes on the inactive X-chromosome. The localised derepression was not associated with the absence of any specific autosomes, nor was the continued repression of the G6PD and PGK genes dependent on the retention of the active X-chromosome. However, this latter fact does not preclude the possibility that the active X-chromosome was responsible for the initiation of inactivity at a number of sites on the inactive X-chromosome. The localised derepression observed may, in fact, be the result of a two step process - a temporary insensitivity of a site regulating the block of genes including the HPRT gene, and an inability to restore inactivity due to the loss of the active X-chromosome. Such a model assumes that the inactivation is due to the production of a regulatory molecule by the active X-chromosome which acts in trans at a number of sites on the inactive X-chromosome. Once initiated, inactivation may be maintained by the inactive X-chromosome itself. The model requires the equivalent sites on the active X-chromosome to be insensitive to the proposed regulatory molecule, perhaps because of some configurational difference. Non-random inactivation, when it occurs, may be due to a configurational difference at the time X-chromosome differentiation occurs. Random inactivation could involve a "flip-flop" event depending solely on a difference in timing of the functioning of the gene coding for the initial regulatory molecule.

In conclusion, it is proposed that the X-chromosome itself monitors its numbers in the cell by a two step regulatory process, initiation and maintenance, involving some regulatory function provided by the active X-chromosome and resulting in autogeneous repression exerted at a number of sites on the inactive X-chromosome. Feedback regulatory mechanisms could be operative in both situations.

ACKNOWLEDGEMENTS

I thank Anne McLaren, Julian Gross and Mary Harper for their friendly encouragement, Mary Harper for preparing the figures, and Hasu Kathuria and Mary Harper for technical assistance.

REFERENCES

ADLER, D.A., WEST, J.D. & CHAPMAN, V.M. (1977) Expression of α-galactosidase in preimplantation mouse embryos. Nature, Lond. 267, 838-839.

BACKMAN, K., PTASHNE, M. & GILBERT, W. (1976) Construction of plasmids carrying the C1 gene of bacteriophage λ. Proc. natn. Acad. Sci. USA 73, 4174-4178.

BEUTLER, E., YEH, M. & FAIRBANKS, V.F. (1962) The normal human female as a mosaic of X-chromosome activity: studies using the gene for G-6PD-deficiency as a marker. Proc. Nat. Acad. Sci. 48, 9-16.

BROWN, S.W. & CHANDRA, H.S. (1973) Inactivation system of the mammalian X-chromosome. Proc. natn. Acad. Sci. USA 70, 195-199.

BURGOYNE, P.S. & BIGGERS, J.D. (1976) The consequences of X-dosage deficiency in the germ line: impaired development *in vitro* of preimplantation embryos from XO mice. Devl. Biol. 51, 109-117.

CATTANACH, B.M. (1962) XO mice. Genet. Res. 3, 487-490.

CATTANACH, B.M. (1970) Controlling elements in the mouse X-chromosome. III. Influence upon both parts of an X divided by rearrangement. Genet. Res. 16, 293-301.

CATTANACH, B.M. (1974) Position effect variegation in the mouse. Genet. Res. 23, 291-306.

CATTANACH, B.M. (1975) Control of chromosome inactivation. Ann. Rev. Genet. 9, 1-18.

CHANDRA, H.S. & BROWN, S.W. (1975) Chromosome imprinting and the mammalian X-chromosome. Nature, Lond. 253, 165-168.

CHAPMAN, V.M. & SHOWS, T.B. (1976) Somatic cell genetic evidence for the X-chromosome linkage of glucose-6-phosphate dehydrogenase, phosphoglycerate kinase and hypoxanthine-phospho-ribosyl transferase. Nature, Lond. 259, 665-667.

CHAPMAN, V.M., WEST, J.D. & ADLER, D.A. (1977) Genetics of early mammalian embryogenesis. In: Sherman, M.I. (Ed.), Concepts in mammalian embryogenesis, MIT Press, Massachusetts and London, pp.95-135.

CHEN, S-H., MALCOLM, L.A., YOSHIDA, A. & GIBLETT, E.R. (1971) Phosphoglycerate kinase: an X-linked polymorphism in man. Am. J. Hum. Genet. 23, 87-91.

COMINGS, D.E. (1966) The inactive X-chromosome. Lancet ii, 1137-38.

COOPER, D.W. (1971) Directed genetic change model for X-chromosome inactivation in eutherian mammals. Nature, Lond. 230, 292-294.

COOPER, D.W., JOHNSTON, P.G., MURTAGH, C.E., SHARMAN, G.B., VANDEBERG, J.L. & POOLE, W.E. (1975) Sex-linked isozymes and sex chromosome evolution and inactivation in kangaroos. In: Markert, L. (Ed.), Isozymes, III. Developmental Biology, Academic, New York, pp. 559-573.

COOPER, D.W., VANDEBERG, J.L., SHARMAN, G.B. & POOLE, W.E. (1971) Phosphoglycerate kinase polymorphism in kangaroos provides further evidence for paternal X-inactivation. Nature (New Biol) 230, 155-157.

DAMJANOV, I. & SOLTER, D. (1974) Experimental teratoma. Curr. Topics Pathol. 59, 69-130.

DAVIDSON, R.G., NITOWSKY, H.M. & CHILDS, B. (1963) Demonstration of two populations of cells in the human female heterozygous for glucose-6-phosphate dehydrogenase variants. Proc. natn. Acad. Sci. USA 50, 481-485.

DeMARS, R., LEVAN, S.L., TREND, B.L. & RUSSELL, L.B. (1976) Abnormal ornithine carbamoyltransferase in mice having the sparse-fur mutation. Proc. natn. Acad. Sci. USA 73, 1693-1697.

DUCIBELLA, T. (1977) Surface changes of the developing trophoblast cell. In: Johnson, M.H. (Ed.), Development in Mammals, Vol. I, North Holland Publishing Company, pp.5-30.

EICHER, E.M. (1970) X autosome translocations in the mouse: total inactivation versus partial inactivation of the X-chromosome. Adv. Genet. 15, 175-259.

EPSTEIN, C.J. (1969) Mammalian oocytes: X-chromosome activity. Science 163, 1078-1079.

EPSTEIN, C.J. (1970) Phosphoribosyltransferase activity during early mammalian development. J. biol. Chem. 245, 3289-3294.

EPSTEIN, C.J. (1972) Expression of the mammalian X-chromosome before and after fertilisation. Science 175, 1467-1468.

EVANS, H.J., FORD, C.E., LYON, M.F. & GRAY, J. (1965) DNA replication and genetic expression in female mice with morphologically distinguishable X-chromosomes. Nature (Lond.) 206, 900-903.

GALTON, M. & HOLT, S.F. (1965) Asynchronous replication of the mouse sex chromosomes. Exp. Cell Res. 37, 111-116.

GARDNER, R.L. & LYON, M. (1971) X-chromosome inactivation studied by injection of a single cell into the mouse blastocyst. Nature 231, 385-386.

GARTLER, S.M. & ANDINA, R.J. (1976) Mammalian X-chromosome inactivation. Adv. Hum. Genet. 7, 99-140.

GARTLER, S.M., ANDINA, R. & GANT, N. (1975) Ontogeny of X-chromosome inactivation in the female germ line. Exp. Cell Res. 91, 454-457.

GARTLER, S.M. & BURT, B. (1964) Replication patterns of bovine sex chromosomes in cell culture. Cytogenetics 3, 135-142.

GARTLER, S.M., CHEN, S-H., FIALKOW, P.J. & GIBLETT, E.R. (1972a) X-chromosome inactivation in cells from an individual heterozygous for two X-linked genes. Nature (New Biol). 236, 149-150.

GARTLER, S.M., LISKAY, R.M., CAMPBELL, B.K., SPARKS, R. & GANT, H. (1972b) Evidence for two functional X-chromosomes in human oocytes. Cell Differentiat. 1, 215-218.

GARTLER, S.M., LISKAY, R.M. & GANT, N. (1973) Two functional X-chromosomes in human fetal oocytes. Exptl. Cell Res. 82, 464-466.

GOLD, L., O'FARRELL, P.Z. & RUSSELL, M. (1976) Regulation of gene 32 expression during bacteriophage T4 infection of *Escherichia coli*. J. Biol. Chem. 251, 7251-7262.

GOLDBERGER, R.F. (1974) Autogenous regulation of gene expression. Science 183, 810-816.

GRZESCHIK, K.H., GRZESCHIK, A.M., BANHOF, S., ROMEO, G., SINISCALCO, M., Van SOMEREN, H., MEERA KHAN, P., WESTERVELD, A., & BOOTSMA, D. (1972) X-linkage of human α-galactosidase. Nature (New Biol). 240, 48-50.

GUIALIS, A., BEATTY, B.G., INGLES, C.J. & CRERAR, M.M. (1977) Regulation of RNA polymerase II activity in α-amanitin-resistant CHO hybrid cells. Cell 10, 53-60.

HERSHEY, A.D. (Editor) (1971) The Bacteriophage Lambda. Cold Spring Harbour.

HOLLIDAY, R. & PUGH, J.E. (1975) DNA modification mechanisms and gene activity during development. Science 187, 226-232.

HORNER, D. & McLAREN, A. (1974) The effect of low concentrations of (^{3}H) thymidine on pre- and post-implantation mouse embryos. Biol. Reprod. 11, 553-557.

ISSA, M., BLANK, C.E. & ATHERTON, G.W. (1969) The temporal appearance of sex chromatin and of the late-replicating X-chromosome in blastocysts of the domestic rabbit. Cytogenetics (Basel) 8, 219-237.

JOHNSTON, P.G., ROBINSON, E.S. & SHARMAN, G.B. (1976) X-chromosome activity in oocytes of kangaroo pouch young. Nature, Lond. 264, 359-360.

KAHAN, B. & DeMARS, R. (1975) Localised derepression on the human inactive X-chromosome in mouse-human cell hybrids. Proc. natn. Acad. Sci. USA 72, 1510-1514.

KAUFMAN, M.H. (1972) Non-random segregation during mammalian oogenesis. Nature (Lond.) 238, 465-466.

KAUFMAN, M.H., GUC-CUBRILO, M. & LYON, M.F. (1978) X-chromosome inactivation in diploid parthenogenetic mouse embryos. Nature, Lond. 271, 547-549.

KINT, J.A. (1970) Fabry's disease - α-Galactosidase deficiency. Science 167, 1268-1269.

KOZAK, L.P., McLEAN, G.K. & EICHER, E.M. (1974) X-linkage of phospho-glycerate kinase in the mouse. Biochem. Genet. 11, 41-47.

KOZAK, C., NICHOLS, E., & RUDDLE, F.H. (1975) Gene linkage analysis by somatic cell hybridisation : assignment of adenine phosphoribosyl transferase to chromosome 8 and α-galactosidase to the X-chromosome. Somatic Cell Genet. 1, 371-382.

KOZAK, L.P. & QUINN, P.J. (1975) Evidence for dosage compensation of an X-linked gene in the 6-day embryo of the mouse. Devl. Biol. 45, 65-73.

LIFSCHYTZ, E. & LINDSLEY, D.L. (1972) The role of X-chromosome inactivation during spermatogenesis. Proc. natn. Acad. Sci. USA 69, 182-186.

LINDER, D. (1969) Gene loss in human teratomas. Proc. natn. Acad. Sci. USA 63, 699-704.

LUSIS, A.J. & WEST, J.D. (1976) X-linked inheritance of a structural gene for α-galactosidase in *Mus musculus*. Biochem. Genet. 14, 849-855.

LUTHARDT, F.W. (1976) Cytogenetic analysis of oocytes and early preimplantation embryos from XO mice. Devl. Biol. 34, 73-81.

LYON, M.F. (1961) Gene action in the X-chromosome of the mouse *(Mus musculus L)*. Nature, Lond. 190, 372-373.

LYON, M.F. (1968) Chromosomal and subchromosomal inactivation. Ann. Rev. Genet. 2, 31-52.

LYON, M.F. (1972) X-chromosome inactivation and developmental patterns in mammals. Biol. Rev. 47, 1-35.

LYON, M.F. (1974) Mechanisms and evolutionary origins of variable X-chromosome activity in mammals. Proc. Roy. Soc. Lond. B. 187, 243-268.

LYON, M.F. & HAWKER, S.G. (1973) Reproductive lifespan in irradiated and unirradiated chromosomally XO mice. Genet. Res. Camb. 21, 185-194.

LYON, J.B. Jr., PORTER, J. & ROBERTSON, M. (1967) Phosphorylase b kinase inheritance in mice. Science 155, 1550-1551.

MAGENIS, R.E., KOLER, R.D., LOVRIEN, E., BIGLEY, R.H., DUVAL, M.C. & OVERTAN, K.M. (1975) Gene dosage : evidence for assignment of erythrocyte acid phosphatase locus to chromosome 2. Proc. natn. Acad. Sci. USA 72, 4526-4530.

MARIMO, B. & GIANELLI, F. (1975) Gene dosage effect in human trisomy 16. Nature 256, 204-206.

MARKS, P.A. & GROSS, R.T. (1959) Erythrocyte glucose-6-phosphate dehydrogenase deficiency: evidence of differences between Negroes and Caucasians with respect to this genetically determined trait. J. Clin. Invest. 38, 2253-2262.

MARTIN, G.R. (1975) Teratocarcinomas as a model system for the study of embryogenesis and neoplasia. Cell 5, 229-243.

MARTIN, G.R., EPSTEIN, C.J., TRAVIS, B., TUCKER, G., YATZIV, S., MARTIN, D.W., CLIFT, S. & COHEN, S. (1978) X-chromosome inactivation during differentiation of female teratocarcinoma stem cells *in vitro*. Nature, Lond. 271, 329-333.

MARTIN, G.R. & EVANS, M.J. (1975) Differentiation of clonal lines of teratocarcinoma cells: formation of embryoid bodies *in vitro*. Proc. natn. Acad. Sci. USA 72, 1441-1445.

McBURNEY, M.W. & ADAMSON, E.D. (1976) Studies on the activity of the X-chromosome in female teratocarcinoma cells in culture. Cell 9, 57-70.

McLAREN, A. (1972) Late labelling as an aid to chromosomal sexing of cultured mouse blood cells. Cytogenetics 11, 35-45.

MIGEON, B.R. (1972) Stability of X-chromosomal inactivation in human somatic hybrids. Nature, Lond. 239, 87-89.

MIGEON, B.R. & JELALIAN, K. (1977) Evidence for two active X-chromosomes in germ cells of female before meiotic entry. Nature, Lond. 269, 242-243.

MONK, M. & HARPER, M. (1978) Evidence for translation of maternal messenger RNA during cleavage in mouse embryos. J. exp. Embryl. Morph. (In press).

MONK, M. & KATHURIA, H. (1977) Dosage compensation for an X-linked gene in pre-implantation mouse embryos. Nature, Lond. 270, 599-601.

MORISHIMA, A., GRUMBACH, M. & TAYLOR, J.H. (1962) Asynchronous duplication of human chromosomes and the origin of sex chromatin. Proc. natn. Acad. Sci. USA 48, 756-763.

MORRIS, T. (1968) The XO and OY chromosome constitutions in the mouse. Genet. Res. 12, 125-137.

MUKHERJEE, A.B. (1976) Cell cycle analysis and X-chromosome inactivation in the developing mouse. Proc. natn. Acad. Sci. USA 73, 1608-1611.

MUKHERJEE, B.B. & SINHA, A.K. (1964) Single-active-X-hypothesis: cytological evidence for random inactivation of X-chromosomes in a female mule complement. Proc. natn. Acad. Sci. USA 51, 252-259.

NESBITT, M.N. (1971) X-chromosome inactivation mosaicism in the mouse. Devl. Biol. 26, 252-263.

NESBITT, M.N. & GARTLER, S.M. (1970) Replication of the mouse sex chromosomes early in the S period. Cytogenetics 9, 212-221.

NESBITT, M.N. & GARTLER, S.M. (1971) The applications of genetic mosaicism to developmental problems. Ann. Rev. Genet. 5, 143-162.

NIELSEN, J.T. & CHAPMAN, V.M. (1978) Electrophoretic variation for sex-linked phospho-glycerate kinase (PGK-1) in the mouse. Genetics. (In press).

ODARTCHENKO, N. & PAVILLARD, M. (1970) Late DNA replication in male mouse meiotic chromosomes. Science 167, 1133-1134.

OHNO, S. (1964) Life history of female germ cells in mammals. 2nd Intern. Conf. Congenital Malformations, 1963, pp. 36-40. Intern. Med. Congr. Ltd., New York.

OHNO, S. (1969) Evolution of sex chromosomes in mammals. Ann. Rev. Genet. 3, 495-524.

OHNO, S. (1976) Major regulatory genes for mammalian sexual development. Cell 7, 315-321.

OHNO, S., KAPLAN, W.D. & KINOSITA, R. (1961) X-chromosome behaviour in germ and somatic cells of *Rattus norvegicus*. Exp. Cell Res. 22, 535-544.

OHNO, S., KLINGER, H.P. & ATKIN, N.B. (1962) Human oogenesis. Cytogenetics 1, 42-51.

PARK, W.W. (1957) The occurrence of sex chromatin in early human and macaque embryos. J. Anat. 91, 369-373.

PIERCE, G.B. & BEALS, T.F. (1964) The ultrastructure of primordial germinal cells and fetal testes and of embryonal carcinoma cells in mice. Cancer Res. 24, 1553-1567.

RACE, R.R. (1965) Identification of the origin of the X-chromosome(s) in sex chromosome aneuploidy. Can. J. Genet. Cytol. 7, 214-222.

RATAZZI, M.C. & COHEN, M.M. (1972) Further proof of genetic inactivation of the X-chromosome in the female mule. Nature, Lond. 237, 393-396.

RAY, M., GEE, P.A., RICHARDSON, B.J. & HAMERTON, J.L. (1972) G6PD expression and X-chromosome late replication in fibroblast clones from a female mule. Nature, Lond. 237, 396-397.

REED, S.I., STARK, G.R. & ALWINE, J.C. (1976) Autoregulation of Simian Virus 40 gene A by T antigen. Proc. natn. Acad. Sci. USA 73, 3083-3087.

RICCIUTI, F.C. & RUDDLE, F.H. (1973) Assignment of nucleoside phosphorylase to D-14 and localisation of X-linked loci in man by somatic cell genetics. Nature, Lond. 241, 180-182.

RIGGS, A. (1975) X-inactivation, differentiation and DNA methylation. Cytogenet. Cell Genet. 14, 9-25.

RUSSELL, L.B. (1961) Genetics of mammalian sex chromosomes. Science 133, 1795-1803.

RUSSELL, L.B. (1963) Mammalian X-chromosome action: inactivation limited in spread and in region of origin. Science 140, 976-978.

RUSSEL, M., GOLD, L., MORRISSETT, H. & O'FARRELL, P.Z. (1976) Translation, autogenous regulation of gene 32 expression during bacteriophage T4 infection. J. Biol. Chem. 251, 7263-7270.

SCAIFE, J. (1976) Bacterial RNA polymerases: the genetics and control of their synthesis. In: Losick, R. & Chamberlin, M. (Eds.) Bacterial RNA Polymerase. Cold Spr. Harb. Symp. Quant, Biol.

SEEGMILLER, J.E., ROSENBLOOM, F.H. & KELLEY, W.N. (1967) Enzyme defect associated with a sex linked neurological disorder and excessive purine synthesis. Science 155, 1682-1683.

SHARMAN, G.B. (1971) Late DNA replication in the paternally derived X-chromosome of female kangaroos. Nature, Lond. 230, 231-232.

SHOWS, T.B. & BROWN, J.A. (1975) Human X-linked genes regionally mapped utilising X-autosome translocations and somatic cell hybrids. Proc. natn. Acad. Sci. USA 72, 2125-2129.

STEELE, M.W. & OWENS, K.E. (1973) Developmental, tissue-specific, and sex differences in activity among three enzymes from human erythrocytes and cultured fibroblasts. Biochem. Genet. 9, 147-162.

STEELE, M.W. & MIGEON, B.R. (1973) Sex differences in activity of G6PD from cultured human fetal cells despite X-inactivation. Biochem. Genet. 9, 163-168.

STEVENS, L.C. (1970) The development of transplantable teratocarcinomas from intratesticular grafts of pre- and post-implantation mouse embryos. Devl. Biol. 21, 364-382.

STEVENS, L.C. & VARNUM, D.S. (1974) The development of teratomas from parthenogenetically activated ovarian mouse eggs. Devl. Biol. 37, 369-380.

TAKAGI, N. (1974) Differentiation of X-chromosomes in early female mouse embryos. Exp. Cell Res. 86, 127-135.

TAKAGI, N. (1976) Stability of X-chromosome differentiation in mouse embryos. Hum. Genet. 34, 207-211.

TAKAGI, N. & OSHIMURA, M. (1973) Fluorescence and Giemsa banding studies of the allocyclic X-chromosome in embryonic and adult mouse cells. Exp. Cell Res. 78, 127-135.

TAKAGI, N. & SASAKI, M. (1975) Preferential inactivation of the paternally derived X-chromosome in the extraembryonic membranes of the mouse. Nature, Lond. 256, 640-642.

VALENTINE, W.N., HSIEH, H.-S., PAGLIA, D.E., ANDERSON, H.M. BAUGHMAN, M.A., JAFFE, E.R. & GARSON, O.M. (1968) Hereditary hemolytic anemia: association with phosphoglycerate kinase deficiency in erythrocytes and leukocytes. Trans. Ass. Am. Physns. 81, 49-65.

VANHA-PERTTULA, T., BARDIN, C.W., ALLISON, J.E., GUMBRECK, L.G. & STANLEY, A.J. (1970) "Testicular feminisation" in the rat: morphology of the testis. Endocrinology 87, 611-619.

WAKE, N., TAKAGI, N. & SASAKI, M. (1976) Non-random inactivation of X-chromosome in the rat yolk sac. Nature, Lond. 262, 580-581.

WEST, J.D., FRELS, W.I., CHAPMAN, V.M. & PAPAIOANNOU, V.E. (1978) Preferential expression of the maternally derived X-chromosome in the mouse yolk sac. Cell. (In press).

WESTERVELD, A., VISSER, R.P.L.S., FREEKE, M.A. & BOOTSMA, D. (1972) Evidences for linkage of 3-phosphoglycerate kinase, hypoxanthine-guanine-phosphoribosyl transferase, and glucose-6-phosphate dehydrogenase loci in Chinese hamster cells studied by using a relationship between gene multiplicity and enzyme activity. Biochem. Genet. 7, 33-40.

ADVANTAGES AND LIMITATIONS OF TERATOCARCINOMA STEM CELLS AS MODELS OF DEVELOPMENT

Gail R. Martin

Department of Anatomy and Cancer Research Institute
University of California
San Franciso, California, 94143

In recent years the concept of using malignant mouse teratocarcinoma cells as models of normal embryonic development has excited considerable interest. Oddly enough, this has perhaps been less true for developmental biologists than for other scientists, who have now found that they may never need to see an embryo to study embryogenesis. The possibility of creating a normal mouse from a teratocarcinoma stem cell with a flick of the micromanipulator has stimulated even greater interest in these tumor cells.

As with any new tool, the extent to which these cells will actually be useful is not yet fully understood, and may even have been overestimated. Part of the problem is that, beyond the fact that the tumor stem cells, like early embryonic cells, are multipotent, we do not really know what specific embryonic cells they resemble. It is therefore difficult to fully appreciate which particular aspects of embryogenesis we can use them to study. However, teratocarcinoma stem cells do offer the advantage of being available in almost unlimited quantities. In addition, they are able to withstand the rigors of *in vitro* cloning, culture and manipulation in ways that normal multipotent embryonic cells are not. This chapter is not intended as a comprehensive survey of the literature. Rather, the main objective is to review the data that reflect on the similarity between teratocarcinoma cells and the cells of the early embryo and to try to assess the extent to which these cells are useful as a model for the study of normal development. Consequently, many interesting studies with teratocarcinoma cells are not discussed. Some of these are covered in recent reviews by Graham (1977) and by Hogan (1977).

ORIGIN OF TERATOCARCINOMA STEM CELLS

Mouse teratocarcinomas are malignant tumors that are characterized by their content of a variety of differentiated cell types including derivatives of all three primary germ layers, and a distinctive cell type known as embryonal carcinoma. The latter are the stem cells of the tumor, and a single embryonal carcinoma cell can give rise to all the differentiated cell types that are observed in the tumors (Kleinsmith and Pierce, 1964). It is the presence of these multipotent undifferentiated embryonal carcinoma cells that is responsible for the malignancy (i.e. progressive growth and transplantability) of these tumors (reviewed by Pierce, 1967; Stevens, 1967; Damjanov and Solter, 1974a). If no embryonal carcinoma cells are present, because they have either differentiated or died, the tumors are benign and are known as teratomas. This latter term, however, is often used to designate both the malignant and benign types of tumor.

Historically there has been a controversy about the origin of these tumors. According to one theory, they arise from multipotent embryonic cells that have escaped the influence of the embryonic organizer. According to the other, they originate from germ cells (see also Heath, this volume). The data suggest that both are correct: teratocarcinomas and teratomas can be produced by grafting a normal embryo to an ectopic site, or they can arise from male primordial germ cells. In discussing this evidence I have, for the sake of brevity, omitted reference to many interesting observations on the genetic and environmental factors that affect the development of these tumors. For more information, readers should consult one of the excellent reviews on the subject of teratocarcinogenesis, such as the one by Stevens (1967) or by Solter *et al.*(1975).

EMBRYO-DERIVED TERATOCARCINOMAS

Teratocarcinomas and teratomas are readily induced by transplanting a normal embryo to an extra-uterine site. The type of tumor which arises, however, is a function of the age of the embryo. Thus, when mouse embryos of 8 or more days of development are transferred to extra-uterine sites they give rise only to teratomas (Damjanov *et al.*,1971). It is thought that this is because, with the exception of germ cells, there are no longer multipotent stem cells present in the embryo after the appearance of the organ primordia (at approximately 7.5 days of development). In contrast, teratocarcinomas containing stem cells can be readily induced experimentally by extra-uterine transplantation of younger embryos (i.e. any stage from the 2-cell embryo up to approximately 7.5 days of development; Stevens, 1968, 1970; Solter *et al.*,1970). This is one indication of the close relationship between embryonal carcinoma cells and the multipotent cells of the normal early embryo.

Experiments involving the grafting of parts of 6-7 day embryos to extra-uterine sites have been carried out to determine which particular cells of the embryo can give rise to teratocarcinomas. Thus, Diwan and Stevens (1976) have shown that the isolated embryonic ectoderm of the 6-day mouse embryo can form a teratocarcinoma and is therefore multipotent, but that the isolated embryonic endoderm is not. Solter and Damjanov (1973) have further demonstrated that the extraembryonic portion of the 7-day mouse egg cylinder apparently lacks the capacity to form a teratocarcinoma and therefore is presumably also not multipotent.

While multipotency of the embryonic cells is necessary for the induction of teratocarcinomas, other factors are also involved. One of these is the genetic background of the host animal, which apparently plays an important role in determining whether an embryo will form a teratoma or a teratocarcinoma. Thus, Damjanov and Solter (1974b; Solter *et al.*,1975) have found that when early embryos of some strains (e.g. C57B1 and AKR) are transplanted to ectopic sites they give rise primarily to teratomas, while in other strains (C3H, CBA and A) teratomas and teratocarcinomas containing stem cells arise in almost equal proportions. However, those embryos which normally give rise to only teratomas in isogeneic hosts (e.g. C57B1 embryo transplanted to a C57B1 host) can give rise to both kinds of tumor when implanted in a histo-compatible host derived from a strain which is permissive for the development of teratocarcinomas (e.g. C57B1 embryo transplanted to a C57B1 x C3H F_1-hybrid host). Although there is as yet no explanation of these data (reviewed by Solter *et al.*,1975), from a practical point of view they indicate that it should be possible to initiate tumors containing stem cells in many strains of mice by choosing the appropriate host.

The embryo which is used to induce a teratocarcinoma or teratoma need not have developed from an egg fertilized by a sperm. Parthenotes, which originate from eggs that are ovulated and begin development without being fertilized, can also be used (Iles *et al.*, 1975).

Embryo-derived tumors can also be obtained spontaneously. In 1974, Stevens and Varnum reported that in the LT strain of mice approximately 50% of all sexually mature females have spontaneous ovarian teratocarcinomas or teratomas. In other mouse strains, such tumors are exceedingly rare. Some of the tumors in the LT mice were found to contain undifferentiated stem cells. The way in which these spontaneous ovarian tumors are formed appears to be analogous to the process of experimental induction by transfer of an embryo to an extra-uterine site. In the case of the spontaneous LT tumors,this site is the ovary. Following parthenogenetic activation of oocytes that have completed the first meiotic division, the haploid eggs undergo a brief period of normal embryonic development (to a stage approximately equivalent to that of the normal 6.5-day embryo) and subsequently continue their growth as a disorganised mass in the ovary (Stevens and Varnum, 1974; Stevens, 1975; Eppig *et al.*,1977). Such spontaneous teratomas and teratocarcinomas might be haploid, but diploidization is possible if there is suppression of the formation of the second polar body or fusion of the second polar body with the ovum after activation. Preliminary data (Eppig *et al.*,1977) suggest that most of the cells of LT tumors are in fact diploid.

MALE GERM CELL-DERIVED TERATOCARCINOMAS

Over the years, Stevens and his co-workers have studied teratocarcinogenesis in male mice (reviewed by Stevens, 1967). In brief, they have found that the primordial germ cells of male strain 129 embryos (and also to some extent A strain embryos) have an unusual propensity for abnormal proliferation. This pro cess, which is somewhat akin to parthenogenic activation of ova, results in the formation of both teratomas and teratocarcinomas. Unlike the ovarian oocytes in the LT mice, tumor formation by these male germ cells apparently does not involve histologically normal embryonic development. Stevens (1975) does point out, however, that one of the earliest stages in the development of these male germ cell tumors is the formation of structures which resemble the isolated embryonic ectoderm. Tumors formed in this way are virtually indistinguishable from the embryo-derived ones. The various ways in which teratocarcinomas and teratomas are obtained are summarized in Table 1.

The observation that teratocarcinomas can arise directly from male germ cells raises the question of whether the progenitors of these tumors are different from the cells that give rise to embryo-derived tumors. There are two hypotheses to explain how one particular cell type could give rise to both germ cell-derived and embryo-derived teratocarcinomas. The first is that an embryo-derived teratocarcinoma can only be obtained if the embryo develops to the point where it contains germ cells *per se*. Although there is as yet no conclusive proof, it is believed that this is not the case (for a review of the evidence, see Graham, 1977; also see Heath, this volume). The alternative hypothesis is that at some time in early development a group of ectodermal cells is "set aside" as germ cell precursors, and these remain in a specific undifferentiated state while the remaining embryonic cells undergo differentiative changes. It would be these undifferentiated cells that could give rise to teratocarcinomas, either when the normal embryonic relationships are perturbed, or in some strains when they become primordial germ cells. Since it is now known that the ectoderm is not homogeneous (see chapter by Snow in this volume), one way of testing this idea would be to determine whether all the cells in the embryonic ectoderm are capable of giving rise to teratocarcinomas with equal efficiency, or whether this ability is

TABLE 1: SUMMARY OF THE WAY IN WHICH TERATOMAS AND TERATOCARCINOMAS CAN BE OBTAINED IN MICE*

	Sex, strain of host	Source of tumor	Time tumors first detectable	Sex of tumor cells
Spontaneous Tumors	♂ 129	primordial germ cells in the fetal genital ridge	approximately 15th day of fetal development	♂
	♀ LT	parthenogenetically activated ovarian eggs	30 days postnatal or older	♀ (haploids possible)
Experimentally Induced Tumors	♂ 129, A	primordial germ cells	shortly after transplant of genital ridge of a 12-13 day embryo to the testis of an adult	♂
	♂ or ♀ many strains	embryo (2-cell to 7.5-days of development)	shortly after transfer of embryo to an ectopic site	♂ pt ♀ (haploids possible if parthenotes grafted)

* Martin, 1977.

limited to only certain ectodermal cells, perhaps those in the proliferative zone.

MORPHOLOGICAL ASPECTS OF TERATOCARCINOMA STEM CELL DIFFERENTIATION

Whatever the origin of the tumor, it is generally accepted that at their inception both teratocarcinomas and teratomas, with the possible exception of those teratomas derived from older embryos, contain multipotent stem cells. It is not yet known why in some cases these cells continue to proliferate in the undifferentiated state and yet in other instances undergo differentiation. However, even when there is a constant or increasing population of undifferentiated stem cells, transplantation of the solid tumor is not an ideal method of maintaining a source of multipotent cells. Under conditions of transplantation, those stem cells which differentiate less frequently are at a selective advantage. It has thus been found that after several transplant generations the tumor stem cells may be "nullipotent", that is, apparently incapable of differentiation (Stevens, 1958). Sometimes the restriction in differentiative capacity is not so extreme, and the stem cells are capable of "limited" differentiation. "Neuro-teratocarcinoma" stem cells are one of the most common examples of such restricted stem cells. Most of the differentiated tissues present in the tumors they form are of a neural type (Stevens, 1958; Damjanov *et al.*,1973).

Another type of restriction which is much more difficult to assess, and therefore often ignored, is one which causes a general decrease in the propensity of the multipotent cells to differentiate. Thus in some tumors the stem cells are classified as multipotent because derivatives of all three germ layers are present, but the actual amount of differentiated tissue they form is small; in other tumors differentiation is so extensive that nests of undifferentiated cells are difficult to find (Figure 1). Whether such differences in the amount or variety of differentiated cell types present in a tumor is a consequence of genetic or of other changes in the stem cell population is not known, but it is important to realize that such differences exist and that not all multipotent cells are necessarily the same.

Obtaining a population of stem cells from a solid teratocarcinoma is difficult because the location and proportion of the stem cells within the tumor is not known. This problem can be overcome by making an "ascitic" conversion of the tumor, which is done by mincing a teratocarcinoma and injecting it intraperitoneally in a histocompatible host (Pierce and Dixon, 1959). In some cases, the stem cells will proliferate in suspension in the ascitic fluid. When this occurs the stem cells can differentiate and form structures known as embryoid bodies (Stevens, 1959; Pierce and Dixon, 1959). When these embryoid bodies implant in the peritoneal wall, or some other surface in the body, multi-differentiated tumors are formed (Stevens, 1960; Pierce *et al.*,1960; Pierce and Verney, 1961). The observation that the stem cells can form embryoid bodies is one of the keys which has led to an understanding of the value of teratocarcinoma stem cells for studies of early mouse development.

EMBRYOID BODY FORMATION IN VIVO

Embryoid bodies are free-floating structures which are so named because of their resemblance to certain stages of normal embryonic development. They are found in two forms, simple and cystic. Embryoid bodies, particularly the cystic ones, often bear

Figure 1 (A-C above opposite): Sections of two teratocarcinomas, stained with hematoxylin, eosin and alcian blue. A. Tumor which contains few embryonal carcinoma cells. Several differentiated cell types are apparent including cartilage (CA) and keratinizing epithelium (KE) with mesodermal cells between them. Approx. 35X. B. Tumor which contains many differentiated cell types, but also a great many embryonal carcinoma cells. Arrows and box indicate areas of such undifferentiated cells. Approx. 35X. C. Detail of embryonal carcinoma cells shown in box in B. The arrows point to particularly clear examples of this cell type: there is relatively little cytoplasm, and a large nucleus usually containing a single distinct nucleolus. These cells are always found in closely packed groups or "nests". Contrast the embryonal carcinoma cells with cartilage (CA) on the right and a "neuroepithelial tubule" on the left. Approx. 150X.

Figure 2 (A-D lower opposite): The similarity between inner cell mass derivatives and simple embryoid bodies. A. Diagram of an "early" mouse blastocyst (approximately 3.5 days p.c.). The dotted line indicates where primitive endoderm will form during the next 24 hours of development. B. Diagram of a "late" or "expanded" mouse blastocyst shortly after implantation (approximately 5 days p.c.). The primitive endoderm has formed. Note that endodermal cells are migrating along the blastocoelic surface of the trophectoderm: these cells will become the "parietal" endoderm that secretes the Reichert's membrane. The endodermal cells that cover the blastocoelic surface of the inner cell mass will form the visceral endoderm. C. Phase contrast photomicrograph of simple embryoid bodies. Arrows point to places where the outer endoderm cell layer is particularly clear. These embryoid bodies most closely resemble the embryonic ectoderm with its outer layer of endoderm, as shown in B. Approx. 200X.
D. Section of a small embryoid body stained with hematoxylin and eosin. The core of embryonal carcinoma cells is surrounded by a layer of visceral endoderm, which is characterized by numerous microvilli and apical vacuoles. Although not shown in this case, parietal endoderm, which has fewer microvilli and vacuoles as well as swollen rough endoplasmic reticulum, often forms part of the outer cell layer. Approx. 250X.

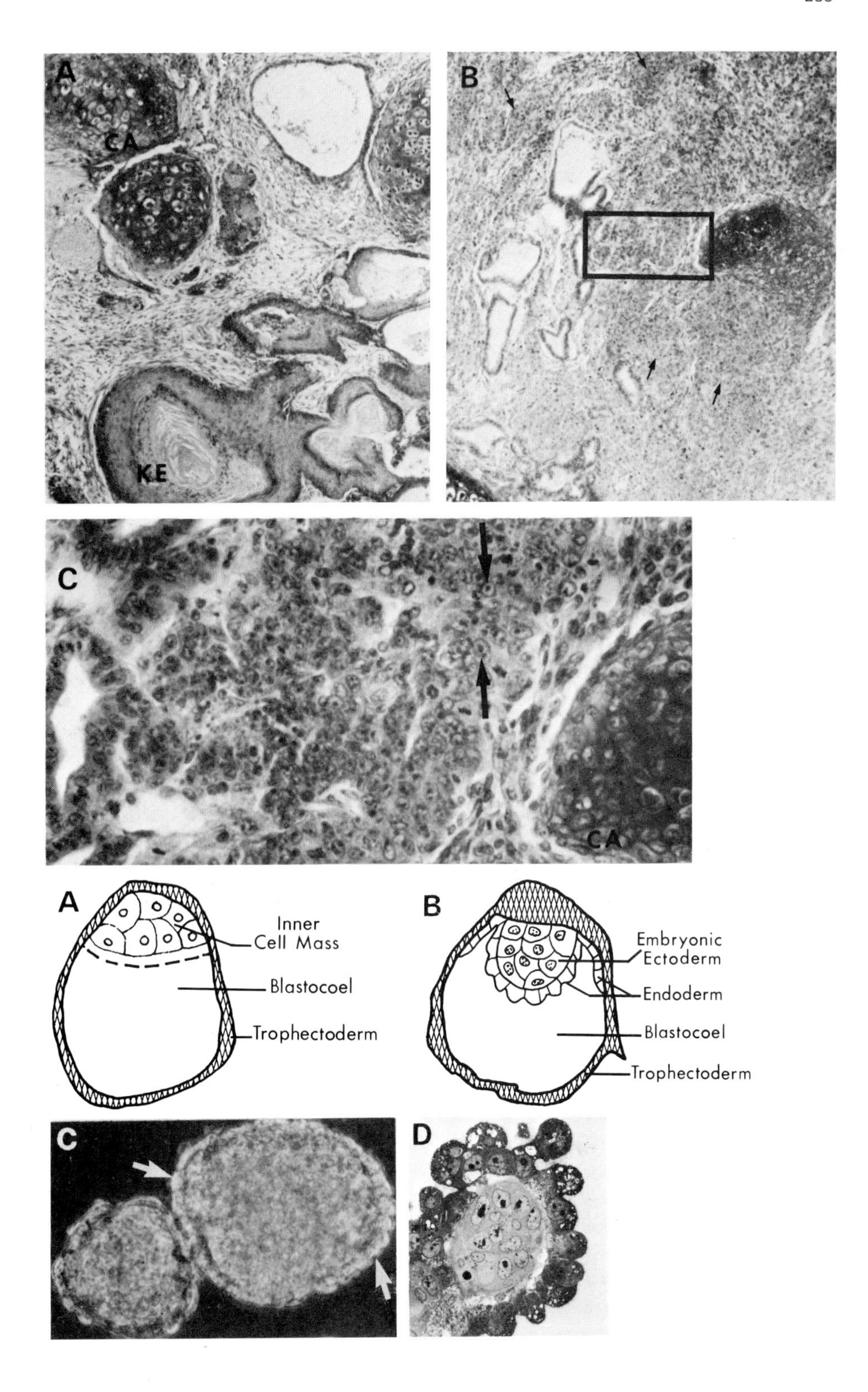
A
CA
KE
B
C
CA
A
Inner
Cell Mass
Blastocoel
Trophectoderm
B
Embryonic
Ectoderm
Endoderm
Blastocoel
Trophectoderm
C
D

a superficial resemblance to mouse blastocysts of approximately 3.5 days of gestation. However, in the blastocyst, the outer cell layer is trophectoderm surrounding an inner cell mass (ICM), while all embryoid bodies have an outer layer of endoderm and do not contain any trophoblastic derivatives.

Simple embryoid bodies consist of a core of undifferentiated embryonal carcinoma cells completely surrounded by an outer layer of endodermal cells. These two-layered structures are most closely similar to the inner cell mass derivatives (endoderm and ectoderm) of a normal embryo at around the time of implantation (approximately 5 days of development; Figure 2). These simple embryoid bodies differ from the normal embryonic inner cell mass in at least two respects. First, in the embryoid bodies the endoderm completely covers the embryonal carcinoma core, in contrast to the situation in the embryo, where the embryonic endoderm does not completely cover the ectoderm (see Figure 2). Second, in the simple embryoid bodies the endoderm which covers the undifferentiated core is often of the parietal type, whereas in the embryo it is of the visceral type. The former is the type of endoderm which is found lining the blastocoelic surface of the trophectoderm (see Figure 2), and which secretes large quantities of the basement membrane substance known as Reichert's membrane (Pierce *et al.*, 1962). It is presumably this type of endoderm which is responsible for the deposition of Reichert's membrane-like material between the outer endoderm and inner embryonal carcinoma core in simple embryoid bodies.

These differences, however, may not reflect intrinsic differences between embryonal carcinoma cells and ICM cells. If the inner cell mass is isolated from the blastocyst and cultured *in vivo* or *in vitro* then a complete outer layer of endoderm is formed (Rossant, 1975; Solter and Knowles, 1975), some of which may be of the parietal type (Strickland *et al.*,1976; Solter, personal communication). In addition, there is evidence that visceral and parietal endoderm may be closely related, and that the former can give rise to the latter in abnormal situations (Solter *et al.*,1974; Diwan and Stevens, 1976).

A second type of embryoid body that is found in the peritoneal cavity is the more complex cystic variety. They are so named because they contain a fluid-filled cyst as well as several differentiated cell types. Overall these structures resemble disorganized inner cell mass derivatives at a stage equivalent to approximately 7-8 days gestation, and they too often possess extra-embryonic parietal endoderm. It was first suggested by Pierce and Dixon (1959) that cystic embryoid bodies develop from simple ones. Hsu and Baskar (1974) have studied this by explanting simple embryoid bodies from the peritoneum and culturing them *in vitro* in parallel with intact mouse blastocysts. They found a striking similarity between the early post-implantation development of the embryo and the formation of cystic embryoid bodies from simple ones. These results emphasize the similarity between the core cells of the embryoid bodies (embryonal carcinoma cells) and the pluripotent cells of the normal embryo and indicate that the former may be "programmed" in the same way as the latter.

PROPERTIES OF EMBRYONAL CARCINOMA CELLS IN VITRO

Multipotent stem cells can be isolated from either solid or ascitic teratocarcinomas, although it is easiest to do so from simple embryoid bodies because the stem cells form a large proportion of the total cell population and their location is known. It was first shown by Kahan and Ephrussi (1970) and Rosenthal, Wishnow and Sato (1970) that the capacity of embryonal carcinoma cells to form multi-differentiated tumors is retained when embryonal carcinoma cells are cloned and cultured *in vitro*. Since then, many embryonal carcinoma cell lines have been isolated and cultured *in vitro* (reviewed by Martin, 1975; Graham, 1977). Some of these cells are multipotent and can differentiate both *in vivo* and *in vitro*, while others have restricted differentiative capacities. For example, the nullipotent cell lines described by Martin and Evans (1975a,b) were isolated from a nullipotent tumor and as yet have not been found to differentiate under any conditions. Other embryonal carcinoma cell lines have been isolated from multi-differentiated tumors but appear to have lost differentiative capacity during *in vitro* culture. One well-known example is the F9 cell line first described by Bernstine *et al.*(1973). These were described as having become nullipotent, although more recently Sherman and Miller (1978) have shown that F9 cells are capable of forming a small amount of endoderm.

It would be difficult to discuss individually all of the many different embryonal carcinoma cell lines that are currently available. They do have a common, highly distinctive morphology which can be taken as characteristic. The cells generally adhere strongly to one another and grow as poorly attached colonies or "epithelioid nests"; in the phase contrast microscope the cell-cell boundaries are very indistinct. The individual cells are rounded or slightly bipolar, with relatively little cytoplasm and a nucleus with one or two prominent nucleoli (Figure 3).

The multipotent embryonal carcinoma cell lines that can differentiate *in vitro* apparently fall into two classes, those that need to be co-cultured with fibroblastic feeder cells* to continue proliferation in the undifferentiated state, and those that do not. Graham (1977) has designated the former

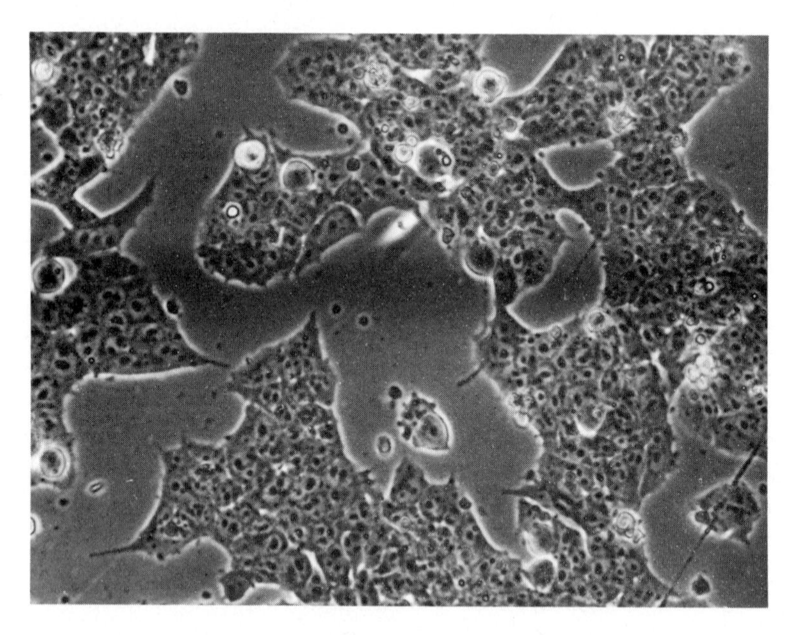

Figure 3: Phase contrast photomicrograph of an embryonal carcinoma cell culture. The boundaries between the individual cells are indistinct. Each cell is rounded or slightly bipolar with a large nucleus containing one or two distinct nucleoli. Even when the cells are carefully dis-associated and are plated as a single cell suspension they rapidly adhere to each other and consequently form colonies. Approx. 250X.

"emancipated" cells and the latter "nursed" cells. Martin and Evans (1975a,b) have demonstrated that in order to maintain such feeder-dependent embryonal carcinoma cells as a proliferating population of undifferentiated cells, they must be subcultured

* Feeders are cells that have been rendered incapable of division as a consequence of X-irradiation or treatment with a drug such as Mitomycin-C. Such cells can metabolize and survive as a confluent monolayer for days or even weeks, providing they are not subcultured. When treated with trypsin, feeder cells tend to disintegrate.

to a feeder layer every 3 or 4 days (Figure 4). If such cells are disaggregated and subcultured two or more times in the absence of feeder cells they die, although they can be rescued by being returned to a feeder layer (Figure 4). The interesting feature of such subcultured embryonal carcinoma cells is that they are capable of forming embryoid bodies *in vitro*, as discussed below. Hogan (1976) has demonstrated that it is possible to select a sub-population of cells whose growth in the undifferentiated state is feeder-independent. However,

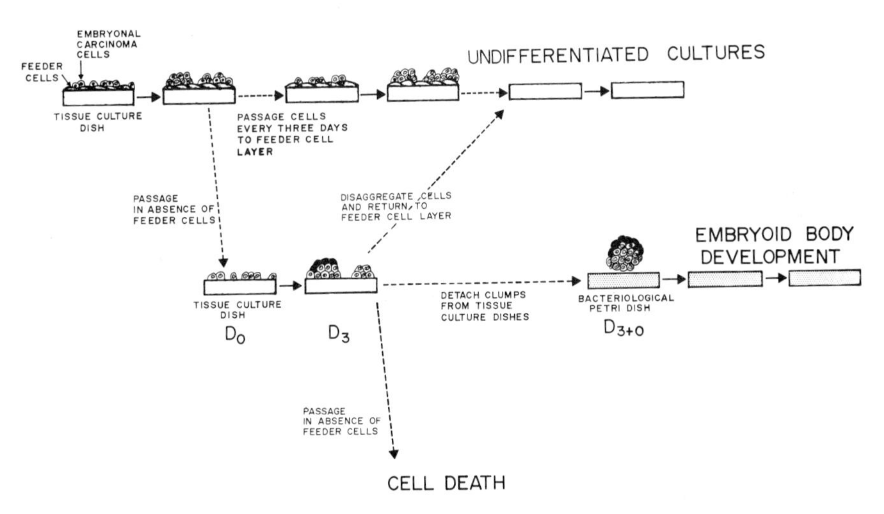

Figure 4: Diagram of culture regime for feeder-dependent embryonal carcinoma cells that form embryoid bodies *in vitro*

cells cultured in this way apparently irreversibly lose the ability to form embryoid bodies, although they may still be capable of *in vitro* differentiation in some other manner. It is of course possible that feeder-independent embryonal carcinoma cell lines capable of forming embryoid bodies *in vitro* will be isolated in the future.

One obvious question is what effect do the feeder cells have on the embryonal carcinoma cells? Although differentiation of the embryonal carcinoma cells is prevented by frequent subculture to a feeder layer, there does not appear to be any specific inhibition of differentiation by the feeder cells.

Thus, embryonal carcinoma cells will differentiate on a feeder layer if they are cultured long enough to form large, dense colonies (Evans and Martin, 1975). Instead, the feeder cells probably nonspecifically inhibit differentiation in two ways. First, they may provide growth factors that stimulate the embryonal carcinoma cells to proliferate more rapidly (Martin and Evans, 1975b). Such rapid growth may in some way be incompatible with a switch to differentiated functions. Second, it has been observed that the embryonal carcinoma cell aggregates assume a more flattened geometry when cultured on a confluent feeder cell layer. If, as discussed below, the process of assuming a more three-dimensional structure in some way triggers differentiation, then a feeder-effect of flattening the clumps would be inhibitory.

There is as yet no explanation of why continuous co-cultivation with feeder cells should enable the embryonal carcinoma cells to retain the ability to differentiate via embryoid body formation. One possibility is that the selection and culture of embryonal carcinoma cells independent of feeder cells may result in cell surface changes that affect the ability of the cells to respond to the normal signals for differentiation.

In Vitro Differentiation of Established Embryonal Carcinoma Cell Lines

Although earlier studies indicated that embryonal carcinoma cells might be capable of some differentiation *in vitro* (Kahan and Ephrussi, 1970; Rosenthal *et al.*,1970), Lehman *et al.*(1974) were the first to demonstrate that an established embryonal carcinoma cell line could undergo extensive differentiation *in vitro:* the cell types that were identified included keratinizing epithelium, cartilage, striated muscle, neuronal cells and endodermal cells. It has since been observed that many embryonal carcinoma cell lines can be induced to differentiate *in vitro* to a greater or lesser extent (see reviews by Martin, 1975; Graham, 1977). From the data that are available, certain generalizations can be made about the factors that affect the differentiation of all teratocarcinoma stem cell lines, regardless of whether or not their growth in the undifferentiated state is feeder-dependent.

The most important of these appears to be cell aggregation: differentiation of embryonal carcinoma cells occurs when the cells are cultured at a high local density. In some cases this is accomplished by allowing single cells or small clumps of cells to attach to a tissue culture surface and culturing them until they become large, tightly rounded colonies (see for example, McBurney, 1976). With other established lines, for example PCC3, the cells may be cultured to high density as a confluent monolayer (Nicolas *et al.*,1975, 1976). Such PCC3 cultures become very dense and then apparently undergo a period of partial lysis, after which numerous differentiated cell types appear. For other embryonal carcinoma cell lines differentiation can be obtained by allowing the cells to aggregate and form clumps in bacteriological petri dishes, to which they do not adhere (Sherman, 1975; Sherman and Miller, 1978).

The second factor that is apparently important in embryonal carcinoma cell differentiation is the attachment of the cells to a substratum. This was first demonstrated by experiments in which embryoid bodies were explanted from the peritoneum and cultured *in vitro*. It was found that if these embryoid bodies were kept in suspension, little if any further differentiation occurred, but if they were allowed to attach to a substratum a great many differentiated cell types appeared (Teresky *et al.*,1974; Levine *et al.*, 1974; Gearhart and Mintz, 1974). Similarly, when embryonal carcinoma cell cultures are aggregated in bacteriological dishes, the number of differentiated cells that appear is small until the clumps are allowed to attach to a tissue culture substratum (Sherman, 1975; Chung *et al.*,1977). There is some evidence that the differentiated cell types in these mass cultures appear in a specific sequence, beginning with endoderm, then neuronal cells, contractile muscle, cartilage, etc. (Nicholas *et al.*,1975). However, since the cultures are usually multi-layered, it is to some extent difficult to quantitate that amount of each of the various cell types that are formed. Taken together, these results suggest that cell-cell contacts and positional information are important in teratocarcinoma stem cell differentiation, as they are in normal embryonic development.

Cystic Embryoid Body Formation and Isolated ICM Development In Vitro

As noted above, unlike "emancipated" embryonal carcinoma cells, certain feeder-dependent cell lines retain the ability to differentiate via the formation of embryoid bodies *in vitro* (Martin and Evans, 1975a,b). Two classes of such cell lines have been isolated: one forms simple embryoid bodies in which the core cells apparently remain undifferentiated as long as these embryoid bodies are kept in suspension: the second type forms simple embryoid bodies that develop into cystic ones when they are kept in suspension (Figure 5). Like the embryoid bodies removed from the peritoneum, both the simple and cystic embryoid bodies formed *in vitro* will differentiate to a very wide variety of cell types if they are allowed to reattach to a culture substratum (Figure 5; Martin and Evans, 1975c).

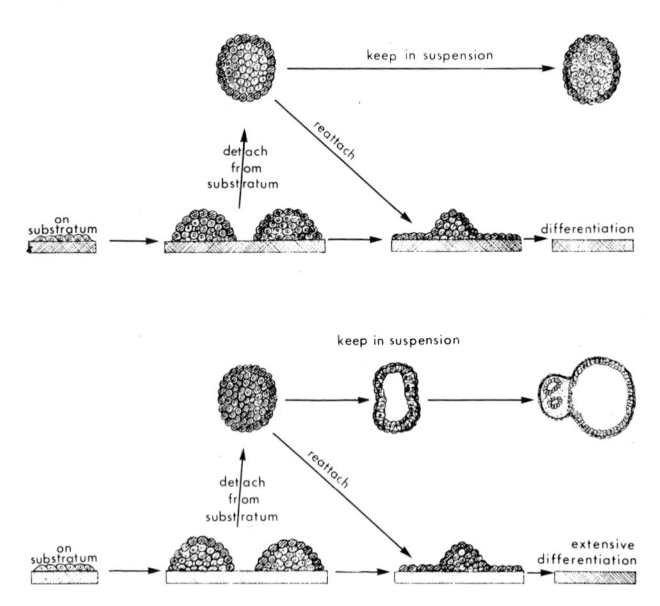

Figure 5: Pattern of differentiation of clonal embryonal carcinoma cell lines that form embryoid bodies *in vitro*. A. Pluripotent cells that form simple embryoid bodies: An endodermal cell layer becomes apparent on the outer surface of clumps formed by these embryonal carcinoma cells. These clumps can be detached from the substratum and kept in suspension; in this case the outer endodermal cell layer completely surrounds the embryonal carcinoma cell core, and the clumps are then known as "embryoid bodies". No further differentiation occurs in suspension. If, however, the cell clumps are not detached from the substratum, or if the embryoid bodies are allowed to reattach, the endodermal cells migrate out, and during the next few weeks the cultures become more complex, multilayered and various differentiated cell types appear. B. Pluripotent cells that form cystic embryoid bodies: These cells behave in the same way as those described above, except that differentiation continues in suspension, and the simple embryoid bodies become cystic. Differentiation of attached cells occurs as described above, but an even greater variety of differentiated cell types form. From Martin (1975).

Such embryonal carcinoma cells can be induced to form embryoid bodies by a very simple culture procedure (see Figure 4): the undifferentiated cultures are disaggregated and seeded in tissue culture dishes that do not contain feeder cells.

Initially the embryonal carcinoma cells attach as flattened aggregates that become rounder and less well-attached to the substratum. This rounding up process appears to trigger the formation of endoderm and is thus the first step in embryoid body formation. The intact cell clumps can then be detached from the substratum and transferred to bacteriological dishes. When this culture regime is followed, the entire cell population synchronously forms a large number of embryoid bodies.

The development of cystic embryoid bodies *in vitro,* like that of the peritoneal embryoid bodies, mimics early embryonic development. A direct comparison of this form of tumor cell differentiation and the behavior of similar multipotent embryonic cells can be made by employing the technique of immunosurgery to isolate the embryonic inner cell mass from the mouse blastocyst (Solter and Knowles, 1975). Both the teratocarcinoma cells and the isolated ICMs can be cultured under comparable conditions *in vitro*. This seems more appropriate than a comparison of embryoid body development with that of the intact embryo. In the latter there may well be some developmental interaction with the trophectoderm and its derivatives, whereas in teratocarcinoma-derived embryoid bodies and isolated inner cell masses there is not.

The similarities and differences in development can be assessed by referring to the work of Martin, Wiley and Damjanov (1977) on teratocarcinoma cystic embryoid body development and of Hogan and Tilly (1978) and Wiley,Spindle and Pedersen (1978) on isolated inner cell mass development *in vitro*. In the case of the embryonal carcinoma cells, embryoid body formation *in vitro* begins once cell aggregates have become three-dimensional structures. The results suggest that if the clumps are attached to a substratum the endoderm forms only on the free surface, and not where the flattened portion of the clumps contact the culture substratum. Within 24 hours of detachment (see Figure 4), a complete layer of endoderm encircles all such undifferentiated cell clumps (Figures 6, 7). The almost identical situation arises when the ICM is isolated from the expanded blastocyst and cultured *in vitro*. At first the endoderm forms a crescent which does not lie over that portion of the ICM surface that was previously covered by trophectoderm (see Figure 2). Within 24 hours after isolation the core cells are completely surrounded by endoderm (Figures 6, 7).

As can be seen from Figures 6,7, the subsequent development of both the teratocarcinoma and ICM-derived structures is remarkably similar: in both, a cavity forms that is similar to the pro-amniotic cavity of the early post-implantation mouse embryo. This cavity subsequently expands and there is a change in cellular organization and morphology, leading to the formation of a columnar epithelial layer which resembles the embryonic ectoderm of the 6.5 day embryo (Figure 7). At later times, some of the structures undergo considerably more expansion and the inner ectodermal cell layer becomes thin and forms a flattened epithelium. In a variable proportion of the teratocarcinoma and ICM-derived embryoid bodies mesoderm forms as a third layer between the outer endoderm and inner columnar ectodermal layers (Figure 7).

In addition to these similarities in the patterns of development, there are also several minor differences. For example, the development of the ICM-embryoid bodies is more rapid than that of the tumor-derived ones. Another difference concerns the type of endoderm formed. In both studies of isolated ICM development *in vitro* no parietal endoderm was observed, whereas the teratocarcinoma cystic embryoid bodies always contained some parietal endoderm. As discussed above, this may not be an important difference. A potentially more significant one stems from the observation by both groups studying *in vitro* ICM development that the isolated ICM can produce extra-embryonic ectoderm. This is a most unexpected observation, since there is good evidence that during normal development the ICM does not give rise to extra-

Figure 6 (opposite): Embryoid bodies obtained by culture of isolated inner cell masses and embryonal carcinoma cells *in vitro*. A. Phase contrast photomicrograph of isolated ICMs after 48 hours of culture. The outer cell layer of visceral endoderm is clear. B. Phase contrast photomicrograph of embryonal carcinoma cell clumps approximately 48 hours after detachment. The outer cell layer of these embryoid bodies is less distinct than in A, possibly because it is a mixture of visceral and parietal endoderm. C. Phase contrast photomicrograph of isolated ICMs after 3 days in culture. These are by now clearly cystic. D. Phase contrast photomicrograph of embryonal carcinoma embryoid bodies after 5 days of culture in suspension. These too are clearly cystic. E. Sections of cystic embryoid bodies formed by isolated ICMs after 48 hours of culture, stained with hematoxylin and eosin. F. Sections of cystic embryoid bodies formed by embryonal carcinoma cells after 4 days of culture in suspension, stained with hematoxylin and eosin. Mag. approximately 120X in A-D; approximately 70X in E and 100X in F. The photographs of cultured ICMs were kindly provided by Dr. Brigid Hogan and were published previously (Hogan and Tilly, 1978). Photographs B and D of teratocarcinoma embryoid bodies have not been published previously; Photograph F is from Martin, Wiley and Damjanov)1977).

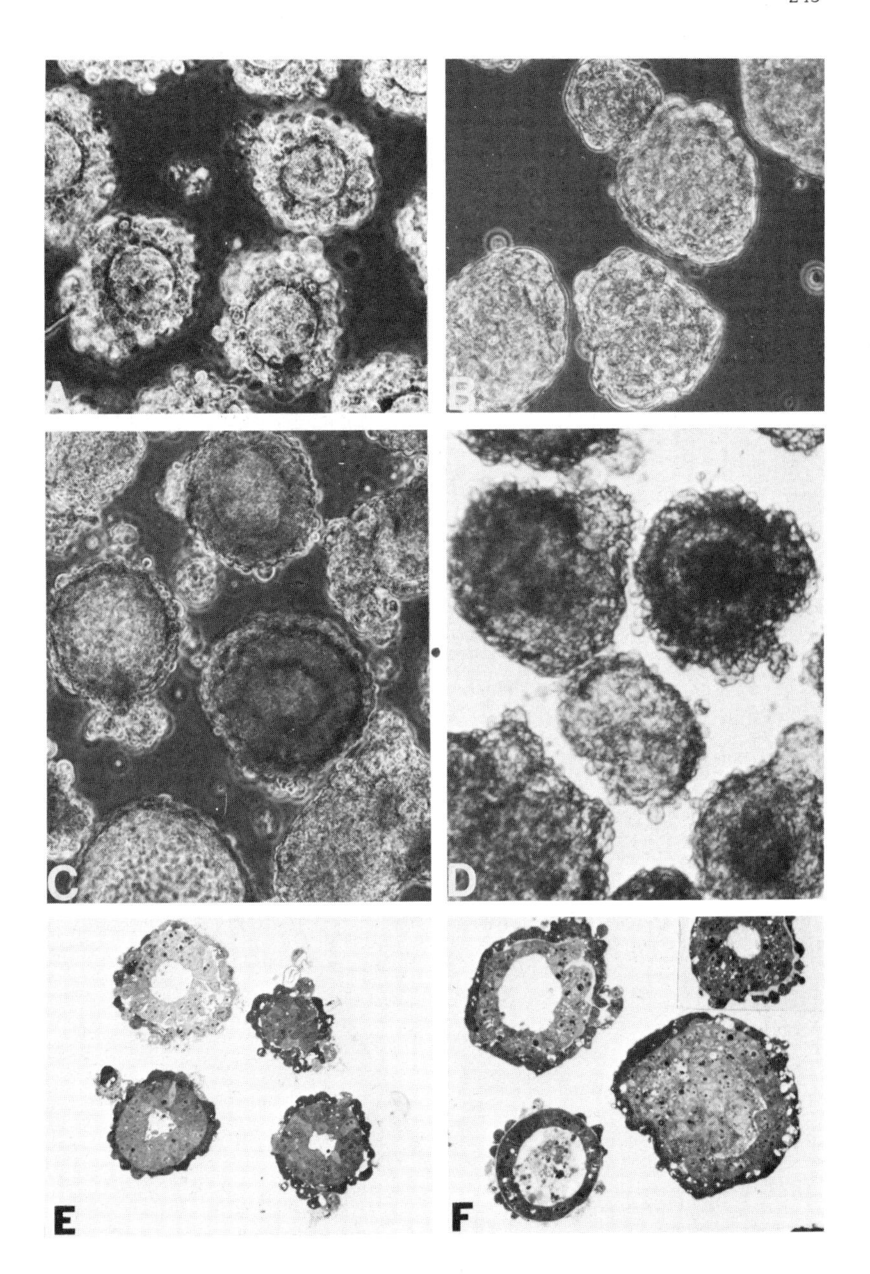
A
B
C
D
E
F

embryonic ectoderm, which is considered a trophectoderm derivative (reviewed by Gardner and Papaioannou, 1975). If the morphological evidence can be substantiated with biochemical data, then there would appear to be a real difference between the teratocarcinoma cystic embryoid bodies and those formed by isolated ICMs, since the former never appear to contain cells that resemble extra-embryonic ectoderm.

In view of the similarities described above, it appears that the embryonal carcinoma cells that form cystic embryoid bodies should provide a useful tool for studying the factors that affect the embryonic processes of endoderm, proamniotic cavity and possibly mesoderm formation. This is especially true in cases where larger quantities of material are necessary than can be obtained by immunosurgery of blastocysts.

GENE EXPRESSION IN TERATOCARCINOMA STEM CELLS

The observations on embryoid body formation suggest that some, if not all, teratocarcinoma stem cells appear to be similar to embryonic cells that no longer have the capacity to form trophectoderm in response to normal signals for differentiation. (It is possible, however, that embryonal carcinoma cells retain the capacity to form trophoblastic derivatives in "abnormal" circumstances. This is suggested by the observation that embryonal carcinoma cell lines can form cells that morphologically resemble trophoblast in dense mass cultures or in tumors). It is important to consider additional criteria in order to determine if there is one particular embryonic cell type to which the embryonal carcinoma cells most closely correspond.

DEVELOPMENTAL STAGE-SPECIFIC PROTEINS

The morphological observations suggest that certain embryonal carcinoma cells are behaviorally similar to the cells of the ICM

Figure 7 (opposite): Similarity between cystic embryoid body development and normal early post-implantation embryonic development. A, B and C are schematic representations of normal embryos at approximately 5, 6 and 7 days of development, respectively. To the left of each are sections of morphologically similar stages in the development of cystic embryoid bodies formed by embryonal carcinoma cells (ECC) or isolated inner cell masses (ICM). The photographs of ECC embryoid bodies are from Martin (1977) and Martin *et al.*(1977); the photographs of ICM embryoid bodies (A, B) are from Wiley *et al.*(1978) and (C) from Hogan and Tilly (1978). Mag. approximately 300X in ECC and ICM - A and B; approximately 200X in ECC - C and approximately 150X in ICM - C.

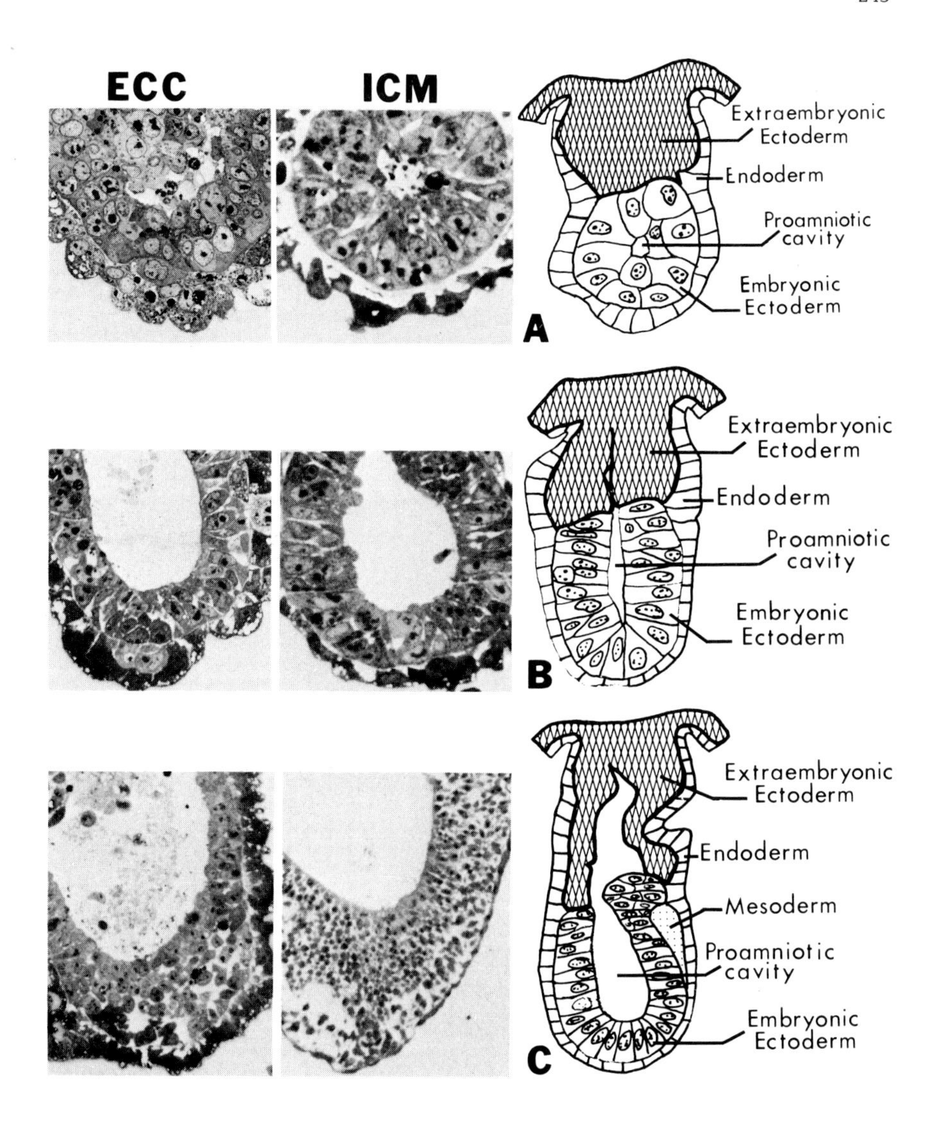
ECC
ICM
Extraembryonic
Ectoderm
Endoderm
Proamniotic
cavity
Embryonic
Ectoderm
A
Extraembryonic
Ectoderm
Endoderm
Proamniotic
cavity
Embryonic
Ectoderm
B
Extraembryonic
Ectoderm
Endoderm
Mesoderm
Proamniotic
cavity
Embryonic
Ectoderm
C

prior to endoderm formation, since both the embryonal carcinoma cell clumps and isolated ICMs begin their differentiation with the formation of an outer layer of endoderm. However, recent studies by Martin, Smith and Epstein (unpublished observations) using one- and two-dimensional SDS-polyacrylamide slab gel electrophoresis, indicate that the pattern of protein synthesis in these undifferentiated teratocarcinoma stem cells is more similar to that of the embryonic ectoderm than that of the ICM prior to endoderm formation. Thus both the embryonal carcinoma cells and the embryonic ectoderm have the same pattern of protein synthesis and synthesize at least one prominent protein (molecular weight approximately 55,000 daltons) that is not detected in significant amounts in embryonic cells at earlier stages of development.

These results indicate that, paradoxically, embryonal carcinoma cells, which begin their differentiation with the formation of an outer layer of endoderm, are biochemically most closely similar to embryonic cells which have already formed a layer of endoderm. This apparent paradox is resolved by the observation of Pedersen, Spindle and Wiley (1977) that the embryonic ectoderm retains the capacity to differentiate an outer layer of endoderm, even after endoderm has already been formed. Therefore, the conclusion that, in protein synthetic terms, embryonal carcinoma cells are closely similar to the cells of the embryonic ectoderm is consistent with other aspects of their behavior. Such results therefore indicate that the best use of the embryonal carcinoma cells is as a model system for the study of the embryonic ectoderm rather than earlier stages of embryonic development. They do not, however, rule out the possibility that the embryonal carcinoma cells are in fact more closely similar to developmentally "older" cells.

Such results emphasize one of the potential uses of embryonal carcinoma cells as a source of embryonic stage-specific molecules. The fact that virtually unlimited quantities of these embryonal carcinoma cells can be obtained by *in vitro* culture means that such molecules can be isolated and their functions ultimately determined. Such a task would be considerably more difficult using only normal embryonic material.

One important question that is as yet unanswered is whether or not *all* embryonal carcinoma cells are "equivalent" to the same type of embryonic cell. This can be considered as two specific questions. The first is whether there are intrinsic "stage-specific" differences among stem cells from tumors of different origins (for example, 3-day embryo-derived as compared with primordial germ cell-derived teratocarcinomas). Part of the reason that it is difficult to resolve this is that most practical methods of comparing embryonal carcinoma cells from different tumors involve culture techniques that may alter the cell population. This raises the second question of whether or not embryonal carcinoma cells, regardless of their origin, can be either altered or selected to be equivalent to different embryonic cell types by *in vitro* culture.

There are several observations that provide some information but do not resolve these questions. The first is that no embryonal carcinoma cell line has been reported to begin its differentiation with the formation of trophectoderm. This suggests that it is unlikely that there are any embryonal carcinoma cells that are equivalent to cleavage stage embryonic cells. Second, in studies using embryonal carcinoma cells isolated from tumors obtained by several different methods of teratocarcinogenesis, Martin, Smith and Epstein (unpublished data) found that each had the same pattern of protein synthesis including the ectoderm-specific autoradiographic band, as determined by one-dimensional gel electrophoresis. Such results suggest that all embryonal carcinoma cells may be basically similar to one particular embryonic cell type. If so, the differences that do exist among embryonal carcinoma cell lines might represent alterations in that cell type rather than similarities to another embryonic cell type.

EMBRYONIC CELL SURFACE ANTIGENS

A sub-class of molecules synthesized by teratocarcinoma cells are expressed on the cell surface and are detected by immunological techniques. It was first shown by Edidin *et al.* (1971) that cells derived from peritoneal embryoid bodies share cell surface antigens with early mouse embryos. By injecting embryonal carcinoma cells of the F9 line into syngeneic mice, Artzt *et al.* (1973) were able to raise a more specific serum and also to demonstrate that the antigen(s) common to teratocarcinoma stem cells and early embryos are not generally expressed by adult tissues, except perhaps for those in immunologically privileged sites. Thus, the anti-F9 serum was shown to react with embryonal carcinoma cells, mouse sperm and cleavage stage mouse embryos, but was reported to have no reactivity against various differentiated cell types. When this anti-F9 serum was tested more fully on embryonic cells, it was found that the F9 antigen is expressed up until the 10th day of development (Jacob, 1977). Other groups have since confirmed that syngeneic sera raised against different embryonal carcinoma cells also react with embryonic cells, but these sera may have somewhat different specificities than the anti-F9 serum (Stern *et al.*,1975; Dewey *et al.*,1977; Gachelin *et al.*,1977).

In a subsequent study, using quantitative absorption of anti-F9 serum with sperm, Artzt, Bennett and Jacob (1974) presented data in support of their hypothesis that the antigen(s) present on the surface of the F9 cells and detected by anti-F9 serum is a product of the *T*-locus, specifically the wild-type allele of the t^{w32} gene. If this were the case, then one might expect that mutant t^{w32}/t^{w32} homozygotes would not express the F9 antigen, but other t mutant homozygotes as well as wild-type embryos would. Kemler *et al.* (1976) tested this and found that, as expected, a significant proportion of the progeny of crosses that produce t^{w32}/t^{w32} homozygotes did not label with the anti-F9 serum. Also, as expected, wild-type embryos and some other t mutant embryos did label. However, an unexpected finding was that, like the t^{w32} homozygotes, t^5 mutant homozygotes also did not label. Kemler *et al.*(1976) suggest that these results indicate that the F9-antigen is probably not the product of any one particular wild-type allele of *t*.

There is now some evidence that whatever gene(s) code for the F9 antigen(s), its product may be important in early development. Thus, Kemler *et al.*(1978), have found that Fab fragments of rabbit anti-F9 IgG can reversibly inhibit cleavage stage embryos and morulae from blastulating.

Taken together, all the results described above emphasize the usefulness of teratocarcinoma cells as immunogens that can be used to obtain sera that detect embryonic cell surface antigens. However, the question of whether these are developmental stage-specific molecules remains open. The data do indicate that some of these antigens may be expressed by cells at various stages of development. If this is the case, it would explain why cells that are neither behaviorally nor biochemically similar to cleavage stage embryos express cell surface antigens in common with them.

Teratocarcinoma stem cells not only express antigens in common with embryonic cells, they also apparently do not express antigens that are difficult to detect on the surface of early embryonic cells. Thus, like early embryonic cells, embryonal carcinoma cells either do not express or express very little H-2 antigen (Artzt and Jacob, 1974; Stern *et al.*,1975). Considering this and the observations that the tumor and embryonic cells share common cell surface antigens, it is clear that differentiating cultures of embryonal carcinoma cells should be useful as a model system for studying changes in antigen expression during early embryonic development.

X-CHROMOSOME INACTIVATION

It is now well known that early in the development of the female embryo a differentiative event occurs which is known as X-chromosome inactivation. This results in the expression of the genes of only one of the two X-chromosomes present in that cell and its descendants (reviewed by Lyon, 1972; Gartler and Andina, 1976; see also Monk, this volume). There is both cytogenetic evidence (Takagi, 1974; Mukherjee, 1976) and data from cell transfer experiments (Gardner and Lyon, 1971; Gardner, personal communication) that suggests that X-inactivation occurs at approximately the time of implantation (when ectoderm formation is occurring) or shortly thereafter.

If female embryonal carcinoma cells are indeed similar to the embryonic ectoderm at a stage prior to X-inactivation, then both of the X-chromosomes in each cell should be producing gene products. If so, when the cells are allowed to differentiate *in vitro* they should undergo X-inactivation, and only one X-chromosome should continue to function in each cell. Martin *et al.*(1978) have found that this appears to be the case, at least for the female embryonal carcinoma cell lines that they tested. They demonstrated this by comparing the levels of X-linked and autosomal gene products in female and male embryonal carcinoma cells both maintained in the undifferentiated state and induced to differentiate by removal from a feeder cell layer as described above (Figure 4). The conclusion they drew from their data was that the undifferentiated female embryonal carcinoma cells contain two genetically active X-chromosomes and that the process of X-chromosome inactivation occurs when the cells are allowed to differentiate *in vitro*. If their interpretation of the data is correct, then one of the main obstacles to molecular studies of the process of X-inactivation, the difficulty in obtaining appropriate samples for biochemical analysis, has been overcome.

Interestingly, not all female embryonal carcinoma cells that have been examined appear to contain two active X-chromosomes.

McBurney and Adamson (1976) concluded that only one X-chromosome was genetically active in the cells they examined. There are two possible explanations for this difference between embryonal carcinoma cell lines. The first involves consideration of the tumors from which the cells were obtained. The cells described by Martin *et al.*(1978) were isolated from a spontaneous ovarian teratocarcinoma, while McBurney and Adamson (1976) examined cells that had been derived from teratocarcinomas obtained by ectopic transfer of embryos of approximately 6.5 days of age. It is possible that the two groups of teratocarcinoma stem cells were derived from embryonic cells at different stages of development, and that those containing only one active X-chromosome are therefore equivalent to embryonic cells at a later, post-X-inactivation stage of development than are those with two active X-chromosomes.

It is perhaps also important to note that the parthenogenetic origin of the LT cells used by Martin *et al.*(1978) may be of significance. Their LT cells presumably arose from haploid ovarian eggs that had already undergone the first meiotic division and then diploidized (Eppig *et al.*,1977). Thus, in these LT teratocarcinoma stem cells both X-chromosomes present have been carried through the process of oogenesis, whereas in the embryonal carcinoma cells derived from a normal embryo, one X-chromosome came from a cell that had undergone oogenesis while the other was from a cell that had undergone spermatogenesis. This abnormal similarity between the two X-chromosomes in the LT cells might possibly account for the observation that they are both genetically active. However, the fact that one of the two X-chromosomes in these cells can apparently undergo inactivation during differentiation *in vitro* suggests that even if both X-chromosomes are active as a consequence of an abnormal situation, their subsequent behavior is normal.

An alternative explanation for the differences among embryonal carcinoma cell lines is that regardless of their origin all embryonal carcinoma cells are similar to the same cells at a pre-X-inactivation stage of development, as discussed above, and that the observed differences in X-chromosome activity are a consequence of *in vitro* culture conditions. Such changes would not necessarily mean that the altered cells are equivalent to developmentally "older" cells in their other properties. Such an explanation is possible in view of the fact that the female cells tested by McBurney and Adamson had been selected for feeder-independent growth, whereas the cells that contain two active X-chromosomes are feeder-dependent: new experiments are needed to determine whether such selection can result in X-inactivation. In addition, it would be interesting to determine whether feeder-dependent female cells isolated from normal embryo-derived teratocarcinomas contain two active X-chromosomes. Only if this is true will it be possible to use teratocarcinoma stem cells to study X-inactivation by means of electrophoretic variants of X-linked markers.

PARTICIPATION OF TERATOCARCINOMA STEM CELLS IN NORMAL DEVELOPMENT

The experiments using teratocarcinoma stem cells that have perhaps attracted the most attention are those that demonstrate the ability of the tumor cells to join with normal embryonic cells in forming an adult mouse. In principle and design these experiments are similar to those which have demonstrated that cells from two genetically different embryos can be mixed and will develop into a single chimeric (mosaic, allophenic) animal (reviewed by McLaren, 1976). With embryonic cells the mixing of the two cell types can be done in two ways. The simplest of these involves the aggregation of cleavage or morula stage cells, which will readily adhere to one another under the appropriate conditions (Tarkowski, 1961; Mintz, 1962). The more complex procedure involves the use of a micromanipulator to inject embryonic cells of one genotype into a host blastocyst of another (Gardner, 1968). In either case the manipulated embryos are placed in pseudopregnant foster mothers and allowed to develop to term. The animals that are born contain varying proportions of cells derived from both initial populations which can be identified by the genetic differences between them.

In studies with teratocarcinoma cells, only the injection technique for combining the tumor and embryonic cells has been successful (Figure 8). The reason that aggregation chimaeras have not been made is that embryonal carcinoma cells apparently do not adhere well to cleavage or morula stage embryonic cells (Mintz *et al.*,1975). This is not surprising since, as discussed above, embryonal carcinoma cells are probably similar to embryonic cells at a later stage of development. Because these experiments require skills that have not as yet been acquired by many scientists, relatively few such experiments have been performed. For this reason, it is difficult to know what the limitations are to this

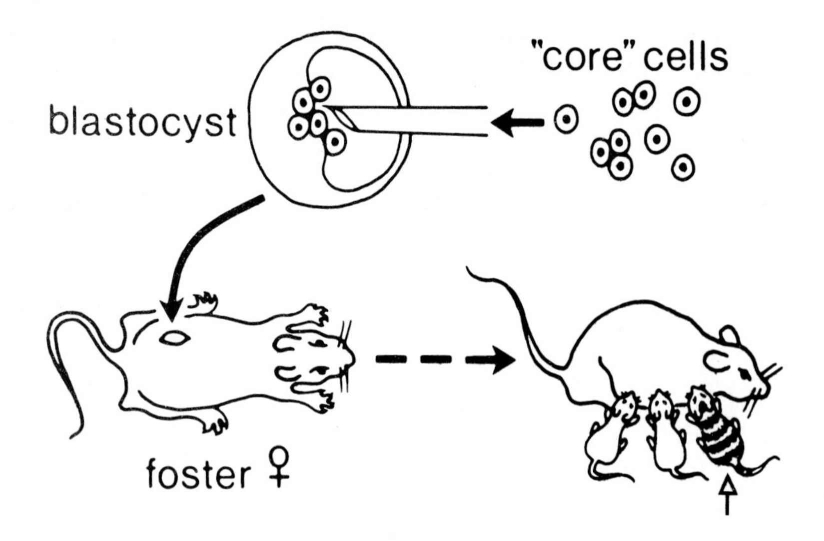

Figure 8: Diagram of experimental protocol for the production of teratocarcinoma-embryo chimeras. Embryonal carcinoma cells isolated from the "cores" of simple embryoid bodies taken from the peritoneum (or established embryonal carcinoma cell lines) are injected into an early mouse blastocyst. The operated embryo is then placed in the uterus of a pseudopregnant foster mother, where it implants. Some of the animals that develop from these embryos have tissues derived from the injected embryonal carcinoma cells as well as the host embryo. The open arrow points to one such animal that shows coat color chimerism. From Mintz and Illmensee (1975).

approach, although some of the problems are now clear, as discussed below.

CELLS TAKEN FROM TUMORS

One group of injection studies was carried out with cells that were taken from the ascitic form of the tumor and which were not cultured *in vitro*. Brinster (1974, 1975) was the first to demonstrate that teratocarcinoma cells could in fact participate in the development of a normal adult mouse. In his study, teratocarcinoma cells derived from a strain 129 SvS1 (agouti) embryo were injected into normal random-bred Swiss (albino) blastocysts. In one case the animal that was born had agouti hairs derived from the injected teratocarcinoma cells. However, since no other readily identifiable genetic differences existed between the injected teratocarcinoma cells and the host embryos, it was impossible to determine the full extent to which the tumor cells had participated in the formation of that or other animals in the experimental series. Nevertheless, these results provided the first indication that teratocarcinoma cells have the capacity to participate in completely normal development.

The full potential of embryonal carcinoma cells for normal embryonic behaviour *in utero* was only appreciated when Mintz and her colleagues carried out experiments using a variety of genetic markers to distinguish the injected embryo-derived teratocarcinoma stem cells from the host embryo (Mintz *et al.*,1975; Mintz and Illmensee, 1975; Illmensee and Mintz, 1976; Mintz, personal communication). They clearly demonstrated that the tumor stem cells could form virtually every tissue of a normal animal. However, their most exciting observation was that in at least two animals, the embryonal carcinoma cells were able to form functional sperm, and to therefore pass their genes to offspring. In another series of experiments, cells taken from a spontaneous ovarian teratocarcinoma were also found to be capable of forming functional oocytes (Mintz, personal communication). Thus, these normal embryo-derived and parthenote-derived teratocarcinoma stem cells that had been carried as malignant tumors for as long as eight years were apparently totipotent and were able to respond to normal developmental signals when placed in the appropriate environment. It is also important to note that none of the chimeric mice they obtained ever developed tumors.

An interesting aspect of their studies (Mintz, personal communication) is that in a few cases teratocarcinomas did form in mice that developed from blastocysts injected with single embryonal carcinoma cells. However, all the normal tissues of these animals were of the host blastocyst genotype. Thus it appears that not all injected embryonal carcinoma cells will necessarily respond to the normal signals for development, even if they have the capacity to do so. One possible reason for this is that they may not become properly integrated into the embryonic inner cell mass. Alternatively, it is possible that not all individuals in the embryonal carcinoma cell population have the same developmental potential.

In another group of experiments conducted in Bar Harbor, single embryonal carcinoma cells with limited developmental potential were obtained from neuroteratocarcinomas and injected into mouse blastocysts (Illmensee, personal communication). The mice that were born from these injected embryos had multiple neuroteratocarcinomas, but none of these animals had normal tissue contributions from the injected cells. In this case, since no mice with chimeric tissues were obtained, it seems likely that the stem cells did not have the ability to respond to the normal embryonic environment.

ESTABLISHED EMBRYONAL CARCINOMA CELL LINES

A somewhat different approach was taken by Papaioannou *et al.*, (1975; 1978), who carried our similar experiments, but who injected embryonal carcinoma cells from four different established cell lines. Their results were the first to demonstrate that cultured embryonal carcinoma cells can participate with normal embryonic cells in the formation of chimeric mice. However, in contrast to the results with cells that had not been cultured *in vitro*, these mice usually developed tumors. In addition, apparently none of the chimeric mice produced from the cultured cells had functional teratocarcinoma-derived germ cells.

It is extremely important to determine the reasons for the differences that are observed between teratocarcinoma stem cells propagated by transplantation *in vivo* and those that have been cultured as established cell lines *in vitro*. If embryonal carcinoma cells are ever to be used as a means of introducing selected mutations into the mouse genome (see below) then teratocarcinoma stem cells cultured *in vitro* must be capable of forming functional germ cells in normal, healthy, tumor-free mice. There is now evidence that some embryonal carcinoma cell lines can form chimeric mice that are tumor-free (Dewey *et al.*,1978).

The reasons why the cell lines tested by Papaioannou and her collaborators formed tumors are not yet known. One possibility is that the tumors were formed as a consequence of the fact that in most of the experiments more than one cell was injected. Thus, while some of the cells injected into a single blastocyst may have participated in normal development the others did not. This could be either because some of the cells did not enter the environment necessary for normal differentiation, or because the cell population was heterogeneous and some individual cells were not capable of responding to the normal signals for development. In either case, such cells could give rise to the tumors that were observed. An alternative explanation is that some of the cell lines that they injected were homogeneous populations with relatively little ability to undergo normal differentiation in an embryonic environment, possibly as a consequence of the way in which they had been cultured *in vitro*. For example, some embryonal carcinoma cells, while classified as pluripotent, have relatively little propensity to differentiate. Such cells form teratocarcinomas in which there may be many differentiated cell types, but there is also a very large amount of undifferentiated embryonal carcinoma (see Figure 1). If the injected cells were of this type, this might account for the high incidence of tumors. This idea could be tested by injecting single cells of these different established lines and determining if the mice that develop contain both tumor and normal tissues of the injected cell genotype.

As noted above, another potential limitation of using cultured embryonal carcinoma cells is that they have not yet been demonstrated to be capable of forming functional germ cells in chimeric mice. At present, the most likely explanation of this is that all of the embryonal carcinoma cell lines that have been found to undergo normal differentiation in an embryonic environment have been aneuploid (McBurney, 1976; Nicolas, *et al.*,1976; Iles and Evans, 1977; Cronmiller and Mintz, unpublished). In some cases, the detectable chromosomal abnormalities have been relatively small. These may, however, have placed any teratocarcinoma-derived germ cells at a selective disadvantage as compared with the host embryo-derived germ cells.

The high incidence of aneuploidy among established embryonal carcinoma cell lines is somewhat surprising. For some time teratocarcinoma stem cells have been known to be unusual as compared with other mouse cell lines, in that they retain a relatively normal chromosomal number even after long periods of *in vitro* culture (reviewed by Martin, 1975; Graham, 1977). Recently, however, when the cell karyotypes have been analysed by the more sophisticated methods of chromosome banding, it has been found that almost all teratocarcinoma stem cell lines have some abnormalities (McBurney, 1976; Martin *et al.*,1978; reviewed by Graham, 1977). It is important to keep this in mind, not only when assessing the results of blastocyst injection experiments, but also of other studies with teratocarcinoma stem cells.

Two groups have now established embryonal carcinoma cell lines that appear to have completely normal karyotypes, as assessed by chromosome banding techniques. One of these is female (Papaioannou, personal communication) and the other is male (Jacob, personal communication). It obviously is important to determine whether these cells can form chimeric mice, and if so, whether any of them can become functional germ cells.

As yet, blastocyst injection studies have been carried out with only the XX-euploid cell line (Papaioannou, Graham and Gardner, unpublished data). Following injection, none of the 287 animals analyzed contained any detectable teratocarcinoma-derived cells. A variety of host embryo strains, injection techniques, and cell treatments were employed, and there is as yet no explanation for the failure of this euploid cell line to form chimeric animals. It has, however, been observed that, in general, teratocarcinoma cells injected into blastocysts produce less chimerism than do normal embryonic cells injected into genetically different blastocysts. This is true both for the total number of chimeras produced and the amount and distribution of chimeric tissue in each animal (discussed by Papaioannou *et al.*, 1978). Whatever the reason for this low frequency of chimera formation, the euploid-XX cell line may have some characteristic, perhaps as a consequence of its *in vitro* culture history, that puts it at an even greater competitive disadvantage against embryonic cells than are other embryonal carcinoma cell lines.

MUTANT EMBRYONAL CARCINOMA CELLS

From the results described above, it is apparent that the tumor cells could provide a means of generating mice with mutant tissues by selecting specific mutant embryonal carcinoma cells and injecting these into blastocysts. The demonstration by Mintz and her colleagues that some embryonal carcinoma cells are capable of forming functional germ cells suggests that it should be possible to go even further and introduce such mutations into the mouse genome. In cases where the specific biochemical lesion that causes a heritable human disorder is known and can be selected for *in vitro*, embryonal carcinoma cells could theoretically be used to create a strain of mice that carry the lesion. These might then serve as a laboratory model of the disease.

Experiments of this nature have been carried out using embryonal carcinoma cells that were deficient in hypoxanthine-guanine phosphoribosyl transferase (HGPRT) activity (Dewey *et al.*,1978). Some of the mice that developed from the injected blastocysts were chimeric, containing both host-derived tissues with normal enzyme activity and teratocarcinoma-derived HGPRT-deficient tissues. These mice were completely normal and showed none of the symptoms that occur in humans with severe HGPRT-deficiency (Lesch-Nyhan Syndrome). Although the experiments did not accomplish the ultimate aim of introducing the HGPRT-deficiency mutation into a mouse strain, they did demonstrate that embryonal carcinoma cells from cultures that have been mutagenized and selected for a specific enzyme deficiency are capable of forming chimeric mice that do not develop tumors.

In view of the remarkable success of all of these studies on teratocarcinoma-injected blastocysts, the question that is most frequently asked is whether teratocarcinoma stem cells can form the whole of a mouse in the absence of normal embryonic cells. However, since embryonal carcinoma cells, even those that differentiate in a manner that closely parallels normal embryogenesis, apparently do not form trophectoderm, they could hardly be expected to implant in the uterus in the normal way. The question must therefore be rephrased: can embryonal carcinoma cells form a whole mouse if they are placed in a trophoblastic vesicle? This question is difficult to answer, since the necessary experiments are technically difficult. In addition, it is unlikely that adequate experiments can be performed until euploid teratocarcinoma stem cells that form chimeras with high frequency are available. Considering the remarkable results that have been reported since 1974 on teratocarcinoma-embryo interactions, it seems likely that the answer to this question will be known relatively soon.

CONCLUSIONS

The data described above can be summarized to a large extent by the diagram shown in Figure 9, which illustrates the relationship between embryonal carcinoma cells and normal embryonic cells. The isolation of pluripotent cell lines directly from normal embryos has yet to be accomplished and is therefore included as a dotted line in the figure. The failure to establish such embryonic cell lines is certainly not due to a lack of effort, since many laboratories have been attempting to do this. The lack of success could be interpreted as an indication that embryonal carcinoma cells are not identical to normal embryonic cells. The latter may have to undergo some "malignant" change before they can become permanently proliferative in the undifferentiated state. On the other hand, it may be that there are very few embryonal carcinoma cell precursors in the embryo and these might have a low efficiency of *in vitro* growth. Thus, when they proliferate *in vivo* to form a tumor, the number of such cells is increased and it therefore becomes less difficult to establish cell lines from the population.

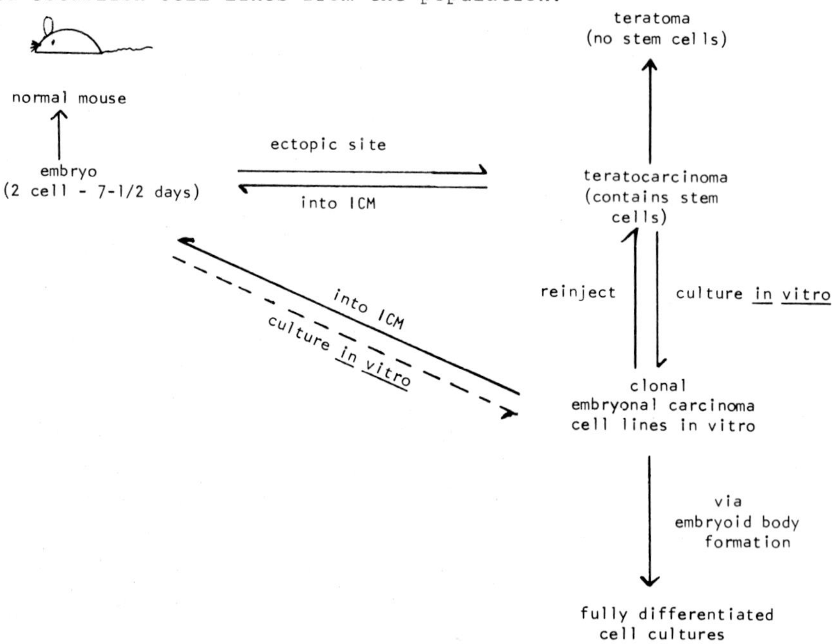

Figure 9: Diagram of the relationship between normal embryonic cells and embryonal carcinoma cells. From Martin (1977).

If a "malignant" change is involved, it is obviously a reversible one, considering the benign nature of the differentiated tumor derivatives of embryonal carcinoma cells and also the relatively high efficiency with which the stem cells can participate in the formation of apparently normal mice. As discussed by Pierce (1967), the malignancy of embryonal carcinoma cells could be the consequence of an epigenetic change in gene expression. Alternatively, it is possible that it could be due to a reversible genetic change such as the insertion or excision of specific DNA sequences (see for example, the work on Crown Gall tumor, Chilton *et al.*,1977). Obviously, more experiments are needed in order to determine whether there are specific changes that must occur before an embryonic cell becomes an embryonal carcinoma cell.

If there are intrinsic differences between embryonal carcinoma cells and their embryonic precursors, these might represent a potential limitation to the usefulness of the tumor cells as models of normal development. Another problem is that embryonal carcinoma cells are probably more often than not altered as a consequence of transplantation *in vivo* or exposure to selective culture conditions. Such genetic and/or epigenetic changes may invalidate to some extent any conclusions about normal development that are based on studies with these cells. Yet another difficulty, as discussed above, is that we still do not know what specific embryonic cells the teratocarcinoma stem cells most closely resemble. The morphological and biochemical studies that have been carried out thus far suggest that some, if not all, embryonal carcinoma cells are perhaps of little use as a model of the earliest differentiative event in mammalian embryogenesis, the formation of inner cell mass and trophectoderm.

On the other hand, the remarkable ability of some embryonal carcinoma cells to mimic the morphological and biochemical events of early mammalian development *in vitro* and also to participate in completely normal development *in vivo*, suggest that these cells will be invaluable in future studies. In many instances these cells represent the only practical means of studying the molecular aspects of the peri-implantation period of development. As we become more aware of their limitations and understand more fully what embryonal carcinoma cells are, it should be possible to use them more effectively to study early development. In addition to the experimental approaches discussed above, there are many other ways in which teratocarcinoma stem cells can be used to study early development. For example, it should be possible to isolate stem cells from tumors derived from mutant embryos: obviously, clonal embryonal carcinoma cell lines homozygous for a mutation that affects early development would be extremely useful. Similarly, human embryonal carcinoma cells that can mimic normal early human development *in vitro* would provide a unique means of studying the early stages of human embryogenesis.

ACKNOWLEDGEMENTS

I am grateful to Drs. G.S. Martin and B. Zetter for their helpful suggestions, and to Mr. D. Akers for his expert help in preparing the photographic material. My own work is supported by the American Cancer Society (grant No. VC-246) and the National Science Foundation (grant No. PCM-7705403).

REFERENCES

ARTZT, K. & JACOB, F. (1974) Absence of serologically detectable H-2 on primitive teratocarcinoma cells in culture. Transplantation 17, 632-634.

ARTZT, K., DUBOIS, P., BENNETT, D., CONDAMINE, H., BABINET, C. & JACOB, F. (1973) Surface antigens common to mouse cleavage embryos and primitive teratocarcinoma cells in culture. Proc. Nat. Acad. Sci. USA 70, 2988-2992.

ARTZT, K., BENNETT, D. & JACOB, F. (1974) Primitive teratocarcinoma cells express a differentiation antigen specified by a gene at the T-locus in the mouse. Proc. Nat. Acad. Sci. USA 71, 811-814.

BERNSTINE, E.G., HOOPER, M.L., GRANDCHAMP, S. & EPHRUSSI, B. (1973) Alkaline phosphatase activity in mouse teratoma. Proc. Nat. Acad. Sci. USA 70, 3899-3903.

BRINSTER, R.L. (1974) The effect of cells transferred into the mouse blastocyst on subsequent development. J. Exp. Med. 140, 1049-1056.

BRINSTER, R.L. (1975) Can teratocarcinoma cells colonize the mouse embryo? In: M. Sherman and D. Solter (Eds.), Roche Symposium on Teratomas and Differentiatiation, Academic Press, New York, p. 51-58.

CHILTON, M.-D., DRUMMOND, M.H., MERLO, D.J., SCIAKY, D., MONTOYA, A.L., GORDON, M.P. & NESTER, E.W. (1977) Stable incorporation of plasmid DNA into higher plant cells: the molecular basis of crown gall tumorigenesis. Cell 11, 263-271.

CHUNG, A.E., ESTES, L.E., SHINOZUKA, H., BRAGINSKI, J., LORZ, C. & CHUNG, C.A. (1977) Morphological and biochemical observations on cells derived from the *in vitro* differentiation of the embryonal carcinoma cell line PCC4-F. Cancer Res. 37, 2072-2081.

DAMJANOV, I. & SOLTER, D. (1974a) Experimental teratoma. Current Topics in Pathology 59, 69-130.

DAMJANOV, I. & SOLTER, D. (1974b) Host-related factors determine the outgrowth of teratocarcinomas from mouse egg-cylinders. Z. Krebsforsch. 81, 63-69.

DAMJANOV, I., SOLTER, D. & SERMAN, D. (1973) Teratocarcinoma with the capacity for differentiation restricted to neuro-ectodermal tissue. Virchows Arch. B. Zellpath. 13, 179-195.

DAMJANOV, I., SOLTER. & SKREB, N. (1971) Teratocarcinogenesis as related to the age of embryos grafted under the kidney capsule. Wilhelm Roux' Archiv. 167, 288-290.

DEWEY, M.J., GEARHART, J.D. & MINTZ, B. (1977) Cell surface antigens of totipotent mouse teratocarcinoma cells grown *in vivo:* their relation to embryo, adult and tumor antigens. Devel. Biol. 55, 358-374.

DEWEY, M.J., MARTIN, D.W.M., JR., MARTIN, G.R. & MINTZ, B. (1978) Mosaic mice with teratocarcinoma-derived mutant cells deficient in hypoxanthine phosphoribosyl transferase. Proc. Nat. Acad. Sci. USA 74, 5564-5568

DIWAN, S.B. & STEVENS, L.C. (1976) Development of teratomas from the ectoderm of mouse egg cylinders. J. Nat. Cancer Inst. 57, 937-942.

EDIDIN, M., PATTHEY, H.L., McGUIRE, E.J. & SHEFFIELD, W.D. (1971) An antiserum to "embryoid body" tumor cells that reacts with normal mouse embryos. In: N.G. Anderson and J.H. Coggin, Jr., (Eds.) Embryonic and Fetal Antigens in Cancer, Oak Ridge, Tennessee: Oak Ridge National Laboratory, p. 239-248.

EPPIG, J.J., KOZAK, L.P., EICHER, E.M. & STEVENS, L.C. (1977) Ovarian teratomas in mice are derived from oocytes that have completed the first meiotic division. Nature 269, 517-518.

EVANS, M.J. & MARTIN, G.R. (1975) The differentiation of clonal teratocarcinoma cell cultures *in vitro.* In: M. Sherman and D. Solter, (Eds.), Roche Symposium on Teratomas and Differentiation, Academic Press, New York, p. 237-250.

GACHELIN, G., KEMLER, R., KELLEY, F. & JACOB, F. (1977) PCC4, a new cell surface antigen common to multipotential embryonal carcinoma cells, spermatozoa and mouse early embryos. Devel. Biol. 57, 199-209.

GARDNER, R.L. (1968) Mouse chimaeras obtained by the injection of cells into the blastocyst. Nature 220, 596-597.

GARDNER, R.L. & LYON, M.F. (1971) X-Chromosome inactivation studied by injection of a single cell into the mouse blastocyst. Nature 231, 385-386.

GARDNER, R.L. & PAPAIOANNOU, V.E. (1975) Differentiation in the trophectoderm and inner cell mass. In: M. Balls and A. Wild (Eds.) The Early Development of Mammals, Cambridge University Press, p. 107-132.

GARTLER, S.M. & ANDINA, R.J. (1976) Mammalian X-chromosome inactivation. Adv. in Human Genet. 7, 99-140.

GEARHART, J.D. & MINTZ, B. (1974) Contact-mediated myogenesis and increased acetylcholinesterase activity in primary cultures of mouse teratocarcinoma cells. Proc. Nat. Acad. Sci. USA 71, 1734-1738.

GRAHAM, C.F. (1977) Teratocarcinoma cells and normal mouse embryogenesis. In: M. Sherman (Ed.), Concepts in Mammalian Embryogenesis, MIT Press, Cambridge, Mass., p. 315-394.

HOGAN, B.L.M. (1976) Changes in the behaviour of teratocarcinoma cells cultivated *in vitro.* Nature 263, 136-137.

HOGAN, B.L.M. (1977) Teratocarcinoma cells as a model for mammalian development. In: J. Paul (Ed.), Biochemistry of Cell Differentiation II, Vol. 15, University Park Press, Baltimore, p. 333-376.

HOGAN, B. & TILLY, R. (1978) *In vitro* development of inner cell masses isolated immunosurgically from mouse blastocysts. I. and II. J. Embryol. exp. Morph. (In press).

HSU, Y.-C. & BASKAR, J. (1974) Differentiation *in vitro* of normal mouse embryos and mouse embryonal carcinoma. J. Nat. Cancer Inst. 53, 177-185.

ILES, S.A. & EVANS, E.P. (1977) Karyotypic analysis of teratocarcinomas and embryoid bodies of C3H mice. J. Embryol. exp. Morph. 38, 77-92.

ILES, S.A., McBURNEY, M.W., BRAMWELL, S.R., DEUSSEN, Z.A. & GRAHAM, C.F. (1975) Development of parthenogenetic and fertilized mouse embryos in the uterus and in extra-uterine sites. J. Embryol. exp. Morph. 34, 387-405

ILLMENSEE, K. & MINTZ, B. (1976) Totipotency and normal differentiation of single teratocarcinoma cells cloned by injection into blastocysts. Proc. Nat. Acad. Sci. USA 73, 549-553.

JACOB, F. (1977) Mouse teratocarcinoma and embryonic antigens. Immunol. Rev. 33, 3-32.

KAHAN, B.W. & EPHRUSSI, B. (1970) Developmental potentialities of clonal *in vitro* cultures of mouse testicular teratoma. J. Nat. Cancer Inst. 44, 1015-1036.

KEMLER, R., BABINET, C., CONDAMINE, H., GACHELIN, G., GUENET, J.L. & JACOB, F. (1976) Embryonal carcinoma antigen and the T/t locus of the mouse. Proc. Nat. Acad. Sci. USA 73, 4080-4084.

KEMLER, R., BABINET, C., EISEN, H. & JACOB, F. (1978) Surface antigen in early differentiation. Proc. Nat. Acad. Sci. USA. (In press).

KLEINSMITH, L.J. & PIERCE, G.B. (1964) Multipotentiality of single embryonal carcinoma cells. Cancer Res. 24, 1544-1552.

LEHMAN, J.M., SPEERS, W.C., SWARTZENDRUBER, D.E. & PIERCE, G.B. (1974) Neoplastic differentiation: characteristics of cell lines derived from a murine teratocarcinoma. J. Cell Physiol. 84, 13-28.

LEVINE, A.J., TOROSIAN, M., SAROKHAN, A.J. & TERESKY, A.K. (1974) Biochemical criteria for the *in vitro* differentiation of embryoid bodies produced by a transplantable teratoma of mice. The production of acetylcholinesterase and creatine phosphokinase by teratoma cells. J. Cell. Physiol. 84, 311-317.

LYON, M.F. (1972) X-chromosome inactivation and developmental patterns in mammals. Biol. Rev. 47, 1-35.

MARTIN, G.R. (1975) Teratocarcinomas as a model system for the study of embryogenesis and neoplasia: Review, Cell 5, 229-243.

MARTIN, G.R. (1977) The differentiation of teratocarcinoma stem cells *in vitro:* parallels to normal embryogenesis. In: M. Karkinen-Jaaskelainen, L. Saxen and L. Weiss (Eds.), Cell Interactions in Differentiation, Academic Press, London, p.59-75.

MARTIN, G.R. & EVANS, M.J. (1975a) The differentiation of clonal lines of teratocarcinoma cells: formation of embryoid bodies *in vitro*. Proc. Nat. Acad. Sci. USA 72, 1441-1445.

MARTIN, G.R. & EVANS, M.J. (1975b) The formation of embryoid bodies *in vitro* by homogeneous embryonal carcinoma cell cultures derived from isolated single cells. In: M. Sherman and D. Solter, (Eds.), Roche Symposium on Teratomas and Differentiation, New York: Academic Press, p. 169-187.

MARTIN, G.R. & EVANS, M.J. (1975c) Multiple differentiation of clonal teratocarcinoma stem cells following embryoid body formation *in vitro*. Cell 6, 230-244.

MARTIN, G.R., WILEY, L.M. & DAMJANOV, I. (1977) The development of cystic embryoid bodies *in vitro* from clonal teratocarcinoma stem cells. Devel. Biol. 61, 69-83.

MARTIN, G.R., EPSTEIN, C.J., TRAVIS, B., TUCKER, G., YATZIV, S., MARTIN, D.W.M., Jr., CLIFT, S. & COHEN, S. (1978) X-chromosome inactivation during differentiation of female teratocarcinoma stem cells *in vitro*. Nature 271, 327-333.

McBURNEY, M.W. (1976) Clonal lines of teratocarcinoma cells *in vitro:* differentiation and cytogenetic characteristics. J. Cell. Physiol. 89, 441-455.

McBURNEY, M.W. & ADAMSON, E.D. (1976) Studies on the activity of the X-chromosomes in female teratocarcinoma cells in culture. Cell 9, 57-70.

McLAREN, A. (1976) Mammalian Chimaeras, Cambridge University Press, London.

MINTZ, B. (1962) Formation of genotypically mosaic mouse embryos. Amer. Zool. 2, 432 (Abstr. 310).

MINTZ, B. & ILLMENSEE, K. (1975) Normal genetically mosaic mice produced from malignant teratocarcinoma cells. Proc. Nat. Acad. Sci. USA 72, 3585-3589.

MINTZ, B., ILLMENSEE, K. & GEARHART, J.D. (1975) Developmental and experimental potentialities of mouse teratocarcinoma cells from embryoid body cores. In: M. Sherman and D. Solter (Eds.), Roche Symposium on Teratomas and Differentiation, New York: Academic Press, p. 59-82.

MUKHERJEE, A.B. (1976) Cell cycle analysis and X-chromosome inactivation in the developing mouse. Proc. Nat. Acad. Sci. USA 73, 1608-1611.

NICOLAS, J.-F., DUBOIS, P., JAKOB, H., GAILLARD, J. & JACOB, F. (1975) Teratocarcinome de la souris: Differenciation en culture d'une lignee de cellules primitives a potentialites multiples. Ann. Microbiol. (Inst. Pasteur) 126A, 3-22.

NICHOLAS, J.-F., AVNER, P., GAILLARD, J., GUENET, J.L., JAKOB, H. & JACOB, H. (1976) Cell lines derived from teratocarcinomas. Cancer Res. 36, 4224-4231.

PAPAIOANNOU, V.E., McBURNEY, M.W., GARDNER, R.L. & EVANS, M.J. (1975) Fate of teratocarcinoma cells injected into early mouse embryos. Nature 258, 70-73.

PAPAIOANNOU, V.E., GARDNER, R.L., McBURNEY, M.W., BABINET, C. & EVANS, M.J. (1978) Participation of cultured teratocarcinoma cells in mouse embryogenesis. J. Embryol. exp. Morphol. (In press).

PEDERSEN, R.A., SPINDLE, A.I. & WILEY, L.M. (1977) Regeneration of endoderm by ectoderm isolated from mouse blastocysts. Nature 270, 435-437.

PIERCE, G.B. Jr. (1967) Teratocarcinoma: model for a developmental concept of cancer. Current Top. Devel. Biol. 2, 223-246.

PIERCE, G.B. & DIXON, F.J. (1959) Testicular teratomas. I. Demonstration of teratogenesis by metamorphosis of multipotential cells. Cancer 12, 573-583.

PIERCE, G.B., Jr., DIXON, F.J. & VERNEY, E.L. (1960) Teratocarcinogenic and tissue-forming potentialities of the cell types comprising neo-plastic embryoid bodies. Lab. Invest. 9, 583-602.

PIERCE, G.B., Jr., & VERNEY, E.L. (1961) An *in vitro* and *in vivo* study of differentiation in teratocarcinomas. Cancer 14, 1017-1029.

PIERCE, G.B., Jr., MIDGLEY, A.R., Jr., RAM, J.S. & FELDMAN, J.D. (1962) Parietal yolk sac carcinoma: a clue to the histogenesis of Reichert's membrane of the mouse embryo. Amer. J. Path. 41, 549-566.

ROSENTHAL, M.D., WISHNOW, R.M. & SATO, G.H. (1970) *In vitro* growth and differentiation of clonal populations of multipotential mouse cells derived from a transplantable testicular teratocarcinoma. J. Nat. Cancer Inst. 44, 1001-1014.

ROSSANT, J. (1975) Investigation of the determinative state of the mouse inner cell mass II: The fate of isolated inner cell masses transferred to the oviduct. J. Embryol. exp. Morph. 33, 991-1001.

SHERMAN, M.I. (1975) Differentiation of teratoma cell line PCC4: aza 1 *in vitro*. In: M. Sherman and D. Solter (Eds.), Roche Symposium on Teratomas and Differentiation, Academic Press, New York, p. 189-205.

SHERMAN, M.I. & MILLER, R.A. (1978) F9 embryonal carcinoma cells can differentiate into endoderm-like cells. Develop. Biol. (In press).

SOLTER, D. & DAMJANOV, I. (1973) Explantation of extraembryonic parts of 7-day-old mouse egg cylinders. Experientia 29, 701.

SOLTER, D. & KNOWLES, B. (1975) Immunosurgery of mouse blastocyst. Proc. Nat. Acad. Sci. USA 72, 5099-5102.

SOLTER, D., SKREB, N. & DAMJANOV, I. (1970) Extrauterine growth of mouse egg cylinders results in malignant teratoma. Nature 227, 503-504.

SOLTER, D., BICZYSKO, W., PIENKOWSKI, M. & KOPROWSKI, H. (1974) Ultrastructure of mouse egg cylinders developed *in vitro*. Anat. Rec. 180, 263-279.

SOLTER, D., DAMJANOV, I. & KOPROWSKI, H. (1975) Embryo-derived teratoma: a model system in developmental and tumor biology. In: M. Balls and A. Wild (Eds.), The Early Development of Mammals, Cambridge University Press, p. 243-264.

STERN, P.L., MARTIN, G.R., & EVANS, M.J. (1975) Cell surface antigens of clonal teratocarcinoma cells at various stages of differentiation. Cell, 6, 455-465.

STEVENS, L.C. (1958) Studies on transplantable testicular teratomas of strain 129 mice. J. Nat. Cancer Inst. 20, 1257-1276.

STEVENS, L.C. (1959) Embryology of testicular teratomas in strain 129 mice. J. Nat. Cancer Inst. 23, 1249-1295.

STEVENS, L.C. (1960) Embryonic potency of embryoid bodies derived from a transplantable testicular teratoma of the mouse. Develop. Biol. 2, 285-297.

STEVENS, L.C. (1967) The biology of teratomas. Adv. Morph. 6, 1-31.

STEVENS, L.C. (1968) The development of teratomas from intratesticular grafts of tubal mouse eggs. J. Embryol. expt. Morph. 20, 329-341.

STEVENS, L.C. (1970) The development of transplantable teratocarcinomas from intratesticular grafts of pre- and post-implantation mouse embryos. Develop. Biol. 21, 364-382.

STEVENS, L.C. (1975) Comparative development of normal and parthenogenetic mouse embryos, early testicular and ovarian teratomas, and embryoid bodies. In: M. Sherman and D. Solter (Eds.), Roche Symposium on Teratomas and Differentiation, Academic Press, New York, p. 17-32.

STEVENS, L.C. & VARNUM, D.S. (1974) The development of teratomas from parthenogenetically activated ovarian mouse eggs. Devel. Biol. 37, 369-380.

STRICKLAND, S., REICH, E. & SHERMAN, M.I. (1976) Plasminogen activator in early embryogenesis: enzyme production by trophoblast and parietal endoderm. Cell, 9, 231-240.

TAKAGI, N. (1974) Differentiation of X-chromosomes in early female mouse embryos. Exp. Cell Res. 86, 127-135.

TARKOWSKI, A.K. (1961) Mouse chimaeras developed from fused eggs. Nature 190, 857-860.

TERESKY, A.K., MARSDEN, M., KUFF, E.L. & LEVINE, A.J. (1974) Morphological criteria for the *in vitro* differentiation of embryoid bodies produced by a transplantable teratoma of mice. J. Cell. Physiol. 84, 319-332.

WILEY, L.M., SPINDLE, A.I. & PEDERSEN, R.A. (1978) Morphology of isolated mouse inner cell masses developing *in vitro*. Devel. Biol. (In press).

MAMMALIAN PRIMORDIAL GERM CELLS

JOHN K. HEATH

Department of Zoology
University of Oxford
South Parks Road,
Oxford OX1 3PS

1. Ultrastructure of PGCs
2. Alkaline phosphatase
3. PGC migration
4. Isolation and culture of PGCs
5. Mutations affecting PGCs
   W, dominant white spotting
   Sl, Steel
6. The origin of PGCs and teratocarcinomas
   Partition
   Interaction
   Germ plasm
7. Conclusions

It is the germ cells which give rise to the gametes in all metazoans. In this chapter, I shall be concerned with the development of these cells prior to the sexual differentiation of the gonad in the mammal (13.5 days post coitum in the mouse). At this stage the cells are usually referred to as 'Primordial Germ Cells' (PGCs).

The origin and development of germ cells has been of interest to embryologists for over 100 years. The interest was stimulated by the conflict between the theories of Waldeyer (1870), who proposed that the germ cells arose by proliferation from the germinal epithelium (or the somatic elements of the gonad), and Weismann (1892) who put forward the idea that the germ line was continuous from generation to generation, and was determined by inheritance of 'germ plasm' from the egg. Evidence has accumulated from a number of sources to support both theories in different systems (reviewed by Eddy, 1975).

Amongst the earliest studies of the germ cell lineage in mammals were those of Rubashkin (1908, 1912) in the guinea pig; Allen (1904) in the pig and rabbit; Fuss (1911, 1912) in the human and guinea pig; De Winiwarter and Saintmont (1909) in the cat; Vannemann (1917) in the armadillo, and Jenkinson (1913) in the mouse. These authors identified a class of cells (PGCs) in the hind gut and associated tissues of the early embryo which migrated into the gonad as it began to form. These cells had a distinctive morphology; large nuclei, prominent nucleoli, lightly staining with conventional stains, and containing prominent perinuclear mitochondria arranged in rods or chains. The exact significance of the cells was disputed. Four different views were put forward (reviewed by Heys, 1931, and Everrett, 1943):

(a) The PGCs migrate into the gonad where they degenerate and are replaced by the definitive germ cells formed by proliferation from the germinal epithelium: in the human (Felix, 1912), in the mouse (Firket, 1920; Kingery, 1917) and in the cat (De Winiwarter and Sainmont, 1909; Kingsbury, 1913).

(b) There was no early segregation of the germ line, and the germ cells arose directly from the germinal epithelium: in the rat (Hargitt, 1925, 1930; Simkins, 1923) and in the human (Simkins, 1928; Willis, 1962). These authors claimed that the distinguishing features of PGCs were vague and could refer to cells about to undergo division or in a state of high metabolic activity.

(c) The PGCs migrate into the gonad where they form germ cells, but their number is periodically added to by proliferation from the germinal epithelium (Allen, 1904).

(d) The PGCs were the strict precursors of the definitive germ cells: in the armadillo (Vanneman, 1917), in the mouse (Jenkinson, 1913) and in the human (Witschi, 1948). Vanneman claimed that the germ line of the armadillo was segregated at the primitive streak stage through the influence of the ectoderm upon the underlying endoderm, at a point where the two layers came in contact.

Experimental evidence bearing upon this controversy came from three types of experiment. Everrett (1945) grafted embryonic gonads under the kidney capsule at a stage before the arrival of the PGCs. He found that when the graft developed it did not contain any recognisable gonocytes. If the grafts were made at a later stage, when the PGCs were present, the gonads did contain gonocytes. This shows that PGCs are required for the formation of definitive gonocytes.

Brambell and Parkes (1927) X-irradiated the ovaries of adult female mice and found, upon analysis some weeks later, that although in some cases all the germ cells were destroyed, there was no evidence for regeneration of germ cells from the germinal epithelium. Mintz (1958) X-irradiated mouse foetuses at the stage of germ cell migration and found that the PGCs were destroyed and that the gonads that developed did not contain germ cells, an observation indicating again that PGCs are required to form gonocytes. Merchant (1975) treated pregnant rats with Busulphan, a drug which destroys dividing PGCs in the early gonad. The gonads of foetuses from treated rats were found to be devoid of germ cells. Although the effects of Busulphan on the somatic component of the gonad are unknown, this observation supports the view that PGCs are the direct precursors of adult germ cells.

A number of workers (Rudkin and Greich, 1962; Peters *et al.*,1962; Kennelly and Foote, 1966; Borum, 1966; Peters and Crone, 1967) have labelled radioactively the DNA of female PGCs prior to meiosis and subsequently found the label in the gonocytes and gametes of the adult, indicating that at least some PGCs form mature gametes.

As stated above, part of the controversy regarding the origin of germ cells stemmed from the fact that the criteria for recognising germ cells were very vague. A considerable advance came from the discovery that the PGCs of many mammalian species had high levels of the enzyme alkaline phosphatase (APase), and so could be distinguished by histochemical procedures (Baxter, 1952). There is now considerable published information on the early history of the germ line using APase as a marker in the human (Baxter, 1953; McKay *et al.*,1953); Pinkerton *et al.*,1961; Falin, 1969, who also used the high glycogen content of human PGCs as a marker); in the bovine (Jost and Prepin, 1966) and in the mouse (Chiquone, 1954; Mintz and Russell, 1957; Mintz, 1960; Odenzki, 1967). The only species so far described where the APase marker seems to be ineffective is the rabbit (Chretien, 1965). The most detailed description of PGC migration using APase histochemistry is that for the mouse, although the information seems generally applicable to other mammalian species.

The earliest appearance of distinctive APase positive PGCs is at about the time of head fold formation (7.5-8d.pc). The cells are found in the caudal end of the primitive streak and the base of the allantois (which is formed by outgrowth of the caudal end of the primitive streak into the exocoelom). A few hours later they are found in the base of the allantois and in the endoderm of the yolk sac adjacent to the allantois. They have also moved laterally away from the mid-sagittal plane of the embryo. By 9.5d.pc the hind gut has been formed by invagination of the endoderm overlying the primitive streak, and the embryo has turned so that its curvature is reversed (Figure 1). The PGCs are now located in the base of the allantois and in the hind gut, which joins directly onto the base of the allantois (Figure 2). The PGCs move caudally along the ventral face of the gut. By 10d.pc the genital ridges have started to form as paired longitudinal thickenings of the epithelium lining the dorsal aspect of the coelom, posterior to the mesonephros. The PGCs, in some cases now as far forward as the full extent of the genital ridges, start to move dorsally and laterally through the dorsal mesentery of the hind gut and around the angle where the mesentery joins the posterior wall into the genital ridges. By day 11, most of the PGCs are located in the genital ridges.

During the process of migration, the number of PGCs increases from about 10-100 on day 8 to 2500-5000 on day 12. Mitotic figures can be seen in PGCs of the hind gut and genital ridges, but no detailed analysis of mitotic rates has been performed. Occasionally ectopic PGCs can be found in the skin epithelium or the extraembryonic regions of the yolk sac adjacent to the placenta.

ULTRASTRUCTURE OF PGCs

The APase marker has proved useful for identifying the migrating PGCs of the mouse at the ultrastructural level (Jeon and Kennedy, 1973; Clark and Eddy, 1975). Other workers (Speiglman and Bennett, 1973; Zamboni and Merchant, 1973) have made use of the fact that migrating PGCs are basophilic and so have a selective affinity for toluidine blue dye.

The migrating PGCs are relatively undifferentiated at the ultrastructural level and bear a strong resemblance to the

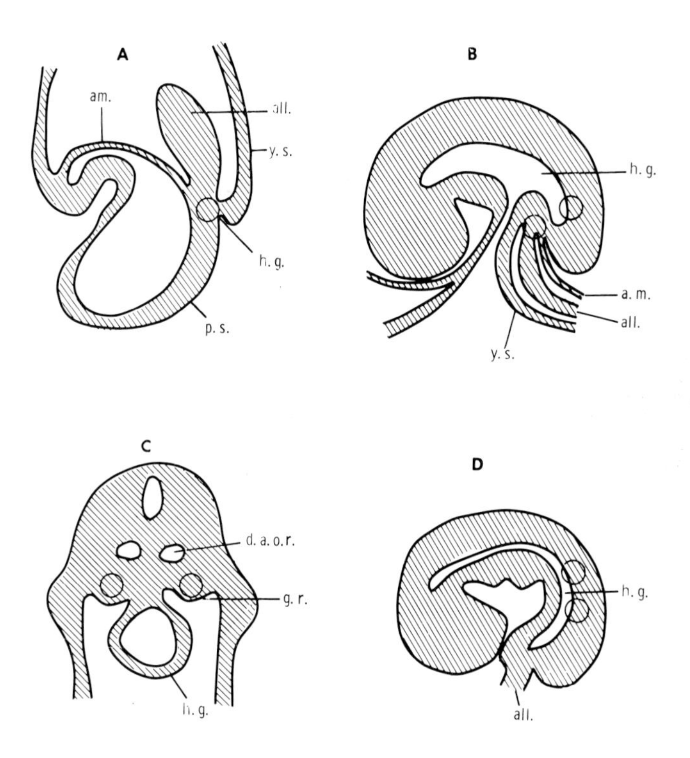

Figure 1: Diagrammatic representation of PGC location in mouse embryos of different ages. Circles represent PGC location. A. 8.5 d.pc sagittal section (am - amnion; all - allantois; y.s. - yolk sac; h.g. - hind gut; p.s. - primitive streak). B. 9.5 d.pc sagittal section. C. 10.5 d.pc transverse section (D.aor - dorsal aorta; g.r. - genital ridge). D. 10.5 d.pc sagittal section

ectodermal and mesodermal cells of the earlier embryo. They contain large amounts of free ribosomes (explaining their basophilic properties) but relatively little endoplasmic reticulum. The mitochondria are large and round or oval in shape with few cristae. The nucleus is large and rounded and has a granulated nucleoplasm, the nucleoli are prominent and reticulate. Occasionally the cytoplasm appears to protrude into the nucleus, either in the form of large indentations or as fine cytoplasmic protrusions (Jeon and Kennedy, 1973).

A feature thought to be unique to PGCs are electron dense granular bodies, often found in association with the nuclear or mitochondrial membranes. Eddy (1974) has proposed that these bodies represent 'nuage' which is observed in the germ cells of a wide range of species (for review see Eddy, 1975).

It is further proposed that this substance represents the remnants of the 'germ plasm' found in insects and amphibia and demonstrated to be involved in germ cell determination in these species (Okada *et al.*,1974; Illmensee *et al.*,1975; Whitingtonand Dixon, 1976). However, the distribution of nuage throughout the germ line or somatic tissue of mammals is not known in detail, and no experimental evidence for such a determinative role for nuage is available. Another feature of migrating PGCs are cytoplasmic vesicles. These could be the result of pinocytotic or exocytotic activity by the cells

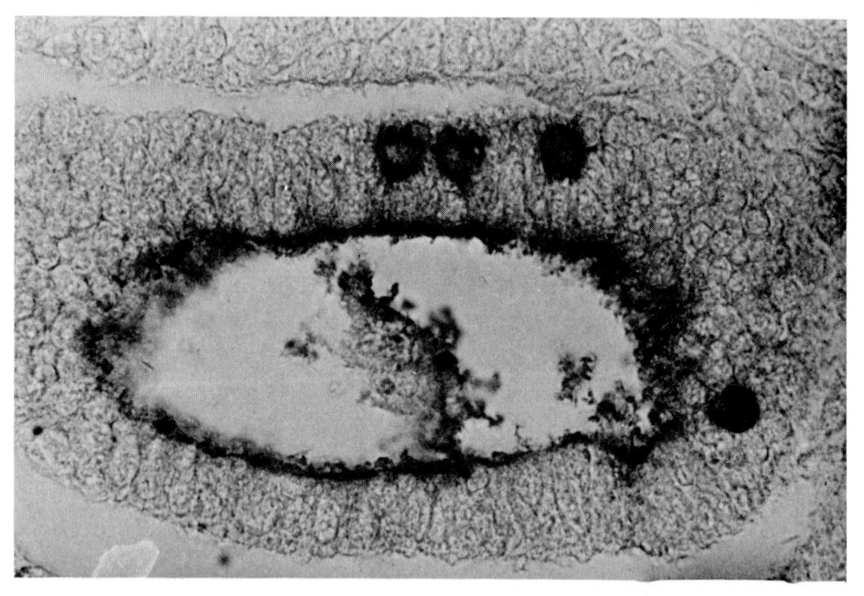

Figure 2: Hind gut of 10 d.pc embryo stained for APase. Note four positively staining PGCs and positively staining epithelium of hind gut. X 300.

(Zamboni and Merchant, 1973), or fixation artefacts resulting from disrupted organelles (Speiglman and Bennett, 1973). Speiglman and Bennett observed that there were portions of endoplasmic reticulum which had one side of the bilaminar membrane missing, creating a 'cisternal space'. These produce a clear crescent-shaped area in the PGCs at the light microscope level. These spaces are probably due to fixation artefacts, and cannot be considered unique to PGCs since they can be seen in ultrastructural studies of other regions of mammalian embryos(Morriss and Steele, 1977).

As the PGCs migrate through the gut they form close associations with the surrounding somatic tissue. This takes the form of membrane interdigitations (Zamboni and Merchant, 1973) or small focal tight junctions (Speiglman and Bennett, 1973).

Electron microscopical studies of migrating human PGCs show considerable similarities with the PGCs of the mouse, with the exception of glycogen storage granules, which appear to be a unique feature of human PGCs (Fukada, 1976; Fujimoto *et al.*, 1977).

Considerable information is available on the ultra-structural appearance of PGCs upon arrival at the genital ridge of the mouse (Pierce and Beals, 1964; Odor and Blandau, 1969; Clarke and Eddy, 1975; Zamboni and Merchant, 1973; Speiglman and Bennett, 1973); of the rat (Eddy, 1974; Merchant, 1975; Franchi and Mandl, 1962, 1964); of the pig (Pelliniemi, 1975a, 1975b, 1976); of the golden hamster (Weakley, 1976); of the rabbit (Gondos and Connor, 1973); and of the guinea-pig (Black and Christiansen, 1969). These studies again reveal a remarkable similarity between the PGCs of this wide range of species. The PGCs at this stage are very similar in appearance to the earlier migrating cells. There is an increase in the amount of endoplasmic reticulum, the golgi apparatus is more developed and the cells are rounded or rectangular in shape. The PGCs are found in small clumps, linked together by cytoplasmic bridges, probably attributable to cell division followed by incomplete cytokinesis (Pierce and Beals, 1964). The cells are again closely associated with the surrounding somatic tissue by means of interdigitating membranes, focal tight junctions and gap junctions (Zamboni and Merchant, 1973; Speiglman and Bennett, 1973).

The ultrastructural work reveals two things. Firstly, PGCs seem to be closely associated with the surrounding tissue during their migration and settlement in the gonad, possibly indicating that they are involved in interactions with their environment. Secondly, the PGCs reveal remarkably little in the way of distinctive ultrastructural features, in fact if anything, they closely resemble the embryonic ectoderm and mesoderm of earlier embryonic stages.

ALKALINE PHOSPHATASE

One of the problems in mouse embryology is the paucity of suitable molecular markers to define specific cell types. It can be seen from the previous section that APase has proved useful in unravelling some of the problems associated with the early history of the mammalian germ line. However, the use of APase as a specific marker for germ cells is subject to a number of important limitations which render its use for a more detailed analysis of the early history of the germ line very difficult for several reasons:

(a) APase is not unique to the germ line. It is found in a number of adult tissues (Fernley, 1971; Fishman, 1974), preimplantation and postimplantation embryos (Damjanov *et al.*, 1977; Solter *et al.*,1973), and in the early stages of organogenesis in a variety of embryonic tissues (Moog, 1965). For example, at the time of the migratory phase of PGCs in the mouse embryo, the epithelium of the gut and the cells of the developing nervous system also stain positively for APase, using conventional histochemical techniques (Figure 2).

Different forms of the enzyme have been identified, on the basis of their electrophoretic properties (Fishman, 1974), immunological properties (Sussman *et al.*,1968), sensitivity to inhibition by various amino acids (Bernstine *et al.*,1973) and heat denaturation properties (Fishman, 1974; Fernley, 1971; Bernstine *et al.*,1973). It is not known which form of APase is expressed in PGCs. It is known that embryonal carcinoma cells express a form of APase, identified by DEAE column chromatography and polyacrylamide gel electrophoresis, which is different from the forms in kidney and placenta (Bernstine and Ephrussi, 1975; Wada *et al.*, 1976). It is possible that this form is shared with PGCs, but the criteria of identification of this form are not readily applicable to the identification of PGCs in sectioned material.

(b) The expression of APase in tissue culture can be modified by a number of exogenous agents. For example, it is possible to induce APase in cultured cell lines by addition of cAMP (Koyama and Ono, 1972); short chain fatty acids (Koyama and Ono, 1976); and both steroid and polypeptide hormones (Melnykovytch *et al.*,1967; Cox and MacLeod, 1961). If the same factors operate *in vivo* as *in vitro*, it is possible that some PGCs might escape detection by APase histochemistry, especially those in ectopic sites or other abnormal environments, due to the absence of suitable inducers.

(c) APase activity varies with the cell cycle in cultured cell lines. The highest level of APase activity occurs during the stationary phase of growth in both embryonal carcinoma/neuroblastoma hybrids (Bernstine and Ephrussi, 1975) and in Hela-23 and Henle embryonic intestinal cells (Melnykovytch *et al.*,1967). This feature might render some migrating and dividing PGCs, especially those in metaphase of mitosis, 'invisible' to APase histochemistry.

(d) The biological role *in vivo* of APase is unknown. A number of functions have been proposed for APase such as trans-membrane transport (Fishman, 1967), mononucleotide pool size regulation (Melnykovytch *et al.*, 1974) and regulation of growth (Sela and Sachs, 1974). It would seem unlikely that APase has a PGC-specific function in view of the widespread tissue distribution and large range of substrates that it can catalyse (Fernley, 1971).

PGC MIGRATION

The means by which PGCs make their way from the yolk sac to the genital ridge are not fully understood. The movement of PGCs falls into three phases: a) the movement from the yolk sac to the hind gut; b) the movement along the hind gut in an anterior direction; c) the movement dorsally out of the hind gut and laterally around the coelomic angles into the genital ridges.

The first phase could be the result of passive movement of the cells into the hind gut by tissue movements in the surrounding cells of the yolk sac and hind gut as the gut invaginates and the embryo turns. The PGCs at this stage do not show any signs of independent motility.

As the PGCs enter the gut, however, they start to show evidence of active motility. The cortical cytoplasm of the cells is often seen to be thrown into bulbous projections, or pseudopodia, characteristic of cells in motion (Witschi, 1948; Zamboni and Merchant, 1973; Speiglman and Bennett, 1973; Clarke and Eddy, 1975). These pseudopodia sometimes display submembranous microfilaments, thought to be necessary for cell motility (Speiglman and Bennett, 1973). Squash preparations of living hind gut cells observed by time lapse cinematography at this time show that the PGCs are indeed actively motile (Blandau *et al.*,1963 ; Blandau, 1969). They migrate in an amoeboid fashion, the pseudopodium forming at the leading edge of the cell, and the nucleus and cytoplasm moving to the opposite end of the cell (Figure 3). The nucleus often becomes elongated in shape. Some authors (Jeon and Kennedy, 1973) have argued that the number of PGCs in the hind gut that show pseudopodia in sectioned material are too few to entirely explain the amount of migration observed. However, pseudopodia in the hind gut are most often orientated in the anterior-posterior axis of the embryo (Clark and Eddy, 1975). Quantitative estimates of the number of PGCs showing pseudopodia in sectioned material, therefore, could depend on the plane of section. Furthermore, if the direction of movement of the PGCs changes, the leading edge of the pseudopodium is withdrawn and the cells return to a rounded shape, before a new pseudopodium is formed in another part of the cell (Blandau, 1969). Therefore, a number of PGCs observed in sections would not be expected to show pseudopodia, were they to be changing direction at the time of fixation.

PGCs observed by time lapse cinematography in squash preparations of later gonads show that the PGCs are still actively moving in the developing gonad. PGCs isolated from 12d-14d.pc gonads and cultured *in vitro* continue to show pseudopodial activity for several days (unpublished observations). This indicates that the PGCs are still capable of active movement although they have reached their target site at the end of their migratory path.

It seems likely, therefore, that the migration from the gut region to the gonad is achieved by means of active amoeboid movement by the PGCs. It then remains to determine what directs the orientation of the PGCs migration from the gut to the gonad. Witschi (1948) proposed that the migration of PGCs was directed by chemotactic factors emanating from the developing genital ridge. Dubois(1968; Dubois and Croiselle,

Figure 3 (top opposite): Squash preparation of 12.5 d.pc genital ridge. PGCs display prominent pseudopodia. x 600.

Figure 4 (lower opposite): Clump of PGCs isolated from squash 13.5 d.pc genital ridge. x 600.

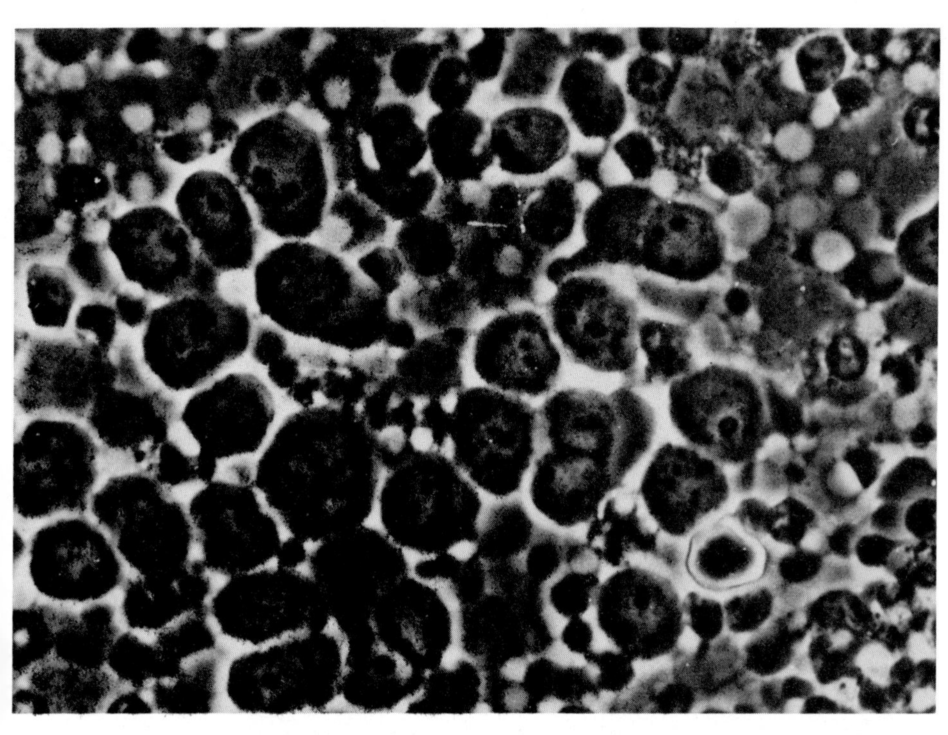

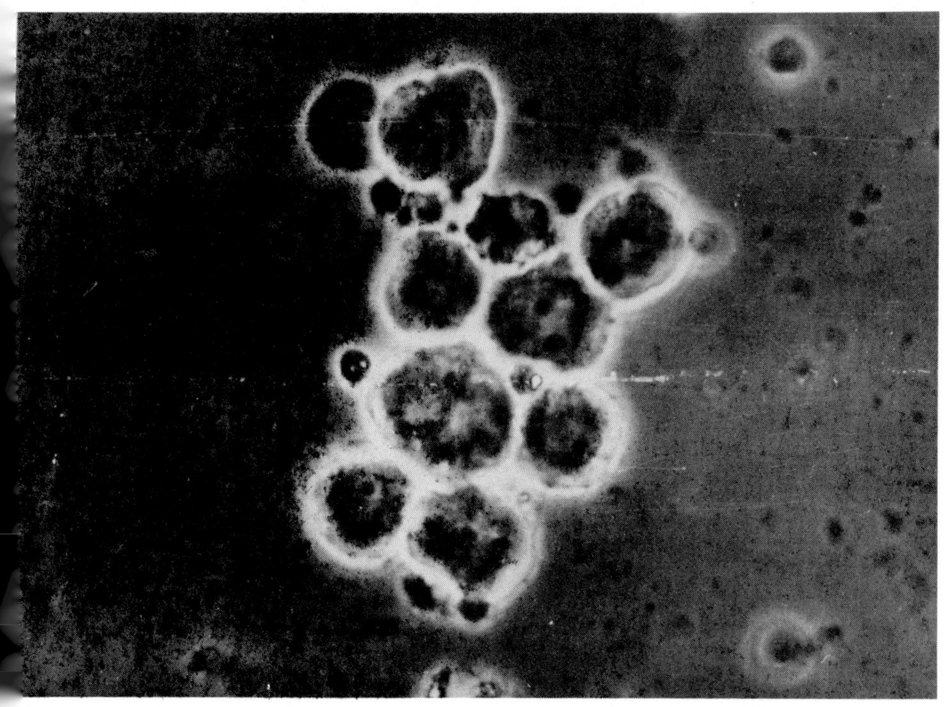

1970; Dubois *et al.*,1976) has produced evidence that the migration of PGCs in the chick is orientated by means of chemotactic factors produced by the gonad. This conclusion was based on the observation that, when radioactively labelled undifferentiated chick gonads were placed next to unlabelled experimentally sterilised gonads, the labelled PGCs could be seen to invade the sterile gonad. If the grafts were separated by an agar barrier or the vitelline membrane of the egg, however, the PGCs were later found to be orientated along the border of the graft next to the permeable barrier. Dubois interpreted this experiment to show that the PGCs of the chick are migrating in response to a factor which can pass through the barrier. Rogulska *et al.*(1971) transplanted the hind guts of mice into the posterior coelomic cavity of chick embryos, next to the chick gonad. Observations on the sections of graft area stained histochemically for APase after the incubation indicated that in a number of cases the mouse PGCs had moved towards the graft/host junction, and in a few cases the PGCs could be seen in the mesonephros and gonad of the host. These experiments were interpreted as showing that the mouse PGCs were migrating in response to the chemotactic factors postulated to be produced by the chick host. However these experiments cannot be considered conclusive evidence for chemotaxis since the incidence of mouse PGCs entering the host gonad was very low (12 grafts out of 90) and the number of PGCs recognisable by APase histochemistry in the graft declined over the time of culture. In addition, no data were presented on the ability of mouse PGCs to invade non-gonadal tissue by grafting to a similar site in a gonadectomised chick host.

Conclusive evidence for chemotaxis in biological systems is difficult to find (Trinkaus, 1969). A chemotactic model of PGC migration in the mouse must necessarily be complex, since there appears to be movement firstly anteriorly along the gut and secondly out of the gut and into the genital ridge. A simpler model of PGC migration might be formulated using the contact guidance theory of Weiss (1945). This would suppose that the germ cells are capable of active motion, and the direction of their movement is dependent upon guidance by the epithelial cells of the gut, mesentery and coelom, or their underlying basement membrane. It is clear from observations on sectioned material that migrating PGCs are often found in close association with the epithelium of the gut and gonad. Where the PGCs come into contact with the epithelium, the epithelial cells lose their usual columnar shape and become rounded or crescenteric in shape (Jeon and Kennedy, 1973; Speiglman and Bennett, 1973; Zamboni and Merchant, 1973). The two cell types are usually separated along the length of their membranes by a basal lamina, but they occasionally show close apposition of naked membranes by means of membrane interdigitations and focal tight junctions. This pattern of association is very similar to that seen between the epiblast cells and the mesenchymal cells of the chick primitive streak during gastrulation, when the orientation of the migrating mesenchymal cells is thought to be influenced by the epiblast cells and their basal laminae (Trelstad *et al.*,1967). Witschi (1948) and Zamboni and Merchant (1973) have proposed that PGCs leave the gut by pushing their way through the basal lamina of the gut epithelium. However, it is possible that the PGCs make use of pre-existing discontinuities in the basal lamina of the gut to leave the gut and enter the dorsal mesentery. The epithelium of the hind gut and coelom form a more or less continuous sheet. In this model, therefore, the migration of the PGCs can be considered as a continuous movement in two dimensions.

The migration of PGCs could be considered to be passive at first, as they are moved out of the yolk sac by movements in the surrounding tissue, and then active as they migrate to the genital ridge, either by means of orientated movement in response to chemotactic substances produced by the gonad, or by means of 'contact guidance' by the epithelium of gut and coelom.

ISOLATION AND CULTURE OF PGCs

The lack of information on PGCs stems in part from the difficulty in securing adequate amounts of material for further characterisation. It is now possible to isolate pure populations of PGCs from mammalian embryos, and many biochemical and serological techniques can be adapted to study small amounts of tissue.

The method routinely used in this laboratory for isolating PGCs from the embryos of mice, rats and voles is based on a two step purification procedure. The isolated gonads containing the PGCs are gently compressed to the point of rupture between a siliconised glass slide and coverslip. The PGCs are released from the somatic tissue by flushing media between the slide and coverslip with a finely pulled micropipette (Figure 4). They are then collected with the aid of a finely pulled siliconised mouth pipette. The cells collected in this manner are about 75-95% PGCs depending on the sex and stage of gonad. The viability of the PGCs is about 70-80% as assessed by Trypan blue dye exclusion. The contaminating somatic cells and dead PGCs can be removed by culturing the cells on tissue culture grade plastic dishes in a serum-containing medium for about two hours. It is found that the somatic contaminants stick to the culture vessel and the PGCs remain floating in the supernatant. The cells collected from the supernatant after this stage are about 99-100% PGCs. Their viability is about 95%, again as assessed by Trypan blue dye exclusion. That these cells are indeed PGCs can be established by morphology, APase production, or by cytological analysis for meiosis at the appropriate stage in the female (Figures 5-7). When gonads from embryos homozygous for the gene W^v, which are severely deficient in PGCs (see next section) are processed in this fashion, the number of PGCs subsequently recovered is also severely reduced. It has been found that enzymic treatment of the gonads is unnecessary, potentially toxic to the PGCs and can be undesirable in certain types of experiments. In experiments where large numbers of cells are required (greater than 10^6), it is judicious to choose a species containing a large number of PGCs. For example, the foetuses from two pregnant rats will yield about $1\text{-}2 \times 10^6$ PGCs when processed in the appropriate manner. PGCs isolated in this fashion are being used in this laboratory to characterise enzymic and serologically-defined characteristics of PGCs at various stages of development.

PGCs can be cultured for at least one week in standard tissue culture media (MEM-based formulations with added foetal

Figure 5 (top left opposite): PGCs after culture. x 1125. Figure 6 (top right opposite): isolated PGC after culture. Note prominent nucleoli, perinuclear mitochondria and high nucleo-cytoplasmic ratio. x 1500. Figure 7 (lower left opposite): binucleate PGC from male genital ridge 13.5 d.pc. x 1500. Figure 8 (lower right opposite): section of testes from W^v/W^v adult male mouse showing limited spermatogenesis and spermatogonial degeneration. x 150.

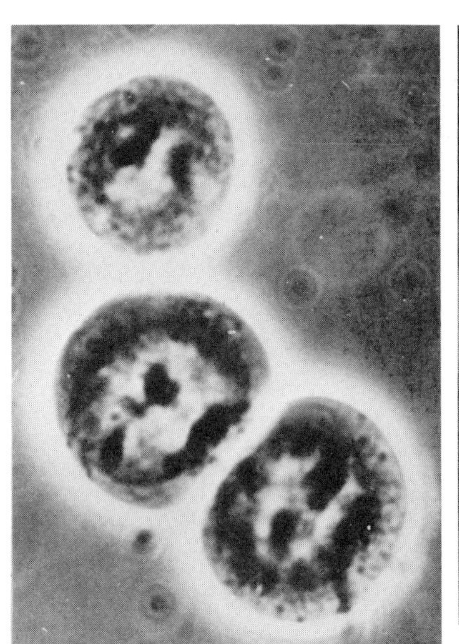

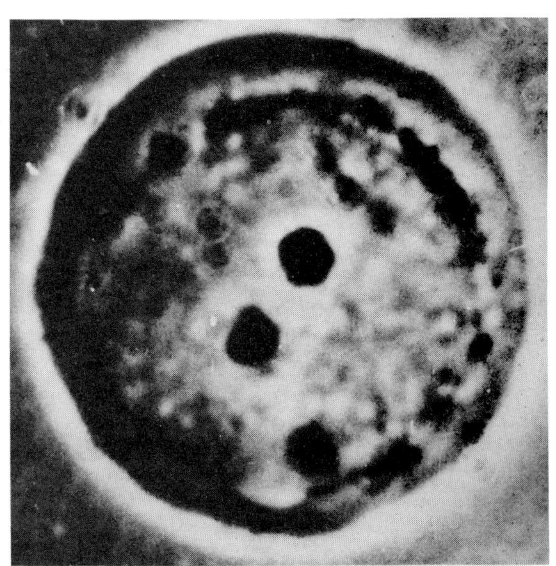

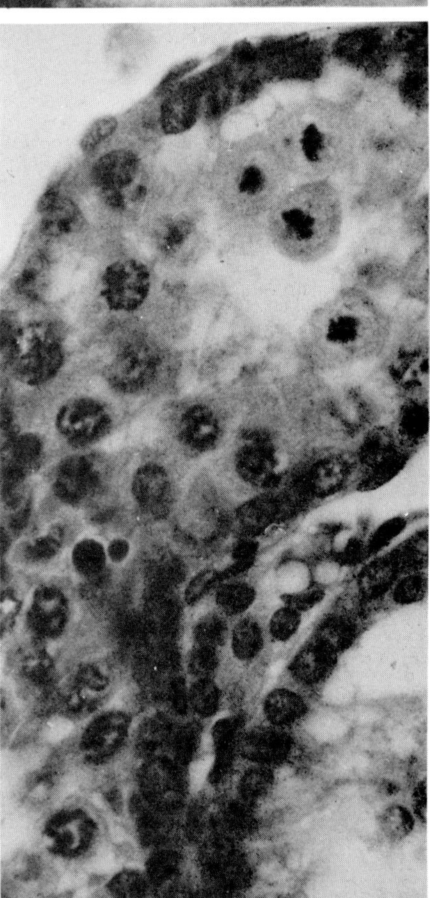

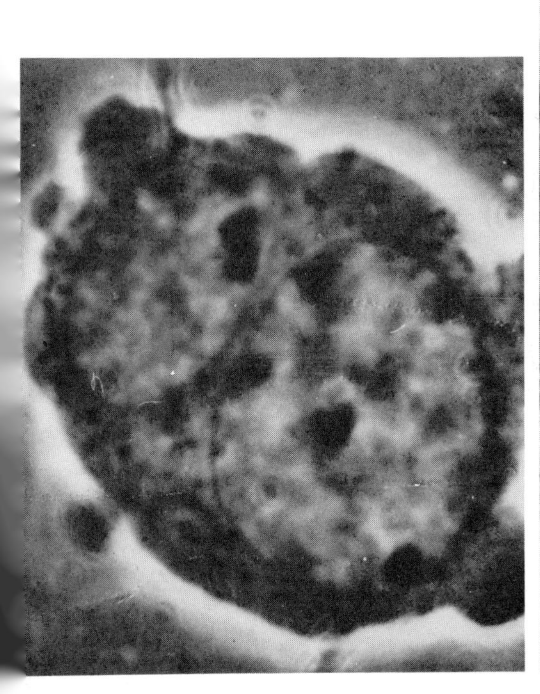

calf serum). The number of viable cells decreases over the period of culture, and PGCs from both sexes and at all stages studied have failed to divide or incorporate ^{3}H-thymidine into their DNA. The use of feeder layers of mitomycin-treated fibroblasts or gonadal somatic cells has so far failed to induce PGCs to divide, but the rate of PGC atresia in culture is decreased. It is clear that more information on the nutrient requirements of PGCs in tissue culture is required.

MUTATIONS AFFECTING PGCs

The use of appropriate mutant genes has proved useful in the analysis of developmental processes in a variety of situations. In the mouse, there are two genetic loci which have effects on PGCs. Analysis of the basis of these defects might help to define extrinsic or intrinsic factors required by PGCs for their growth and differentiation, and the role played by PGCs in the differentiation of the gonad.

W, DOMINANT WHITE SPOTTING

This is a series of semi-dominant alleles located on chromosome 5 with pleiotropic effects, including spotting, macrocytic anaemia and sterility (reviewed Russell, 1970).

The degree of spotting can vary between alleles. Animals heterozygous for some mutant alleles have only a white belly patch, whereas animals heterozygous for other alleles have an almost completely white coat except for a hood of coloured hair around the face. Viable homozygotes have a white coat with black eyes, the pigment being confined to the retina.

Viability of the homozygous mutant individuals varies over the series from lethality in late foetal and early neonatal life in some cases, to normal viability in others. The viability of the mice is probably inversely related to the degree of anaemia present. Some alleles show no anaemia in the homozygotes where other alleles produce anaemia in both homozygotes and heterozygotes.

The fertility of the homozygous mutant individuals also varies over the series from complete sterility in some cases to normal fertility in others. Sterile homozygous mutant individuals have a marked reduction in the number of germ cells, and a reduction in the size of the gonad.

The anaemia is due to some defect in the haemopoietic stem cells of the foetal liver and adult bone marrow. The erythrocyte number is reduced and the mean blood cell volume increased (Russell and Fondal, 1951). The foetal liver and adult bone marrow show a preponderance of immature erythroid cells compared to normal littermates (Russell *et al.*,1953), indicating some defect in blood cell maturation. The rate of erythrocyte production is affected. If labelled haem precursors are injected into normal mice, labelled erythrocytes appear in the circulation after about three days compared to about ten days for affected individuals (Altman and Russell, 1964). The anaemia depends entirely on the genotype of the haemopoietic stem cells,since it is possible to completely cure affected mice by injection of isologous normal bone marrow cells (Bernstein and Russell, 1959). If bone marrow stem cells from anaemic mice of the *W* series are injected into lethally irradiated normal hosts, they fail to produce macroscopically visible colonies in the spleen (McCulloch *et al.*, 1964). This is consistent with a defect in either the proliferation or maturation of haemopoietic stem cells.

Harrison and Cherry (1975) have shown that suitably marked bone marrow transplants into anaemic individuals can colonise both lymphoid and haemopoietic systems indicating that the defect might reside in the putative common precursor cells of the lymphoid and haemopoietic systems.

The homozygous coat colour defect appears to be at least partly the result of a specific defect in the melanocytes. Mayer and Green (1968) showed that if portions of 9d. neural crest from embryos homozygous for the allele W^v are grafted to the testes alongside normal 11 d.pc skin, the mutant melanocytes fail to produce pigment in the grafted skin. However, normal neural crest cells can produce pigment in W^v/W^v skin in a similar graft. In addition, normal neural crest cells can produce pigment in skin from 16 d.pc W^v/W^v skin. The melanocytes have normally arrived in the skin by 16 d.pc and further melanocyte invasion is prevented (Mayer, 1970). This indicates that either W^v/W^v melanocytes fail to reach the skin or reach the skin but die before 16 d.pc. Deol (1973) has argued that the distinctive distribution of melanocytes in heterozygotes might involve a tissue environment effect as well as a melanocyte defect.

The germ cell defect in mutant individuals has been less studied than other aspects, possibly due to the lack of available techniques. Mintz and Russell (1957) counted the number of PGCs in a series of embryos obtained from the mating of heterozygous $W^V/+$ mice. They observed that by 11 d.pc a distinct class of embryos, amounting to about 29% of the total had significantly fewer PGCs than their littermates, as assessed by use of APase histochemistry. This class was presumed to represent the homozygous mutant individuals.

The few PGCs that were observed in these individuals followed the usual course of migration, but the total number of observable PGCs failed to increase over the course of migration. It is not clear whether this is due to failure to divide or degeneration *en route*. The deficiency of PGCs in homozygous mutant mice almost certainly accounts for the deficiency of mature germ cells in the adult gonad (Coulombre and Russell, 1954). The few germ cells which are found in adult W^V/W^V male mice undergo a limited amount of differentiation, but seem to degenerate before mature gametes are produced (Figure 8).

The only method currently available for deciding whether the PGC deficiency is intrinsic to the mutant PGCs or is a function of some defect in the environment of the PGCs is by production of embryo-derived chimaeras between mutant and normal mice. If the PGC deficiency could be rescued in the chimaera and viable gametes produced, it would imply that the defect in affected mice lay in some feature of the PGC environment. A number of such animals have been produced, but there is currently no evidence for transmission of mutant gametes (unpublished observations).

Sl, STEEL

This is a series of semi-dominant mutations, located on chromosome 10, with a variety of pleiotropic effects closely mimicking those of the *W* locus. However, despite the superficial similarity between the two mutants, the underlying bases of the defects appear to be different. Attempts at curing the anaemia by transplantation of isologous bone marrow cells have proved unsuccessful. Indeed it is possible to cure the *W* anaemia by transplantation of *Sl/Sl* bone marrow cells (McCulloch *et al.*,1965). *Sl/Sl* bone marrow cells also produce normal spleen colonies in lethally irradiated recipients (McCulloch *et al.*, 1965). This implies that the anaemia is caused by some feature of the bone marrow haemopoietic stem cell environment (reviewed Cole, 1975).

Mayer and Green (1968) in a similar series of experiments to those outlined for *W* showed that *Sl/Sl* neural crest could produce pigment in normal skin, but normal neural crest cells could not produce pigment in *Sl/Sl* skin. This implies some deficiency in the melanocyte environment in *Sl/Sl* mice.

Counts of PGCs in embryos from matings of *Sl/+* mice show a class of embryos amounting to 25% of the total which are deficient in PGCs by 11d.pc (Bennett, 1956; McCoshan and McCallion, 1975). Those PGCs observed in the deficient embryos follow the normal path of migration, but fail to increase in number. No further experimental analysis has been attempted.

THE ORIGIN OF PGCs AND TERATOCARCINOMAS

Teratocarcinomas are neoplastic growths, usually found in the gonads of a variety of mammalian species, which are characterised by the presence of differentiated derivatives of all three embryonic germ layers. Teratocarcinomas contain transplantable pluripotent stem cells called embryonal carcinoma cells (EC Cells) (Kleinsmith and Pierce, 1964). The benign form of the tumour, which lacks EC cells, is called a teratoma. Most of the experimental analysis of these tumours has been performed in the mouse (reviewed by Stevens, 1967; Damjanov and Solter, 1974; Graham, 1977; Martin, this volume).

Teratocarcinomas occur rarely in normal mouse populations. However, there are two strains of mouse which are characterised by a high incidence of spontaneously occurring teratomas. Stevens and Little (1954) noted that the incidence of testicular teratomas in the male mice of 129J strain was about 1%. The incidence could be raised to about 10% in 129J mice heterozygous for the gene *Steel* (see previous section; Stevens and Mackensen, 1961), and further raised to about 32% in a subline of 129J called 129J/Ter Sv (Stevens, 1973). About 50% of female mice of the strain LT are found to have spontaneous ovarian teratomas by three months of age (Stevens and Varnum, 1974). This strain is also characterised by a high incidence of spontaneous parthenogenetic activation.

It is also possible to experimentally induce teratomas and teratocarcinomas. Stevens (1964, 1966, 1970a) found that 75% of male genital ridges grafted to the testes of isogeneic recipients formed teratomas. The ability to form teratomas was highest for ridges from 11 or 12 d.pc foetuses. It was possible to induce these teratomas in both 129J and A/He, a strain which has a very low incidence of spontaneous teratomas. There are strains, for example C3H, in which it has so far proved impossible to produce teratomas experimentally by grafting genital ridges. Stevens (1970a) noted that the incidence of teratoma induction was highest in genital ridges grafted to the scrotal testes, indicating a temperature effect on the process of induction.

Mouse embryos aged 2-7d.pc and of either sex can also form teratomas when grafted to ectopic sites in the adult (Stevens, 1968, 1970b; Solter *et al.*,1970; Skreb *et al.*,1972; Solter *et al.*,1974; Ilses, 1977). Embryo-derived teratomas can be induced in a number of strains which are not susceptible to teratoma formation in grafted genital ridges. Teratomas are only formed in transplants containing the embryonic ectoderm of the egg cylinder. Diwan and Stevens (1976) divided the 7th day embryo into three parts: primitive endoderm, embryonic ectoderm and extraembryonic ectoderm. When the various parts were grafted separately to isogeneic hosts, it was found that only the portion containing embryonic ectoderm gave rise to teratomas.

Teratocarcinomas and teratomas seem to arrive directly from PGCs in spontaneous testicular teratomas and grafted genital ridges. Support for this view comes from the observation of Stevens (1967) that transplanted genital ridges from foetuses homozygous for the gene *Steel* have a greatly reduced incidence of teratoma formation. Histological analysis of early spontaneous teratomas has revealed that the earliest stages form inside the seminiferous tubules of the testis (Stevens, 1962). This implies that the teratoma arises either from the Sertoli cells or the germ cells. Pierce and Beals (1964) showed that there was a very close ultrastructural similarity between the germ cells of the foetal testes and the EC cells of the tumour. The earliest tumours observed in transplanted genital ridges took the form of small nests of undifferentiated cells which were again very similar ultrastructurally to PGCs and EC cells (Pierce *et al.*, 1967).

Embryo-derived and spontaneous ovarian teratomas and teratocarcinomas are thought to arise directly from the embryonic ectoderm of the 7 day pc mouse embryo. The spontaneous parthenogenetic embryos of LT mice go through the normal course of development up to the egg cylinder stage (Stevens and Varnum, 1974), before their growth becomes disorganised and the teratoma forms. Grafted embryos develop upto the egg cylinder stage in a similar fashion before their growth becomes disorganised and the teratoma is formed (Stevens, 1968).

The ability to form teratomas is a specific property of certain cell types in the embryo, including PGCs. The origin of PGCs and teratomas are therefore closely related problems (see also Martin, this volume). There are three main views on the extragonadal origin of PGCs in the mammalian embryo, and it is useful to consider the implications of these theories for understanding the origin of teratomas (Figure 9).

PARTITION

This theory was first proposed in a general way by Rubashkin (1912). It suggests that the PGCs are a small group of embryonic ectoderm cells partitioned away from the rest of the embryonic ectoderm as it begins to differentiate. PGCs retain most of the features of embryonic ectoderm cells, including the ability to form teratomas. This ability is lost upon differentiation into mesoderm and endoderm in the case of embryonic ectoderm or into gametes in the case of PGCs.

Damjanov and Solter (1974) have argued that the EC stem cells of embryo-derived and PGC-derived teratomas are identical. The partition theory would also state that the progenitor cells in the normal embryo of both types of EC cells are also equivalent. Teratoma formation would therefore be a unique property of a single set of cells. This theory makes a number of testable predictions. (a) There should be no detectable germ line distinct from embryonic ectoderm before 7 days pc. There are in fact very few discernible differences between embryonic ectoderm, EC cells and PGCs. The ultrastructural resemblances between the three cell types has already been remarked upon. PGCs, EC cells and embryonic ectoderm cells contain high levels of APase (Bernstine *et al.*,1973; Damjanov *et al.*,1977). Embryonic ectoderm and PGCs are both derivatives of the non-endodermal cells of the 4.5 d.pc inner cell mass (Gardner and Lyon, unpublished observations). A syngeneic

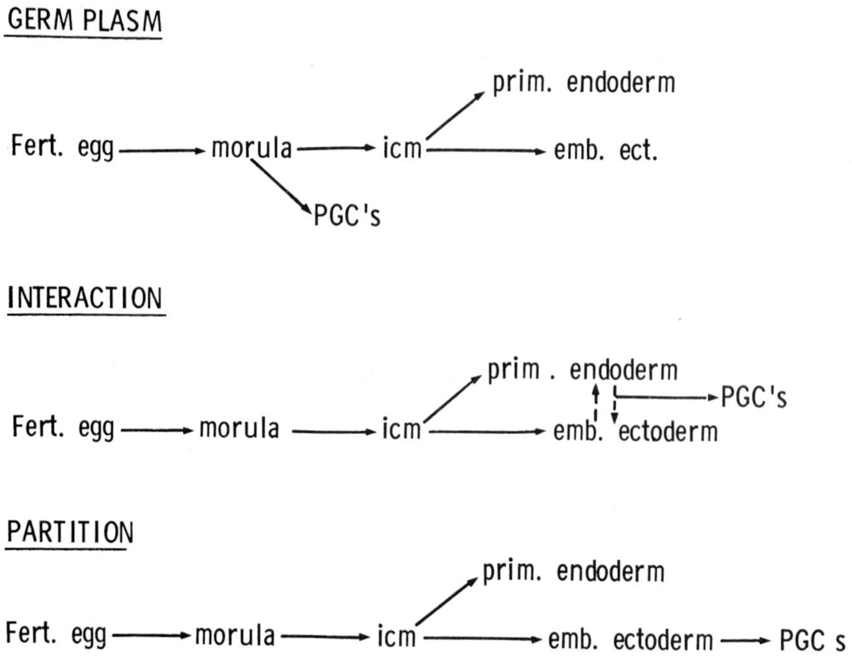

GERM PLASM

prim. endoderm
Fert. egg → morula → icm → emb. ect.
PGC's

INTERACTION

prim . endoderm
PGC's
Fert. egg → morula → icm → emb. ectoderm

PARTITION

prim. endoderm
Fert. egg → morula → icm → emb. ectoderm → PGC s

Figure 9: diagrammatic representation of three theories of germ cell origin in the mouse. Solid lines indicate cell lineages, broken lines indicate cell interaction.

antiserum raised against EC cells of the tumour F9 is detectable on embryonic ectoderm cells and PGCs as well as EC cells from a variety of different teratocarcinomas (Jacob, 1976). PG-1 a xenogeneic antiserum raised against 13.5 d.pc PGCs cross reacts with EC cells from a variety of terato-carcinoma cell lines (unpublished observations). (b) It should only be possible to produce teratomas from tissues containing PGCs or embryonic ectoderm. Experimental foetec-tomy in the rat can lead to the formation of teratomas in the portion of the yolk sac trapped outside the uterus after the operation (Sobis and VanDerPutte, 1975). It is claimed that these growths do not arise from PGCs in the yolk sac because they also occur in pregnant rats treated with busulphan (Sobis and VanDerPutte, 1976). However, the time course of PGC migration in the rat is not known in detail, and it is not known whether busulphan treatment of the pregnant mother has any effect on PGCs in the yolk sac of the foetus. The most favoured extragonadal sites of teratoma formation in the human are at the base of the spine and at the base of the buccal cavity behind the skull (Willis, 1966). If one assumes that these tumours are not the result of metastases from a primary gonadal tumour which subsequently regresses, these locations are consistent with the location of ectopic PGCs which have either failed to migrate dorsally and continued along the gut, or which have lost contact with the epithelium after leaving the dorsal mesentery and continued to migrate dorsally into the notochord. A detailed study of ectopic PGCs in the human foetus could help to resolve this point. (c) Embryonic ectoderm alone should be capable of giving rise to PGCs in the appropriate conditions. This could be tested by growing embryonic ectoderm and other embryonic tissues in culture, in grafts or by injecting embryonic ectoderm cells into genetically marked blastocysts, and testing for the appearance of marked cells during gametogenesis.

INTERACTION

This theory was first proposed by Vanneman (1917). This suggests that PGCs arise as the result of some kind of interaction between embryonic ectoderm and primitive endoderm. The PGCs are therefore formed *de novo* as the result of the interaction, and are different from embryonic ectoderm and primitive endoderm. This view is supported by studies in other species. Kocher-Becker and Tiedemann (1971) and Satsuraja and Nieuwkoop (1974) have shown that PGC formation in the urodeles is the result of an interaction between ectoderm and endoderm in the gastrula. Endoderm, which does not normally produce PGCs, can form them when cultured in contact with appropriate ectoderm. In the mouse, PGCs would be expected to form from the embryonic ectoderm as Gardner and Lyon (unpublished observations) have shown that only nonendodermal cells of the 4.5 d.pc ICM will colonise the germ line of the foetus. Although this theory does not make any specific predictions about the origin of teratomas, it could be distinguished from the partition model since it predicts that a specific interaction between two cell types is required to form PGCs. Embryonic ectoderm should only be capable of giving rise to

PGCs therefore if cultured or grafted in the presence of primitive endoderm.

GERM PLASM

This view comes from the work of Weismann (1892) and has been more recently supported for the mammal by Eddy (1974). The hypothesis suggests that the germ line is formed very early in development, and is dependent on the presence of cytoplasmic factors in the egg which are segregated to a small group of cells which are thereby determined to become germ cells. The cells in this small group do not appreciably increase in numbers until they are recognised as distinct APase-positive cells on day 8 pc. This view has support from work on a number of species where a distinct germ plasm can be recognised by its staining properties and so followed through development (Beams and Kessel, 1975; review by Whitington and Dixon, 1976). In the insects, the role of this substance in germ cell formation can be demonstrated by transplantation into ectopic sites in the blastoderm (Illmensee *et al.*,1975). There is, however, no distinctive cytoplasmic marker of the germ line known in the mammal, and in the absence of such a marker a conclusive test of this theory is difficult to devise. The putative cytoplasmic substance must be found in all cells of the early embryo upto at least the 8 cell stage, since Kelly (1975) has shown that all the cells of the 8 cell embryo are equally capable of giving rise to adult tissues, including functional gametes. The implications of this theory for the origin of teratomas is less clear. It is possible to say that the only cells capable of giving rise to teratomas are members of the germ line. The germ plasm theory would then predict that only that portion of the embryonic ectoderm containing the germ cells could form teratomas. This could be tested by examining the incidence of teratoma formation in small portions of embryonic ectoderm grafted to ectopic sites. It is also possible to say, however, that teratoma formation is a property of two different types of tissue. Stevens (1970b) has argued that the sex and strain restrictions on PGC-derived teratoma formation compared to embryo-derived teratomas indicate an independent origin for each type of tumour. The failure to produce teratomas from female grafted genital ridges might be due to the fact that the PGCs in these grafts are entering meiotic prophase. The strain restrictions might reflect the greater resistance of the PGCs in the graft environment to neoplastic change. The susceptibility to neoplasia in other situations can be genetically restricted, but this restriction can be overridden by appropriate experimental techniques. For example, neoplasms of the somatic component of the ovary occur spontaneously in some strains of mice (Ce, C3HeB/FeJ, RIII/J, Murphy, 1966) but can be induced in non-susceptible strains by grafting the ovary to an ectopic site (Li and Gardner, 1949; Hummell, 1954). It is possible therefore, that the strain restrictions on teratoma formation might be overridden in an appropriate experimental situation.

CONCLUSIONS

The origin and differentiation of the germ line is of interest not only from its relevance to gonadogenesis, but also from its relevance to problems of tumour formation and cell deployment in the early embryo. Our understanding of the origin of the germ line in mammals has not advanced in many ways from the ideas of the early embryologists. However, the technology is now becoming available for one to attempt to answer this question, one of the oldest problems in mammalian embryology.

ACKNOWLEDGEMENTS

I would like to thank Drs. R.L. Gardner, C.F. Graham and V.E. Papaioannou for valuable discussions. I am supported by an M.R.C. research studentship.

REFERENCES

ALLEN, B.M. (1904) The embryonic development of the ovary and testes in mammals. Am. J. Anat. 9, 117.

ALTMAN, K.M. & RUSSELL, E.S. (1964) Heme synthesis in normal and genetically anaemic mice. J. cell. comp. Physiol. 64, 293-301.

BAXTER, R. (1952) Alkaline phosphatase in the primordial germ cells of a 10mm. human embryo. Int. Anat. Congr. Oxford, p. 17-18.

BEAMS, H.W. & KESSEL, R.G. (1975) The problem of germ cell determinants. Int. Rev. Cytol. 39, 413-479.

BENNETT, D. (1956) Developmental analysis of a mutation with pleiotropic effects in the mouse. J. Morph. 98, 199-234.

BERNSTEIN, S.E. & RUSSELL, E.S. (1959) Implantation of normal blood-forming tissue in genetically anaemic mice without X-irradiation of the host. Proc. Soc. exp. Biol. Med. 101, 769.

BERNSTINE, E.G. & EPHRUSSI, B. (1975) Alkaline phosphatase activity in embryonal carcinoma and its hybrids with neuroblastoma. In: M.I. Sherman and D. Solter (Eds.), Teratomas and differentiation, Academic Press, New York, pp. 271-287.

BERNSTINE, E.G., HOOPER, M.L., GRANDCHAMP, S. & EPHRUSSI, B. (1973) Alkaline phosphatase activity in mouse teratoma. Proc. natn. Acad. Sci. U.S.A. 70, 3899-3902.

BLACK, V.H. & CHRISTIANSEN, A.K. (1969) Differentiation of interstitial cells in the fetal guinea pig testes. Am. J. Anat. 124, 211-232.

BLANDAU, R.J. (1969) Observations on living oogonia and oocytes from human embryonic and fetal ovaries. Am. J. Obstet. Gynec. 104, 310-319.

BLANDAU, R.J., WHITE, B.J. & RUMERY, R.E. (1963) Observations on the movement of the living primordial germ cells in the mouse embryo. Fert. Steril. 14, 482-489.

BORUM, K. (1966) Oogenesis in the mouse: A study of the origin of mature ova. Expl. Cell Res. 45, 39-47.

BRAMBELL, F.W. & PARKES, A.S. (1927) Changes in the ovary of the mouse following exposure to X-rays: 3. Irradiation of the non-parous adult. Proc. Roy. Soc. B. 101, 36.

CHIQUONE, A.D. (1954) The identification, origin and migration of the primordial germ cells of the mouse embryo. Anat. Rec. 118, 135-146.

CHRETIEN, F.Ch. (1965) Etude de l'origine, de la migration et de la multiplication des cellules germinales chez l'embryon de Lapin. J. Embryol. exp. Morph. 16, 591-607.

CLARK, J.M. & EDDY, E.M. (1975) Fine structural observations on the origin and associations of the primordial germ cells of the mouse. Devl. Biol. 47, 136-155.

COLE, R.J. (1975) Regulatory functions of microenvironment and hormonal factors in prenatal haemopoetic tissue. In: M. Balls and A.E. Wild (Eds.) The early development of mammals, Cambridge University Press, London, pp. 335-358.

COULOMBRE, J.L. & RUSSELL, E.S. (1954) Analysis of the pleiotropism at the W locus in the mouse. J. exp. Zool. 126, 277-295.

COX, R.P. & MacLEOD, C.M. (1962) Alkaline phosphatase content and the effects of prednisilone. J. gen. Physiol. 45, 439-485.

DAMJANOV, I. & SOLTER, D. (1974) Experimental teratoma. Current topics in pathology, 59, 69-130.

DAMJANOV, I., CUTLER, L.S. & SOLTER, D. (1977) Ultrastructural localisation of membrane phosphatases in teratocarcinoma and early embryos. Am. J. Path. 87, 297-304.

DEOL, S. (1973) The role of tissue environment in the expression of spotting genes in the mouse. J. Embryol. exp. Morph. 30, 483.

DE WINIWARTER, H. & SAINMONT, G. (1909) Nouvelles recherches sur l'ovogenese et organogenese de l'ovaire des mammifieres (chat). Arch. Biol., Paris, 24. 1.

DIWAN, S. & STEVENS, L.C. (1976) The development of teratomas from the ectoderm of mouse egg cylinders. J. natn. Cancer Inst. 57, 937-942.

DUBOIS, R. (1968) La colonisation des ebauches gonadiques par les cellules germinales de l'embryon de poulet en culture in vitro. J. Embryol. exp. Morph. 20, 189-213.

DUBOIS, R. & CROISELLE, Y. (1970) Germ cell line and sexual differentiation in birds. Phil. Trans. Roy. Soc., London, 259, 73-79.

DUBOIS, R., CUMINGE, D. & SMITH, J. (1976) Interpretation of some recent results in experimental embryology and the problem of the germ line. In: M. Balls and M.A. Monnickendam (Eds.) Organ Culture in Biomedical Research, Cambridge University Press, London, pp. 61-93.

EDDY, E.M. (1974) Fine structural observations on the form and distribution of nuage in the germ cells of the rat. Anat. Rec. 178, 731-758.

EDDY, E.M. (1975) Germ plasm and the differentiation of the germ cell line. Int. Rev. Cytol. 43, 229-280.

EVERRETT, N.B. (1943) Observational and experimental evidences relating to the origin and differentiation of the definitive germ cells in mice. J. exp. Zool. 92, 49-92.

EVERRETT, N.B. (1945) The present status of the germ cell problem in vertebrates. Biol. Rev. 20, 45-55.

FALIN, L.I. (1969) The development of the genital glands and the origin of the germ cells in human embryogenesis. Acta Anat. 72, 533-537.

FELIX, W. (1912) Manual of human embryology. F. Keibel and F.P. Mall, Philadelphia, p. 752.

FERNLEY, H.N. (1971) Mammalian alkaline phosphatases. In: P.D. Boyer (Ed.) The enzymes, Academic Press, New York, p. 417.

FIRKET, J. (1920) On the origin of germ cells in higher vertebrates. Anat. Rec. 18, 309.

FISHMAN, W.H. (1974) Perspectives on alkaline phosphatase isozymes. Amer. J. Med. 56, 617.

FRANCHI, L.L. & MANDL, A.M. (1962) The ultrastructure of oogonia and oocytes in the foetal and neonatal rat. Proc. Roy. Soc. B., 157, 99-114.

FRANCHI, L.L. & MANDL, A.M. (1964) The ultrastructure of germ cells in the foetal and neonatal male rat. J. embryol exp. Morph. 12, 289.

FUJIMOTO, T., MYAYAMA, Y. & FUJITA, M. (1977) The origin, migration and fine morphology of human primordial germ cells. Anat. Rec. 118, 315-330.

FUKADA, T. (1976) Ultrastructure of primordial germ cells in the human embryo. Virch. Arch. B. Cell. Path. 20, 85-89.

FUSS, A. (1911) Uber extraregionaire geschlechts zellen bei einem menschlichen embryo von 4 wochen. Anat. Anz. 39, 407.

FUSS, A. (1912) Uber die geschlechts zellen des menschen und der saugitiere. Arch. mikroskop. Anat. EntwMech. 81, 1-12.

GRAHAM, C.F. (1977) Teratocarcinoma cells and normal embryogenesis. In: M.I. Sherman (Ed.) Concepts in mammalian embryogenesis. MIT Press, London, pp. 315-394.

GONDOS, B. & CONNOR, L.A. (1973) Ultrastructure of developing germ cells in fetal rabbit testes. Am. J. Anat. 136, 23-42.

HARGITT, G.T. (1925) The formation of the sex glands and germ cells of mammals. 1: The origin of germ cells in the female rat. J. Morph. 40, 517.

HARGITT, G.T. (1930) The formation of the sex glands and germ cells of mammals. 3: The history of the female germ cells in the female rat to the time of sexual maturity. J. Morph. 49, 277.

HARRISON, D.E. & CHERRY, M. (1975) Survival of marrow allografts in W/W^v anaemic mice: Effect of disparity at the Ea-2 locus. Immunogenetics 2, 219-229.

HEYS, F. (1931) The problem of the origin of the germ cells. Q. Rev. Biol. 6, 1-45.

HUMMELL, K.P. (1954) Induced ovarian and adrenal tumours. J. natn. Cancer Inst. 15, 711-715.

ILLMENSEE, K., MAHOWOLD, A.P. & LOOMIS, M.R. (1975) The ontogeny of germ plasm during oogenesis in Drosophila. Devl. Biol. 49, 40-65.

ILSES, S.A. (1977) Mouse teratomas and embryoid bodies: their induction and differentiation. J. Embryol, exp. Morph. 38, 63-75.

JACOB, F. (1976) Mouse teratocarcinoma and embryonic antigens. Immunological Reviews. 33, 3-32.

JENKINSON, J.W. (1913) Vertebrate embryology, Oxford, p. 29.

JEON, K.W. & KENNEDY, J.R. (1973) The primordial germ cells in early mouse embryos: light and electron microscopic study. Devel. Biol. 31, 275-284.

JOST, A. & PREPIN, J. (1966) Donnes sur la migration germinales primordial du foetus du veau. Arch. Anat. Micr. Morph. exp. 55, 161-186.

KELLY, S.J. (1975) Studies on the potency of early cleavage blastomeres of the mouse. In: M. Balls and A.E. Wild (Eds.) The early development of mammals, Cambridge University Press, London, pp. 97-105.

KENNELLY, J. & FOOTE, K.H. (1966) Oocytogenesis in rabbits: The role of neogenesis in the formation of the definitive ova and the stability of oocyte DNA measured with tritiated thymidine. Am. J. Anat. 118, 573-590.

KINGERY, H.M. (1917) Oogenesis in the white mouse. J. Morph. 30, 261.

KINGSBURY, B.F. (1913) Morphogenesis of the mammalian ovary: Felis Domestica. Am. J. Anat. 15, 354.

KLEINSMITH, L.J. & PIERCE, G.B. (1964) Multipotentiality of single embryonal carcinoma cells. Cancer Res. 24, 1544-1552.

KOCHER-BECKER, W. & TIEDEMANN, H. (1971) Induction of mesodermal structures and primordial germ cells in Triturus by a vegetalising factor from chick embryos. Nature, London, 233, 65-66.

KOYAMA, H. & ONO, T. (1972) Induction of alkaline phosphatase by cyclic AMP and its dibutyryl derivatives in a hybrid cell line between mouse and chinese hamster in culture. Biochem. Biophys. Res. Commun. 46, 305-311.

KOYAMA, H. & ONO, T. (1976) Induction by short chain fatty acids of alkaline phosphatase in cultured mammalian cells. J. Cell. Physiol. 88, 49-56.

LI, M.H. & GARDNER, W.V. (1949) Further studies on the pathogenesis of ovarian tumours in mice. Cancer Res. 9, 35-41.

MacCOSHAN, J.A. & MacCALLION, D.J. (1975) A study of primordial germ cells during their migratory phase in Steel mutant mice. Experientia 31, 589-590.

MAYER, T.C. (1970) A comparison of pigment cell development in albino, steel and dominant spotting mutant mouse embryos. Devl. Biol. 18, 62-75.

MAYER, T.C. & GREEN, M.C. (1968) An experimental analysis of the pigment defect caused by mutation at the W and Sl loci in mice. Devl. Biol. 9, 269-286.

McCULLOCH, E.A.L., SIMINOVITCH, L., TILL, J.E.A., RUSSELL, E.S. & BERNSTEIN, A. (1965) The cellular basis of the genetically determined hemapoietic defect in anemic mice of genotype Sl/Sl. Blood 26, 399-410.

McKAY, D.G., HERTIG, A.T., ADAMS, E.C. & DANZIGER, S. (1953) Histochemical observations on the germ cells of human embryos. Anat. Rec. 117, 407-408.

MELNYKOVYTCH, G., BISHOP, C.F. & SWAYZE, M.A.B. (1967) Fluctuation of alkaline phosphatase activity in synchronised heteroploid cell cultures: effects of prednisilone. J. Cell. Physiol. 70, 231-236.

MERCHANT, H. (1975) Rat gonadal and ovarian organogenesis with and without germ cells: an ultrastructural study. Devl. Biol. 44, 1-21.

MINTZ, B. (1958) Irradiation of primordial germ cells in the mouse embryo. Anat. Rec. 130, 341 (Abs.).

MINTZ, B. (1960) Embryological phases of gametogenesis. J. cell. comp. Physiol. (Supp.1), 56, 31-48.

MINTZ, B. & RUSSELL, E.S. (1957) Gene induced embryological modifications of primordial germ cells in the mouse. J. exp. Zool. 134, 207.

MOOG, F. (1965) Enzyme development in relation to functional differentiation. In: R. Weber (Ed.), Biochemistry of animal development. Academic Press, New York, p.307.

MORRISS, G.M. & STEELE, C.E. (1977) Comparison of the effects of retinol and retinoic acid on post-implantation rats in vitro. Teratology. 15, 109-119.

MURPHY, E.D. (1966) Characteristic tumours of the mouse. In: Jackson laboratory Staff (Eds.), Biology of the laboratory mouse, 2nd Ed. MacGraw-Hill, New York, pp. 521-562.

OKADA, M., KLEINMAN, I.A. & SCHNEIDERMAN, H.A. (1974) Restoration of fertility in sterilised Drosophila eggs by transplantation of polar cytoplasm. Devl. Biol. 37, 43.

ODOR, D.L. & BLANDAU, R.J. (1969) Ultrastructural studies on fetal and postnatal mouse ovaries. 2: Cytodifferentiation. Am. J. Anat. 125, 177-216.

ODZENSKI, W. (1967) Observations on the origins of primordial germ cells in the mouse. Zool. Pol. 17, 367-379.

PELLINIEMI, L.J. (1975a) Ultrastructure of gonadal ridge in male and female pig embryos. Anat. Embryol. 147, 19-34.

PELLINIEMI, L.J. (1975b) Ultrastructure of early ovary and testes in pig embryos. Am. J. Anat. 144, 89-112.

PELLINIEMI, L.J. (1976) Ultrastructure of indifferent gonad in male and female pig embryos. Tissue and Cell. 8, 163-174.

PETERS, H. & CRONE, M. (1967) DNA synthesis in oocytes of mammals. Arch. Anat. Microscop. 56, 160-170.

PETERS, H., LEVY, E. & CRONE, M. (1962) Deoxyribonucleic acid synthesis in oocytes of mouse embryos. Nature, London, 195, 915-916.

PIERCE, G.B. & BEALS, T.F. (1964) The ultrastructure of primordial germ cells in the fetal testes and embryonal carcinoma cells in mice. Cancer Res. 24, 1553-1567.

PIERCE, G.B., STEVENS, L.C. & NAKANE, P.K. (1967) Ultra-structural analysis of the early development of terato-carcinoma. J. natn. Cancer Inst. 39, 755-773.

PINKERTON, J.H.M., McKAY, D.G., ADAMS, E.C. & HERTIG, A.T. (1961) Development of the human ovary using histochemical techniques. Obstet. Gynec. 18, 165-181.

ROGULSKA, T., ODSENSKI, W. & KOMAR, A. (1971) Behaviour of mouse primordial germ cells in the chick embryo. J. Embryol. exp. Morph. 25, 155-164.

RUBASHKIN, W. (1908) Zur frage von der entethung der keimzellen bei saugetierembryon. Anat. Anz. 32, 222.

RUBASHKIN, W. (1912) Zur celue von der keimbahn bein Saugertieren uber die entwickleung der keimsdrusan. Anat. Hefte. Arb. Anat. inst. 46, 345-411.

RUDKIN, G. & GREICH, H.A. (1962) On the persistence of oocyte nucleii from fetus to maturity in the laboratory mouse. J. Cell Biol. 12, 169-175.

RUSSELL, E.S. (1970) Abnormalities of erythropoiesis associated with mutant genes in the mouse. In: A.S. Gordon (Ed.), Regulation of haematopoiesis,Appleton Century Crofts, New York, pp. 649-675.

RUSSELL, E.S., SNOW, C.M., MURRAY, L.M. & CORMIER, J.P. (1953) The bone marrow in inherited macrocytic anemia in the house mouse. Acta. haemat. 10, 247-259.

RUSSELL, E.S. & FONDAL, E.L. (1951) Quantitative analysis of the normal and four alternative degrees of an inherited macrocytic anemia in the house mouse. 1: Number and size of erythrocytes. Blood, 6, 892-905.

SATSURAJA, L.A. & NIEUWKOOP, P.D. (1974) The induction of primordial germ cells in the urodeles. Wilhelm Roux Arch. EntwMech. Org. 175, 199-220.

SELA, B-A. & SACHS, L. (1974) Alkaline phosphatase activity and the regulation of growth in transformed mammalian cells. J. cell. comp. Physiol. 83, 27-34.

SIMKINS, C.S. (1923) On the origin of the so called primordial germ cells in mouse and rat. Acta. zool. Stockh. 4, 241.

SIMKINS, C.S. (1928) Origin of the sex cells in man. Am. J. Anat. 41, 249.

SKREB, N., DAMJANOV, I. & SOLTER, D. (1972) Teratomas and teratocarcinomas derived from rodent egg shields. In: R. Harriss and D. Viza (Eds.), Cell Differentiation, Munksgaard, Copenhagen, pp. 151-155.

SOBIS, H. & VANDERPUTTE, M. (1975) Yolk sac derived study of teratomas from displaced yolk sac. Devl. Biol. 45, 267-290.

SOBIS, H. & VANDERPUTTE, M. (1976) Yolk sac derived teratomas are not of germ cell origin. Devl. Biol. 51, 320-323.

SOLTER, D., SKREB, N. & DAMJANOV, I. (1970) Extrauterine growths of mouse egg cylinders result in malignant teratoma. Nature, London, 227, 503-504.

SOLTER, D., DAMJANOV, I. & SKREB, N. (1973) Distribution of hydrolytic enzymes in early rat and mouse embryos:- a reappraisal. Z. Anat. EntwGesch. 139 119-126.

SOLTER, D., BICZYKOW, W.W. & KOPROWSKI, H. (1974) Ultrastructure of mouse teratomas developed in vitro. Anat. Rec. 180, 263-280.

SPEIGLMAN, M. & BENNETT, D. (1973) A light and electron microscopic study of primordial germ cells in the early mammalian embryo. J. Embryol. exp. Morph. 30, 97-118.

STEVENS, L.C. (1962) Testicular teratomas in fetal mice. J. natn. Cancer Inst. 28, 247-268.

STEVENS, L.C. (1964) Experimental production of testicular teratomas in mice. Proc. natn. Acad. Sci., U.S.A. 52, 654-661.

STEVENS, L.C. (1966) Development of resistance to teratocarcinogenesis by primordial germ cells in mice. J. natn. Cancer Inst. 37, 859-867.

STEVENS, L.C. (1967) The origin of testicular teratomas from primordial germ cells in mice. J. natn. Cancer Inst. 38, 549-552.

STEVENS, L.C. (1968) The development of teratomas from intra-testicular grafts of tubal mouse eggs. J. Embryol. exp. Morph. 20, 329-341.

STEVENS, L.C. (1970a) The development of transplantable teratomas from pre- and post- implantation mouse embryos. Devl. Biol. 21, 364-382.

STEVENS, L.C. (1970b) Experimental production of testicular teratomas in mice of strain 129, A/He and their F1 hybrids. J. natn. Cancer Inst. 44, 929-932.

STEVENS, L.C. (1973) A new inbred subline of mice (129/terSv) with a high incidence of spontaneous congenital testicular teratomas. J. natn. Cancer Inst. 50, 235-242.

STEVENS, L.C. & LITTLE, C.C. (1954) Spontaneous testicular tumours in an inbred strain of mouse. Proc. natn. Acad. Sci., U.S.A. 40, 1080-1087.

STEVENS, L.C. & MacKENSEN, J.A. (1961) Genetic and environmental influences on teratocarcinogenesis in mice. J. natn. Cancer Inst. 27, 443-453.

STEVENS, L.C. & VARNUM, D.S. (1974) The development of teratomas from parthenogenetically activated ovarian mouse eggs. Devl. Biol. 37, 369-380.

SUSSMAN, H.H., SMALL, P.A. & CUTLOVE, E. (1968) Human alkaline phosphatase. Immunochemical identification of organ specific isozymes. J. Biol. Chem. 243, 160-166.

TRELSTAD, R.L., HAY, E.D. & REVEL, J-P. (1967) Cell contact during early morphogenesis in the chick embryo. Devl. Biol. 16, 78-106.

TRINKAUS, J.P. (1969) Cells into Organs. Prenticehall, New Jersey, p. 38.

VANNEMANN, A.S. (1917) Early history of the germ cells of the armadillo: Tatsuia novomentica. Am. J. Anat. 22, 341.

WADA, H.G., VANDENBURG, S.R., SUSSMAN, H.H., GROVE, W.E. & HERMAN, M.M. (1976) Characterisation of two different alkaline phosphatases in mouse teratoma: partial purification, electrophoretic and histochemical studies. Cell, 9, 37-44.

WALDEYER, W. (1870) Eirstock und Ei: Beitrag zur anatomie und entwicklungsgeschite der sexualorgane. Leipzeig.

WEAKLEY, B.S. (1976) Variations in mitochondrial size and ultrastructure during germ cell development. Cell Tissue Res. 169, 531-550.

WEISS, P. (1945) Experiments on cell and axonal orientation in vitro: the role of extracellular exudates in tissue organisation. J. exp. Zool. 100, 353-386.

WEISMANN, A. (1892) Das Keimplasma. Eine theorie der vereburg. Gustav Fischer, Jena.

WHITINGTON, P. McD. & DIXON, K.E. (1976) Quantitative studies of germ plasm and germ cells during early embryogenesis of Xenopus laevis. J. Embryol. exp. Morph. 33, 57-74.

WILLIS, R.A. (1962) The pathology of tumours, 4th ed. Butterworth, London, Washington.

WILLIS, R.A. (1966) The borderland of embryology and pathology, 3rd ed. Butterworth, London, Washington.

WITSCHI, E. (1948) Migration of the germ cells of human embryos from the yolk sac to the primitive gonadal folds. Contrib. Embryol. Publs. Carnegie Inst. 32, 67-80.

ZAMBONI, L. & MERCHANT, H. (1973) The fine morphology of mouse primordial germ cells in extragonadal locations. Am. J. Anat. 127, 299-336.

THE INTERACTION BETWEEN THE GERMINAL AND SOMATIC CELLS IN GONADAL DIFFERENTIATION AND DEVELOPMENT

Dr. Wai-sum O

Department of Physiology,
Laboratory of Human Reproduction and Reproductive Biology,
Harvard Medical School,
45 Shattuck Street,
Boston, Ma 02115, U.S.A.

1. Sex and Soma
   Origin of germ cells
   Somatic cell differentiation in the gonad
2. Experimental and natural sex reversal of mammalian gonads
3. Role of mesonephros and the rete ovarii in gonadogenesis
   Initiation and control of meiosis
4. Germ cell differentiation in the gonad
   Mammalian chimaeras and the fate of germ cells
5. Conclusion

The mechanism by which an individual develops into a male or a female has interested scientists since the time of Aristotle (384-320 B.C.), yet the present state of our knowledge on gonadal sex differentiation is still so fragmentary that there remain fascinating and fundamental problems for developmental biologists. In the past, emphasis on gonadal sex differentiation has centered on descriptive histological studies. Relatively little attention has been devoted to the complex cellular relationships involved or to the possible dependence of gamete differentiation upon specific organisation within the gonad. The aim of this chapter is to discuss the recent experimental investigations against the conventional descriptive histological approach.

SEX AND SOMA

The embryonic origin of different gonadal components has been a controversial subject, but the triple origin of cell types in the gonad has been well recognised: the gonad consists of germ cells, coelomic epithelium and mesenchymal tissue. The latter two are the somatic elements of the gonad, contributing to its structural and endocrine function, while the germ cells give rise to the gametes.

The germ cells are a unique cell line in the organism because they are capable of undergoing both mitosis and meiosis, whereas somatic cells can only divide mitotically. Another characteristic of germ cells is that in female mammals, both X chromosomes are active while only one X chromosome is functional in all female somatic cells (Lyon, 1974).

The most fundamental event in early embryonic development is the segregation of the germ cells from the soma. Before considering the origin of the germ cells, an answer has to be given to the old question of whether somatic cells can become germ cells and whether germ cells can become somatic cells. It is now well established that germ cells are segregated from somatic cells early in development. The primordial germ cells of the embryo appear to give rise to the entire germ cell population of the adult. If germ cells are destroyed, e.g. by busulphan (Merchant, 1975), or fail to survive, as in the Wj mice (Mintz and Russell, 1957), they will not be replaced and gonads without germ cells result (see review by Franchi *et al.*, 1962). Although somatic cells cannot become germ cells after germ line segregation, there is evidence to suggest that germ cells can become somatic cells. The teratocarcinoma cells of germ cell origin have been shown to be capable of becoming somatic cells. About 20-40 teratocarcinoma cells when injected into mouse blastocysts, resulted in mosaic mice with several tissues contributed by the injected neoplastic germ cells (Papaioannou *et al.*, 1975; see also discussion by Heath, this volume).

The germ and soma segregation in early embryogenesis does not appear to involve irreversible changes in the genes. This has been demonstrated in amphibia by the transplantation of living nuclei with suitable genetic markers into fertilized or unfertilized eggs. In *Xenopus laevis*, fertile adult frogs have been produced by transplanting nuclei from intestinal cells of feeding tadpoles to enucleated eggs (Gurdon, 1968). However, similar results have not been obtained for mammalian cell nuclei transplantation using cell fusion techniques (Graham, 1971).

Origin of germ cells

The exact mechanism of delimiting a clone of germ cells which is capable of both mitotic and meiotic divisions is not well known. In amphibia and insects, there are morphological observations which demonstrate a substance in the cytoplasm segregated into specific cells during blastulation and determinative for the progenitor cells of the germ cell line during subsequent development. The substance has been termed germ plasm in amphibians and pole plasm in insects (for review, see Eddy, 1975). Decisive evidence for the morphogenetic importance of germ plasm has been provided by injecting vegetal cytoplasm containing the special germ plasm into eggs sterilized by ultraviolet light. The resulting embryos that would have been sterile, become fertile. This type of experiment has been done by Smith (1966) in *Rana pipiens* and subsequently by Illmensee and Mahowald (1974) in *Drosophila*. Although germ plasm and pole plasm have been shown to be RNA-rich protein fractions of the cytoplasm, the actual mechanism by which these structures specify the germ cell line has remained unsolved since Weissmann (1885) proposed the theory of continuity of germ cells.

The volume of the germ plasm has been measured in *Xenopus* by Whittington and Dixon (1975) who found no change during cleavage. The germ plasm is segregated in a few cells by the normal process of cleavage, subsequently taking up a perinuclear position. The cells containing germ plasm undergo a small number of cloning divisions. This is contemporaneous with the first sign of differentiation.

In mammals, a structure similar to germ plasm has been observed in the cytoplasm of oogonia, oocytes (Eddy, 1974), premeiotic and meiotic spermatogonia and spermatids (Fawcett *et al.*,1970). The term *nuage* has been given to this structure by Andre and Rouiller (1957). Although *nuage* has been found to be nucleolar in origin, the precursor of this structure and its rate of turnover is not known and there is no definitive proof of its direct control of germ cell line segregation.

The delimiting of the germ cell line has still to take place in the 4½ day mouse embryo, since a single ectodermal cell from a 4½ day embryo injected into a host blastocyst can give rise to both germinal and somatic cell elements in the resulting chimaeras (Gardner and Rossant, 1976). It is perhaps more remarkable that the teratocarcinoma cells of germ cell origin can dedifferentiate, since when injected into normal blastocysts, these cells colonise the normal gonad of the chimaeric individual giving rise to normal fertile gametes (Illmensee and Mintz, 1976).

The earliest mammalian embryonic stage in which germ cells have been demonstrated is in the 7½ day mouse embryo, where they are found in the primitive streak (or the mesoderm) (Ozdzenski, 1967). Later, these primordial germ cells are found in the endoderm of the yolk sac. This indicates that Witschi's observation of primordial germ cells in the endodermal yolk sac of 13-somite human embryo (Witschi, 1948) is only a secondary observation. However, the controversy as to whether the germ cells arise in the extraembryonic endoderm or mesoderm would seem unimportant if the germ cells are segregated before the development of these layers (see discussion by Heath, this volume).

Once migration of the germ cells has commenced, they move up the stalk of the yolk sac and enter the embryo caudally near the allantois (Witschi, 1948). They subsequently pass up the gut mesentry and enter the developing gonadal primordia (Figure 1). The various theories which have been proposed for the mode of migration of primordial germ cells are fully reviewed by Franchi *et al.*(1962) and Zuckerman and Baker (1977). The primordial germ cells could migrate by (i) active amoeboid movements, (ii) passive movement, (iii) passive movement via the blood stream, or (iv) under the influence of a chemical inductor. Primordial germ cells cultured *in vitro* are capable of amoeboid movements (Blandau *et al.*,1963). Electron microscope studies on the morphology of the primordial germ cells in extragonadal sites have confirmed that they are capable of migrating by active amoeboid movement. The study of the fine structure shows that the energy required for locomotion is dependent on the close association with somatic cells (Zamboni and Merchant, 1973). On arrival at the genital ridges, the cells lose their amoeboid features and assume a

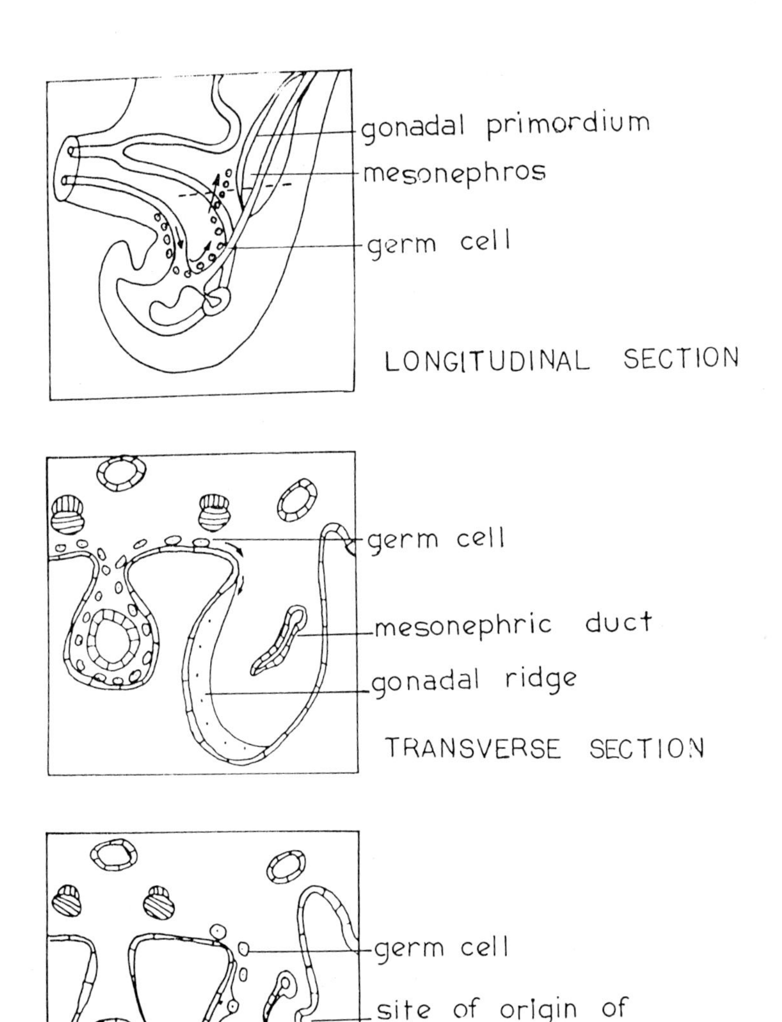

Figure 1: Diagrams to illustrate the migration of primordial germ cells.

structural organisation much simpler than that of the preceding stage, indicating a short stage of quiescence which is followed by an active mitotic phase. Once these germ cells colonise the gonadal primordia, sex differentiation takes place. The events are summarised in Figure 2.

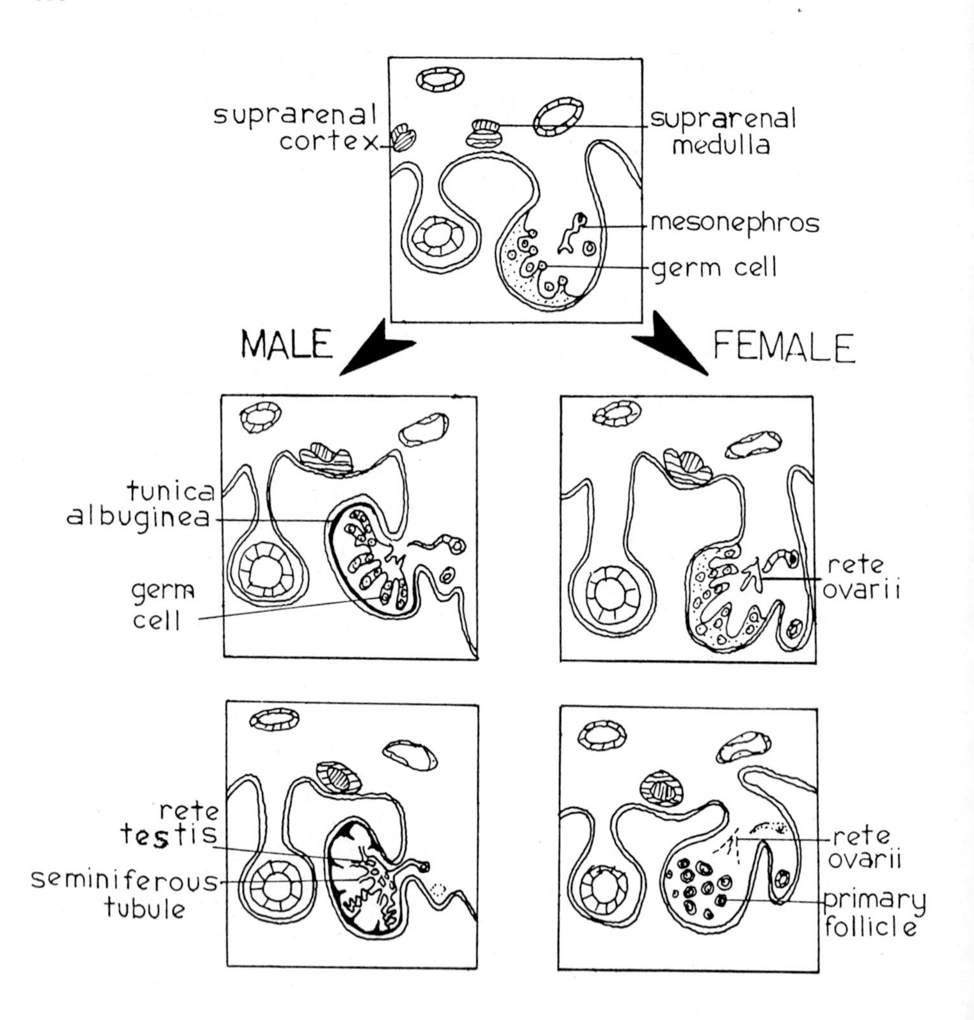

Figure 2: Schematic sections to illustrate the differentiation of the indifferent gonads into testes or ovaries.

Somatic Cell Differentiation in the Gonad

The concept that the embryonic gonads in vertebrates are developmentally bipotential, having two discrete primordia, a medullary component which develops into a testis, and a cortical component which becomes an ovary has been widely accepted. This simple view may be incorrect: mammalian gonads may be structurally bipotential, yet the medulla and the cortex may not have equal developmental potentials. Under natural or experimental conditions, spontaneous sex reversal can occur in either direction in fishes and amphibians (for review, see Chan, 1970). Among the amniotes, the bipotentiality in the embryonic gonads is not so obvious and the medullary and cortical components do not have an equal developmental potential. In birds, sex reversal usually involves masculinisation of a female and only rarely does the reverse occur. In mammals, the freemartin condition in cattle and some anomalous cases in humans, pigs and goats are the rare naturally occurring instances of developmental expressions of medullary cords in the ovary. Experimental attempts to bring about gonadal sex reversal of any sort become more and more difficult in higher vertebrates, suggesting that the great lability shown by gonads of lower forms has progressively diminished with evolution. Thus, in mammals, although the structural basis of bipotentiality may still exist, the ability of the structure to respond to factors that cause morphological changes has largely been lost.

Recently, immunological work (for review, see Silver and Wachtel, 1977) has suggested that there is a single testis-determining gene residing on the Y-chromosome, which acts by specifying a plasma membrane protein, the H-Y antigen. Since the expression of the H-Y antigen has been claimed to be present in 50% of the 8-cell mouse embryos (Krco and Goldberg, 1976), it is probably Y-dependent rather than merely a product of testicular differentiation. However, this is a weak antigen that is difficult to detect and more confirmatory evidence is needed if we are to accept its fundamental role in gonadal differentiation.

EXPERIMENTAL AND NATURAL SEX REVERSAL OF MAMMALIAN GONADS

In amphibians, steroid hormones can cause transformation of germ cells as well as the somatic tissue of gonads. In mammals, a similar effect has been observed only in marsupials. Administration of a low dose of oestradiol diproprionate to male newborn opossums (Burns, 1956) and wallabies (Alcorn, 1975) resulted in differentiation of the presumptive testes into ovotestes or ovaries. In eutherian mammals, attempts to reverse experimentally the course of gonad differentiation by male hormones have been completely unsuccessful.

There are, however, reports that ovotestes and rudimentary testes have been obtained from kidney capsule grafts of rat gonads judged to be genetically female. Moore and Price (1942) have found that ovaries grafted subcutaneously may convert into testes. Similarly, Mangoushi (1977) has shown that scrotal grafts of rat fetal ovaries also differentiate into ovotestes. Intraocular grafts of ovaries have been transformed into testes with limited degree of development (Torrey, 1950). The authors suggest that this is because of environmental inhibition in the cortical development, in contradiction to medullary inhibition of cortical development, as shown by the well known example of the freemartin.

In mammals, it is known that the testis can produce some factors that are capable of acting on the developing ovary. This is shown in transplantation experiments where testicular grafts influence the development of an ovary grown in close proximity. The only successful masculinisation of somatic tissue as well as sex-reversal of the germ cells in embryonic rat and mouse ovaries has been claimed by Turner and Asakawa (1962) and Turner (1969). A limited degree of masculinisation such as the retention and hyperdevelopment of the rete ovarii and Pflunger tubules, both of which contain atretic follicles has been observed by several authors (MacIntyre, 1956; Holyoke, 1949; Ozdzenski, 1972; Ozdzenski *et al.*,1976; O, 1977). The results are comparable to the masculinisation of freemartin gonads where inhibition of cortical growth and the subsequent development of seminiferous tubules occur (Jost *et al.*, 1975). These effects are believed to be caused by a diffusible masculinising factor which passes from the male co-twin to the freemartin via placental vascular anastomosis. Before day 45 p.c., ligature of the blood vessels between the female and the male co-twin prevents the freemartin condition from occurring although XX/XY chimaerism is evident (Vigier *et al.*,1976). The identity of this masculinising factor is still unknown and requires further investigation.

Although the masculinising factor from the testis inhibits cortical development and stimulates medullary differentiation, initiation of meiosis in the female germ cells is not impaired in freemartin gonads or in the ovarian portion of the heterosexual gonadal grafts. However, the subsequent development of the female germ cells is jeopardized by the testicular environment. Therefore, the factor responsible for gonadal organisation is probably different from the meiotic inductor substance.

ROLE OF MESONEPHROS AND THE RETE OVARII IN GONADOGENESIS

The role of the mesonephros in gonadogenesis has been studied in several submammalian species with varying results. The general theme behind the experiments depends on the fact that the development of the mesonephros from its early blastema is induced by the primitive ureter which, if prevented from reaching the mesonephric area, results in lack of differentiation of the nephric structure. In the newt, *Pleurodeles waltii*, when there is complete mesonephric agenesis, no gonad is formed. The primordial germ cells simply accumulate beneath the coelomic epithelium (Houillon, 1956).

In the chick embryo, stopping the growth of the Wolffian duct produces mesonephric agenesis, but does not prevent small ovaries and testes from being formed (Bishop-Calame, 1966). These gonads lacked definite mesonephric connections but received medullary structures from a primordium common to the adrenal cortex. Similar results have been briefly reported in the frog, *Rana dalmatins* (Cambar and Mesnage, 1963). These experiments supported the concept that the presence of a medullary component is a prerequisite for the development of the gonad, although it may be difficult to assess from which cells in the mesonephros-adrenal-gonadal blastema it stems.

There is now evidence available in mammals to indicate the involvement of the mesonephros in early gonadogenesis. Many embryologists considered the rete blastema to be mesonephric in origin (Waldeyer, 1870; Wallart, 1928; Witschi, 1951). In the testis, the rete testis forms the structural part connecting the seminiferous tubules with the vas deferens. In the female, the rete ovarii (which is homologous to the rete testis) does not become the reproductive duct system and its function in adult life is not well known. The rete ovarii takes up several morphological appearances in different species of mammals. It is quite large and compact in many species e.g. in one genus of bat, *Uroderma*, it differentiates into a rete type of interstitial gland which resembles an endocrine gland whose exact function is not known (Mossman and Duke, 1973). In the jumping mouse (*Zapus*), the guinea pig, and the white-tailed deer (*Odocoileus virginianous)* the rete epithelium is columnar and the lumen is wide, indicating an exocrine secretion into the rete. There is recent evidence showing that rete ovarii produces a factor that initiates and controls meiosis in the mouse (Byskov, 1974), and the hamster ovary (O and Baker, 1976) (for details, see next section).

There are, however, other embryologists who believe that the rete ovarii does not originate from the mesonephros, but it develops from the mesenchymal tissue (Wichman, 1916; Sauramo, 1954) or from the surface epithelium (Coert, 1898; Politzer, 1933; Gillman, 1948). In organ culture of rat ovaries there is a suggestion that it is the mesonephros rather than the rete ovarii that is essential for the initiation of meiosis (Rivelis *et al.*,1976). If this question is to be settled, the origin of the rete ovarii cells must be reinvestigated using such methods as radioactive labelling of different cell types.

Initiation and control of meiosis

One major developmental difference between the male and the female is that male germ cells commence meiosis at puberty whereas female germ cells enter meiotic prophase during fetal life. This predetermines a definite number of germ cells in the ovary early in fetal development. Large numbers of male germ cells can be produced by the testis through life since 'stem' cells persist in the seminiferous tubules. This difference might be due to meiosis being actively induced in the fetal ovary, or inhibited in the prepubertal testis. Testicular inhibition could be active, or due to the lack of an appropriate stimulus, or to a combination of both effects. This uncertainty has been partially resolved recently.

In the ovary, the early meiosis in the female germ cells is not an autodifferentiational event. Byskov (1974) transplanted fetal ovaries into nude mice and showed that meiosis in female germ cells is only 'triggered' in an ovary if the rete ovarii is present in the transplant: in the absence of the rete, meiosis failed to occur. Studies of the histology of cat, mink and ferret ovaries (Byskov, 1975) showed that there is cellular contact between rete and germinal cells in the medullary region of the ovary, where meiosis is first observed, and that further proliferation of the rete could account for the subsequent induction of meiosis in peripheral germ cells. The actual meiotic inductor substance (MIS) has not been identified but may be related to the PAS-positive granules found in the cells of the rete (Byskov, 1975). Recent studies employing organ culture of ovaries with and without the rete ovarii and/or mesonephros (which may have a common origin) have confirmed the vital role of MIS (O and Baker, 1976; Revelis *et al.*,1976).

The initiation of meiosis by a diffusible factor secreted from the rete might explain the situation in rodents, where the onset of meiosis of germ cells takes place synchronously. However, in mammals with longer gestational periods, meiosis in female germ cells occurs asynchronously over a long period of time (about 7 months in cow and man) and it is difficult to visualise how a diffusible factor could selectively affect germ cells which lie close together. This may suggest that the meiotic inducing factor must be a very weak inductor which acts only on the oogonia in close proximity to the rete cells. As intercellular bridges occur between adjacent oogonia as a result of incomplete cytokinesis during mitotic division (Gondos and Zamboni, 1969), the meiotic inducing substance could influence individual syncytial groups of germ cells without necessarily affecting neighbouring groups.

In the male, meiosis normally does not occur until puberty. It has been repeatedly shown, however, that the rete ovarii can induce meiosis in testicular germ cells. When fetal testes and ovaries (with rete) are co-cultured on opposite sides of a Millipore filter of suitable pore size, germ cells in both the ovary and testis enter meiosis, suggesting the presence of a diffusible factor (mouse: Byskov and Saxen, 1976; O and Baker, 1976). The failure of male germ cells to enter meiosis during fetal life could be due to the lack of the MIS stimulus or to an active inhibition of meiotic activity. Byskov and Saxen (1976) also observed that fetal mouse testes containing testicular cords, secrete a meiosis preventing substance (MPS) which can arrest female germ cells within the meiotic prophase. However, this MPS effect has not been observed in hamsters (O, 1977).

In fetal ovary-testis co-culture experiments, disorganisation of the seminiferous tubules, resulting in extratubular germ cells, is always observed in testes in which germ cells have entered meiosis (O and Baker, 1976) (Figure 3). It remains unknown whether such disorganisation is the primary or secondary effect of MIS. Another interesting point in these experiments is that male germ cells, when induced to undergo

TABLE 1

TIME REQUIRED BY THE MALE AND THE FEMALE GERM CELLS TO REACH DIPLOTENE STAGE OF MEIOTIC PROPHASE

Species	Days to develop from preleptotene to diplotene			
	Male germ cells		female germ cells	
	in vivo	in vitro (with rete)	in vivo	in vitro (with rete)
mouse	13+	7*	4++	4*
hamster	20+	8**	8**	8**

+ Courot *et al.*,1971
++ Baker & O, 1976
* estimated from Byskov & Saxen, 1976
** estimated from O & Baker, 1976

precocious meiosis, take a shorter time than normal to reach the diplotene stage of meiotic prophase (Table 1).

Whether the secretion of MIS depends on the presence of two X chromosomes, or whether the presence of a Y chromosome inhibits the secretion, remains to be determined. By considering various naturally occurring situations, as shown in Table 2, we may get some clues to the answer. Both X chromosomes may not be necessary to cause the rete to secrete MIS because XO mice and human fetuses with Turner's syndrome have ovaries in which germ cells enter meiosis. Does the presence of the Y chromosome prevent the secretion of the MIS? In

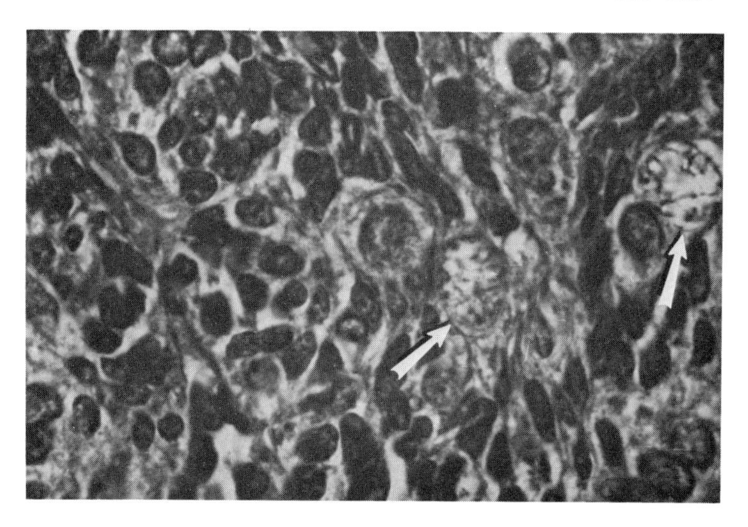

Figure 3: Day 14 p.c. hamster testis cultured for 12 days with an ovary of the same age. Note the large numbers of somatic cells and the male germ cells (arrowed) undergoing meiosis. X 262

normal males, this appears to be the case. However, in the case of the female wood lemming (normal XY), the rete cells must be secreting MIS even in the presence of a Y chromosome. This baffling situation can be explained because we believe that there is an X-linked Y suppressor gene, so that it is as if there is no functional Y chromosome (Fredga *et al.*,1975). This idea is supported by the absence of the H-Y antigen in the XY female wood lemming (Wachtel *et al.*,1976). The inhibition of MIS production by the Y chromosome may also be blocked by the X-suppressor gene. In the case of XO, Sxr (sex-reversed) and XX, Sxr mice where the Y chromosome is absent, the rete

TABLE 2

RELATIONSHIP BETWEEN THE SEX CHROMOSOMES OF SOMA AND GERM CELL FATE

Genotype of somatic cells	Gonadal sex	Germ cells entering meiosis	Fate	Reference
XX	♀	+	oocyte	Normal development
XY	♂	+	spermatocyte	Normal development
	♀	+	oocyte	Wood lemming: Fredga *et al.*,1976
XO	♀	+	oocyte	mouse: Welshon & Russel, 1956
	♂	+	spermatocyte	man: Kjessler, 1966
XX,Sxr	♂	-*	-	mouse: Cattanach, 1975
XO,Sxr	♂	+	spermatocyte	mouse: Cattanach, 1975
XXY	♂	-*	-	man: Ferguson-Smith, 1966
				mouse: Cattanach, 1961
				pig: Breeuwsma, 1970

* Prenatal atresia

testis should secrete the MIS. But Wachtel *et al.* (1975) showed that the cells from sex reversed mice are H-Y positive. This either implies that there has been an autosomal mutation, so that the Sxr can mimic the Y expression of H-Y antigen, or there may have been a Y autosomal translocation (Cattanach, 1975).

Table 2 also shows the difference between XO,Sxr and XX,Sxr mice. Germ cells from the former become spermatozoa although they are immotile, while those from the latter become atretic at the spermatocyte stage. This indicates that XX germ cells cannot survive in a testicular environment, hence the absence of germ cells in Klinefelter syndrome (XXY). This also shows that although the initiation of meiosis in the germ cells (XX or XY) is under somatic influence, the genotype of the germ cells also determines whether they can survive in a particular environment. This point will be further discussed in the following section.

GERM CELL DIFFERENTIATION IN THE GONAD

Although the prime function of germ cells in the gonad is to produce gametes, the germ cells also play an important role in gonadal development. So far, the evidence to suggest that gonads can differentiate in the total absence of germ cells is not decisive. Busulphan treatment can eliminate a large number of germ cells during their migration into the primordial gonad (Merchant, 1975) but it is not known whether any reach the gonad and initiate development before they die. In explaining the freemartin condition in cattle, Ohno *et al.*, (1976) have suggested that the XY germ cells from the male co-twin reach the female gonad and produce enough H-Y antigen to initiate masculinisation. This idea no longer seems attractive in view of the experiment (see earlier) by Vigier *et al.* (1976) who demonstrated conclusively that cellular chimaerism itself cannot be the cause of the freemartin gonadal hypoplasia.

As far as the testis is concerned, endocrine function can proceed normally in the absence of male germinal tissue. This is illustrated in the XXY males who lose the germinal cells at some stage prior to puberty and in whom androgen secretion of the gonad is not necessarily impaired. In the ovary, the development of the Graafian follicle, the endocrine apparatus of the ovary, is dependent on the presence of the oocyte (for review see O and Short, 1977). It has been shown that in the busulphan treated mice, the presence of germ cells in the epithelial cords at birth is essential for the structural differentiation of somatic cells into steroid synthesising tissue in the ovary (Merchant-Larois, 1976).

Mammalian chimaeras and the fate of germ cells

The production of artificial chimaeric mice by aggregating cleavage stages of mouse embryos *in vitro* (Tarkowski, 1961; Mintz, 1962) made it possible to create a mammalian organism composed of both XX and XY cells. Experiments show that hermaphroditism is rare in these XX/XY chimaeras (for review see McLaren, 1976). In fact, most sex chimaeras become normal fertile males which have a mixed population of XX and XY cells present throughout the body (Mystokowski and Tarkowski, 1970; Mintz, 1968). This indicates the dominant role of the male-directing, morphogenetic factor in the differentiation of the gonad.

Although it has been claimed that XX germ cells may enter the fetal testis in secondary chimaeras from heterosexual twin pregnancies in cattle and marmosets (see Tarkowski, 1970), there is no evidence that such cells complete meiosis. Extensive progeny-testing and cytogenetic analysis show that in XX/XY chimaeric male mice, only the XY germ cells give rise to the spermatozoa (McLaren, 1975). Similarly, extensive progeny testing and meiotic studies of like-sexed cattle twins sharing a placental vascular anastomosis, and of bull calves born twin to freemartins have failed to reveal any evidence of germ cell exchange (Ford and Evans, 1977). Evidence from both primary and secondary chimaeras thus makes it most unlikely that XX germ cells can be induced to undergo spermatogenesis. Other lines of evidence, reviewed by Short (1972) and O and Short (1977), lead to a similar conclusion for other mammals.

The fate of the XY germ cells in the ovary is much more difficult to determine. So far, only two fertile XX/XY female chimaeric mice have been reported (Ford *et al.*,1975). Twenty-two of the twenty-three progeny they produced were proven to be derived from the XX germ cell component of the mother. The remaining male was XXY and the Y had come through the oocyte from the XX/XY mother. To what extent XY or XXY germ cells can be transformed into oocytes when they enter an ovary remains an open question. However, there is some recent evidence to show that functional sex reversal of XY germ cells in the female XX/XY chimaeras is possible. Recently, a single XY oocyte in meiosis has been found in an XX/XY chimaeric female mouse (Evans *et al.*,1977). However, there is no evidence available as to whether the somatic cells surrounding the germ cells affect their differentiation. The limitation of the mouse chimaera as a model of gonadal differentiation is that the 'patch size' for each cell type cannot be determined (see contribution by West in this volume).

Techniques permitting the separation of germinal and somatic cells from gonads are now available (Steinberger and Steinberger, 1966; Yamada *et al.*,1973). By reaggregating germ cells of one sex with somatic cells of the opposite sex, an 'artificial' chimaeric gonad can be made and cultured, thereby making it possible to study the fate of germ cells in a genetically 'inappropriate' somatic environment. Details of the method of making the chimaeric gonads have been reported (O and Baker, 1978) and a brief summary is given here. Day 12 post-conception hamster fetuses can be sexed by examining for the presence of a sex chromatin body in liver cells. Gonads of the same sex are then dissected and pooled together, and trypsinised, giving a suspension of germinal and somatic cells. The cells are washed and cultured overnight. In this way, the somatic cells become attached to the plastic Petri dish while the germ cells remain free-floating in the medium. The germinal and somatic cells are then separately recovered. By mixing the germinal cells with the somatic cells of the opposite sex, and reaggregating by centrifugation, an 'artificial' chimaeric gonad is prepared for culture (for summary, see Figure 4). The results indicate that XY germ cells can survive in the XX somatic environment and are induced to enter meiosis precociously while XX germ cells fail to survive in an XY somatic environment. This method provides a new system for investigating the interaction between germinal and somatic cells in relation to initial gonadal differentiation and can certainly be applied to other mutants in mice so as to investigate unanswered problems in gonadal sex differentiation.

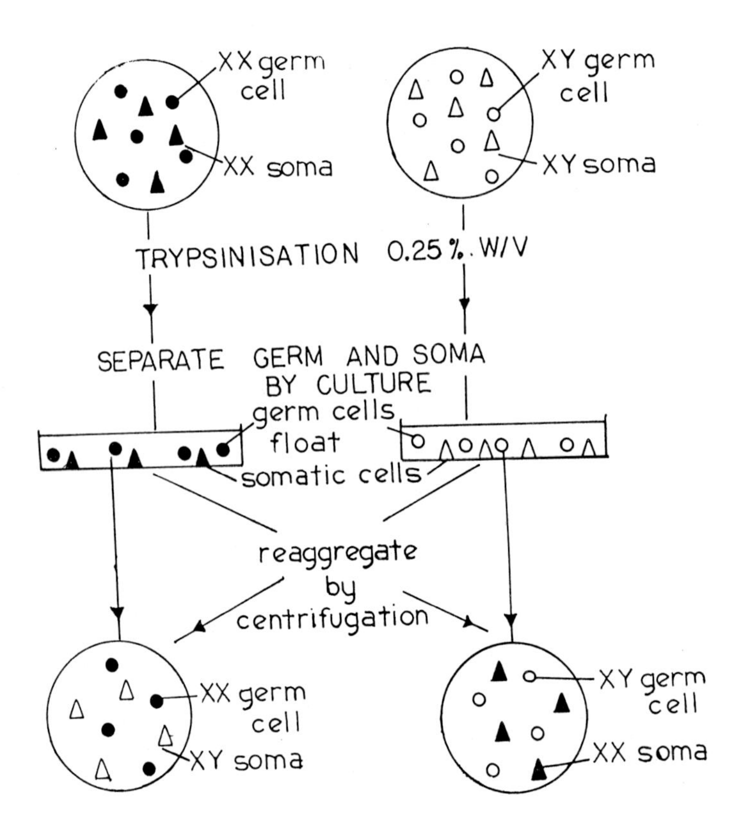

Figure 4: Diagrammatic flow sheet for protocol for preparation of chimaeric gonads.

CONCLUSION

The germinal and somatic elements of the gonads are segregated during early embryogenesis and are under separate genetic control. They subsequently acquire a high degree of functional interdependence. It has long been known that, in mammals, femaleness corresponds to an inherent trend of development. Application of such reasoning to gonadal differentiation has to be reconsidered since the entry of female germ cells into meiosis during early development has been conclusively shown not to be autodifferentiation but actively induced by the somatic element of the ovary.

Both naturally occurring and experimental findings show that XX germ cells cannot differentiate contrary to their genetic sex while XY germ cells can be sex-reversed. The behaviour of germ cells is partly determined by their genetic sex and in part by the phenotypic environment in which they find themselves.

Differentiation of somatic tissue indicates the dominant role of the male-directing morphogenetic factor in the organisation of the gonad. The development of the endocrine activity in the soma of an ovary is dependent on the presence of germ cells. The somatic cells of the testis can develop normally and function as an endocrine organ in the absence of any germ cells. A variety of techniques are now available which may help to solve some of the basic problems of gonadal differentiation. The 'artificial' chimaeric gonad system could be particularly useful in terms of finding the specific requirements for germ cell differentiation since various mutant germinal and soma somatic cell lines can be combined.

REFERENCES

ANDRE, J. and ROUILLER, C. (1957) L'ultrastructure de la membrane des ovocytes de l'araignee (*Tegenaria domestica* Clark). In: Electron Microscopy Conference Proceeding, Stockholm, 1956, pp. 162-164.

ALCORN, G.T. (1975) Ovarian development in the tammar wallaby *Macropus eugenii*. Ph.D. Thesis, MacQuarie University.

BAKER, T.G. and O, W.S. (1976) Development of the ovary and oogenesis. In: M.C. MacNaughton & A.D.T. Govan (Eds.), Clinics in Obstetrics and Gynaecology, Vol.3, Saunders Co. Ltd., London, pp. 3-26.

BISHOP-CALAME, S. (1966) Etude experimentale de l'organogenese du systeme urogenital de l'embryon de poulet. Arch. Anat. microsc. Morphol. exp. 55, 215-309.

BLANDAU, R.J., WHITE, B.J. and RUMERY, R.E. (1963) Observations on the movements of the living primordial germ cells in the mouse. Fert. Ster. 14, 482-489.

BREEUWSMA, A.J. (1970) Studies on intersexuality in pigs. Drukkerij Bronden - Offset, N.V. Rotterdam.

BURNS, R.K. (1956) Transformation du testicule embryonnaire de l'opposum en ovotestis ou en "ovaire" sous l'action de l'hormone femelle, le diproprionate d'oestradiol. Arch. Anat. microsc. Morphol. exp. 45, 173-202.

BYSKOV, A.G.S. (1974) Does the rete ovarii act as a trigger for the onset of meiosis? Nature, Lond. 252, 396-397.

BYSKOV, A.G. (1975) The role of the rete ovarii in meiosis and follicle formation in different mammalian species. J. Reprod. Fert. 45, 201-209.

BYSKOV, A.G. and SAXEN, L. (1976) Induction of meiosis in fetal mouse testis *in vitro*. Develop. Biol. 52, 193-200.

CAMBAR, R. and MESNAGE, J. (1963) L'agenesie experimentale du mesonephros n'influence par le developpement de la grande genitale chez les Amphibien Anoures. C. r. hebd. Seanc. Acad. Sci., Paris, 257, 4021-4023.

CATTANACH, B.M. (1961) XXY mice. Genet. Res. 2, 156-160.

CATTANACH, B.M. (1975) Sex reversal in the mouse and other mammals. In: M. Balls & A.E. Wild (Eds.), The early development of mammals, Cambridge University Press, London, pp. 305-317.

CHAN, S.T.H. (1970) Natural sex reversal in vertebrates. Phil. Trans. Roy. Soc. Ser. B 259, 59-71.

COERT, H.T. (1898) Over de entwikkeling en den bouw van de geschlachtsklier bix de Zoodieren meer en het bejzonder van den eierstok. Proefschrift, Lieden Thesis.

COUROT, M., HOCHEREAU-DE REVIERS, M. and ORTAVANT, R. (1971) Spermatogenesis. In: A.D. Johnson, W.R. Gomes & N.L. Van Demark (Eds.), The Testis, Vol.1, Academic Press, New York, pp. 339-432.

EDDY, E.M. (1974) Fine structural observations on the form and distribution of nuage in germ cells of the rat. Anat. Rec. 178, 731-758.

EDDY, E.M. (1975) Germ plasm and the differentiation of the germ cell line. Int. Rev. Cytol, 43, 229-280.

EVANS, E.P., FORD, C.E. and LYON, M.F. (1977) Direct evidence of the capacity of the XY germ cells in the mouse to become an oocyte. Nature, Lond. 267, 430-431.

FAWCETT, D.W., EDDY, E.M. and PHILLIPS, D.M. (1970) Observations on the fine structure and relationships of the chromatoid body in mammalian spermatocyte. Biol. Reprod. 2, 129-153.

FERGUSON-SMITH, M.A. (1966) X-Y chromosome interchange in the aetiology of true hermaphroditisms and of XX Kleinefelter's syndrome. Lancet 2, 475-476.

FORD, C.E. and EVANS, E.P. (1977) Cytogenetic observations on XX/XY chimaeras and a reassessment of the evidence for germ cell chimaerism in heterosexual twin cattle and marmoset. J. Reprod. Fert. 49, 25-33.

FORD, C.E., EVANS, E.P., BURTENSHAW, M.D., CLEGG, H.N., TUFFREY, M. and BARNES, R.D. (1975) A functional 'sex-reversed' oocyte in the mouse. Proc. Roy. Soc. B 190, 187-197.

FRANCHI, L.L., MANDL, A.M. and ZUCKERMAN, S. (1962) The development of the ovary and the process of oogenesis. In: S. Zuckerman, A.M. Mandl & P. Eckstein (Eds.), The Ovary, Vol.1, 1st ed., Academic Press, London, pp. 1-88.

FREDGA, K., GROPP, A., WINKLING, H. and FRANK, I. (1975) Fertile XX- and XY- type females in the wood lemming *Myopus schisticolor*. Nature, Lond. 261, 225-226.

GARDNER, R.L. and ROSSANT, J. (1976) Determination during embryogenesis. In: Symposium on Embryogenesis in Mammals, London, 1975, Ciba Foundation, North-Holland, pp. 4-25.

GILLMAN, J. (1948) The development of the gonads in man, with a consideration of the role of fetal endocrine and histogenesis of ovarian tumour. Contrib. Embryol. 32, 83-131.

GONDOS, B. and ZAMBONI, L. (1969) Ovarian development: the functional importance of germ cell interconnections. Fert. Steril. 20, 176-189.

GRAHAM, C.F. (1971) Virus assisted fusion of embryonic cells. Karolinska Symposium in Reproductive Endocrinology 3, 154-165.

GURDON, J.B. (1968) Changes in somatic cell nuclei inserted into growing and maturing amphibian oocytes. J. Embryol exp. Morphol. 20, 401-414.

HOLYOKE, E.A. (1949) The differentiation of embryonic gonads transplanted to the adult omentum in the albino rat. Anat. Rec. 103, 675-699.

HOUILLON, C. (1956) Recherches experimentales sur la dissociation medulli-corticale dans l'organogenese des gonades chez le triton *Pleurodeles waltlii michah.* Bull. biol. Fr. Belg. 90, 359-455.

ILLMENSEE, K. and MAHOWALD, A. (1974) Transplantation of posterior pole plasm in Drosophila. Induction of germ cells at the anterior pole of the egg. Proc. Natn. Acad. Sci. U.S.A. 71, 1016-1020.

ILLMENSEE, K. and MINTZ, B. (1976) Totipotency and normal differentiation of a single teratocarcinoma cells cloned by injection into blastocysts. Proc. Natn. Acad. Sci. U.S.A. 73, 549-551.

JOST, A., PERCHELLET, J.P., PREPIN, J. and VIGIER, B. (1975) The prenatal development of bovine freemartins. In: R. Reinboth (Ed.), Intersexuality in the Animal Kingdom, Springer, Heidelberg, pp. 392-406.

KJESSLER, B. (1966) Karyotype, meiosis and spermatogenesis in a sample of men attending infertility clinics. In: S. Karger (Ed.), Monograph in Human Genetics, Vol.2, New York.

KRCO, C.J. and GOLDBERG, E.H. (1976) H-Y (male) antigen: detection on eight-cell mouse embryos. Science 193, 1134-1135.

LYON, M. (1974) Evolution of X-chromosome inactivation in mammals. Nature, Lond. 250, 651-653.

MANGOUSHI, M.A. (1977) Contiguous allografts of male and female gonadal primordia in the rat. J. Anat. 123, 407-413.

MACINTYRE, M.N. (1956) Effects of the testis on ovarian differentiation in heterosexual embryonic rat gonad transplants. Arch. Anat. microsc. Morphol. exp. 48, 141-153.

McLAREN, A. (1975) Sex chromosome and germ cell distribution in a series of chimaeric mice. J. Embryol. exp. Morphol. 33, 205-216.

McLAREN, A. (1976) Mammalian Chimaeras. Cambridge University Press, Cambridge.

MERCHANT, H. (1975) Rat gonadal and ovarian organogenesis with and without germ cells. An ultrastructural study. Devel. Biol. 44, 1-21.

MERCHANT-LAROIS, H. (1976) The role of germ cells in the morphogenesis and cytodifferentiation of the rat ovary. In: N. Muller-Berat (Ed.), Proceedings of the 2nd International Conference on Differentiation, North Holland Publishing Co., Amsterdam, pp. 453-462.

MINTZ, B. (1962) Formation of genotypically mosaic mouse embryos. Amer. Zool. 2, 432 (Abstract).

MINTZ, B. (1968) Hermaphroditism, sex chromosomal moasicism and germ cell selection in allophenic mice. J. Anim. Sci. (Suppl. 1) 27, 51-60.

MINTZ, B. and RUSSELL, E. (1957) Gene induced embryological modifications of primordial germ cells in the mouse. J. exp. Zool. 134, 207-239.

MOORE, C.R. and PRICE, D. (1942) Differentiation of embryonic reproductive tissues of the rat after transplantation into post-natal host. J. exp. Zool. 90, 229-265.

MOSSMAN, H.W. and DUKE, K.L. (1973) Comparative morphology of the mammalian ovary. The Wisconsin University Press: Wisconsin.

MYSTOKOWSKI, E.T. and TARKOWSKI, A.K. (1970) Observations in CBA-p/CBA-T6T6 mouse chimaeras. J. Embryol. exp. Morphol. 20, 33-52.

O, W.S. (1977) Gonadal sex determination and differentiation in rats and hamsters. Edinburgh University, Ph.D. Thesis.

O, W.S. and BAKER, T.G. (1976) Initiation and control of meiosis in hamster gonads *in vitro*. J. Reprod. Fert. 48, 399-401.

O, W.S. and BAKER, T.G. (1978) Germinal and somatic cell interrelationships in gonadal sex differentiation. Ann. Biol. anim. Bioch. Biophys. 18 (in press).

O, W.S. and SHORT, R.V. (1977) Sex determination and differentiation in mammalian germ cells. In: R.J. Blandau & Bergsma D. (Eds.), Morphogenesis and Malformation of the Genital System, Alan R. LISS, Inc., New York, pp. 1-12.

OHNO, S., CHRISTIAN, L.C., WACHTEL, S.S. and KOO, G.C. (1976) Hormone-like role of H-Y antigen in Bovine freemartin gonad. Nature, Lond. 261, 597-599.

OZDZENSKI, W. (1967) Observations on the origin of primordial germ cells in the mouse. Zool. Pol. 17, 367-379.

OZDZENSKI, W. (1972) Differentiation of the genital ridges of mouse embryos in the kidney of adult mice. Arch. Anat. microsc. Morph. exp. 61, 267-278.

OZDZENSKI, W., ROGULSKA, T., BATAKIER, H., BRZOZAWSKA, M., BEMBISZEWSKA, A. and STEPINAKA, U. (1976) Influence of embryonic and adult testis in the differentiation of embryonic ovary in the mouse. Arch. Anat. microsc. Morph. exp. 65, 285-294.

PAPAIOANNOU, V.E., McBURNEY, M.W., GARDNER, R.L. and EVANS, M.J. (1975) Fate of teratocarcinoma cells injected into early mouse embryos. Nature, Lond. 258, 70-73.

POLITZER, G. (1933) Die Keinbajn des Menschen. J. Anat. 100, 331-361.

RIVELIS, C., PREPIN, J., VIGIER, B. and JOST, A. (1976) Prophase meiotique dans les cellules germinales de l'ebanche ovarienne de Rat cultivee *in vitro* en milieu anhormonal. C.r. hebd. Seanc. Acad. Sci., Paris, 282, 1429-1432.

SAURAMO, H. (1954) Development, occurrence, function and pathology of the rete ovarii. Acta Obstet. Gynecol. Scand. 33 Suppl. 2. 29-46.

SHORT, R.V. (1972) Germ cell sex. In: R.A. Beatty & S. Glucksohn-Waelsch (Eds.), Genetics of the Spermatozoon, Copenhagen, pp. 325-345.

SILVER, W.K. and WACHTEL, S.S. (1977) H-Y antigen: Behaviour and function. Science 195, 956-960.

SMITH, L.D. (1966) The role of a 'Germinal plasm' in the formation of primordial germ cells in *Rana pipiens*. Devel. Biol. 14, 330-347.

STEINBERGER, A. and STEINBERGER, E. (1966) *In vitro* culture of rat testicular cells. Expl. Cell Res. 44, 443-452.

TARKOWSKI, A.K. (1961) Mouse chimaeras developed from fused eggs. Nature, Lond. 190, 857-860.

TARKOWSKI, A.K. (1970) Germ cells in natural and experimental chimaeras in mammals. Phil. Trans. Roy. Soc. Ser. B. 259, 107-111.

TORREY, T. (1950) Intraocular grafts of embryonic gonads of the rats. J. Exp. Zool. 37-58.

TURNER, C.D. (1969) Experimental reversal of germ cells. Embryologia 10, 206-230.

TURNER, C.D. and ASAKAWA, H. (1962) Experimental reversal of germ cells in ovaries of fetal mice. Science, 143, 1344-1345.

VIGIER, B., LOCATELLI, A., PREPIN, J., DU MESUIL DU BUISSON, F. and JOST, A. (1976) Les premieres manifestations du freemartinisme chez le foetus du veau ne dependent pas du chimaerisme chromosomique XX/XY. C.r. hebd. Seanc. Acad. Sci., Paris, 282, 1355-1358.

WACHTEL, S.S., OHNO, S., KOO, G.C. and BOYSE, E.A. (1975) Possible role for H-Y antigen in primary determination of sex. Nature 257, 235-236.

WACHTEL, S.S., KOO, G.C., OHNO, S., GROPP, A., DEV, V.G., TANTRAVAHI, R., MILLER, D.A. and MILLER, O.J. (1976) H-Y antigen and the origin of XY female wood lemmings (*Myopus schisticolor)*. Nature, Lond. 264, 638-639.

WALDEYER, W. (1870) Eierstok und Ei. Enzelmann: Leipzig.

WALLART, J. (1928) Contribution a l'etude des origines du rete ovarii. Bull. Hist. appl. Physio. Path. 5, 181-190.

WEISMANN, A. (1885) Die Continuitatdes Keimplasmas als Grundlage einer Theorie der Vererbung. Jena.

WELSHONS, W.J. and RUSSEL, L.B. (1959) The Y-chromosome as the bearer of male-determining factors in the mouse. Proc. Natn. Acad. Sci. U.S.A. 45, 560-566.

WHITTINGTON, P.McD. and DIXON, K.E. (1975) Quantitative studies of germ plasm and germ cells during early embryogenesis of *Xenopus laevis*. J. Embryol. exp. Morphol. 33, 57-74.

WICHMAN, S.E. (1916) Das Epoophoron seine Anatomie und Entwicklung bein Menschen von der Embryonalizeit bis ins Greisenalter. Thesis, Helsinki.

WITSCHI, E. (1948) Migration of the germ cells of human embryos from the yolk sac to the primitive gonadal folds. Contrib. Embryol. 32, 67-80.

WITSCHI, E. (1951) Embryogenesis of the adrenal and reproductive gland. Recent Prog. Horm. Res. 6, 1-27.

YAMADA, M., YASUE, S. and MATSUMOTO, K. (1973) Formation of C21-17-hydroxysteroids and C19-steroid from 3β-hydroxypreg-5-en-20-one and progesterone *in vitro* by germ cells from immature rat testes. Endocrinology, 93, 81-89.

ZAMBONI, L. and MERCHANT, H. (1973) The true morphology of mouse primordial germ cells in extragonadal locations. Am. J. Anat. 137, 299-316.

ZUCKERMAN, S. and BAKER, T.G. (1977) The development of the ovary and the process of oogenesis. In: S. Zuckerman & B.J. Weir (Eds.), Ovary, Vol.1, 2nd ed., Academic Press, New York, pp.42-68.

THE ROLE OF MULLERIAN INHIBITING SUBSTANCE IN MAMMALIAN SEX DIFFERENTIATION

Patricia K. Donahoe, M.D.* and David A. Swann, Ph.D**

**The Pediatric Surgical Research Laboratory and the Division of Surgery, Massachusetts General Hospital, Harvard Medical School, and*
***The Department of Biological Chemistry Harvard Medical School*

1. Ultrastructure of the Regressing Mullerian Duct and the Sertoli Cell.
2. Mullerian Inhibiting Substance Production After Birth.
3. Biochemical Isolation of Mullerian Inhibiting Substance.
4. Gonadotropin Control of Mullerian Inhibiting Substance.
5. Speculation

The early embryogenesis of the mammalian genital structures is autonomous in the female but subject to hormonal control in the male. In the absence of a gonad, or in the presence of an ovary, the Mullerian duct structures develop into uterus, fallopian tube and upper vagina, and the common genital anlagen develops into clitoris, labia minora and labia majora; the Wolffian duct regresses. In the presence of a testis, Wolffian duct structures develop into the vas deferens, epididymis and seminal vesicals, and the common genital anlagen develop into glans, shaft of penis and scrotum; the Mullerian duct regresses. In the late 1940's, Jost elucidated these phenomena in a series of *in vivo* experiments on rabbit embryos in which undifferentiated gonads were removed and replaced with ovarian or testicular grafts early in embryonic development before the Wolffian or Mullerian ducts had differentiated. If whole testis was replaced with testosterone alone, then Wolffian ducts were stimulated, but Mullerian ducts did not regress, leading Jost to postulate the existence of a second testicular hormone named Mullerian Inhibiting Substance, which is directly responsible for regression of the Mullerian ducts (Jost, 1946a, 1946b, 1947). It is this intriguing substance, which directly causes the active regression of an embryonic organ system, that is the subject of this report.

Proper sex differentiation depends upon a number of factors, of which Mullerian Inhibiting Substance plays only a small role; these factors include chromosomal endowment, gonadal differentiation, hormonal production and receptor response to gonadal hormones. The germ cells must receive a normal chromosomal endowment and then must migrate properly from the allantois (Moore, 1974), along the hind gut and over the retroperitoneum (Figure 1) to the urogenital ridge, where they take residence. In the presence of a Y chromosome, or merely the short arm of the Y, the gonad undergoes differentiation into seminiferous tubules. The mechanism by which the Y chromosome directs testicular morphogenesis is obscure, but

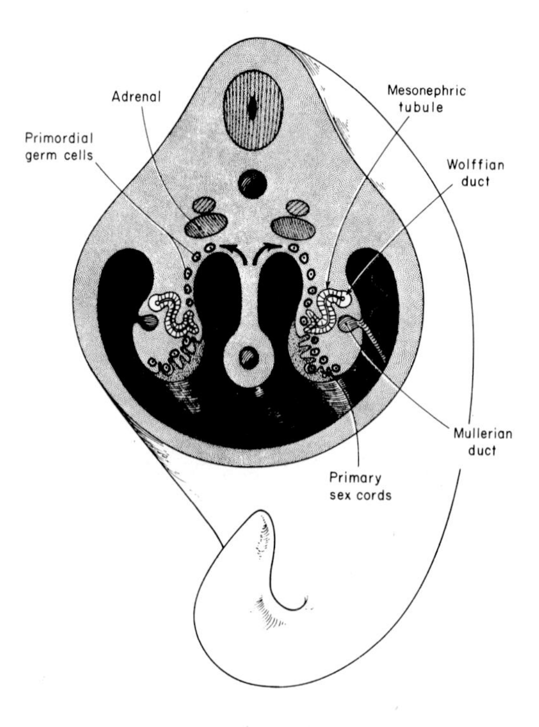

Figure 1: Germ cells migrate from the hindgut over the retroperitoneum to the urogenital ridge. Y gene products induce testicular morphogenesis and the testis then produces testosterone which stimulates Wolffian duct and causes regression of Mullerian duct. In the absence of Y gene products, the gonad differentiates as an ovary; the Wolffian duct regresses passively; and the Mullerian duct develops autonomously.

recent cytogenetic and immunologic studies of 46,XX true hermaphrodites and 46,XX sex-reversed males who underwent testicular organogenesis and developed a male phenotype in the apparent absence of a Y chromosome have brought some understanding to this phenomenon. Wachtel and his colleagues (1976) detected the presence of H-Y antigen in these 46,XX individuals.

H-Y antigen was postulated to be the gene product of the short arm of the Y (Ohno, 1976) after Eichwald and Silmer (1955) observed that inbred female mice rejected transplanted tissue taken from males of the same strain. Antibodies presumably formed to a Y product, subsequently referred to as H-Y antigen, since the remainder of the chromosomes were common to both sexes. The amount of detectable H-Y antigen is directly related to the number of determining genes present, since individuals with two Y chromosomes (Wachtel *et al.*,1975) i.e. XXYY or XYY, possess more H-Y antigen than normal males with one Y, and normal males express more H-Y antigen than true hermaphrodites or XX sex-reversed males (Wachtel *et al.*,1976; Saenger *et al.*, 1976). It appears that the short arm of the Y chromosome alone is required for expression of the H-Y antigen (Koo *et al.*, 1976) and for male phenotypic development (Devictor-Vuillet *et al.*, 1971). Even if the short arm of the Y is translocated to another chromosome, then H-Y antigen can still be detected, testicular differentiation can take place, and male determination, partial or complete, can occur.

ULTRASTRUCTURE OF THE REGRESSING MULLERIAN DUCT AND OF THE SERTOLI CELL

The testis differentiates first the seminiferous tubules at day 13 in the rat, and day 43 to 50 in humans (Jirasek, 1971), and later the interstitium at 15 days in the rat, and 60 days in the human (Jost *et al.*, 1973), from whence Leydig cells produce testosterone. Just after the testis differentiates into a recognizable tubular structure, the Sertoli cell manufactures Mullerian Inhibiting Substance, which acts on the Mullerian ducts causing them to regress. Mullerian duct responsivity to Mullerian Inhibiting Substance is limited in both males and females to a short period of embryonic development, until late day 15 in the rat (Picon, 1969), before day 60 in the human (Josso, 1974), and until approximately day 62 in the calf (Jost *et al.*, 1973).

Regression of the Mullerian duct under the influence of Mullerian Inhibiting Substance is characterized ultrastructurally by an interesting sequence of events. The nuclei of the Mullerian duct epithelial cells of the male rat appear euchromatic or stimulated, rather than pycnotic throughout the period of dissolution from day 14 to day 17. On day 15, electron-dense cytoplasmic particles appear (Price *et al.*, 1977). Since these particles are rich in acid phosphatase, they are probably lysosomes. Mesenchymal cells then condense around the basement membrane. The Mullerian duct cells are soon engulfed by phagocytes which cross the basement membrane of the epithelial duct from the surrounding mesenchyme. This sequence of autodigestion and autophagocytosis characterizes the process of programmed cell death of the Mullerian duct in response to the Mullerian Inhibiting Substance signal.

Evidence that the Sertoli cells manufacture Mullerian Inhibiting Substance was provided by Josso (1973), who separated fetal calf seminiferous tubules from interstitium. When separately cultured with agonadal urogenital ridge from 14 day rat foetuses, the Mullerian duct regressed under the influence of seminiferous tubules, but not when cultured with interstitium. Stripped tubules free of interstitium were depleted of germ cells and then grown in tissue culture. Clumps of cells, thought to represent predominantly Sertoli cells (Blanchard & Josso, 1974) caused regression of the Mullerian duct when co-cultured with 14 day fetal rat urogenital ridge. More indirect evidence to implicate the Sertoli cell as the cell responsible for the production of Mullerian Inhibiting Substance was provided by an electron microscopic study of the newborn calf testis at a time when Mullerian Inhibiting Substance activity is high (Donahoe *et al.*, 1977d). These Sertoli cells, which predominate in the tubule, have the dense rough endoplasmic reticulum and swollen cisternae characteristic of a highly active protein producing cell (Fawcett, 1966).

MULLERIAN INHIBITING SUBSTANCE PRODUCTION AFTER BIRTH

Contrary to the expectations that Mullerian Inhibiting Substance was a protein unique to fetal life, it was found that the rat continues to produce Mullerian Inhibiting Substance after birth for 21 days (Donahoe *et al.*, 1976b), after which it

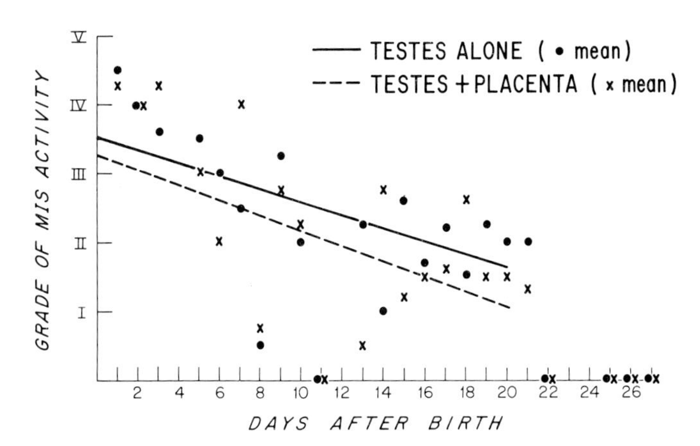

Figure 2: Mullerian Inhibiting Substance activity (0 to V) observed when small fragments of RAT testes were incubated with the urogenital ridge of the 14 day female rat embryo. High Mullerian Inhibiting Substance activity gradually diminished until 21 days after birth, after which it could not be detected. Incubation of testes with placenta did not significantly vary the regression observed with testis alone.

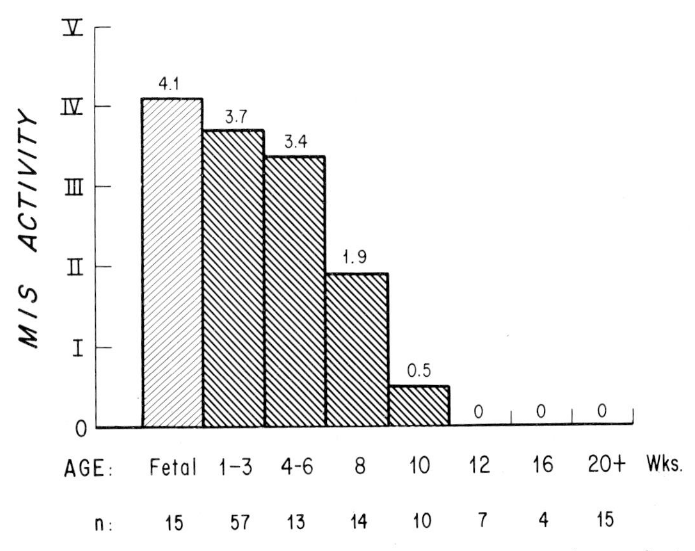

Figure 3: Mullerian Inhibiting Substance activity (0 to V) observed when small fragments of CALF testes were incubated with the urogenital ridge of the 14 day female rat embryo. High Mullerian Inhibiting Substance activity gradually diminished until 8 to 10 weeks after birth, after which it could not be detected.

may no longer be detected by presently available techniques (Figure 2). Similarly, the calf continues to produce Mullerian Inhibiting Substance after birth for up to 8 to 10 weeks (Figure 3), as determined by organ culture techniques (Donahoe, 1977d). This serendipitous finding led to the discovery of a large source of raw material for the biochemical isolation and characterisation of Mullerian Inhibiting Substance (Swann, Donahoe, Ito, Morikawa & Hendren, unpublished). The human testis also continues to manufacture Mullerian Inhibiting Substance after birth (Figure 4), up to age two years, after which

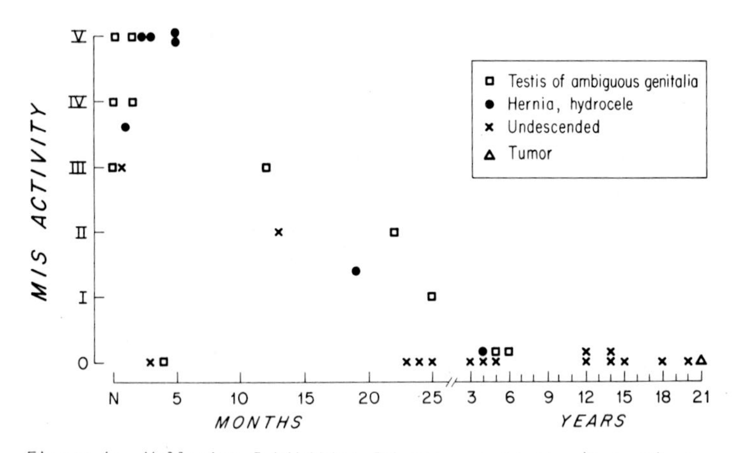

Figure 4: Mullerian Inhibiting Substance activity (0 to V) observed when small fragments of HUMAN testes were incubated with the urogenital ridge of the 14 day female rat embryo. High Mullerian Inhibiting Substance activity gradually diminished until 2 years of age, after which it could not be detected.

it can no longer be detected by present methodology (Donahoe *et al.*, 1977a). Absence of Mullerian duct regression may result either from poor production of Mullerian Inhibiting Substance as occurs in the dysgenetic testis of Mixed Gonadal Dysgenesis, or from failure of end organ responsiveness, a possible mechanism which could explain the otherwise normal male with retained Mullerian duct structures (Figure 5). It is of interest that these males commonly have cryptorchidism, which initiates their presentation for evaluation (Brook *et al.*, 1973). Because of these observations, we compared Mullerian

Inhibiting Substance activity in patients of matched ages (less than 2 years). Those with undescended testes always had less Mullerian Inhibiting Substance than those with descended testes, causing us to speculate that Mullerian Inhibiting Substance may play a role in descent of the testis (Donahoe *et al.*, 1977a).

BIOCHEMICAL ISOLATION OF MULLERIAN INHIBITING SUBSTANCE

Newborn calf testes with a high activity of Mullerian Inhibiting Substance (Donahoe *et al.*, 1977b) when cocultured as intact fragments, maintain high biologic activity after prolonged extraction, dialysis, and mild centrifugation (Swann *et al.*, unpublished data). If this active supernatant is subjected to sedimentation on a caesium chloride density gradient, the regression activity is limited to a protein fraction. The active protein fraction obtained by density gradient has been further fractionated by chromatography. Gel filtration consistently yielded a single active fraction, as did ion exchange chromatography. A positive dose response curve was obtained when increasing concentrations of active fractions were added to the assay system. The fractions isolated by both of these procedures comprised between 82 and 89% amino acids, had similar carbohydrate compositions, and contained several protein constituents with the same electrophoretic mobility when analyzed by gel electrophoresis. An antiserum prepared to the extract supernatant and absorbed with fetal calf serum 1) demonstrated specific precipitin bands on double diffusion, 2) localized to the newborn calf testis by immunofluorescence, and 3) blocked high biologic activity of an extract supernatant in the organ culture assay. Thus, analysis of the active fraction indicates that Mullerian Inhibiting Substance is a protein or a glycoprotein, and the immunofluorescence studies support the contention that the protein or glycoprotein is a testis-derived constituent. It is not yet clear which of the testis constituents extracted is Mullerian Inhibiting Substance or whether duct regression is caused by more than one constituent. It is also not clear how the active constituent or constituents cause duct regression at a distance from their site of origin. More refined biologic and immunologic techniques should allow isolation and identification of this embryonic hormone.

GONADOTROPIN CONTROL OF MULLERIAN INHIBITING SUBSTANCE

Evidence has recently been obtained to suggest that production of Mullerian Inhibiting Substance may not be autonomous to the testis, but is controlled from some other site. Since production wanes soon after birth or at the time of weaning, it was anticipated that chorionic gonadotropin or prolactin might act as a modulator. Placental fragments co-cultured with postnatal rat testes *in vitro* failed to affect Mullerian Inhibiting Substance activity (Donahoe *et al.*, 1976a). Prolactin added *in vitro* in increasing concentrations to the incubating media of 10 day postnatal rat testes, failed to vary Mullerian Inhibiting Substance production. Neither of these observations conclusively eliminates chorionic gonadotropin or prolactin as potential modulators. A high titre antiserum to Luteinizing Hormone Releasing Hormone (LHRH) when given to mothers on day 13 and 20 of pregnancy, caused an elevation of Mullerian Inhibiting Substance in the testes of male pups assayed both on embryonic day 17 and on day 6 after birth when compared to testis of pups from control mothers receiving normal rabbit serum. Testis weight and diameter were reduced in the experimental pups, indicating that endogenous LH was blocked by LHRH antiserum as had been demonstrated previously (Bercu *et al.*, 1977a) at a later stage of development. The contrasting observation that Mullerian Inhibiting Substance activity increased, indicated that one of the gonadotropins, FSH or LH, inhibited Mullerian Inhibiting Substance activity (Bercu *et al.*, 1977b). In order to differentiate between LH and FSH, pups of mothers receiving antiserum to LHRH were given daily injections of exogenous LH or FSH for 5 days after birth; Mullerian Inhibiting Substance activity was assayed on day 6. FSH replacement after birth restored Mullerian Inhibiting Substance activity to lower control levels. These studies indicate that the gonadotropins modulate Mullerian

Figure 5: Intraoperative photo (Provided by Dr. Charles Klippel, Medical College of Ohio at Toledo) of a child with bilateral undescended testes in whom retained Mullerian duct structures were found (fallopian tube, uterus).

Figure 6: Intraoperative findings in a child with True Hermaphroditism. The ovotestis on the right caused ipsilateral regression of Mullerian duct.

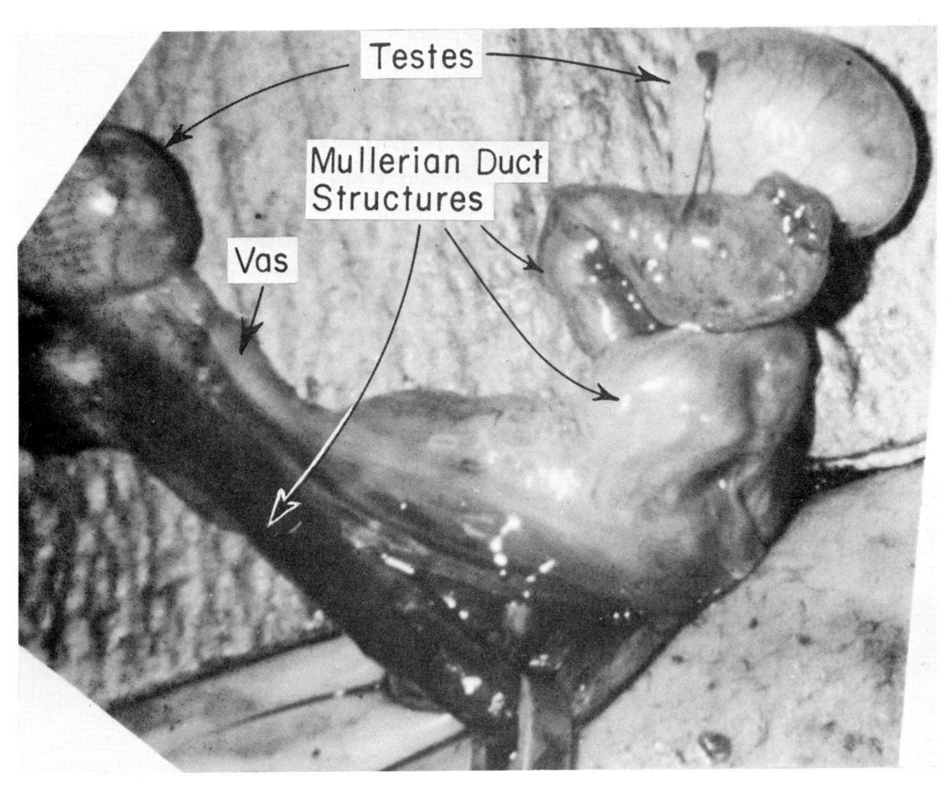
Testes
Mullerian Duct Structures
Vas

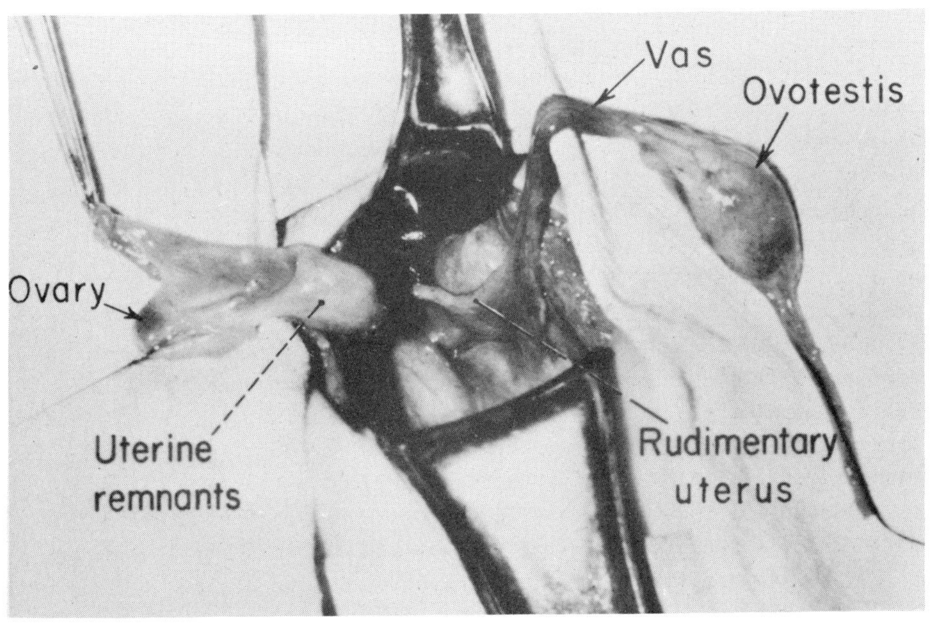
Vas
Ovotestis
Ovary
Uterine remnants
Rudimentary uterus

Inhibiting Substance production, and specifically, that FSH inhibits Mullerian Inhibiting Substance production. FSH modulation of Mullerian Inhibiting Substance production has not been documented in any other mammal. Gonadotropin inhibition may provide an explanation for the disappearance of Mullerian Inhibiting Substance after birth, since the pituitary matures soon thereafter.

SPECULATION

Since the production of Mullerian Inhibiting Substance continues after Mullerian duct regression, the question arises of other roles for the substance. We suspect that Mullerian Inhibiting Substance may be involved in descent of the testis. Evidence to date, however, is inferential. Elger, Richter and Korte (1976), failed to detect androgen dependence of descent of the testis in fetal rabbits, mice, or monkeys, using androgen antagonists. In contrast, we detected high Mullerian Inhibiting Substance activity in human testes after birth with progressive diminution in regression until two years of age, after which no further activity could be detected by the organ culture method (Donahoe *et al.*, 1977a,c). Comparison of Mullerian Inhibiting Substance activity in descended and undescended testes revealed that, for matched ages, Mullerian Inhibiting Substance was lower in patients with undescended testes.

Patients with dysgenetic testes as in Mixed Gonadal Dysgenesis, lack both testosterone and Mullerian Inhibiting Substance (Donahoe *et al.*, 1977a). Male pseudohermaphrodites (Donahoe *et al.*, 1977c) with normal testicular histology, who produce both testosterone and Mullerian Inhibiting Substance, lack receptors for testosterone or dihydrotestosterone, but demonstrate normal Mullerian duct regression. True hermaphrodites (Donahoe *et al.*, 1977b,d), who undergo sufficient testicular organogenesis under the direction of the short arm of the Y chromosome, produce both testosterone and Mullerian Inhibiting Substance, but the latter is only ipsilaterally active on the side of the testis or ovotestis (Figure 6). In the female pseudohermaphrodite or adrenogenital syndrome female, excessive testosterone production due to enzymatic deficiencies in the pathway of metabolism of cholesterol to cortisol leads to stimulation of the external genitalia, but Mullerian structures develop normally due to the absence of Mullerian Inhibiting Substance in these otherwise normal females (Hendren & Crawford, 1972). Thus Mullerian Inhibiting Substance clearly does play a role in the sex differentiation of human and other mammalian males after testicular organogenesis has occurred.

REFERENCES

BERCU, B.B., JACKSON, I.M.D., SAFAII, H. and REICHLIN, S., (1977a) Permanent impairment of testicular development after transient immunological blockade of endogenous LHRH in the neonatal rat. Endocrin. 101, 1871-1879.

BERCU, B.B., MORIKAWA , Y., JACKSON, I. and DONAHOE, P.K., (1977b) Increased secretion of Mullerian Inhibiting Substance after immunological blockade of endogenous luteinizing hormone releasing hormone in the rat. Pediat. Res. (In press).

BLANCHARD, M. and JOSSO, N., (1974) Source of the anti-Mullerian hormone synthesized by the fetal testis: Mullerian Inhibiting activity of fetal bovine Sertoli cells in tissue culture. Pediat. Res. 8, 968-971.

BROOK, C.G., WAGNER, H., ZACHMAN, M., PRADER, A., ARMENDARES, S. FRENKS, S., ALEMAN, P., NAJJAR, S.S., SLIM, M.S., CONTON, M. and BOZIE, C., (1973) Familial occurrence of persistent Mullerian structures in otherwise normal males. Brit. Med. J. 1, 771-773.

DEVICTOR-VUILLET, M., LUCIANI, J., CARLON, N. and STAHL, A., (1971) Anomalies de structure et role du chromosome Y chez l'homme. Pathol. Biol. 19, 231-249.

DONAHOE, P.K., CRAWFORD, J.D. and HENDREN, W.H., (1977a) Management of the neonate with male pseudohermaphroditism. J. Pediat. Surg. 12, 1045-1058.

DONAHOE, P.K., CRAWFORD, J.D. and HENDREN, W.H., (1977b) True Hermaphroditism - a clinical description and a proposed function for the long arm of the Y chromosome. J. Pediat. Surg. (In press).

DONAHOE, P.K. and HENDREN, W.H., (1976a) Evaluation of the newborn with ambiguous genitalia. Pediat. Clin. of N. Amer. 23, 361-370.

DONAHOE, P.K., ITO, Y., MARFATIA, S., HENDREN, W.H., (1976b) The production of Mullerian Inhibiting Substance by the fetal neonatal and adult rat. Biol. of Reprod. 15, 329-333.

DONAHOE, P.K., ITO, Y., MORIKAWA, Y. and HENDREN, W.H., (1977c) Mullerian Inhibiting Substance in human testes after birth. J. Pediat. Surg. 12, 323-329.

DONAHOE, P.K., ITO, Y., PRICE, J.M. and HENDREN, W.H., (1977d) Mullerian Inhibiting Substance activity in bovine fetal, newborn, and prepubertal testes. Biol. of Reprod. 16, 238-243.

EICHWALD, E. and SILMSER, C., (1955) Skin. Transpl. Bull. 2, 148-149.

ELGER, W., RICHTER, J. and KORTE, R., (1976) Failure to detect androgen dependence of the descensus testiculorum in foetal rabbits, mice and monkeys. Bierich, J.R., Rager, K. and Ranke, M.B. (Eds) Maldescensus Testis, Colloquium at Tubingen pp.187-191.

FAWCETT, D., (1966) The Cell, W.B. Saunders, Philadelphia.

HENDREN, W.H. and CRAWFORD, J.D., (1972) The child with ambiguous genitalia. Curr. Prob. in Surg. 1-64.

JIRASEK, J.E., (1971) (ed. M. Michael Cohen) Development of genital system and male pseudohermaphroditism. John Hopkins Univ. Press, Baltimore.

JOSSO, N., (1973) In vitro synthesis of Mullerian Inhibiting hormone by seminiferous tubules isolated from the calf fetal testis. Endoc. 93, 829-834.

JOSSO, N., (1974) Fetal sexual differentiation in Mammals. Ped. Ann. 3, 67-79.

JOST, A., (1946a) Sur la differentiation sexuelle de l'embryon de lapin experiences de paraboise. C. R. Soc. Biol. 140, 463-464.

JOST, A., (1946b) Sur la differentiation sexuelle de l'embryon de lapin remarques au sujet de certaines operations chirurgical. C. R. Soc. Biol. 140, 460-461.

JOST, A., (1947) Sur les derives mulleriens d'embryons de lapin des deus sexes castres a 21 jours. C. R. Soc. Biol. 141, 135-136.

JOST, A., VIGIER, B., PREPIN, J. and PETCHELLET, J.P., (1973) Studies on Sex Differentiation in Mammals. Rec. Prog. in Hormone Res. 29, 1-41.

KOO, G., WACHTEL, S., BREG, W. and MILLER, O., (1976) Mapping the locus of the H-Y antigen. Birth Defects 12, 175-180.

MOORE, K., (1974) Before We Are Born. Basic Embryology and Birth Defects. W.B. Saunders, Philadelphia.

OHNO, S., (1976) Major regulatory genes for mammalian sexual development. Cell 7, 315-321.

PICON, R., (1969) Action du testicule foeatal sur le development in vitro des canaux de Muller le rat. Arch. Anat. Micro. Exp. 58, 1-19.

PRICE, J.M., DONAHOE, P.K., ITO, Y. and HENDREN, W.H., (1977) Programmed cell death in the Mullerian duct induced by Mullerian Inhibiting Substance. Am. J. Anat. 149, 353-376.

SAENGER, P., LEVINE, L., WACHTEL, S., KORTH-SCHUTZ, S., DOBERNE, Y., KOO, G., LAVENGOOD, R.W., GERMAN, J.L. and NEW, M.I., (1976) Presence of H-Y antigen and testis in 46,XX true hermaphroditism, evidence for Y chromosomal function. J. Clin. Endoc. and Metab. 43, 1234-1239.

WACHTEL, S., KOO, G., BREG, W., ELIAS, S., BOYSE, E. and MILLER, O.J., (1975) Expression of H-Y antigen in human males with two Y chromosomes. N.E.J. of Med. 293, 1070-1072.

WACHTEL, S., KOO, G., BREG, W., THALER, H., DILLARD, G., ROSENTHAL, I., DOSIK, H., GERALD, P., SAENGER, P., NEW, M., LIEBER, E. and MILLER, O., (1976) Serologic detection of a Y-linked gene in XX males and XX true hermaphrodites, N. E. J. of Med. 95, 750-754.

PROLIFERATIVE CENTRES IN EMBRYONIC DEVELOPMENT

Michael H.L. Snow

MRC Mammalian Development Unit
Wolfson House (University College London)
4 Stephenson Way, London NW1 2HE

In words it is fairly easy to define a proliferative centre; it is 'a locality in a developing tissue at which cell division takes place at a significantly elevated rate compared to the tissue as a whole'. A consequence of such a centre is that a disproportionate number of cells may be generated at the site which in turn could cause morphogenetic changes, such as tissue spreading or localised thickening. On the other hand, it is in practice sometimes very difficult to positively identify a proliferative centre for a variety of technical reasons which will be discussed later.

Historically the notion of proliferative centres has its origin in the embryo studies of von Baer in the 1820s which were formulated by His (1874) into a hypothesis suggesting that the epibolic tissue movements of gastrulation were brought about, at least in part, by the pushing effects of growth centres. Local increases in tissue volume (cell number) provided a simple mechanism whereby cell layers could spread or change shape. Nevertheless, the studies of mitotic activity in the gastrulae of lower vertebrates (fish: Richards, 1935; Richards and Porter, 1935; Richards and Schumacher, 1935; Pasteels, 1936; Self, 1937. Amphibia: Bragg,

1938; Pasteels, 1943) presented apparently equivocal and somewhat uncritical data which led Holtfreter (1943) to conclude that 'the principle of cell division can be dismissed as a causative or guiding factor in early Amphibian morphogenesis'. Pasteels (1943) went even further to state that, with the exception of the mouse and rat, gastrulation in the vertebrates studied (three fish, two Amphibians, two birds and two mammals) could not be associated with regional differences in mitotic index or the presence of proliferative centres.

The considered opinions of these two authors seemed to stifle almost all further interest in any causal relationship between growth centres and morphogenetic movements although it was not until nearly 20 years later when Kessel (1960) showed that gastrulation in the fish *Fundulus* continued in the presence of colchicine, that direct experimental evidence dissociated the two events. Since then, similar reports of continued gastrulation in the presence of mitotic inhibitors have been made for *Xenopus laevis* (Cooke, 1973a, b). In addition, Diwan (1966) treated chick embryos of the definitive streak and headfold stages with colchicine and found that the younger embryos developed many abnormalities especially in the brain, and also showed a shorter longitudinal axis whereas the headfold stage embryos developed normally. Since considerable morphogenesis had occurred between the primitive streak stage and the detection of brain abnormalities, it would appear that the chick embryo does not depend upon cell division for these developmental processes.

The negation of His' hypothesis associating morphogenetic movements with cell proliferation so adequately demonstrated by Kessel (1960) and Cooke (1973a, b) does not necessarily deny the existence of proliferative centres but indicates that, if and where they exist, their biological significance is in a different context. This article will therefore consider three questions.

a) Do proliferative centres exist during gastrulation?
b) What is their role in morphogenesis?
c) Are they found in later development?

The literature on lower vertebrates will be re-appraised to provide a background against which current knowledge on mammalian development can be assessed.

PROLIFERATIVE CENTRES - FACT OR FICTION?

The identification of a proliferative centre is heavily dependent upon the methods of analysis used which can be complicated by difficulties in orientation of tissues lacking any structural features. For instance, the simple observation of a high mitotic index in a restricted area of a single individual is often misleading as it may only indicate close synchrony within a group of cells fortuitously caught in division. This was the essence of Pasteels' (1943) dismissal of the data on fish presented by Richards and his colleagues (Richards, 1935; Richards and Porter, 1935; Richards and Schumacher, 1935) and that on Amphibia by Bragg (1938). Even similar observations on several specimens at the same stage of development need not indicate a proliferative centre but merely reflect clonal development and control of the timing of cell division based upon some circadian rhythm (Pilgrim *et al.*, 1963; Pilgrim and Maurer, 1965). Adjacent cells may be programmed to divide at a different time of day.

Even where care has been taken to avoid such pitfalls as these, it is sometimes impossible to compare the conclusions of reports in which the authors have used slightly different forms of topographical analysis. Figure 1 illustrates the nature of this problem. It shows the distribution of dividing cells in four samples of a tissue consisting of 2,000 cells. In column A the tissue is mapped into 20 squares, each containing 100 cells (as could result, for example, from scoring with an eyepiece graticule), and in column B the subdivision is into 25 areas of 80 cells each (in this case the sides of the tissue have been divided proportionately). At the foot of each column the areas are assigned their average mitotic index (%) calculated from the sums of the numbers of divisions and cells for each area. Quite different impressions are created by these two slightly different analyses. The map generated in B shows a single proliferative centre with a mitotic index double that of other regions, whereas map A suggests a rather even distribution of dividing cells. This particular artefact means that provided adequately controlled experiments have been carried out, a single report showing a proliferative centre may have a much greater significance than several failing to show it and seeking to deny its existence. This confusion has beset the debate about mitotic centres in avian embryos.

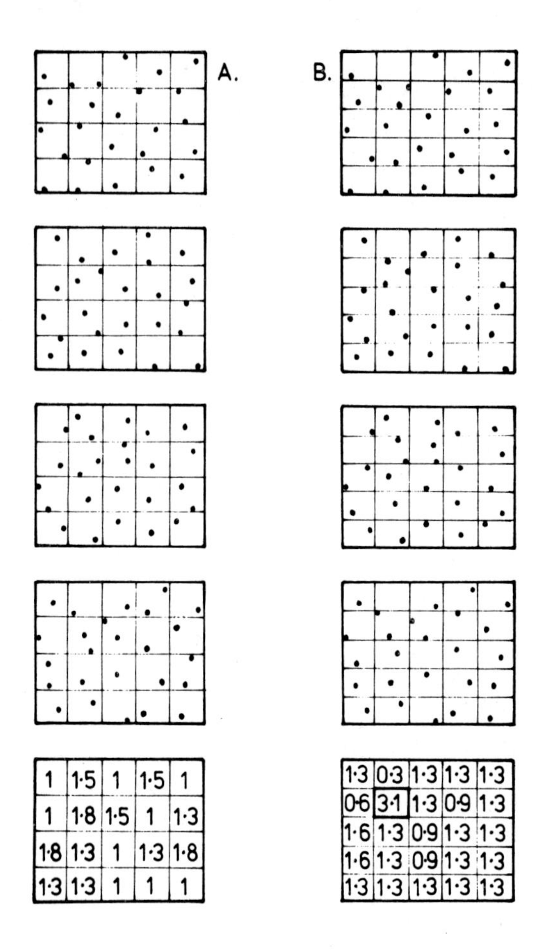

Figure 1: The effect of a small variation in topographical analysis upon the detection of a proliferative centre in a tissue composed of 2,000 cells. Each dot represents a dividing cell (see text for details).

With the foregoing considerations in mind, does the early literature support the notion of proliferative centres?

FISH

The mechanisms of teleost gastrulation only became clearly understood in the 1960s through the work of Ballard (1966a,b,c; 1968), and as all the studies of mitotic activity in these embryos preceded this work, it was carried out against a background of an improper understanding of the processes involved in gastrulation. Nevertheless, the analyses were of such a general nature that the observations and conclusions are not invalidated by that lack of knowledge. No definite conclusion can be drawn, principally because the analyses have not been carried out in sufficient detail. Richards and Porter (1935, working on *Fundulus*), Richards and Schumacher (1935, on *Coregonus*) and Self (1937, on *Gambusia*) all report that in the early primitive streak blastoderm the highest mitotic activity is in the anterior regions but each of these reports is based upon analysis of a single embryo. Only Self (1937) examined the embryo region by region for variation in mitotic activity and although he records patches of high mitotic activity, it is of course impossible to estimate their significance. Pasteels (1943) compared two Trout embryos at both the early and late gastrula stages and scored mitotic index in two small regions of the embryo, one posterior and one anterior. No difference in mitotic activity was found either between regions or between stages.

AMPHIBIA

The embryos of three Anurans have been studied and at the gastrula stage all show a similar, high mitotic index in the ectoderm (Pasteels, 1943, in *Discoglossus*; Bragg, 1938, in *Bufo*; and Maleyvar and Lowery, 1972, in *Xenopus*). Regional variation of mitotic activity was found in ectoderm of *Bufo* and *Xenopus* embryos but not investigated in *Discoglossus*. At the onset of gastrulation in *Bufo*, Bragg (1938) identified a cap of micromeres at the animal pole with a markedly elevated mitotic index, and Maleyvar and Lowery (1972), analysing slightly older *Xenopus* embryos show localisation of a region of high mitotic activity at six different stages of gastrulation. In this latter study, the peak of mitotic activity is observed in a progressively more anterior position at each stage and the authors interpret their data as showing a wave of mitotic activity passing along the embryo. As they do not demonstrate that an entirely different population of cells is involved in the mitoses of different stages, the alternative explanation, i.e. that there is a group of cells undergoing extremely rapid proliferation around which the embryo changes shape, would account for the observations equally well.

BIRDS

In a detailed study on ducks, Chen (1932) reports an examination of primitive streak formation in which the numbers of mitoses were scored for anterior, middle and posterior regions of the embryo. The mitoses in a single histological section per region were counted in nine embryos of comparable stage. The pooled data indicates greater mitotic activity in an anterior region of the ectoderm. In the absence of information about cell size and number, however, no conclusions can be drawn about cell proliferation in the respective regions.

A later independent study of the distribution of mitotic index in chick embryos suggests that Chen's (1932) figures may well reflect proliferative activity. Derrick (1937) plotted regional variation in mitotic index of three chick embryos of two different strains. Her maps show that although in any individual two or three small regions may show similar indices, comparison of embryos indicates that only one area is consistently high, namely in the mid-line in an anterior position, around the end of the primitive streak.

The significance of these data was challenged by Pasteels (1937) who reported that there was no difference between the mitotic indices of the primitive streak and its surrounding ectoderm, in either duck or chick embryos. The controversy was increased by Spratt (1966) whose diagrams located the most mitotic activity in the streak itself and around its posterior margin. Nevertheless his paper includes a footnote observing a very high incidence (greater than any area on his maps) of metaphase and anaphase figures around the anterior end of the definitive primitive streak. Since Emanuelsson (1961) presents data essentially similar to Derrick's (1937) locating a peak of mitotic activity in that part of the mid-line which includes Hensens node it would seem incorrect to conclude that a proliferative centre is absent in avian embryos. The consensus from all the papers that have investigated the distribution of mitoses in the ectoderm suggests that the area around the anterior end of the primitive streak is proliferating at a

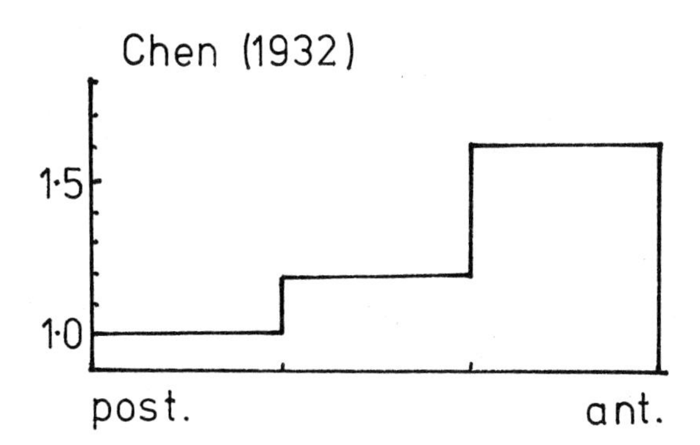

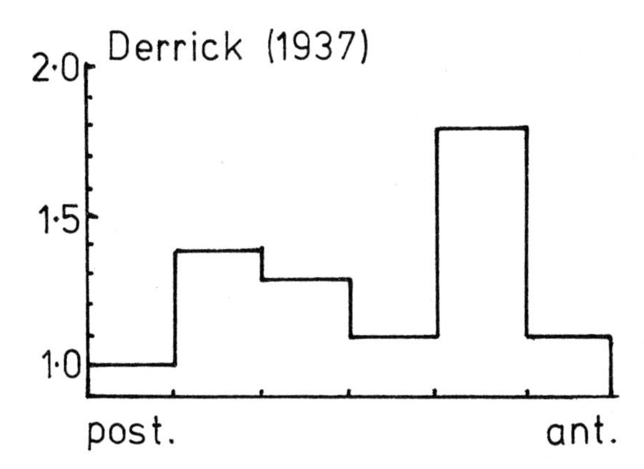

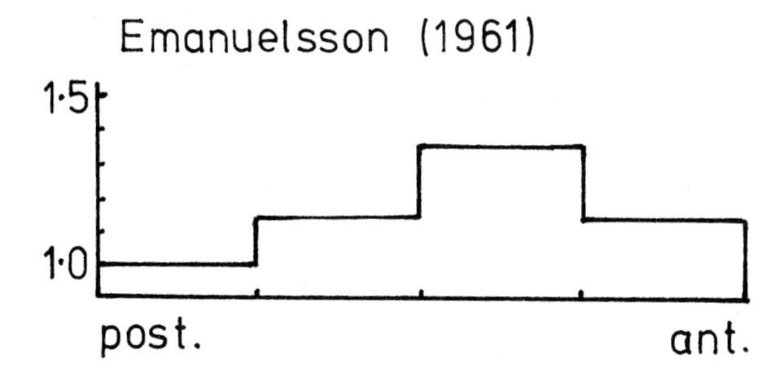

Figure 2: The relative mitotic activity along the posterior-anterior axis in the mid-line of chick primitive streak stage embryos. The region with the lowest score is assigned the value 1.0. Chen (1932) scored numbers of mitoses and Derrick (1937) and Emanuelsson (1961) scored mitotic index.

greater rate than any other region. Figure 2 plots the relative mitotic activity along the mid-line according to the studies of Chen (1932), Derrick (1937) and Emanuelsson (1961).

MAMMALS

Preto (1939) compared the mitotic indices (MIs) of the three germ layers in late primitive streak stage rat embryos (two individuals) and found that the ectodermal index was double that of the mesoderm and some 7 to 8 times that in the endoderm. Pasteels (1943) made similar observations on mice (three embryos) but went further to show that the MI in the primitive streak was only about half that of the ectoderm (epiblast). A decade later, the extensive report by Corliss (1953) seemed to discredit these earlier studies. Corliss calculated mitotic density (number of divisions per unit area of tissue) in the various germ layers, for various regions of the embryo. In order to facilitate statistical analysis he sought to increase the number of metaphases using colchicine as a mitotic inhibitor. The data showed no consistent differences in mitotic density between ectoderm and mesoderm, nor between regions within either tissue. Corliss considered that the colchicine-enhanced mitotic density figures reflected the mitotic rate in the various regions and concluded that differential proliferation of cells was not involved in rat gastrulation. A number of assumptions are made in this interpretation, all of which appear to be invalid. Firstly it is assumed that colchicine is not immediately toxic and serves solely to block cells in metaphase and does not interfere in other processes. Some years earlier, however, both Van Dyke and Richey (1947), and Kerr (1947) demonstrated that a single injection of a 'physiological' concentration (to the mother, presumably) of colchicine is lethal to embryos of rats and mice. Kerr (1947) observed degenerative changes in the 6½ day mouse embryo (early primitive streak stage) only a few hours after injection. More recently, Williams (1968, 1970) and Williams and Carpentieri (1967) have shown that antimitotic agents such as colchicine, appear to inhibit later stage rat embryo cells during their DNA synthesis phase. If this is true also for egg cylinder stages, then it exposes a flaw in Corliss' second assumption, that cells in the different tissues and regions will have accumulated to the same extent. While this may be true for the initial accumulation rate, it becomes invalid in circumstances where cells are prevented from leaving the 'S' phase and progressing to division if the time between injection of colchicine and scoring metaphases is longer than the shortest G_2 period. Then, the maximum number of divisions scored merely reflects the number of cells in G_2, and variation in G_2 duration will significantly affect these figures. The autoradiographic studies of Solter and Skreb (1968) and Solter *et al.* (1971) show that the G_2 period for mouse ectoderm and mesoderm are 0.6h and 0.75h respectively. Since Corliss allowed 5 to 7h between colchicine injection and fixation, it might be anticipated that up to 25% more cells would accumulate in division in mesoderm than in ectoderm. Finally, for mitotic density to reflect similar mitotic rates in different regions, the cell size in the compared tissues must be the same. At least for the mouse, mesoderm cells during early primitive streak stages are smaller than ectoderm cells (Snow, 1976a). Both of these latter factors will lead to an overestimate of division rate in the mesoderm with respect to ectoderm and thus obscure such differences between the tissues as were observed by Preto (1939) and Pasteels (1943). The more recent work on mice confirms these two earlier studies. The autoradiographic data of Solter *et al.*(1968, 1971) indicate cell cycle times in epiblast and mesoderm of 6.5 and 8h respectively and my own data (Snow, 1976a, 1977) on cell number increase fits well with their figures, requiring an *average* cell cycle time of around 6h in the epiblast and considerably longer in mesoderm for the 24h period from 6½ to 7½ days *post coitum*. Detailed mapping of cell divisions in these mouse embryos (Snow, 1977, Figures 3 and 4) reveals a non-uniform distribution of mitotic activity with a proliferative centre located in an anterior region of the mid-line in a position which roughly corresponds to the future anterior extremity of the primitive streak or in other words the primitive knot. This corresponds to Hensens Node in chick embryos. At this site, the level of mitotic activity is about 2½ times greater than any other comparable area in the epiblast at all primitive streak stages in the mouse. This estimate is based upon analysis of 41 embryos from 17 litters and two different strains of mice. A similar study on rabbit embryos by Daniel and Olsen (1966) had previously shown a precisely similar map of mitotic activity with a proliferative centre identifiable in a small area which includes the primitive knot. In this case, the mitotic index is less spectacularly high being a little less than double that of any other areas.

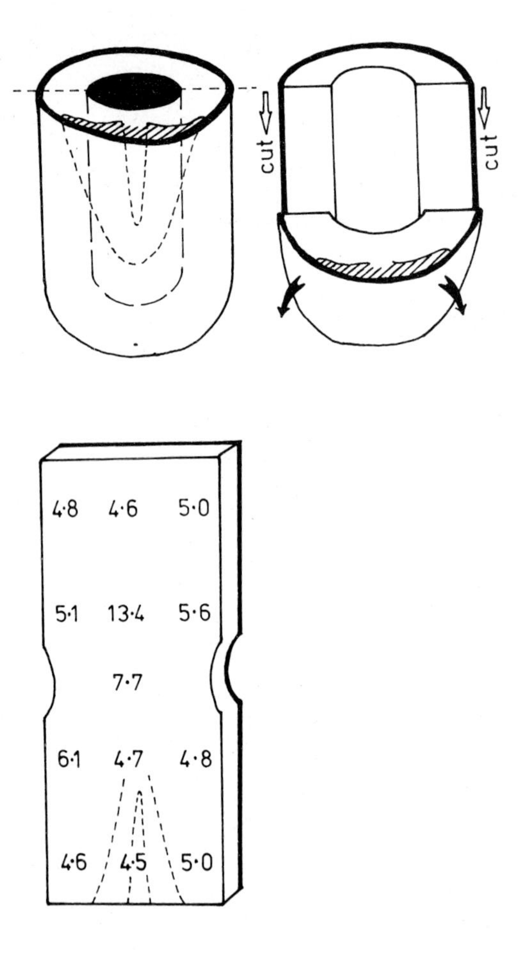

Figure 3: The mapping of metaphase/anaphase index in the epiblast of mouse egg cylinders. The distribution of M/A index is compiled from camera lucida drawings of serial transverse sections and mapped onto a plan of the embryo as shown in the lower part of figure. This plan supposes that the cylindrical embryo has been cut down its lateral margins and opened out flat as in top left and right (see Snow, 1977). ▭ = epiblast, ▨ = mesoderm, ■ = primary endoderm.

CONCLUSIONS

It is quite clear that in all the major vertebrate classes studied, where the analyses have considered regional variation, that the anterior part of gastrulating embryos exhibits a higher cell proliferation rate than the posterior regions. Although the data for fish and amphibia do not permit more specific comment than that, in birds and mammals it would appear that the elevation of mitotic activity in this anterior region is due to the presence of a rapidly dividing proliferative centre located in the mid-line at a site which corresponds to the most anterior point reached by the primitive streak.

WHAT ROLE DO PROLIFERATIVE CENTRES PLAY IN GASTRULATION?

Several studies appear to establish that cell division during gastrulation is not essential for the process, at least in lower vertebrates. Trifonowa (1934) observed epibolic movements similar to those associated with gastrulation in parthenogenetic fish embryos in which cell division had ceased. Furthermore, in some embryos he observed movement in a mass of acellular tissue overlying the egg yolk. More recently Kessel (1960) showed that gastrulation of *Fundulus* embryos was morphologically normal in the presence of colchicine at a concentration which he showed effectively blocked cell division. Cooke (1973 a,b) came to a similar conclusion for *Xenopus* finding that both mitomycin C and colcemid, which inhibit cell division by different mechanisms, allow completion of gastrulation if the inhibition of division commences at the early gastrula stage. At this stage there are apparently sufficient numbers of cells to undergo gastrulation as there was no increase in cell number throughout the experimental period. Cell number can be a limiting factor according to Cooke (1973b), as embryos blocked at the mid-blastula stage complete only the early invagination movements of gastrulation. This observation may have significance when considering the development of mammalian, particularly rodent embryos.

Chick embryos of the headfold stage develop normally in the presence of colchicine but definitive primitive streak stages develop a variety of abnormalities (Diwan, 1966). Many of the observed defects, such as absence or reduction in size of somites, absence of optic vesicle and reduction in length of the embryo are easily attributable to paucity of cells, while others, such as malformation of the brain and neural tube, and of the vitelline veins, although possibly due also to lack of cells, may be the result of a different type of lesion. Nevertheless the presence of colchicine does not appear to inhibit morphogenetic movements. It is noted that the most obvious lesions in the chick involve the neural ectoderm.

In chicks, therefore, the response to mitotic arrest for the most part is qualitatively similar to that in fish and amphibia except that there seems to be a requirement for further cell production during gastrulation.

In the mouse (Kerr, 1947) and rat (Van Dyke and Ritchey, 1947), a single injection of colchicine into the pregnant female during or very shortly before the primitive steak stage is invariably lethal but during organogenesis stages may only result in developmental retardation. Both these papers indicate that degeneration of the embryo commences very soon after colchicine injection, thus not permitting significant progress through gastrulation.

Judged from these data, gastrulation in rodents appears to be dependent upon continuing cell division. But is it required simply to provide cells for the process or does it generate motive force from the proliferative centre identified by Snow (1977)?

Following a single injection into a pregnant female, colchicine is unlikely to be generally available to cells for more than 1h and most of it will have been absorbed into cells or excreted before that. In that time, the drug could be expected only to affect those cells entering metaphase in its presence. If this were the case, then a comparatively small proportion of cells in the embryo would be arrested. According to Snow (1976a, 1977) the overall average cell cycle time in primitive streak stage mouse embryos is around 6.5h so a reduction in cell number of some 15% would result if cell proliferation were uniformly distributed. It is already known that the mouse embryo will complete gastrulation successfully when cell numbers are reduced at least by 50% and possibly by more than 75% (Tarkowski, 1959a,b; Snow, 1973, 1975, 1976b) so paucity of cells would seem an unlikely explanation for the effects of colchicine. However, in the proliferative centre of the mouse epiblast (proliferative zone of Snow, 1977) cell cycle times of less than 3h are proposed for the 50 or so cells involved at the initiation of gastrulation. Colchicine would have a more severe effect upon this small population. Even so, based upon the kinetics of cell production described by Snow (1977), the loss of cells resulting from a one hour blockage to division, i.e. a reduction of the proliferative zone by one third and of the remaining epiblast by one seventh would only lower cell number in the post-gastrulation embryo by 30%. Nevertheless the disproportionate interference with the proliferative zone could upset the geometry of gastrulation in an unacceptable manner. This interpretation would suppose that the proliferative zone is a necessary adjunct to gastrulation which, if not providing the motive force for tissue movement, is required to supply cells to fill the space that, in the absence of tissue stretching, the migration of cells posteriorly and out through a primitive streak would tend to create.

On balance, it would be unwise to draw firm conclusions from the above arguments on their own, especially as Williams (1968, 1970) and Williams and Carpentieri (1967) impute an interference with DNA synthesis in rat embryos to stathmokinetic agents. To what extent their findings reflect direct action on the embryo or unidentified effects upon maternal physiology is not known. Notwithstanding this note of caution, recent experiences with the growth *in vitro* of mouse embryos over the primitive streak period further implicate an important role for accelerated cell proliferation in gastrulation. When mouse blastocysts are cultured in medium facilitating inner cell mass development, the structures produced fall into two classes. The larger class exhibit fairly slow growth throughout and produce a variety of vesicular structures that histologically resemble yolk sac, (composed of mesoderm and endoderm) whereas the smaller group, although also growing slowly in their initial stages, undergo a transition to a rapidly growing structure, which shows many features of normal embryonic development, including formation of primitive streak, neural folds, beating hearts and of some other organ primordia (Hsu, 1971, 1972, 1973; McLaren and Hensleigh, 1975; Wiley and Pedersen, 1977). Our own observations (Hensleigh, Tam and Snow, unpublished) illustrate some interesting parallels between the *in vitro* grown embryos and their *in vivo* counterparts which seem to hinge upon the formation of a proliferative centre. The growth profile of the *in vitro* embryos showing the switch to rapid growth is very similar to that of embryos *in vivo* (see Goedbloed, 1972; Kohler *et al.*,1972; McLaren, 1976). The switch in growth rate *in utero* not only coincides with the appearance of the primitive streak and proliferative zone (Snow, 1977) but also with changes in chromosome replication behaviour (Takagi, 1974; Snow, unpublished, Table 1). *In vitro* this characteristic is confined to the rapidly growing embryos, although the data at present are not extensive.

TABLE 1: CHANGES IN CHROMOSOME REPLICATION BEHAVIOUR IN MOUSE EMBRYOS DURING PRIMITIVE STREAK STAGES

Age (days)	No. embryos	No. showing* late label			No. metaphases scored	No. labelled	No. late* labelled
		♀	♂	?			
6½	22	7	5	2	94	81	34 (42%)
7½	9	4	3	1	75	59	34 (58%)
8½	12	5	7	0	>400	166	136 (82%)

* Includes late labelled sex chromosomes and/or autosomal late label.

Of 12 slow growing embryos, all were negative, while 5 out of 6 of the rapidly growing ones were positive for late replication patterns. When embryos are dissected from the uterus between 6 and 6½ days *post coitum* (prior to the formation of a proliferative zone) and cultured *in vitro* most grow slowly into vesicles whereas if isolated 6h later when a proliferative zone is present their growth is rapid and includes the formation of a primitive streak, neural folds and eventually beating hearts (Table 2). It would seem that culture conditions which permit and support the acceleration of growth in the mouse embryo also facilitate gastrulation. Of course these facets of development may be independently triggered and therefore separable in theory but at present their co-existence seems somewhat more than mere coincidence.

CONCLUSIONS

In fish, amphibia and birds continuing cell division is not required for the morphogenetic movements of gastrulation to occur, and in the lower vertebrates the evidence suggests that the late blastula/early gastrula is fully equipped with sufficient cells to carry out the process normally. In birds, however, there is the report that mitotic inhibition by colchicine, leading to a paucity of cells, results in malformation in the post-gastrulation embryo (Diwan, 1966). These abnormalities are primarily in the brain and associated neural structures and it is interesting to note that these tissues have their origin from the epiblast around Hensens Node, an area already identified as a centre of high proliferative activity by Derrick (1937), Emanuelson (1961) and Spratt (1966), and which would be expected to be highly sensitive to antimitotic agents.

TABLE 2: CHANGES IN THE *IN VITRO* DEVELOPMENTAL POTENTIAL OF MOUSE EMBRYOS OVER THE PRIMITIVE STREAK PERIOD (TAM, unpublished)

Age (days)	Primitive streak	Number cultured	Number showing development of: Endoderm	Ext.emb. tissues	Embryonic ectoderm	Mesoderm	Somites, organs
6¼	absent	28	20	11	0	0	0
6½	absent	79	38	26	0	0	0
	formed	47	27	27	17	24	0
7+	formed	170	86	84	84	82	82

In the rat and mouse, the best studied mammals, the situation is somewhat confused with a number of unanswered questions concerning the action of colchicine on embryos *in vivo*. Nevertheless, the comparative studies of embryos *in vivo* and *in vitro* strongly suggest that the onset of gastrulation is associated with, and seems dependent upon, a sudden acceleration in cell proliferation and that, certainly *in vivo*, this is largely brought about by the establishment of a proliferative centre.

There is no compelling reason to believe that this proliferative centre provides the motive power behind the tissue movements of gastrulation but since the slow growing embryos *in vitro* do continue cell proliferation and may eventually achieve a mass comparable to that reached by their faster, gastrulating fellows, it seems certain that the centre contributes more to gastrulation than a simple increase in cell number. In support of this inference, there is circumstantial evidence from a *t*-locus mutant (Snow and Bennett, 1978).

In the homozygous mutant t^{w18}/t^{w18} embryos which die shortly after primitive streak formation because of abnormalities in their mesoderm, disturbances in mitosis (blockage in metaphase, disorientation of cleavage planes) can be detected in the epiblast in all areas except the proliferative zone (Figure 4). Since the ectoderm of the mutant can develop normally in ectopic sites (Artzt and Bennett, 1972), it is tempting to infer that the proliferative zone is a site of ectoderm production. The kinetics of cell production at this site would permit the generation of almost all the ectoderm

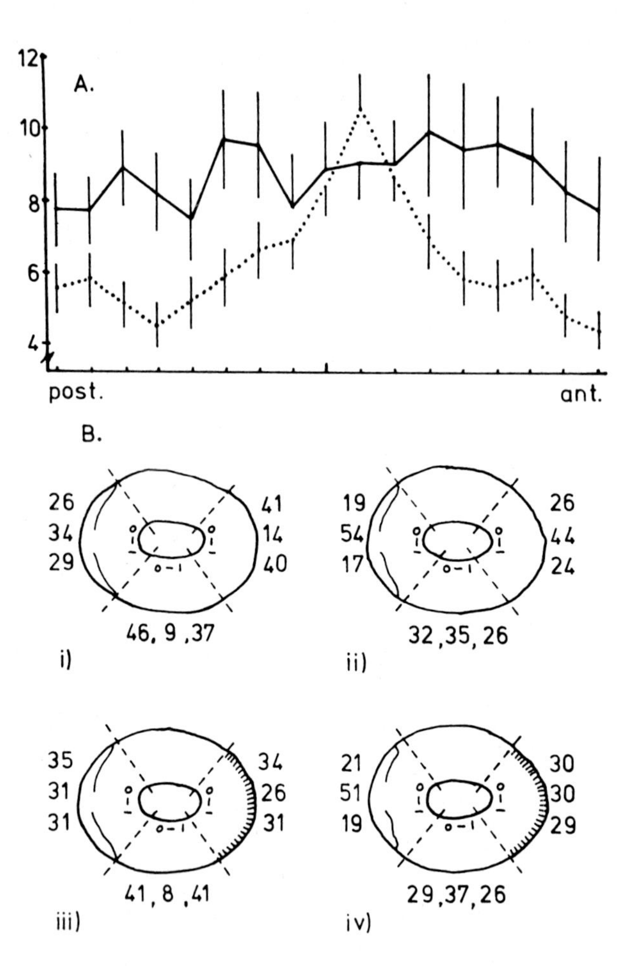

Figure 4: Comparison of some mitotic characteristics in normal, and homozygous t^{w18}/t^{w18} embryos, at the early primitive streak stage. (A) Metaphase/Anaphase Index along the midline of normal (......, n=23) and mutant (———, n=8) embryos. Vertical bars indicate ± one s.e. (B) The direction and frequency (%) of cell cleavage planes in primitive streak, lateral and head-fold sectors of the epiblast. O represents a division perpendicular to the plane of section i.e. a metaphase plate seen thus ⁂, whereas the short bars (–, ı) indicate the equatorial plate of the spindle of divisions in the plane of the section, thus . Between 5% and 10% of divisions cannot be assigned accurately to any group. (i) and (ii) represent the overall distributions for normal and t^{w18}/t^{w18} respectively, and (iii) and (iv) the comparable figures for the level including the proliferative centre (in the head-fold sector, marked).

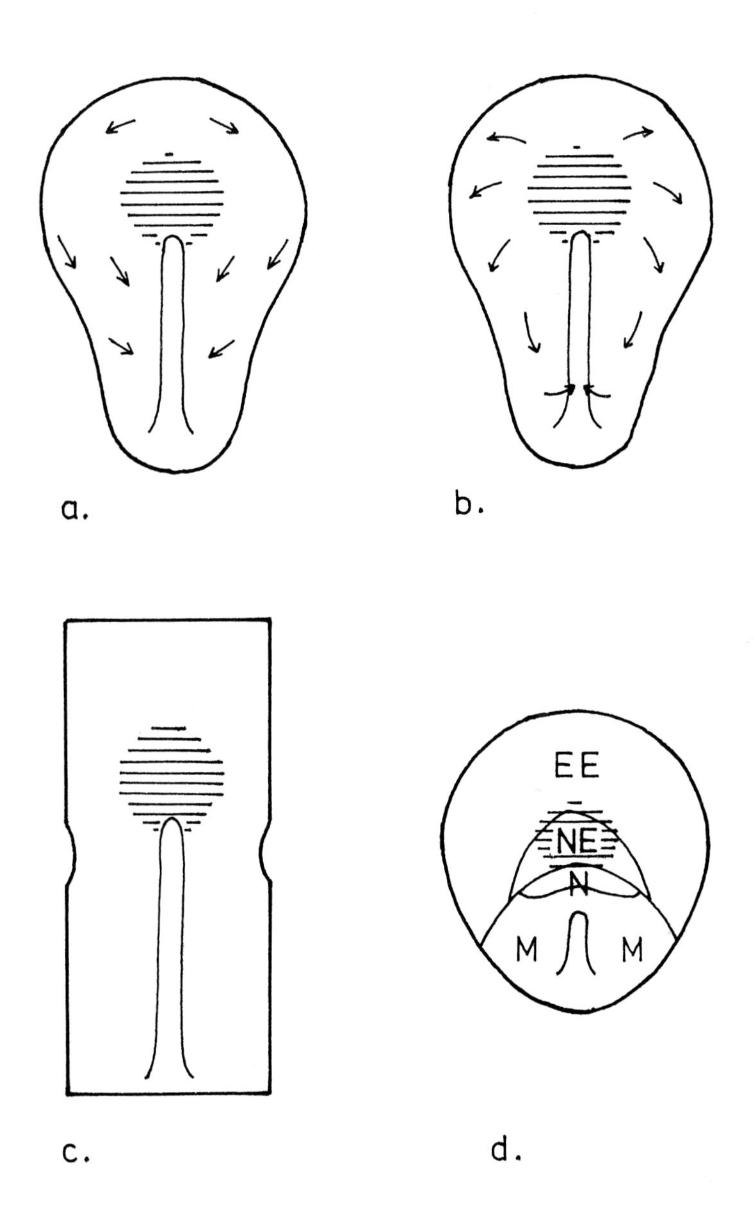

Figure 5: Location of proliferative centres and, where known, cell movements in the epiblast of (a) the rabbit (after Daniel and Olson 1966), (b) birds (after Derrick, 1937; Emanuelsson, 1961; Spratt, 1966; Rosenquist, 1966), and (c) the mouse (Snow, 1977). For comparison, a crude fate map of the early chick embryo indicating the position of the proliferative centre is shown in (d). EE = epidermal ectoderm; NE = neural ectoderm; N = notochord; M = mesoderm.

of the headfold stage embryo (Snow, 1977). Such a scheme would suggest that the mouse embryo is very similar to the chick in which neural ectoderm structures have their origin from a proliferative centre situated in an identical location in the embryo (Pasteels, 1937; Rudnick, 1944; Rosenquist, 1966). Figure 5 shows diagrammatic representations of the epiblast of the chick, mouse and rabbit depicting early primitive-steak stages and the locality of the proliferative centres in each.

ARE PROLIFERATIVE CENTRES FOUND IN LATER DEVELOPMENT?

One of the more impressive features of vertebrate development is the speed with which organ primordia become established. Even in mammals, whose embryos are regarded as developing comparatively slowly, the transition from a featureless tissue to a well sculptured region with an organ primordium in an advanced state of development may be a matter of only a few hours. The initial change in many organ systems is an increase in cell population density producing what are called condensations. The term implies a concentrating of an existing cell population generally assumed to be either by cell movement or by tissue contraction. However, Trinkaus (1965) on the subject of morphogenetic movements admits 'These condensations could result from aggregative cell movements, contraction of the whole cellular mass, or a local increase in cell division. The importance and ubiquity of such condensations is undeniable, but evidence on the mechanisms of their formation is largely lacking'. One of the problems facing investigators of this phenomenon is the identification of the site of a future condensation. Consequently, in a more recent review, Trinkaus (1976) offers no further explanation for the origin of the condensations.

In discussing the growth control of organs, Bellairs (1971) suggests that there is a burst of rapid mitotic activity in many tissues just prior to the onset of differentiation, but cites only Wessells (1965) on early feather formation as evidence for high cellular proliferation, going on principally to illustrate that in the chick lens (Modak *et al.*,1968), in myoblasts (Okasaki and Holtzer, 1966) and in erythroblasts (Marks and Kovach, 1966) the onset of differentiation coincides with an abrupt cessation of mitotic activity. Trinkaus (1965) also cites Wessells (1965) as being exceptional in providing evidence for a mechanism of condensation formation, namely differential mitotic activity.

Since Wessells (1965) constitutes the only quantitative investigation of the involvement of differential mitotic rate in very early organ formation, it is worth brief re-examination of the data. The gross morphological horizons of feather formation used by Wessells (1965) are broadly similar to those involved in the development of other organs in which epithelial/mesenchyme interactions are involved. The early stages are (1) no regional variations in cell density or shape in either epidermis or dermis; (2) epidermal placodes present, formed by localised changes in cell shape; and (3) dermal condensations are present beneath these placodes. By scoring the incidence of mitoses and of cells labelled with H^3-thymidine, it was found that initially proliferation was randomly distributed in both tissues, but that in stage (3), cells within the placodes and dermal condensations were unlabelled and non-dividing. There was, however, a concentration of labelled and dividing cells around the periphery of the dermal condensations implicating a contribution to the primordium by cell division. Since cells labelled with H^3-thymidine in stage (1) showed a very uniform grain count in autoradiographs, Wessells argued that if the dermal condensations arose from differential rates of cell proliferation, the reduction in labelling intensity associated with cell division should be greater in these regions, and went on to demonstrate significantly lower grain counts per cell in the dermal condensations. He concluded, quite reasonably, that the primordium arose by rapid cell proliferation. In the absence of identification of the sites of mitosis, the data do not exclude the possibility, however unlikely, of migration into the condensation of cells generated from divisions occurring elsewhere. The working hypothesis, that the initial formation of the dermal condensation is by rapid multiplication of a small colony (clone?) of cells *in situ*, is supported by the observations of large numbers of mitoses at the site of primordium formation in bone (Hale, 1956), mouse hair and vibrissae (Wessells and Roessner, 1965), adrenal glands (Waring, 1935), pituitary (Kerr, 1944), pancreas (Wessells, 1964), thyroid (Rudnick, 1932), and in formation of deciduoma in pregnant mice (Zhinkin and Samoshkina, 1967).

In more advanced stages in organ development, variations in the distribution of mitotic activity, which appear to have a sculpturing effect, are reported for the heart (DeHaan, 1965), retina (Coulombre, 1965), lung (Sorokin *et al.*,1959), central nervous system (Kallen, 1965; Watterson, 1965) and chick limb (Hornbruch and

Wolpert, 1970). The best studied of these are the development of the brain and the chick limb. In the brain, although cell streaming and gross changes in tissue dimensions contribute considerably to morphogenesis, the process known as neuromery seems to be a consequence of localised increases in mitotic activity. At about the time of neural tube closure, a series of transverse bulges, apparently segmentally arranged, can be seen in the anterior end of the neural tube of many species, including birds, reptiles and mammals (see Källen, 1962; Watterson, 1965). These have been described as artefacts but this now seems unlikely as they are clearly visible in living chick embryos. They are transitory structures, apparent only in the early phases in the thickening of the brain wall, a process in which they are probably an important factor. Each neuromeric bulge develops around a region of high mitotic activity (Källen, 1956a,b,c, 1962; Watterson, 1965).

During the growth of the chick-wing from stage 18 to 30 (Hamburger and Hamilton, 1951), the region of highest mitotic activity has been located in the distal end of the elongating limb (Hornbruch and Wolpert, 1970; Summerbell and Wolpert, 1972; Summerbell and Lewis, 1975). However, the autoradiographic experiments of Janners and Searls (1970, 1971), and Searls and Janners (1971) seem to show essentially uniform indices of proliferating cells throughout the limb during stages 18 to 22 of development, with a tendency for the mesenchyme of the tip of the limb to show a somewhat higher incidence only in older limbs (stages 23 to 25). This apparent conflict is resolved, however, if the variations in cell density, volume and nuclear size are taken into account. Summerbell and Wolpert (1972) show that during stages 18 to 22 cell density is at its greatest in the tip of the limb and at its least in the mesenchyme at the proximal end. This distribution of cell density is reversed in stages 23 to 30. They also show that nuclear diameter is linearly related to cell density with the largest nuclei found in the least dense regions. The relationship between nuclear size, section thickness and range of the beta particles from tritium is crucial in estimating the incidence of labelled nuclei in autoradiographs of a tissue (Modak *et al.*,1973), and can be taken into account in an analysis, such as that of Janners and Searls (1970). According to the data of Summerbell and Wolpert (1972), the regional variations in nuclear size expected in stage 18 to 22 limb buds will have led Janners and Searls (1970) to underestimate the incidence of labelled cells in the distal tip of the limb (their subridge area) to a greater extent than elsewhere, thus bringing their observations more into accord with those of Hornbruch and Wolpert (1970). In addition, Summerbell and Lewis (1975) report an autoradiographic study carried out using sections 1 to 1.5 μm thick, thus avoiding the criticism described above, and report that mesenchyme labelling indices of around 98% are found in the distal tip up to stage 28. Other regions of intense mitotic activity may occur transiently in the limb, according to Hornbruch and Wolpert (1970) who report such centres in the elbow region of stage 21 and wrist region of stage 23 limbs.

CONCLUSIONS

There is sufficient evidence, albeit mostly circumstantial, to suggest that the early formation of many organ primordia involves the creation of a small proliferative centre in an hitherto uniformly dividing, homogeneous tissue. The existence of these centres in most cases seems short-lived, giving way fairly rapidly to aggregative cell or tissue movements. It is tempting to consider the cells of these proliferative centres in relationship to stem cells and clonal development but no evidence establishing such a link is available. Indeed, current knowledge of cell populations in adult organ systems indicates that stem cells of renewing tissues generally replicate fairly slowly (Cheng and Leblond, 1974; Potten *et al.*,1974; Cairnie *et al.*,1976).

In later organogenesis, regional variations in mitotic activity are widely reported. In these instances, however, it seems more likely to have been brought about by a lowering of proliferative activity in regions embarking upon differentiation rather than localised acceleration of cell division.

REFERENCES

ARTZT, K. & BENNETT, D. (1972) A genetically caused embryonal ectodermal tumor in the mouse. J. Natl. Canc. Inst. 48, 141-158.

BALLARD, W.W. (1966a) The role of the cellular envelope in the morphogenetic movements of teleost embryos. J. exp. Zool. 161, 193-200.

BALLARD, W.W. (1966b) Origin of the hypoblast in *Salmo*. I. Does the blastodisc edge turn inward? J. exp. Zool. 161, 201-210.

BALLARD, W.W. (1966c) Origin of the hypoblast in *Salmo* II. Outward movement of deep central cells. J. exp. Zool. 161, 211-220.

BALLARD, W.W. (1968) History of the hypoblast in *Salmo*. J. exp. Zool. 168, 257-272.

BELLAIRS, R. (1971) Developmental processes in higher vertebrates. Logos Press Ltd., London.

BRAGG, A.N. (1938) The organisation of the early embryos of *Bufo cognatus* as revealed especially by the mitotic index. Zeit. Zellforschung mikr. Anat. 28, 154-178.

CAIRNIE, A.B., LALA, P.K. & OSMOND, D.G. (1976) Stem cells of renewing cell populations. Acad. Press, New York.

CHEN, B.K. (1932) The early development of the duck egg, with special reference to the origin of the primitive streak. J. Morph. 53, 133-187.

CHENG, H. & LEBLOND, C.P. (1974) Origin, differentiation and renewal of the four main epithelial cell types in the mouse small intestine V. Unitarian theory of the origin of the four epithelial cell types. Am. J. Anat. 141, 537-562.

COULOMBRE, A.J. (1965) The eye. In: R.L. DeHaan and H. Ursprung (Eds.) Organogenesis. Holt, Rinehart & Winston, New York, London, pp. 219-252.

COOKE, J. (1973a) Morphogenesis and regulation in spite of continued mitotic inhibition in *Xenopus* embryos. Nature 242, 55-57.

COOKE, J. (1973b) Properties of the primary organisation field in the embryo of *Xenopus laevis*. IV. Pattern formation and regulation following early inhibition of mitosis. J. Embryol. exp. Morph. 30, 49-62.

CORLISS, C.E. (1953) A study of mitotic activity in the early rat embryo. J. exp. Zool. 122, 193-227.

DANIEL, J.C. & OLSON, J.D. (1966) Cell movement, proliferation and death in the formation of the embryonic axis of the rabbit. Anat. Rec. 156, 123-128.

DEHAAN, R.L. (1965) Morphogenesis of the vertebrate heart. In: R.L. DeHaan and H. Ursprung (Eds.) Organogenesis. Holt, Rinehart & Winston, New York, London, pp. 377-420.

DERRICK, G.E. (1937) An analysis of the early development of the chick by means of the mitotic index. J. Morph. 61, 257-284.

DIWAN, B.A. (1966) A study of the effects of colchicine on the process of morphogenesis and induction in chick embryos. J. Embryol. exp. Morph. 16, 245-257.

EMANUELSSON, H. (1961) Mitotic activity in chick embryos at the primitive streak stage. Acta. physiol. scand. 52, 211-233.

GOEDBLOED, J.F. (1972) The embryonic and postnatal growth of rat and mouse. Acta. Anat. 82, 305-336.

HALE, L.J. (1956) Mitotic activity during the early differentiation of the scleral bones of the chick. Quart. J. Microscop. Sci. 97, 333-353.

HAMBURGER, V. & HAMILTON, H. (1951) A series of normal stages in the development of the chick embryo. J. Morphol. 88, 49-92.

HIS, W. (1874) Unsere Körperform und des physiologische Problem ihrer Entstehung. Leipzig (Cited by Holtfreter, 1943).

HOLTFRETER, J. (1943) A study of the mechanics of gastrulation. J. exp. Zool. 94, 261-318.

HORNBRUCH, A. & WOLPERT, L. (1970) Cell division in the early growth and morphogenesis of the chick limb. Nature 226, 764-766.

HSU, Y.-C. (1971) Post-blastocyst differentiation *in vitro*. Nature 231, 100-102.

HSU, Y.-C (1972) Differentiation *in vitro* of mouse embryos beyond the implantation stage. Nature 239, 200-202.

HSU, Y.-C (1973) Differentiation *in vitro* of mouse embryos to the stage of early somite. Devel. Biol. 33, 403-411.

JANNERS, M.Y. & SEARLS, R.L. (1970) Changes in the rate of cellular proliferation during the differentiation of cartilage and muscle in the mesenchyme of the embryonic chick wing. Devel. Biol. 23, 136-165.

JANNERS, M.Y. & SEARLS, R.L. (1971) Effect of removal of the apical ectodermal ridge on the rate of cell division in the subridge mesenchyme of the embryonic chick wing. Devel. Biol. 24, 465-476.

KÄLLÉN, B. (1956a) Contribution to the knowledge of the regulation of the proliferation processes in the vertebrate brain during ontogenesis. Acta. Anat. 27, 351-360.

KÄLLÉN, B. (1956b) Experiments on neuromery in *Ambystoma punctatum* embryos. J. Embryol. exp. Morph. 4, 66-72.

KÄLLÉN, B. (1956c) Studies on the mitotic activity in chick and rabbit brains during ontogenesis. Kgl. Fysiograf. Sallskap. Lund. Handl. 26, 171-184.

KÄLLÉN, B. (1962) Mitotic patterning in the central nervous system of chick embryos; studied by a colchicine method. Z. Anat. Entwick. 123, 309-319.

KERR, T. (1944) The development of the pituitary of the laboratory mouse. Quart. J. Microscop. Sci. 86, 5-29.

KERR, T. (1947) On the effect of colchicine treatment on mouse embryos. Proc. Zool. Soc. Lond. 116, 551-564.

KESSEL, R.G. (1960) The role of cell division in gastrulation of *Fundulus heteroclitus*. Exp. Cell Res. 20, 277-282.

KOHLER, E., MERKER, H.-J., EHMKE, W. & WOJNOROWICZ, F. (1972) Growth kinetics of mammalian embryos during the stage of differentiation. Naunyn-Schmiedeberg's Arch. Pharmacol. 272, 169-181.

McLAREN, A. (1976) Growth from fertilization to birth in the mouse. In: Embryogenesis in Mammals. Ciba Foundation Symposium 40. Elsevier, Amsterdam.

McLAREN, A. & HENSLEIGH, H.C. (1975) Culture of mammalian embryos over the implantation period. In: M. Balls and A.E. Wild (Eds.), The Early Development of Mammals. Cambridge University Press, Cambridge, London, New York, pp. 45-60.

MALEYVAR, R.P. & LOWERY, R. (1972) The patterns of mitosis and DNA synthesis in the presumptive neurectoderm of *Xenopus laevis*. In: M. Balls and F.S. Billett (Eds.), The Cell Cycle in Development and Differentiation. Cambridge University Press, Cambridge, London, New York, pp. 249-255.

MARKS, P. & KOVACH, K. (1966) Development of mammalian erythroid cells. In: A. Moscona and A. Monroy (Eds.), Current Topics in Developmental Biology 1, Academic Press, New York and London, pp. 213-252.

MODAK, S.P., MORRIS, G. & YAMADA, T. (1968) DNA synthesis and mitotic activity during the early development of chick lens. Devel. Biol. 17, 544-561.

MODAK, S.P., LEVER, W.E., THERWATH, A.M. & UPPULURI, V.R.R. (1973) Estimation of the proportion of cell nuclei in tissue sections falling within tritium-autoradiographic range. Exp. Cell Res. 76, 73-78.

OKASAKI, K. & HOLTZER, H. (1966) Myogenesis, fusion, myosin synthesis and the mitotic cycle. Proc. Natl. Acad. Sci. 56, 1484-1490.

PASTEELS, J. (1936) Etudes sur la gastrulation des vertébrés méroblastiques. I. Téléostéens. Arch. Biol. 47, 205-312.

PASTEELS, J. (1937) Etudes sur la gastrulation des vertébrés méroblastiques. III. Oiseaux. IV. Conclusions générales. Arch. Biol. 48, 381-488.

PASTEELS, J. (1943) Proliférations et croissance danse la gastrulation et la formation de la queue des vertebrés. Arch. Biol. 54, 1-51.

PILGRIM, C., ERB, W. & MAURER, W. (1963) Diurnal fluctuations in the number of DNA synthesizing nuclei in various mouse tissues. Nature 199, 863.

PILGRIM, C. & MAURER, W. (1965) Autoradiographiche untersuchung über die Konstanz de DNS-verdopplungsdauer bei zellarten von maus und ratte durch doppelmarkierung mit H^3- and C^{14}-thymidin. Exp. Cell Res. 37, 183-199.

POTTEN, C.S., KOVACS, L. & HAMILTON, E. (1974) Continuous labelling studies on mouse skin and intestine. Cell Tiss. Kinet. 7, 271-283.

PRETO, V. (1939) Analisi della distribuzione dell'attivata mitotica in giovani embrioni di ratto. Arch. Ital. Anat. Embriol. 41, 165-206.

RICHARDS, A. (1935) Analysis of early development of fish embryos by means of the mitotic index. Am. J. Anat. 56, 355-363.

RICHARDS, A. & PORTER, R.P. (1935) Analysis of early development of fish. II. Mitotic index in pre-neural tube stages of *Fundulus heteroclitus*. Am. J. Anat. 56, 365-393.

RICHARDS, A. & SCHUMACHER, B.L. (1935) Analysis of early development of fish. III. Mitotic index of young embryos of *Coregonus clupeiformis*. Am. J. Anat. 56, 395-408.

ROSENQUIST, G.C. (1966) A radioautographic study of labelled grafts in the chick blastoderm: development from primitive streak stages to stage 12. Contr. Embryol. Carneg. Inst. 38, 71-110.

RUDNICK, D. (1932) Thyroid-forming potencies of the early chick blastoderm. J. exp. Zool. 62, 287-318.

RUDNICK, D. (1944) Early history and mechanics of the chick blastoderm. Quart. Rev. Biol. 19, 187-212.

SEARLS, R.L. & JANNERS, M.Y. (1971) The initiation of limb bud outgrowth in the embryonic chick. Devel. Biol. 24, 198-213.

SELF, J.T. (1937) Analysis of the development of fish embryos by means of the mitotic index. IV. The process of differentiation in the early embryos of *Gambusia affinis affinis*. Zert. fur Zellforsch. Mikr. Anat. 26, 673-695.

SOLTER, D. & SKREB, N. (1968) La durée des phases du cycle mitotique dans differentes régions du cylindre-oeuf de la souris. C.R. Hebd. Séances Acad. Sci. Ser. D. Sci Nat. 267, 659-661.

SOLTER, D., SKREB, N. & DAMJANOV, I. (1971) Cell cycle analysis in the mouse egg cylinder. Exp. Cell Res. 64, 331-334.

SNOW, M.H.L. (1973) Tetraploid mouse embryos produced by cytochalasin B during cleavage. Nature 244, 513-515.

SNOW, M.H.L. (1975) The functional competence of the inner cell mass. In: D. Neubert and H.J. Merker (Eds.), New approaches to the evaluation of abnormal embryonic development. George Thieme. Stuttgart, pp. 394-407.

SNOW, M.H.L. (1976a) Embryo growth during the immediate post-implantation period. In: M. O'Connor (Ed.), Embryogenesis in mammals. Ciba Foundation Foundation Symposium 40. Elsevier Excerpta Medica, North-Holland, Amsterdam, pp. 53-70.

SNOW, M.H.L. (1976b) Embryonic development of tetraploid mice during the second half of gestation. J. Embryol. exp. Morph. 34, 707-721.

SNOW, M.H.L. (1977) Gastrulation in the mouse: growth and regionalisation of the epiblast. J. Embryol. exp. Morph. 42, 293-303.

SNOW, M.H.L. & BENNETT, D. (1978) Gastrulation in the mouse: establishment of cell populations in the epiblast of t^{w18}/t^{w18} embryos. J. Embryol. exp. Morph. (In press).

SOROKIN, S.P., PADYKULA, H.A. & HERMAN, E. (1959) Comparative histo-chemical patterns in developing mammalian lungs. Devel. Biol. 1, 125-151.

SOROKIN, S.P. (1961) A study of development in organ cultures of mammalian lungs. Devel. Biol. 3, 60-83.

SPRATT, N.T. (1966) Some problems and principles of development. Am. Zool. 6, 9-19.

SUMMERBELL, D. & WOLPERT, L. (1972) Cell density and cell division in the early morphogenesis of the chick wing. Nature New Biology 239, 24-26.

SUMMERBELL, D. & LEWIS, J.H. (1975) Time, place and positional value in the chick limb-bud. J. Embryol. exp. Morph. 33, 621-643.

TAKAGI, N. (1974) Differentiation of X chromosomes in the early mouse embryos. Exp. Cell Res. 86, 127-135.

TARKOWSKI, A.K. (1959a) Experiments on the development of isolated blastomeres of mouse eggs. Nature 184, 1286-1287.

TARKOWSKI, A.K. (1959b) Experimental studies on regulation in the development of isolated blastomeres of mouse eggs. Acta. Theriologica. 3, 190-267.

TRIFONOWA, A. (1934) Parthenogenese der Fische. Acta. Zool. 15, 183-213.

TRINKAUS, J.P. (1965) Mechanisms of morphogenetic movements. In: R.L. DeHaan and H. Ursprung (Eds.), Organogenesis. Holt, Rinehart & Winston, New York. London, pp. 55-104.

TRINKAUS, J.P. (1976) On the mechanisms of metazoan cell movements. In: G. Poste and G.L. Nicolson (Eds.), Cell Surface Reviews 1, 225-329.

VAN DYKE, J.H. & RICHEY, R.G. (1947) Colchicine influence during embryonic development of the rat. Anat. Rec. 97, 375 (abstr.).

WARING, H. (1935) The development of the adrenal gland of the mouse. Quart. J. Micr. Sci. 78, 329-366.

WATTERSON, R.L. (1965) Structure and mitotic behaviour of the early neural tube. In: R.L. DeHaan and H. Ursprung (Eds.), Organogenesis. Holt, Rinehart & Winston, New York. London, pp. 129-159.

WESSELLS, N.K. (1964) DNA synthesis, mitosis and differentiation in pancreatic acinar cells *in vitro*. J. Cell Biol. 20, 415-433.

WESSELLS, N.K. (1965) Morphology and proliferation during early feather development. Devel. Biol. 12, 131-153.

WESSELLS, N.K. & ROESSNER, K.D. (1965) Nonproliferation in dermal condensation of mouse vibrissae and pelage hairs. Devel. Biol. 12, 419-433.

WILEY, L.M. & PEDERSEN, R.A. (1977) Morphology of mouse egg cylinder *in vitro:* a light and electron microscope study. J. exp. Zool. 200, 389-402.

WILLIAMS, J.P.G. (1968) Inhibition of embryonic deoxyribonucleic acid synthesis by colcemid. Eur. J. Pharmacol. 3, 337-340.

WILLIAMS, J.P.G. (1970) Selective inhibition of embryonic deoxyribonucleic acid synthesis by vinblastine. Cell Tiss. Kinet. 3, 155-159.

WILLIAMS, J.P.G. & CARPENTIERI, U. (1967) The different responses of embryonic and adult rats to demicolcin. Life Sci. 6, 2613-2620.

ZHINKIN, L.N. & SAMOSHKINA, N.A. (1967) DNA synthesis and cell proliferation during the formation of deciduomata in mice. J. Embryol. exp. Morph. 17, 593-605.

AN APPROACH TO CRANIAL NEURAL CREST CELL MIGRATION AND DIFFERENTIATION IN MAMMALIAN EMBRYOS

Gillian M. Morriss* and Peter V. Thorogood**

** Department of Human Anatomy*
*and ** Department of Zoology*
South Parks Road,
Oxford, U.K.

Neural crest (NC) cells originate in the most lateral part of the neural plate, and are at first morphologically indistinguishable from neighbouring neuroepithelial cells. Their separation and migration from this position involves conversion from an epithelial to a mesenchymal organization analogous to the conversion of epiblast to mesoderm (primary mesenchyme) in the primitive streak. NC-derived mesenchyme is sometimes referred to as *mesectoderm* in order to emphasise its distinction from primary mesenchyme.

Studies on the migration and subsequent differentiation of NC cells in avian and amphibian embryos have demonstrated that the migration pathways are well-defined, and specifically related to the nature of the differentiated end product. Although there is general agreement that mammalian NC cells behave in a manner essentially similar to those of other vertebrates, little experimental work has been carried out to verify this assumption. As a result, discussion of the aetiology of human craniofacial anomalies which have been attributed to abnormal neural crest cell migration, has relied largely on experimental evidence from studies on non-mammalian vertebrates.

While the basic principles of NC cell migration and differentiation may be common to all vertebrates, differences in the pattern of the *adult* structure between vertebrate classes suggest that significant developmental differences exist also. This is particularly important in relation to the craniofacial skeleton, where the elements (though not their shape) show consistent similarities within the class Mammalia, and clear distinctions between mammals and other vertebrates. The purpose of this article is to consider some observations on normal and abnormal migration and

Figure 1 (upper, opposite): Scanning electron micrograph of an 8-somite rat embryo (early day 10), dorsal view. The line through the otic region of the hindbrain shows the position of the section illustrated in Figure 2. Neural tube closure is extending anteriorly from somite 4, and posteriorly from somite 8. The hindbrain extends from just anterior to the transverse groove (preotic sulcus) to the posterior edge of somite 2. Torn edges of the amnion and yolk sac form the edges of the specimen. Note that the hyoid arch is adjacent to the otic region of the hindbrain. M-mandibular arch; H-hyoid arch; S1-first somite. (Bar = 200μ). Figure 2 (lower, opposite): Slightly oblique transverse section through otic (left) and preotic (right) region of hindbrain neural folds of an 8-somite rat embryo. Neural crest cells have just begun to break the smooth line formed by the basal lamina on the left, and have begun to migrate between surface ectoderm and dorsal aorta on the right. M-mesoderm of mandibular arch; N-notochord; O-otic pit; P-pharyngeal endoderm. 1μm section glutaraldehyde/ osmium tetroxide fixed, embedded in Spurr, stained toluidine blue. (Bar = 50μ).

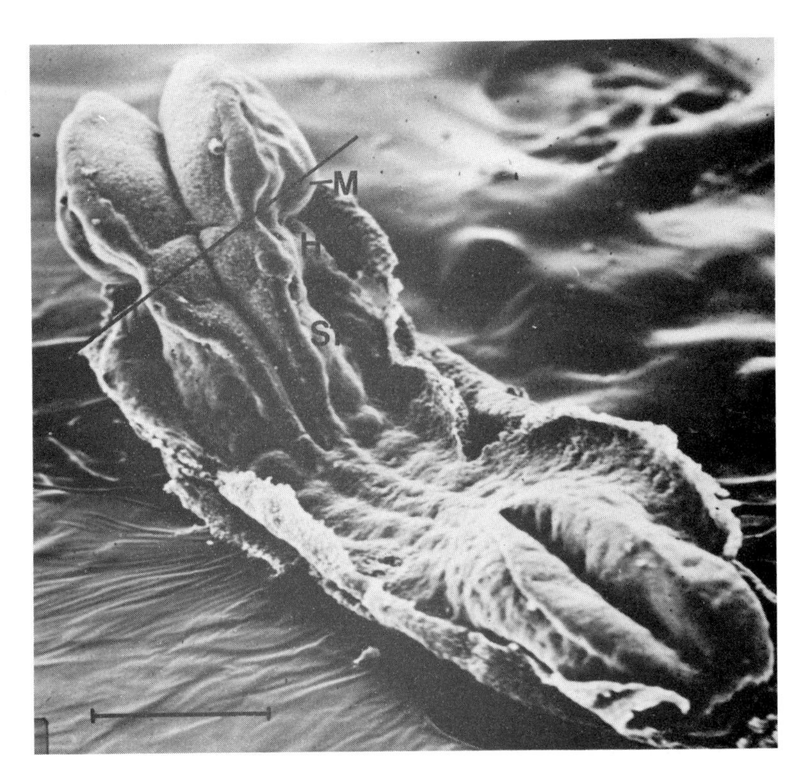
M
H
S

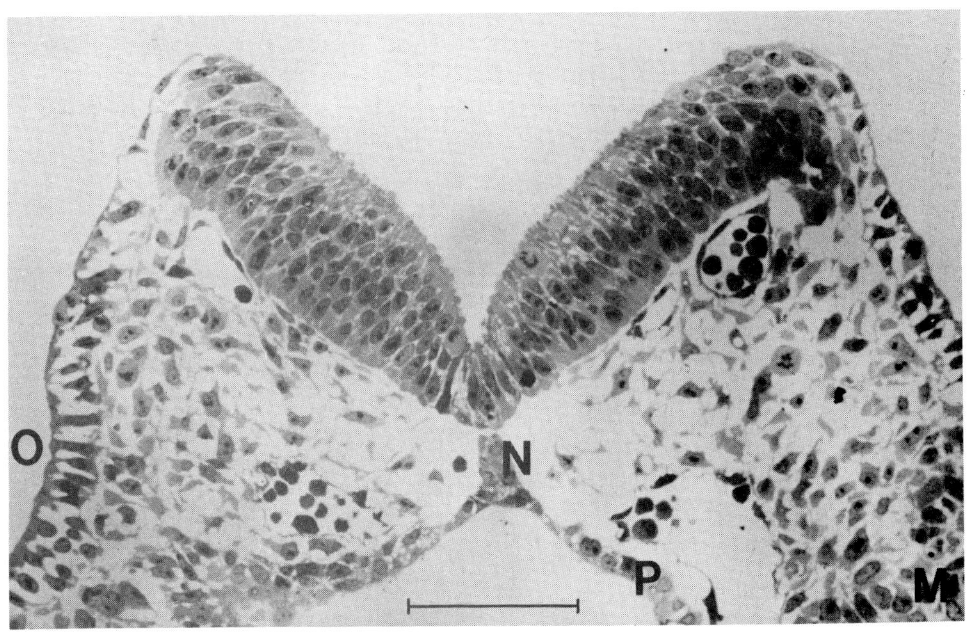
O
N
P
M

differentiation of cranial NC cells in mammalian embryos in relation to the more detailed information which is now available from studies on both cranial and trunk regions in non-mammalian vertebrates. In considering mechanisms of migration, we will also refer to studies of non-neural crest cells *in vitro*.

We have used three approaches to the study of cranial NC cells in a mammalian embryo (rat): first, morphological observations of the early stages of migration in normal embryos *in vivo;* secondly, morphological observation of normal and abnormal crest cell migration in whole embryos *in vitro;* thirdly, comparison of the normal and abnormal morphogenesis of pharyngeal arch cartilages and membrane bones of the face in embryos developing *in vivo*. These three approaches will be used to assess how far it is possible to correlate early disturbances in the pattern of cranial NC cell migration in mammalian embryos with subsequent abnormalities of the craniofacial skeleton. Finally, we will consider possibilities for more direct observations on the role of neural crest in mammalian embryos.

CRANIAL NEURAL CREST CELL MIGRATION

THE ONSET OF MIGRATION

Studies on avian and amphibian embryos have demonstrated that the onset of NC cell migration occurs at, or soon after, the time of closure of the neural tube (Hörstadius, 1950; Johnston, 1966). In avian embryos these events occur in an anteroposterior sequence (see cover illustration of Wessells, 1977). In mammalian embryos, neural tube closure begins in the upper cervical region and progresses anteriorly as well as posteriorly (Hamilton and Mossman, 1972, and Figure 1). The later closure of the cranial neural tube is probably a consequence of the differentiation of an extremely broad cranial neural plate (Morriss and Solursh, 1978a). Migration of the preotic cranial NC begins considerably earlier than closure of this region of the neural tube (Figure 2), suggesting that the

Figure 3 (upper, opposite): Scanning electron micrograph of a section through the upper cervical region of a 9-somite rat embryo. Neural crest cells are beginning to migrate out from the dorsal part of the neural tube. The large tube underlying the neural tube is the foregut, with notochord dorsal to it and splanchnic mesoderm ventrally. (Bar = 40μ). Figure 4 (lower, opposite): Higher magnification of neural crest cells shown in Figure 3, showing their "flask" shape. The leading edge of the cells contain the nucleus, and an elongated filament stretches out from this to, or close to, the apical surface of the neural epithelium. The cells have finer filamentous processes (microspikes) on their surfaces. (Bar 10μ).

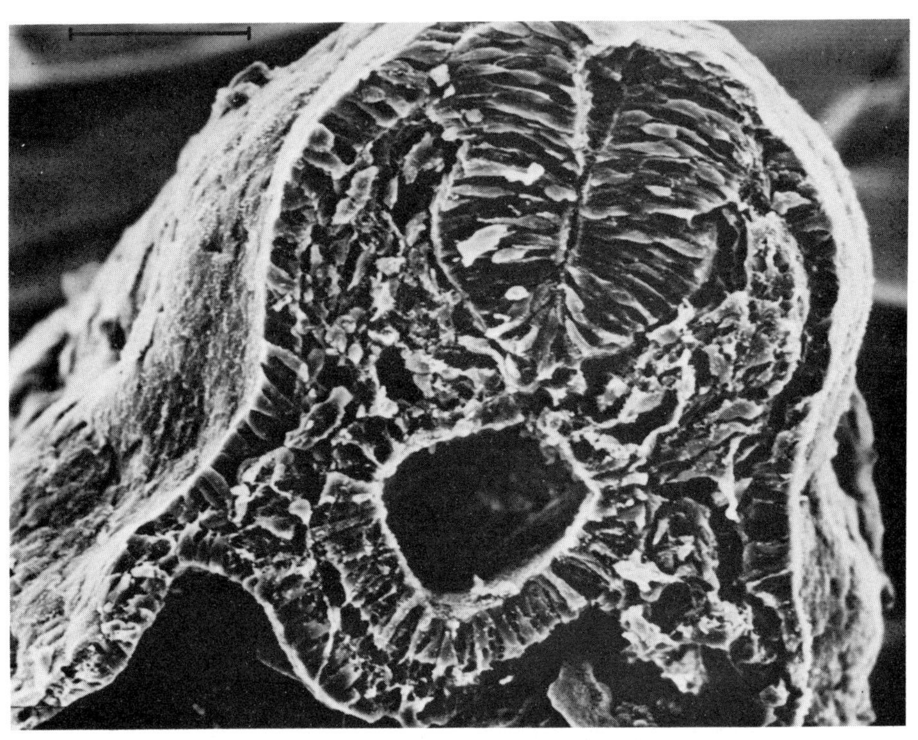

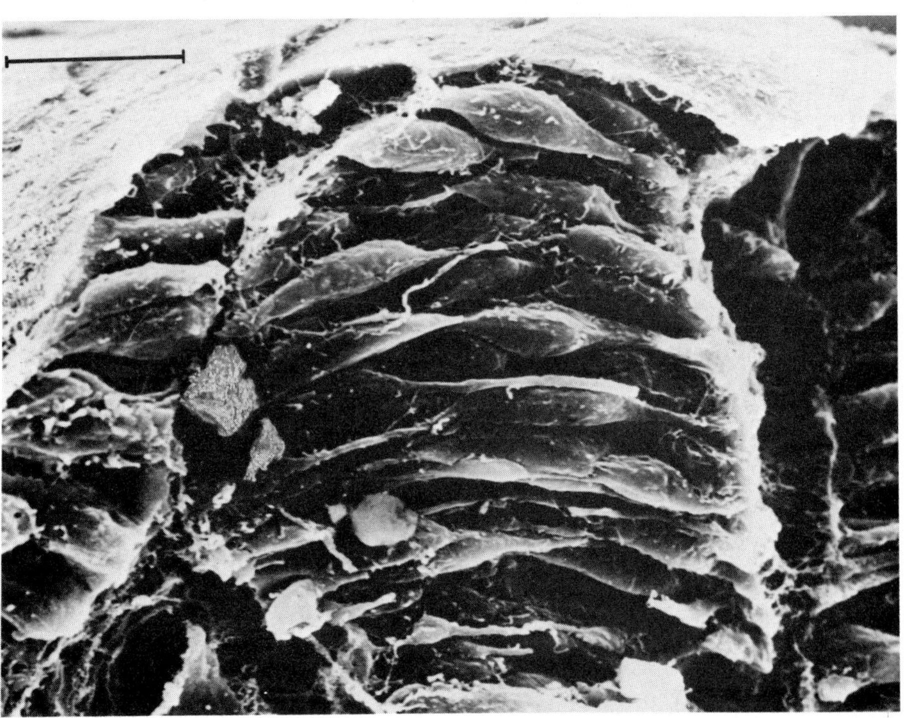

onset of migration is a function of the age of the neural epithelium rather than a component of the process of neural tube closure. Migration of the mammalian trunk NC conforms to the avian pattern, beginning at the time of fusion of the neural folds of the same region (Figures 3 and 4).

The onset of migration involves breakdown of the most lateral part of the neuroectodermal basal lamina; the crest cells become flask-shaped, resembling primitive streak epiblast cells. The necks of the cells become greatly elongated before contact with the apical neuroepithelial surface is lost (Figure 4). In chick embryos, NC cells have been observed to retain their flask-shape during migration, and to contact each other and adjacent structures by fine filopodia, in contrast to the much broader interconnections of adjacent mesoderm (Steffek *et al.*,1975). Scanning electron microscopy of ferret embryos suggests that migrating cranial NC is similarly distinguishable from mesoderm in mammals (Johnston and Steffek, personal communication).

Marchase *et al.*(1976) have suggested that migration may be initiated by a loss of the cells' adhesion to their original environment, but there is no experimental evidence for this idea. The role of cell adhesiveness in normal and abnormal cell migration is discussed further in a later section.

POSSIBLE MECHANISMS OF ORIENTED MIGRATION

As in other vertebrates, mammalian cranial NC cells follow two quite distinct pathways. In the preotic region of the hindbrain, the first cells move between the dorsal aorta and ectoderm (Figure 2), remaining subectodermal. By analogy with studies on chick embryos (Johnston, 1966; Noden, 1975), these cells probably represent the population which moves down into the mandibular arch. Later, a second population of crest cells migrates over the basal lamina of the neural tube, subsequently differentiating to form neurones of the cranial sensory ganglia (visible as denser regions in Figure 8).

Studies on the mechanism of directionality of migration have been carried out both *in vitro* and *in vivo*. NC cells migrate readily from an explant cultured on a glass or plastic surface (Epperlein, 1974; Greenberg and Pratt, 1977). In these conditions, the direction of migration is random, even in the presence of tissue (pharyngeal endoderm) with which the cells will interact prior to their differentiation (Epperlein, 1974; Epperlein and Lehmann, 1975).

This observation suggests that the pharyngeal endoderm does not have a chemotactic influence on the migration pathway.

Further evidence that chemotaxis is not involved in cranial NC cell migration comes from work in amphibian embryos (Hörstadius, 1950, p.60, Figure 31): the left hindbrain neural crest (containing cells which would normally migrate into the pharyngeal arches to form arch cartilages) was removed, stained with Nile Blue sulphate, and implanted in a horizontal position on the right ventral aspect of the head. The right neural crest was left *in situ,* and stained with Neutral Red. Stained neural crest cells could easily be seen as they migrated: tongues of blue cells from the implant moved dorsally (i.e. away from the pharyngeal arches), forming an alternating pattern of colours with the unoperated NC cells as the latter moved ventrally. The orientation of the two migration pathways was the same, but the direction was different. One possible explanation for this is that material in the substratum may be involved in a "contact guidance" phenomenon. Indeed, Löfberg (1976, 1978) has observed orientation of extracellular matrix fibrils in the spaces through which neural crest cells migrate in axolotl larvae.

Most of our understanding of the role of contact guidance in the orientation of cell migration comes from studies on fibroblasts and other non-neural crest cells *in vitro*. Although these studies seem far removed from the problem of neural crest *in vivo* the same basic principles of cell movement must be involved. Dunn and Heath (1976) observed that migrating chick heart fibroblasts *in vitro* orientate lengthwise on glass fibres of radius less than 100μm, whereas they orientate randomly on fibres of greater radius. Their directional movement is also affected (deflected or inhibited) where there is a ridge in the substratum whose angle is greater than about 4°. Elsdale and Bard (1972) have shown that when hydrated collagen gels are drained to produce a lattice of oriented fibres, cells placed on them orientate with their long axis parallel to the direction of the fibres. Ebendal (1976) explanted embryonic spinal ganglia onto an oriented collagen lattice prepared in this way, and observed that the cells preferentially migrated out from the explant in a direction parallel to the fibres.

What evidence do we have that the migratory behaviour of crest cells *in vivo* is influenced by contact guidance? Unfortunately, very little. Using scanning electron microscopy, Wessells (1977, p.220) illustrates oriented fibrils on the basal lamina of chick embryo somites prior to NC migration; Johnston and Steffek (personal communication) have shown a similar alignment of sub-ectodermal fibrils in the preotic region of both chick and ferret embryos. But other workers (Ebendal, 1977; Bancroft and Bellairs, 1976) have looked in vain for fibril alignment appropriate to the crest cell migration pathway in chick embryos, even though matrix fibrils are a conspicuous feature of the environment through which the cells migrate. These results are surprising in view of the observations of Weston (1963): he found, using ^{3}H-thymidine as a label, that migration of the neural tube surface population of trunk NC cells was uniform until it reached the level of the somites; thereafter, migration was confined to the somitic boundaries, and did not occur in the intersomitic regions, so that a segmental pattern of migration was established. Transplantation of labelled neural tube (with neural crest) to a region of unsegmented mesoderm abolished the segmental effect, although the cells still migrated out as two separate (dorsal and ventral) populations. Weston (1963) suggested that environmental factors - discontinuities, or "influences" from the neural tube and somites - impart a segmental pattern on the ventral NC cell population, prior to its differentiation into segmental structures (dorsal root and sympathetic ganglia).

Further evidence for environmental influences on migration pathways has been obtained through heterochronic grafting, using time differences of up to 12 hours (Weston and Butler, 1966). "Old" NC cells from labelled donors, grafted into "young" hosts at an early stage of migration of the host's own NC cells, contributed to the full range of the host's NC derivatives; in contrast, "young" grafts failed to contribute to the host's sympathetic ganglia when the time difference was maximal, suggesting that the existence of migration pathways is temporally restricted to the normal period of NC cell migration through them.

In sum, the above evidence suggests that physical channels of migration exist, but that the phenomenon of contact guidance plays, at most, a minor role in directing cells along them.

"Contact inhibition" has been suggested by Weston (1963) to be important in initiating migration. This phenomenon, as defined by Weiss (1958), occurs when adhesion between cell and substrate is stronger than between cells ("substrate" may be a different cell type - Weston and Roth, 1969). Thus cells migrate out radially from an explant *in vitro*, forming a monolayer (Abercrombie and Heaysman, 1954), and their intercellular contacts are minimized.

If the behaviour of NC cells leaving the neural tube is interpreted in relation to observations on contact inhibited cells *in vitro*, we can see that they would be unable to move anteriorly or posteriorly without contacting other cells in the same sheet, but that their movement in a lateral or ventral direction would only be limited by the presence of the ectoderm dorsolaterally, the surface of the neural tube medioventrally, and mesoderm (somitic or cranial mesenchyme) between the two. The subectodermal and neural tube surface pathways are thus the only two available, and migration into them does not require any substantial change in direction of the cells from their initial orientation within the neural epithelium. (Figures 2-4 show the orientation of the NC cells at the time of onset of migration of the subectodermal population; the cells which migrate over the neural tube surface leave after increase in diameter of the neural tube has changed the orientation of its basal surface relative to the orientation of the more ventral NC cells).

We have suggested above that the two initial pathways taken are the only two available. But prior to NC cell migration, the mesoderm itself is in contact with the surface ectoderm on one side and the neural tube on the other. A space for the crest cells to migrate into must therefore be created by separation or at least loosening of these mesodermal contacts. Pratt *et al.* (1975) have shown that migration of the subectodermal population of cranial NC cells in the chick embryo is preceded by the formation of a hyaluronate-rich cell-free space, into which the cells subsequently move. Much of this hyaluronate is synthesized by the overlying surface ectoderm (Pratt *et al.*,1976), but Greenberg and Pratt (1977) have recently shown that significant amounts are also secreted by the crest cells themselves.

We do not know whether hyaluronate is important in mammalian NC cell migration, but the absence of a significant subectodermal cell-free space in the equivalent place and time in rat embryos (Figure 2) suggests that it is not required in such a large quantity as in the chick. This mammalian-avian difference in quantity of extracellular matrix around migrating cells also occurs in the primary mesenchyme (Morriss and Solursh, 1978b). The function of hyaluronate around migrating cells may be related not only to migration itself, but to a postponement of differentiation, since differentiation following migration has been correlated with local production of hyaluronidase and consequent closer contact between cells (Toole, 1973, 1976).

MIGRATION PATHWAYS

Migration pathways of cranial NC cells have been worked out in some detail in avian and amphibian embryos. However, with the exception of a single unpublished observation (Johnston and Krames, see below), we have no direct evidence of migration pathways in mammalian embryos. Our interpretation of normal and abnormal craniofacial development to be described later will therefore rely heavily on non-mammalian studies, some of which will now be described.

Chibon (1967) and Noden (1975) have used similar techniques to study cranial NC migration in an amphibian (*Pleurodeles waltii*) and in an avian (chick) embryo. They labelled whole embryos with ^{3}H-thymidine, then grafted pieces of neural tube or neural fold from these donors into unlabelled hosts, using autoradiography to visualise the position of the grafted crest cells during migration and after they had reached their final position. Chibon (1967) was thus able to confirm and amplify the fate map of amphibian NC worked out in *Amblystoma* by Hörstadius and Sellman (1946) with particular attention to the contribution of NC to the cranial and visceral arch skeleton. He showed that the cells move directly laterally from their point of origin (see above).

Noden (1975) used a similar approach to the elucidation of migration pathways in chick embryos. The greater sophistication of his techniques, and the closer relationship of his material to mammalian embryos, make his findings of particular relevance to our present purpose. ^{3}H-thymidine-labelled neural fold grafts were implanted orthotopically into unlabelled hosts, followed by autoradiographic analysis and reconstruction of the head from serial sections. The normal migration pathways and destinations of the crest cells were thereby defined: cells from the midbrain region consisted of the subectodermal population only, and migrated to the facial region; in the hindbrain region both populations were present, the neural tube-adjacent group giving rise to cranial sensory ganglia, and the subectodermal population migrating into the mandibular and hyoid arches. The pathways relevant to craniofacial skeletal development are summarized in Figure 5. Heterotopic grafting of labelled neural folds was carried out in order to discover whether the characteristic migratory patterns for each region were determined in the cells before their migration, or whether a particular pattern was imposed on the cells by the environment through which they migrate. One class of experiment selected from the extensive series is sufficient to illustrate the point. A labelled graft from area 4 (see Figure 5), which normally contributes crest cells to the mandibular arch mesenchyme, was implanted into areas 2 and 3 - normally periocular mesenchyme and maxillary process. Subsequent reconstruction of such embryos from autoradiographed sections revealed large numbers of labelled cells associated with the optic cup, optic stalk and maxillary mesenchyme, i.e. a distribution indistinguishable from an orthotopic region 2/3 control graft.

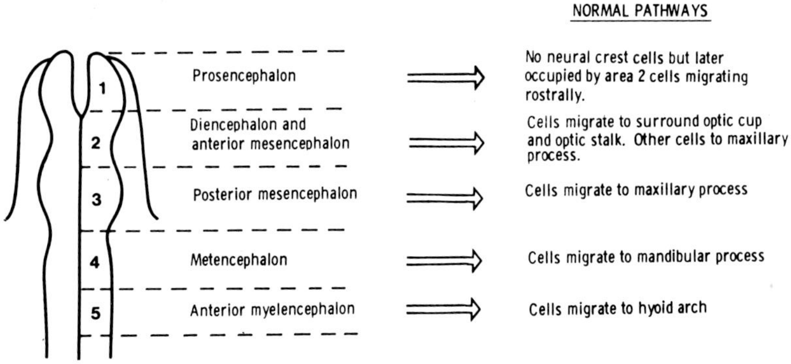

Figure 5: Migration pathways of cranial neural crest cells in the chick embryo (after Noden, 1975). Pathways leading to neural derivatives are omitted. The head is shown in profile, with major regions of the brain designated.

Radioactive markers, as used by Chibon (1967) and Noden (1975) are diluted with time as a result of cell division. This means that it is not possible to follow the differentiative fate of the labelled cells. Le Douarin (1973) found a potentially permanent graft marker in the nucleolus of the quail: most vertebrate nucleoli have only very small quantities of heterochromatin associated with them, and therefore do not stain with the Feulgen reaction; the quail is peculiar in that a large amount of heterochromatin is associated with the nucleolus, which is therefore Feulgen-positive and unusually large in size. The quail nucleolus can be identified in this way in all embryonic and adult tissues. Using chick-to-quail and quail-to-chick orthotopic and heterotopic grafting procedures, Le Douarin and her colleagues (Le Douarin and Teillet, 1973, 1974; Le Lievre, 1974; Le Lievre and Le Douarin, 1975; Le Douarin, 1976) have demonstrated that the differentiation of NC cell types is related to the migration pathway, and not to the initial position of the crest prior to grafting. An important exception to this is the differentiation of chondrocytes from cranial NC grafted into the trunk (Le Douarin and Teillet, 1974), which will be referred to in appropriate places in subsequent parts of this article. Other details of the chick-quail system in relation to differentiation will also be discussed later.

The observation of Johnston and Krames referred to at the beginning of this subsection indicates that migration of cranial NC cells in mammals conforms to the pattern of other vertebrates (Johnston, personal communication; Johnston and Listgarten, 1972). The hindbrain cranial neural fold on one side of a day 9 rat embryo was labelled with ^{3}H-thymidine by means of a fine paintbrush. After 24 hours of culture (using the dish culture technique of New, 1966) the embryo was fixed and sectioned for autoradiography; labelled cells were found in the mandibular arch in the position characteristic of neural crest cells in avian embryos, i.e. surrounding the unlabelled mesoderm cells of the arch. Although this experiment has not been repeated, it is an important observation for two reasons: first, it indicates that certain extrapolations from avian to mammalian embryos may be valid; secondly, it indicates that although techniques of manipulation and labelling of mammalian embryos *in vitro* are still difficult procedures, they have considerable potential for future exploitation. This is particularly so now that the rat embryo culture technique has been improved to allow perfectly normal development *in vitro* for up to 48 hours after initiation of cranial NC cell migration (New *et al.*,1976a). This point will be discussed further in the final section of this article.

DIFFERENTIATION

INTRODUCTION

Much of our basic knowledge about cell and tissue types derived from the neural crest stems from work carried out in the 1930's and 1940's using amphibian species. Generally these investigations were based on morphological analysis of migration of NC cells in intact embryos, or on neural fold and/or neural plate ablation experiments, and in both cases the results were usually evaluated by histological means (see reviews by de Beer, 1947; Hörstadius, 1950). Approaches such as these have considerable shortcomings if they are employed to trace the derivatives of the NC. Analysis of normal migration and subsequent differentiation by morphological means is complicated by the fact that after the initial stage of movement away from the neural epithelium, NC cells do not necessarily migrate as an integrated sheet of cells but often as a dispersed population; mesectodermal cells in many species may appear no different morphologically from any other mesenchyme cell in the environment through which they are migrating. Prior to migration, the topographical limits of the NC at the ectodermal-neuroepithelial junction are often arbitrarily defined and this, coupled with the fact that the NC is capable of regulation (Hörstadius, 1950; Noden, 1975), suggests that interpretation of ablation experiments was often rather equivocal. Furthermore, the realization that differentiation of NC cells often (always?) involves some type of tissue interaction suggests that absence of a particular tissue or structure following NC ablation does not necessarily mean that the direct cellular antecedents have been removed by that extirpation. Such a result can alternatively be explained by the possible deletion of a component in a 'multi-factorial' interaction system. However, the early work often produced results that have since proved to be essentially correct and, as Weston (1970) has pointed out, the confusion and controversies that did arise can usually be ascribed to ambiguities in the method of analysis.

Explantation techniques were also applied to the problem of NC differentiation but these reveal only the potentialities of cells under certain defined culture conditions and little about the *in vivo* fate of a population of cells. To follow the normal differentiative fate of NC cells, *in vivo* experimentation must be developed in conjunction with a cell marker technique that permits orthotopic and heterotopic grafting and subsequent mapping of the distribution of labelled cells.

Such an approach was first introduced by Raven (1931) who developed a method of xenoplastic grafting between amphibian genera and employed nuclear morphology and size as criteria to distinguish graft-derived cells from host tissue. The later development of more sophisticated cell labelling and marking techniques by Weston, Noden and Le Douarin in birds has enabled more reliable and precise experimental analyses to be made of NC migration and differentiation in higher vertebrates instead of dependence upon extrapolation from amphibian work. Largely as a result of the use of cell marker techniques used in a variety of grafting experiments, differentiative fates of NC cells are now thought to be determined by micro-environmental factors encountered during migration and acting on a cell population which is, to a greater or lesser degree, pluripotential. Possibly some 'restriction' of these ectoderm-derived cells has occurred as a result of 'primary induction', but nevertheless the crest still gives rise to an impressively wide range of cell types.

Thus neural crest cell differentiation is not a function of a cell's position along the anteroposterior axis of the embryo but is a reflection of the environments through which it migrates and where it finally resides. Experimental evidence to support this claim can be drawn from many sources but one example using heterotopic grafting within the cervico-truncal region of avian embryos demonstrates the point convincingly. Le Douarin and Teillet (1974) grafted segments of quail neural tube with associated NC into host chick embryos. Control studies, using orthotopic grafts, showed that enteric (parasympathetic) ganglion cells were derived from the 'vagal' NC (at the level of somites 1-7) and the sacral NC (somites 28+), whereas the intervening trunk NC (at the level of somites 8-28) gave rise to sympathetic ganglia and adrenomedullary cells, the latter arising specifically from the crest between somites 18 and 24. Heterotopic grafting of appropriate quail tissue from the 'adrenomedullary' region of crest into the 'vagal' region resulted in normal enteric ganglion cells derived from the graft and associated with the oesophagus. The corresponding graft of vagal NC into the adrenomedullary region resulted in normal adrenomedullary tissue carrying the quail nucleolus marker indicative of its graft origin. It is evident from this that the quail NC cells differentiate according to site of grafting rather than site of graft origin.

This finding, together with others similar in nature, provides support for the contention that NC cells are not committed or determined when migration from the dorsal side of the neural epithelium commences, and that the unique route, final location and differentiative fates characterizing the migration of NC cells from any point along the anteroposterior axis can be attributed to the particular tissue environment which they encounter. (This is not to say that NC cells at any point along the anteroposterior axis have the same range of potentiality; see discussion in Weston and Butler, 1966.) The unique character of a migratory route might be said to be due to the sum total of local microenvironmental factors. Do we know anything about such factors operative in determining that a particular NC cell might, for instance, become chondrogenic rather than myogenic? To broaden the question - what do we know about the causal mechanisms involved in determination of neural crest cells and in the expression of their differentiated state?

DERIVATIVES OF CRANIAL NEURAL CREST AND CAUSAL MECHANISMS INVOLVED IN THEIR DIFFERENTIATION

In higher vertebrates virtually all of the experimental work tracing NC derivatives and analyzing their differentiation has been carried out on avian embryos. The results of these studies are summarised in Table 1. The few experimental analyses of mammalian NC, and the reasons for the relative paucity of mammalian studies, are discussed later.

The experimental evidence for the NC origin of all the tissues listed in Table 1 has been reviewed elsewhere (Weston, 1970; Le Douarin and Teillet, 1974; Le Lievre and Le Douarin, 1975), but the experimental analysis of causal mechanisms in higher vertebrates is restricted largely to chick cranial NC mesenchymal derivatives.* In mammals, a similar range of tissues is surmised to arise from the NC but there is very little experimental evidence for this and such conjecture is based largely on extrapolation from studies of the chick embryo. Since an experimental manipulation of postulated NC mesenchymal tissues in the rat embryo is described in the next section, it is appropriate now to review briefly some relevant analyses of cranial mesectodermal tissues in higher vertebrates.

Odontogenesis

The dental papilla mesenchyme from which odontoblasts are derived has been demonstrated to be NC in origin in Amphibia (Mangold, 1936; Balinsky, 1947); consequently odontoblasts and dentine are assumed to be NC products in higher vertebrates. The enamel epithelium, which subsequently differentiates into ameloblasts, is believed to be largely ectodermal - at least anteriorly, although

* A major exception to this generalization is the range of analyses of chick melanogenesis and neurogenesis. This has been discussed extensively elsewhere, e.g. Cowell & Weston (1970), Cohen (1972).

TABLE 1: SUMMARY OF PRINCIPAL NEURAL CREST DERIVATIVES IN BIRDS

TISSUE TYPE	TRUNK	NECK/ HEAD
NEURAL:		
sensory	+	+
sympathetic	+	-
parasympathetic	+	+
supportive (glial and sheath)	+	+
PIGMENT CELLS		
(of various types)	+	probably only a very small contribution
ENDOCRINE:		
cells producing:		
calcitonin	-	+
ACTH	-	+
MSH	-	+
carotid body	-	+
MESENCHYMAL:		
dental papilla	-	+
membrane bone	-	+
cartilage	-	+
connective tissue	-	+
meninges	+	-
smooth muscle	-	+

posteriorly it may be an endodermal epithelium (discussed in Hörstadius, 1950). In normal tooth development in mammals, there is an interaction between the epithelial and mesenchymal components as a result of which the latter differentiate into odontoblasts and produce a dentine matrix, and the former differentiate into ameloblasts and release enamel (Koch, 1972; Slavkin, 1972). Superficially this appears to be a single event, but results from a wide range of tissue isolation and heterotypic recombination experiments using mammalian material have established that it is in fact a sequence of interactions, each step being vital if the sequence is to be completed and normal tooth morphogenesis to ensue.

The odontoblasts apparently differentiate slightly in advance of the enamel organ epithelium and it is only after the early dentine matrix is first secreted that the epithelial cells differentiate into ameloblasts and an enamel matrix appears. Initially there is a polarity in this interaction, with the mesenchyme influencing the differentiation of the epithelium (Koch, 1972) and it has, appropriately, been termed a 'directive' interaction (Thesleff, 1977). This and later aspects of tooth germ development have received relatively more attention than earlier aspects, possibly because the former involve the synthesis, release and accumulation of products of the differentiated state (i.e. specific matrices), whereas any earlier interactions result in less tangible developmental changes such as restriction of potential and acquisition of competence. If we examine earlier stages, the epithelial component is found to be nonspecific, and other epithelia (e.g. from embryonic foot pad) will substitute satisfactorily in tissue recombinations with dental papilla mesenchyme (Kollar and Baird, 1970). However, the mesenchymal component is specific in that mesenchyme from any other source tested will not become odontogenic in tissue recombination with the enamel organ epithelium. More recent studies indicate that the active constituent of the epithelium in these interactions is most likely to be the collagenous component of the basal lamina (Kollar, 1972; Galbraith and Kollar, 1976). In this respect, interaction of the papilla mesenchyme with an epithelial component resembles interactions in early corneal development (Hay, 1973; Meier and Hay, 1975). The lack of specificity of the epithelial component but the need for the presence of a basal lamina and associated material for progression from mesenchyme to the odontoblast stage may therefore be interpreted as a 'permissive' interaction (Thesleff, 1977). The specificity of dental papilla mesenchyme mentioned above is, of course, based upon its ability, in a 'permissive' environment, to respond in a unique fashion by becoming odontogenic. An earlier developmental event must have occurred to produce in NC-derived cells a competence to respond in this way, and experiments on Amphibia published over thirty years ago demonstrate that in one group of vertebrates at least, this event is based upon an interaction between stomodeal endoderm and the NC cells during their migration (Sellman, 1946). The nature of this interaction has not been investigated, nor has an equivalent event in mammals been demonstrated although it is generally assumed to occur.

Chondrogenesis

It is generally believed that in both birds and mammals the greater part of the facial and visceral skeleton is mesectodermal in origin. This has been experimentally demonstrated in birds using labelled implants (Johnston, 1966) and the quail/chick system (Le Lievre, 1974), and some of the causative mechanisms have been analyzed in considerable detail. However, the most familiar developmental system in this respect (but not necessarily the best understood) comes not from higher vertebrates but from the differentiation of the mesectodermal branchial arch skeleton in Amphibia. Hörstadius and Sellman (1946) and later Okada (1955) found that ablation of pharyngeal endoderm in urodele neurulae resulted in the absence of branchial arch cartilages. These authors suggested that during normal branchial (=pharyngeal) development there was an association between mesectoderm and endodermal epithelium which caused some cells of the former tissue to become chondrogenic and contribute to the visceral skeleton. However, the evidence was circumstantial given the limitations of ablation experiments, and unequivocal proof of such a relationship was later provided by *in vitro* investigations (e.g. Holtfreter, 1968; Epperlein, 1974; Epperlein and Lehmann, 1975). Epperlein and his colleagues, using *Triturus*, found that an explant of head neural fold produced a radial outgrowth comprising epidermal cells, melanocytes and undifferentiated mesenchymal cells; an explant of pharyngeal endoderm alone expanded to form an epithelioid monolayer; cultures of adjacent explants of the two tissues resulted in radial outgrowth of cells from the neural fold explant as described above, but wherever these cells contacted endoderm cells they became chondrogenic and developed into cartilaginous nodules.

Interaction, if it does occur *in vitro*, is apparently the result of chance encounters between cells migrating randomly from the explants (see section 1).

The nature of this interaction between mesectoderm and pharyngeal endoderm has never been analyzed and indeed, the term 'contact' is used as a light-microscope criterion: whether this entails a cell contact- or matrix-mediated interaction is uncertain. From the limited experimental evidence available, an interaction with pharyngeal endoderm does not appear to be essential in higher vertebrates: the transplantation of quail cranial NC into the chick trunk produced a chondrogenic progeny in spite of those cells having never encountered endoderm (Le Douarin and Teillet, 1974).

The second example of cranial chondrogenesis is the formation of scleral cartilage in the chick embryo, a phenomenon without equivalent in mammals, but important in the present context because it is a system which has been analysed in considerable detail. Many nonmammalian vertebrates possess a hyaline cartilage layer in the sclera in addition to layers of fibrous tissue. In the chick, recognizable cartilage appears on days 8-9 of development (Romanoff, 1960); it arises from the periocular mesenchyme which is derived from mesencephalic NC cells migrating rostrally and 'arriving' in the periocular region at 2½ days (Johnston *et al.*, 1973; Noden, 1975). From ectopic grafting experiments, Reinbold (1968) postulated that the differentiation of scleral cartilage from mesenchyme was dependent upon the presence of the pigmented layer of the retina at a particular developmental stage. This predicted relationship was confirmed by grafting various combinations of cranial mesenchymal and optic cup tissues onto the chorioallantoic membrane (CAM) of the chick embryo (Newsome, 1972). It was concluded that a tissue interaction of some type is responsible for the ensuing chondrogenesis, and that this event is completed by 3½ days of development, i.e. approximately five days before cartilage appears. The retinal pigmented epithelium is normally associated with a collagenous, PAS-positive, extracellular layer separating it from periocular mesenchyme *in vivo*, and can secrete such a layer of extracellular material *in vitro*. It has now been established that the interaction is mediated by this extracellular material and not by cell contact (Newsome, 1976). Clones of retinal pigmented epithelial cells were cultured on Millipore filters and deposited a matrix on the filter surface. The cells were lysed and washed off, leaving a cell-free matrix on which cloned NC or periocular mesenchymal cells were grown; these subsequently became chondrogenic, as judged by morphological criteria and by the ratio of $\alpha1:\alpha2$ collagen chains (Newsome, 1975). Other matrices, such as rat tail collagen, were tested but none had the same effect; the possibility of a diffusible agent being involved was also eliminated by a variety of transfilter experiments. The results indicated that direct contact with the extracellular material deposited *in vitro* by retinal pigmented epithelium will cause competent periocular mesenchyme cells to become chondrogenic: the active factors are not diffusible and living epithelial tissue need not be present. Further analysis of the matrix is anticipated with interest particularly in the light of recent findings that a number of tissues (e.g. neural retina - Smith *et al.*,1976) go through transient phases of type II collagen synthesis - a collagen type previously believed to be unique to differentiated cartilage, and to the notochord during 'induction' of sclerotome chondrogenesis (e.g. von der Mark *et al.*,1976). Scleral chondrogenesis and its dependence upon retinal pigmented epithelium may prove to be analogous to the relationship between sclerotome and notochord but until further information is forthcoming it is well to bear in mind that differences may exist between mesectodermal and mesodermal cartilages (Chiakulas, 1957). If these cartilages are 'non-equivalent' (Lewis and Wolpert, 1976) then factors operative in their differentiation may be dissimilar.

Osteogenesis

The calvaria of birds and mammals are laid down as membrane bone, and ossification of the mesenchyme concerned is dependent upon an influence from the underlying neural epithelium (Schowing, 1968). In the absence of that epithelium, the membrane bones comprising the roof of the cranium fail to form.

It has recently been demonstrated that the overlying surface ectodermal epithelium exerts a similar influence in the formation of membrane bones in the mandible of chick embryos (Tyler and Hall, 1977). Meckel's cartilage, the pharyngeal cartilage of the mandibular arch, is replaced by bone only at its proximal end, forming the articulare of reptiles and birds, and the malleus of mammals (see section 3). The mandible develops initially as membrane bone(s) closely investing Meckel's cartilage. The NC derivation of all of these skeletal elements in birds has been unequivocally demonstrated using the quail nucleolar marker system (Le Lievre,

1974). Tyler and Hall (1977) used organ culture and CAM grafting techniques to grow intact mandibular processes or separated ectodermal and mesenchymal components, obtaining essentially the same result with both approaches: intact mandibular processes of all stages developed moderately well, and both Meckel's cartilage and membrane bone differentiated. Isolated mesenchyme (irrespective of stage) formed Meckel's cartilage, but only mandibular mesenchyme, isolated from its epithelium at 4½ days or later, subsequently produced membrane bone in addition to cartilage (membrane bone in the mandible *in vivo* appears at 7 days). Heterotypic combinations have demonstrated that wing or leg bud ectoderm can be substituted for mandibular arch ectoderm (Hall, personal communication) indicating that the epithelial component is not specific. This analysis has recently been extended to the maxillary process where the dependence upon a maintained epithelial presence for *in vitro* ossification of mesectoderm ceases after 4 days, that is 3½ days before membrane bones form in this region (Hall and Tyler, personal communication). The disparity in stage-dependence between the maxillary and mandibular processes, 4 days in the former and 4½ days in the latter, possibly reflects differences in the time of onset and distance of migration of NC-derived cells. Cells migrate further to reach the mandibular process and consequently reach it later than more anterior NC cells migrating earlier to the maxillary region (Noden, 1975). The mandibular epithelial-mesenchymal interaction may therefore be completed later in development.

These experiments indicate that an epithelial involvement is essential for mesectoderm cells to form membrane bone, but the precise role of the epithelium remains to be elucidated. The apparently 'permissive' character of the interaction arising from lack of epithelial specificity is reminiscent of the early interaction in odontogenesis discussed previously. Has mesectoderm in the maxillary and mandibular processes, like dental papilla mesenchyme, acquired a 'competence' to respond specifically (in this case by osteogenesis) as the result of an earlier influence during migration?

Periosteal cells around maxillary membrane bones also have chondrogenic potential which is expressed in the development of secondary or adventitious cartilage (Murray and Smiles, 1965), in the repair of experimental fractures (Hall and Jacobsen, 1975), and in explant culture (Thorogood, unpublished and Figures 6 and 7).

Further analysis should be aimed at establishing whether this epithelio-mesenchymal interaction produces a solely osteogenic response or a 'skeletogenic' response, the exact nature of which is modulated between chondrogenic and osteogenic expression depending on microenvironment.

Summary

As we have shown above, the neural crest origin of a wide range of craniofacial tissues has been established in birds by reliable experimental means. The migration pathways from the dorsal side of the neural tube to the final site at which the differentiated state is expressed have been investigated, but our current understanding of these events is limited. It is evident that microenvironmental factors influence, or even determine, the differentiation of a population of ostensibly pluripotent cells. At present we understand these 'factors' to be cell-to-cell or cell-to-matrix associations and the interactions thereby generated may be considered as three types:

(i) isotypic cell association - quantitative differences in this type of association are operative in neurogenesis and melanogenesis *in vitro*. Experiments from NC-derived cells from early chick embryo sensory ganglia demonstrate an *in vitro* correlation between cell dispersion/melanogenesis and cell aggregation/neurogenesis (Cowell and Weston, 1970). The implications of this in relation to clonal analysis will be discussed later.

(ii) heterotypic cell association - as seen in epithelio-mesenchymal interactions such as those underlying the development of membrane bones. Most remain to be analyzed at the cellular and molecular level and some might prove to be equivalent to:

(iii) cell-matrix association - as shown to occur in the formation of scleral cartilage. It appears initially to be an interaction

Figure 6 (above, opposite): Cell outgrowth from a 14-day explant of 10-day chick embryo membrane bone periosteum, viewed in the living state by inverted phase microscopy. Some polygonal epithelioid cells and spherical cells are surrounded by a refractile matrix which stains metachromatically (see Figure 7) and are therefore interpreted as chondrogenic. (Bar = 50μm). Figure 7 (lower, opposite): Explant outgrowth similar to that shown in Figure 6; stained May-Grunwald and Giemsa. A chondrogenic nodule of cells with stained lacunae (arrow) is present. (Bar = 30μm).

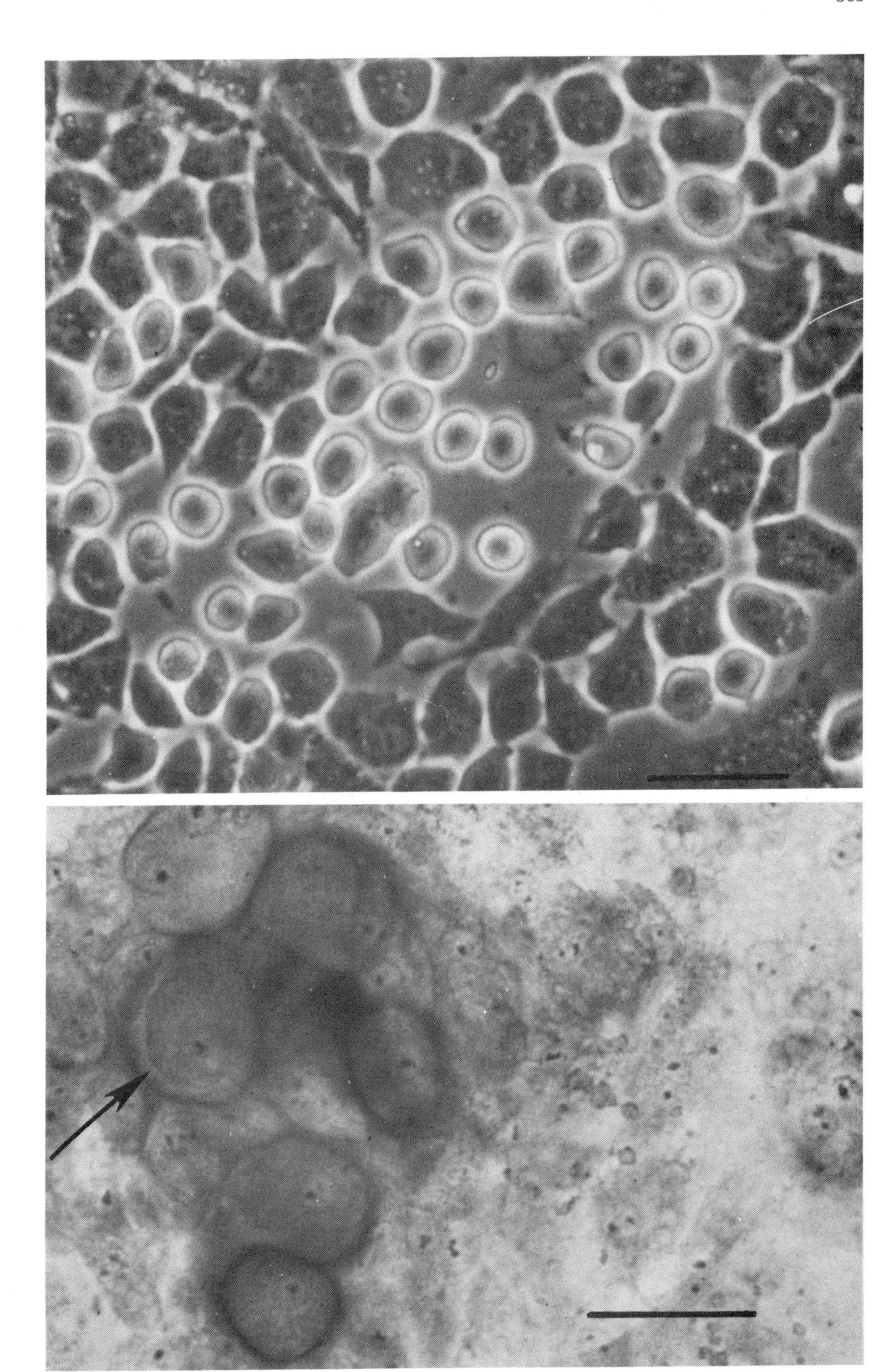

between an epithelium and mesenchyme but in fact the interaction is mediated through a matrix secreted by the former. Given the presence of the matrix for the mesenchyme cells to interact with, the living epithelium need not be present for differentiation.

STUDIES ON NEURAL CREST DIFFERENTIATION IN MAMMALIAN EMBRYOS

In mammals, cranial (and trunk) NC derivatives are not so well-defined. Of the relatively few accounts of NC in mammals most are morphological descriptions of the crest prior to or during early migration (e.g. Kallen, 1953; Bartelmez, 1960) or histochemical surveys of the differentiation of presumed mesectodermal tissues in the head (Milaire, 1959). Nevertheless, evidence for the NC origin of craniofacial tissues equivalent to those discussed earlier in birds has been obtained from the study of mutants and teratological studies.

There is a considerable degree of homology between the heads of bird and mammalian embryos, and in spite of the pitfalls of assuming developmental equivalence, there has been much extrapolation from amphibian and bird NC studies to mammals. Although many of these suppositions will probably be verified, at present unequivocal evidence of NC derivation remains to be established and virtually no experimental investigation of NC differentiation has been carried out. The several exceptions involve non-mesenchymal and non-cranial derivatives. The elegant experiments performed by Rawles (1947) established beyond doubt that melanocytes in the mouse are of neural crest origin. Transplants of tissue, including neural fold or somitic and lateral plate tissue before and after predicted NC 'invasion', were placed into the coelom of White Leghorn chick embryos. Subsequent recovery of the grafts demonstrated that only those containing neural fold or NC cells developed pigmented hair. Not only was the NC derivation of melanocytes shown, but the use of small tissue samples from different regions of embryos of different ages enabled the rate of migration to be plotted. This early analysis of melanogenesis facilitated a genetic approach to NC differentiation using the well-documented range of mutants with coat colour abnormalities (e.g. 'steel', Mayer, 1975 and see Heath, this volume; 'piebald' , Mayer, 1977. The achievements and potential of this approach will be discussed in section 4. The heterospecific grafting experiments reported by Pugin (1973) in which mouse neural tube and associated crest were grafted axially into chick embryos of approximately equivalent developmental stages, and produced neural and supporting tissue of graft origin, is another (modified) application of the approach adopted by Rawles, and indicates that its usefulness may extend to NC derivatives other than melanocytes.

NORMAL AND ABNORMAL MORPHOGENESIS OF CRANIAL NEURAL CREST-DERIVED STRUCTURES IN RAT EMBRYOS

In this section, we will describe the morphogenesis of a pattern of craniofacial anomalies induced experimentally in rat embryos. We will then assess how far the abnormal developmental events described can be attributed directly to abnormal migration and differentiation of NC cells, and how far our interpretation must still rely on evidence derived from the studies on non-mammalian vertebrate embryos described in previous sections.

EFFECTS OF EXCESS VITAMIN A ON NEURAL CREST CELL MIGRATION

Abnormal morphogenesis of the facial skeleton can be induced in rat embryos by maternal administration of excess vitamin A (retinyl palmitate) within the period of development beginning just prior to differentiation of the neural ectoderm and ending at the time of cranial neural tube closure (day 8 to day 10 of pregnancy when day of mating = day 0) (Morriss, 1972, 1973; Poswillo, 1975). By examining live or cleared embryos at day 11 and day 12, it is possible to observe groups of migrating neural crest cells as less translucent regions (Johnston *et al.*,1977). Comparison of control and vitamin A-affected embryos prepared according to this technique shows that the preotic hindbrain NC cells migrate later and/or more slowly in vitamin A-affected embryos; furthermore, their migration pathway has shifted anteriorly, with cells moving into the maxillary region at a time when this cannot be seen to

Figures 8-11 (overleaf): 11- and 12-day rat embryos fixed in 1/4 strength Bouins fluid, dehydrated and cleared in 3:1 methyl salicylate: benzyl benzoate. (Bar = 0.5mm; same for each figure). Figure 8 (top, left): Day 11 embryo, control. Groups of preotic neural crest cells (arrowed) remain in the position of the cranial sensory ganglia (nerves V, VII and VIII) and have also moved into the mandibular and hyoid arches. The otocyst lies adjacent to the hyoid arch. Figure 9 (lower, left): Day 11 embryo from dam given 100,000 i.u. vitamin A palmitate s/c on day 8 of pregnancy. Preotic neural crest cells (arrowed) are in a more dorsal position than those of controls, and have not yet reached the pharyngeal arches (compare density of these structures with those in Figure 6). The otocyst lies adjacent to the mandibular arch. Figure 10 (top, right): Day 12 embryo, control. Figure 11 (lower, right): Day 12 embryo from day 8 - vitamin A-treated dam. A group of neural crest cells can be seen in the proximal region of the maxillary process (arrow). These cells cannot be seen in the control embryo (Figure 8).

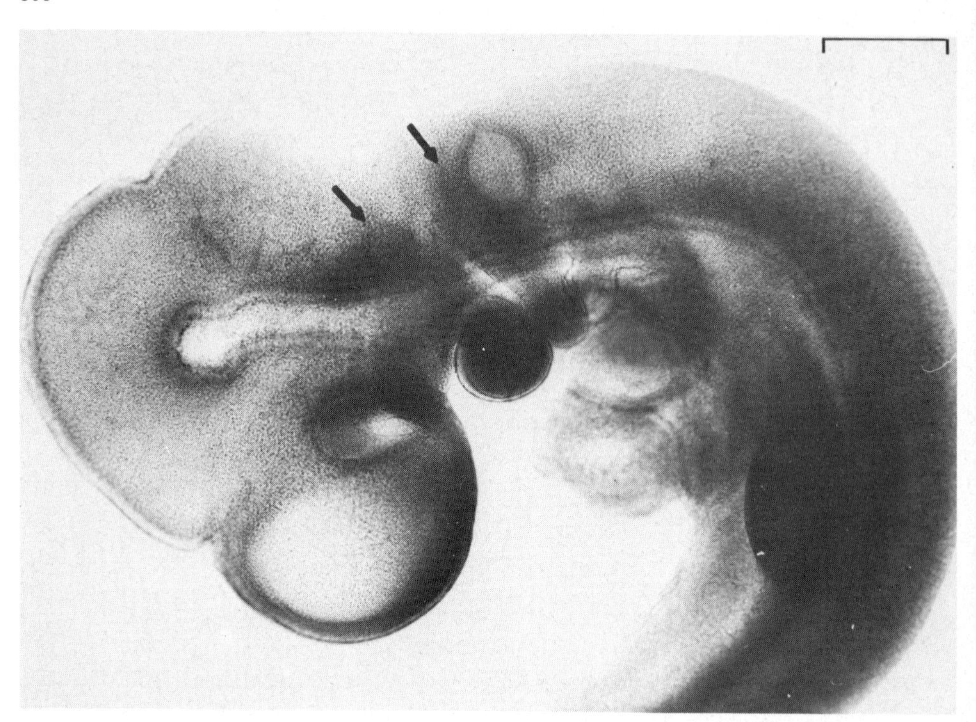

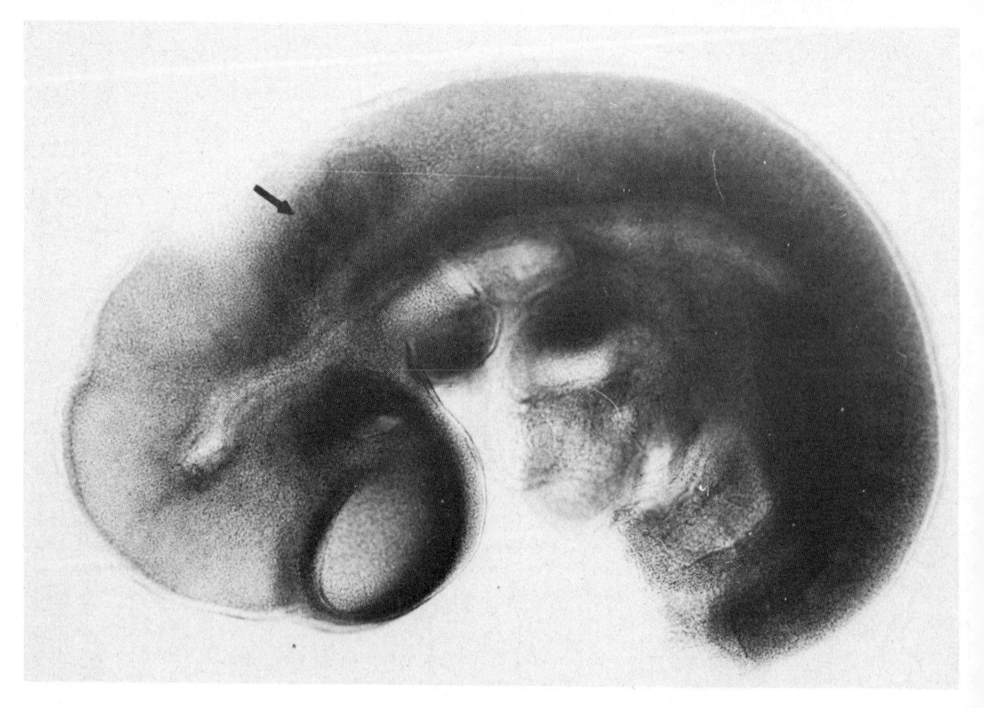

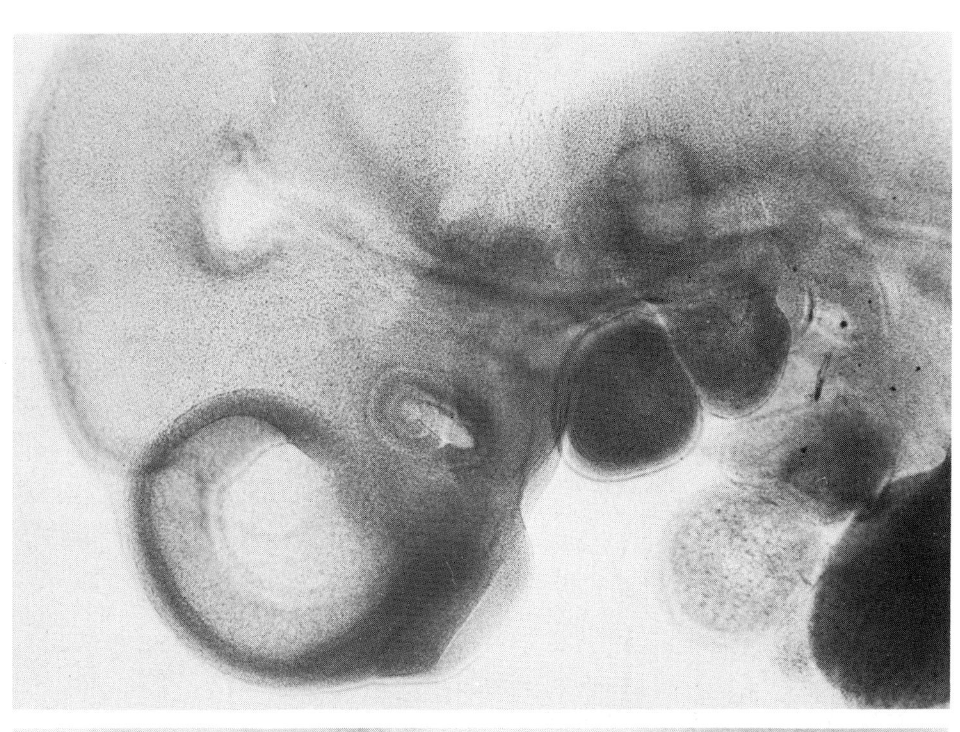

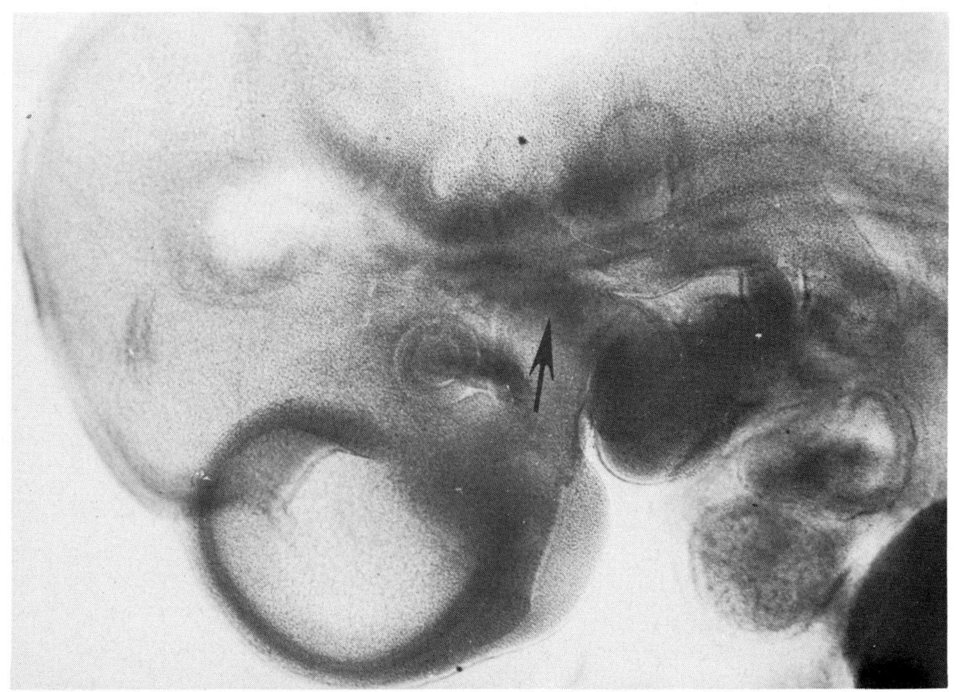

occur in the controls (Figures 8-11).

Late or slower NC cell migration has also been demonstrated in intact rat embryos *in vitro*. Embryos were explanted at day 9 (early cranial neural plate stage) and cultured for 48 hours by the improved technique of New *et al.* (1976a) in serum containing added retinol or retinoic acid (Morriss and Steele, 1977). Preotic neural crest cells were observed close to the otocyst in experimental embryos at a time when they had reached the mandibular arch in embryos in the control cultures. These experiments strongly suggest that vitamin A affects the time of onset or rate of migration of neural crest cells in intact embryos.

Evidence in favour of a direct effect of vitamin A on cell migration comes from studies on migrating mesenchyme cells *in vitro* (Morriss, 1975; Kwasigroch and Kochhar, 1975). Morriss (1975) explanted primary mesenchyme from day 9 embryos into hanging drop cultures. After 24 hours, successful explants were transferred to medium containing 0.1 or 0.01 μg/ml retinol, or the diluent (ethanol) alone. Cells in control medium moved in an essentially similar way to other locomoting cells on a flat surface (e.g. chick-heart fibroblasts, Abercrombie and Heaysman, 1954), having a leading lamella with ruffled membrane, and a trailing edge. However, in the presence of 0.01 μg/ml retinol, they did not form a typical leading lamella; instead, blebs formed at apparently arbitrary regions of the surface, being alternately extended and withdrawn in a manner which suggested a failure to adhere normally to the substratum. Locomotion was much slower than in the control cultures. In the presence of 0.1 μg/ml retinol the blebbing phenomenon was exaggerated, followed by a rounding up of the cells, which all died within four hours.

These observations on the effect of vitamin A on locomotory cells *in vitro* suggest that the retardation of neural crest cell migration observed in intact embryos in the presence of vitamin A *in vivo* and *in vitro* may involve similar phenomena. Transmission electron microscopy of primary mesenchyme in vitamin A-treated embryos has shown blebbing of cells and some cell death. Unfortunately, similar studies have not been carried out at the stage of neural crest cell migration. The actual mechanism of the locomotory abnormality is not understood, but may involve vitamin A-induced expansion of the plasma membrane (Daniel *et al.*,1966) and/or an effect on glycosylation of the integral glycoproteins of the cell membrane (De Luca and Yuspa, 1974). The possibility that the mechanism may involve glycosylation of membrane glycoproteins is intriguing in the present context, since evidence is now strong that cell-cell and cell-substrate adhesiveness involves a reaction between glycoproteins and glycosyltransferases (Shur and Roth, 1975) or by bridging molecules between the carbohydrate chains of surface glycoproteins (Rees *et al.*,1977). Furthermore, Weston (1970) suggested that cell-substrate adhesiveness may be important in determining the NC cell migration pathway. (It should be stressed here that this is a separate phenomenon from that of contact guidance, which involves only physical aspects of the substratum: see Dunn and Heath, 1976, and section 1 of this article).

NORMAL AND ABNORMAL DIFFERENTIATION OF PHARYNGEAL ARCH CARTILAGES

Later consequences of developmental disturbances can only be followed in *in utero*-affected embryos, because the embryo culture technique is at present limited to the period of yolk sac-mediated nutrition. The following summary of normal and abnormal pharyngeal arch cartilage differentiation in control and vitamin A-affected embryos is based on the assumption that these structures are derived from preoptic hindbrain cranial neural crest cells (i.e. the groups of cells observed in the cleared specimens, Figures 8-11); however, it should be borne in mind that this has not yet been unequivocally demonstrated to be true for mammalian embryos. For a more detailed account of these structures and the method of preparation, see legends to Figures 12-15.

The first and second arch cartilages appear first in day 15 embryos. Chondrification of the malleus, incus, and stapes occurs at the same time as the chondrification of the periotic capsule. The three future ear ossicles develop within a single

Figures 12-15 (next page): Sagitally-halved rat embryos, stained for cartilage with Alcian blue 8GX and subsequently prepared as for Figures 6-9 (Gurr, 1973). E-eye; C-cochlea; M-Meckel's cartilage; S-semicircular canals. Figure 12 (top, left): Day 15 embryo, control. Cartilaginous rudiments of the malleus, incus, and stapes can be seen between Meckel's cartilage and the upper cochlear region. (Bar = 1mm). Figure 13 (lower, left): Day 15 embryo from day 8-vitamin A-treated dam. Although the cartilaginous capsule of the inner ear is not as well differentiated as that in Figure 10, it can clearly be seen that the proximal end of Meckel's cartilage is curved downwards to a position ventral to the chochlea. (Bar = 1mm). Figure 14 (top, right): Day 17 embryo, control. Malleus and incus cartilages can be distinguished at the proximal end of Meckel's cartilage, and the stapes cartilage lies adjacent to the oval window. There is no cartilage in the maxillary region. (Bar = 1mm). Figure 15 (lower, right): Day 17 embryo from day 8-vitamin A-treated dam. The relationship between the proximal end of Meckel's cartilage and the inner ear is as in Figure 11, and there are no ear "ossicle" cartilages. An ectopic "maxillary" cartilage lies immediately below the eye. (Bar = 1mm).

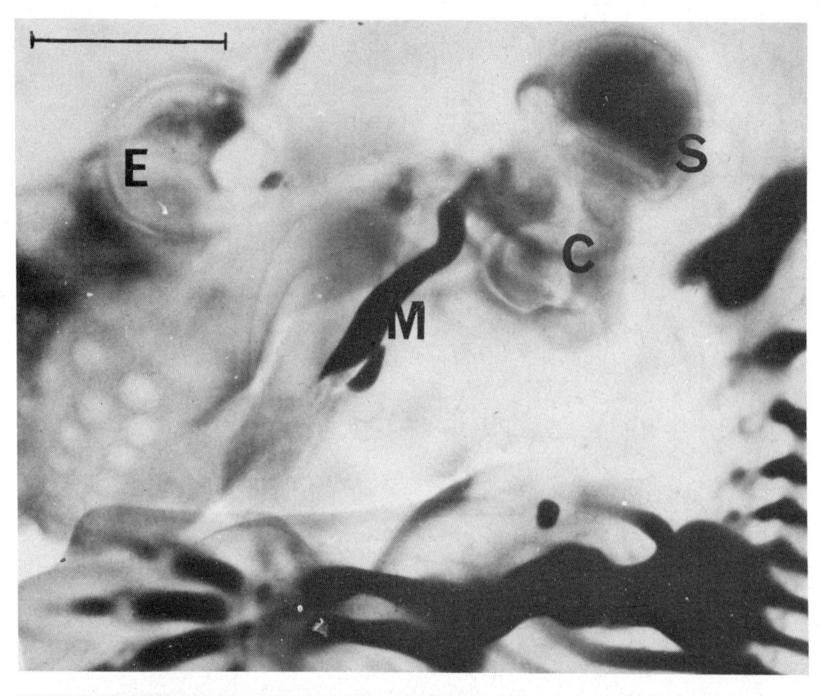
E
S
C
M

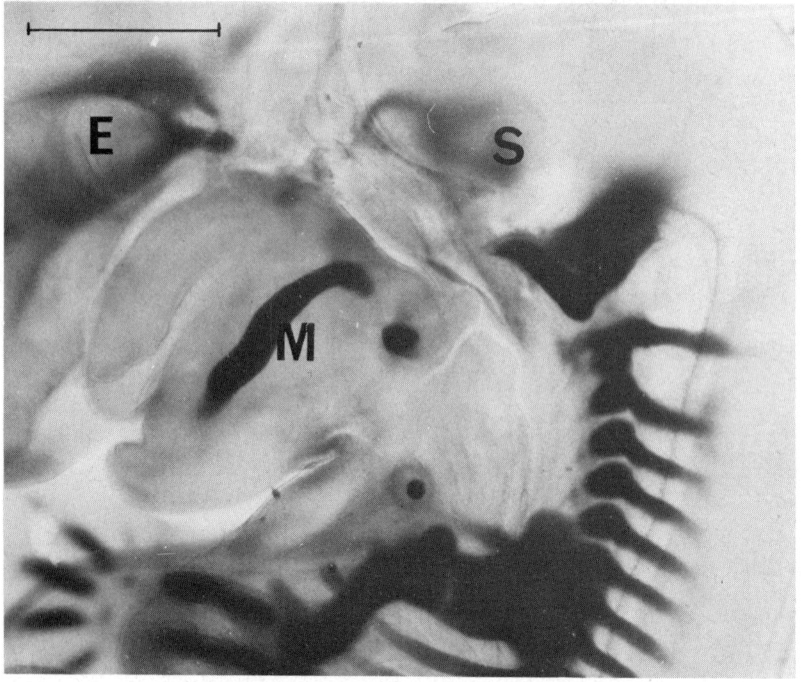
E
S
M

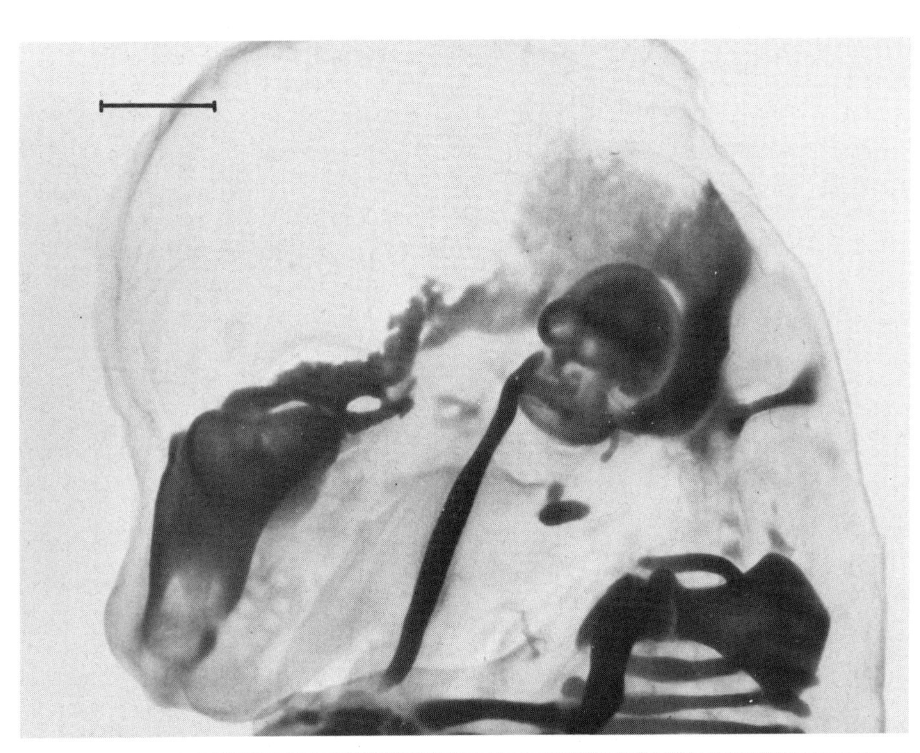

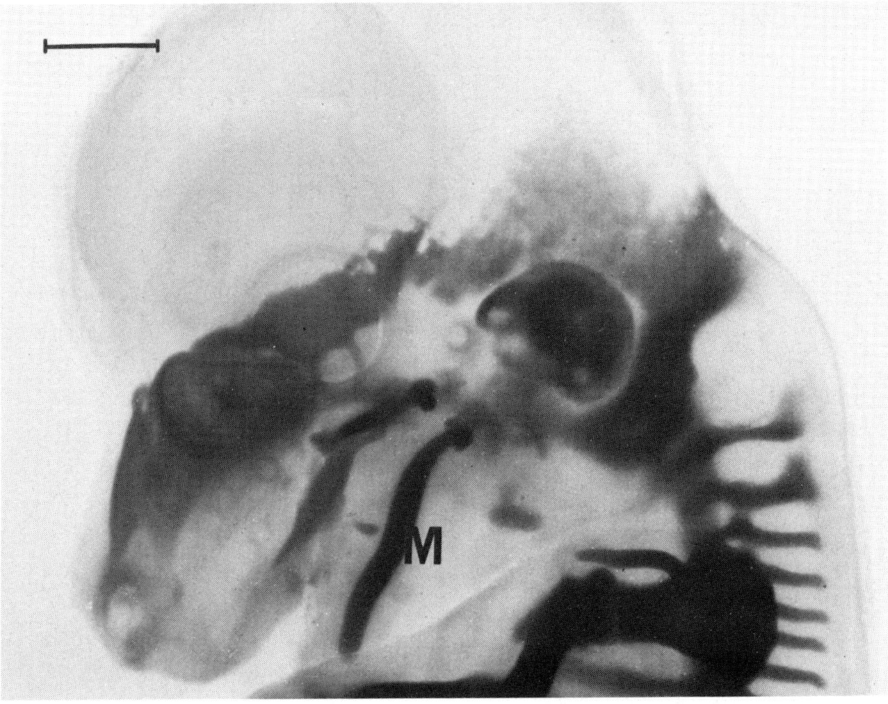
M

mesenchymal condensation, and the malleus is clearly in continuity with Meckel's cartilage until day 18 or 19, when a region of degenerating cells begins the process of separation (not shown). The stapes develops in the region of the developing oval window (Figures 12 and 14).

In vitamin A-affected embryos, the proximal end of Meckel's cartilage does not develop the complex shape and relationships described above, and there is no clear differentiation of malleus, incus, and stapes in the region adjacent to the oval window. The actual pattern of cartilage differentiation shows some variation between specimens, and some of those examined at day 18 or later show small fragments of cartilage situated between the proximal end of Meckel's cartilage and the periotic capsule. A curious and constant feature of these abnormal embryos is the presence of a "maxillary cartilage" (Figure 15). Interpretation of the possible origin of this structure is discussed below.

POSSIBLE CAUSES OF THE FORMATION OF A "MAXILLARY" CARTILAGE

The palatoquadrate cartilage of avian and reptilian embryos represents the epibranchial component of the mandibular arch (Goodrich, 1958). It is normally represented in mammalian embryos only by the incus (see above) although the greater wing of the mammalian sphenoid bone (alisphenoid) may be homologous with the epipterygoid part of this cartilage in reptiles. The most plausible explanation for the formation of a "maxillary" or "palatoquadrate" cartilage in vitamin A-affected embryos is that a population of neural crest cells, which would normally enter the mandibular arch or come to lie in the middle ear region, has migrated along an abnormally anterior pathway to reach the maxillary process. There it has differentiated into a chondrogenic phenotype as it would normally do, forming an ectopic cartilage.

There is some evidence for this interpretation if the cleared day 11 and day 12 embryos (Figures 8-11) are interpreted in relation to the pathways of migration elucidated in non-mammalian vertebrates. Chibon (1967) and Hörstadius and Sellmann (1946) observed that migration of neural crest from the midbrain and hindbrain regions in amphibian embryos was along a plane directly transverse to the neural tube. Noden (1975) showed that crest cells reaching the maxillary region of chick embryos originate in the mesencephalic (midbrain) region, whereas those reaching the mandibular arch originate in the metencephalon (upper hindbrain). One of the earliest effects of vitamin A, preceding neural crest migration, is the differentiation of a cranial neural plate which is shortened along its anteroposterior axis (Morriss, 1972). This results in the otocyst differentiating in a position level with the mandibular arch instead of the 2nd (hyoid) arch (Figures 2-4, 8,9). Thus, neural crest cells migrating from the preotic region of the hindbrain arise from a point anterior to their normal level with respect to the pharyngeal endoderm.

There is, as yet, no evidence that the mammalian pharyngeal endoderm is involved in an inductive interaction with neural crest cells prior to differentiation of the arch cartilages. However, if this does prove to be the case, the formation of the "maxillary" cartilage would suggest that the absence of a palatoquadrate cartilage in mammals is due to a different pattern of cell migration from that of other vertebrates, and not to an inherent inability of the upper pharyngeal endoderm to stimulate chondrogenesis. Alternatively, it may be that the subectodermal neural crest cell population emanating from the hindbrain requires only contact with the subectodermal basal lamina in order to express its potentiality of chondrogenesis. This interpretation can also be applied to the results from Le Douarin and Teillet (1974) where hindbrain neural crest cells transplanted to the trunk produced chondrogenic progeny.

NORMAL AND ABNORMAL DIFFERENTIATION OF MEMBRANE BONES OF THE FACE AND EAR REGIONS

Alizarin staining of bones in day 20 normal and vitamin A-affected rat embryos shows that malformation of the membrane bones of the face and ear region always occurs subsequent to abnormal cartilage differentiation. The severity of membrane bone abnormalities can be lessened by reducing the amount of vitamin A administered (not shown), or by later administration (Figures 16 to 18).

The membrane bone of the mandible forms around and lateral to Meckel's cartilage, in close but not apparently intimate juxtaposition to it. This observation is based on light microscopy, and the actual nature of the association is unknown. Interestingly, the small membrane bones that form below the orbit in the vitamin A-affected embryos (Figures 17 and 18) show a similar association with the "maxillary" cartilage (Figure 19). This observation suggests some functional relationship between the osteogenic cells (which by homology with avian embryos are probably neural crest in origin - see earlier) and the first arch cartilages in both normal and abnormal embryos. This suggestion does not conflict with the results of Tyler and Hall (1977) showing that in chick embryos differentiation of membrane bone from mesectoderm of the mandibular and maxillary process requires previous contact with the overlying

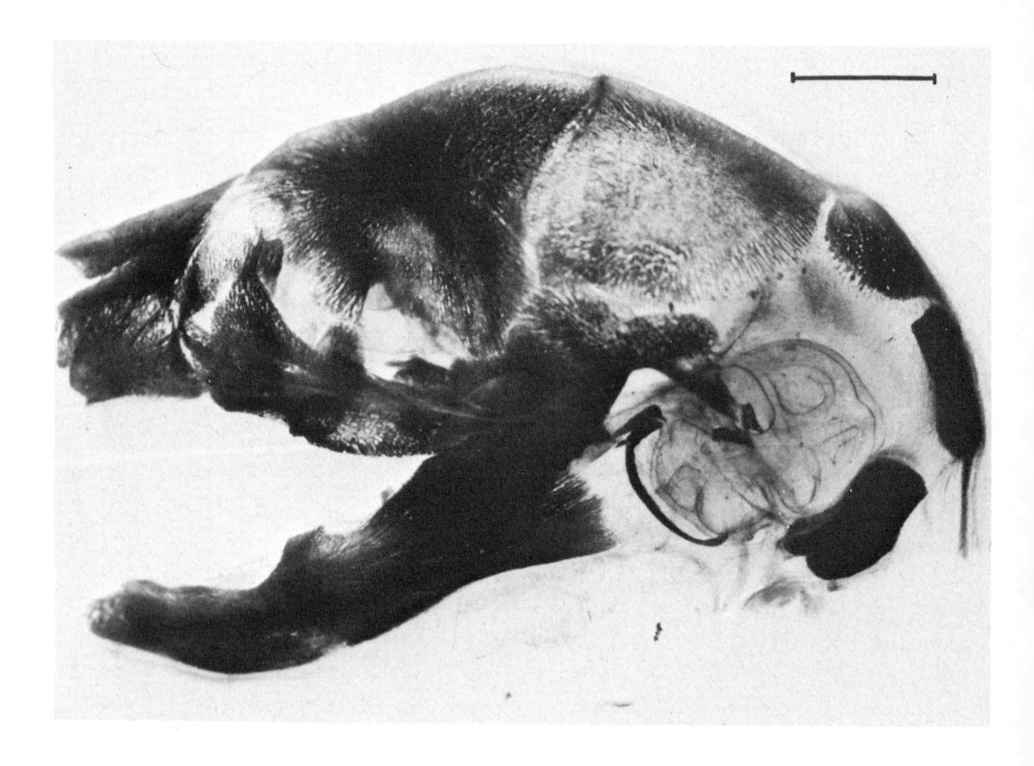

Figures 16-18: Heads of day 20 embryos stained for bone by the Alizarin technique. (Bar = 2mm throughout). Figure 16 (above): Head of control embryo, for comparison with Figures 15 and 16. (The periotic capsule can be seen as an unstained structure, being still cartilaginous at this stage). Figure 17 (upper, opposite): Head of embryo from dam given 100,000 i.u. vitamin A s/c on day 9 of pregnancy. Extra membrane bones are present in the region of the zygomatic arch (i.e. below the eye), and the tympanic and squamous temporal bones are reduced in size. Figure 18 (lower, opposite): Head of embryo from dam given 100,000 i.u. vitamin A s/c on day 8 of pregnancy. The malformations noted in Figure 15 are present in exaggerated form: a large number of small membrane bones have formed in the zygomatic region, and form an elongated area of fusion with the bone of the mandible. The tympanic bone is absent.

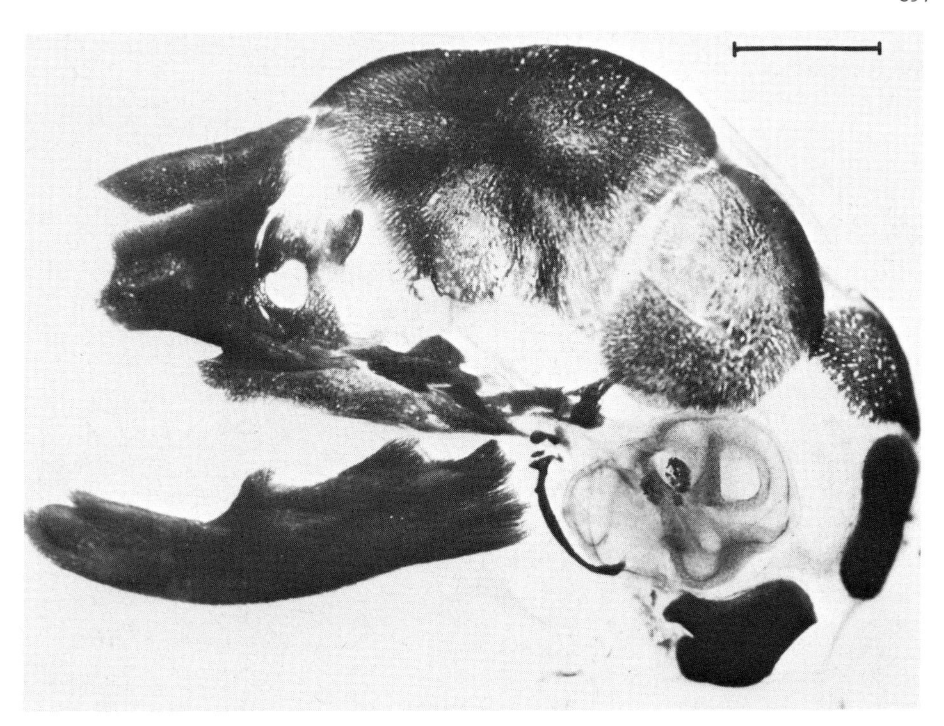

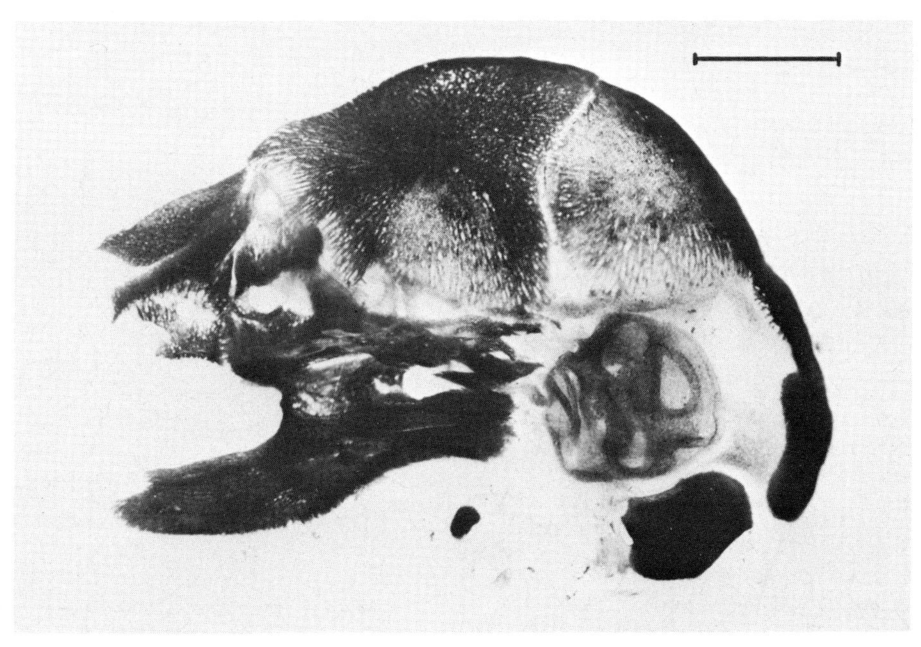

ectodermal epithelium (see earlier). Rather, it suggests that in addition to the interaction with overlying surface ectoderm, a second interaction involving the first arch cartilage may also be involved. Absence or reduction of the tympanic bone following reduction of the proximal end of Meckel's cartilage (Figures 13, 15, 17 and 18) is consistent with this interpretation.

SUMMARY

In this section we have described a complex sequence of morphogenetic events which indicate that abnormal neural crest cell migration in a mammalian embryo represents an early stage in the process of abnormal craniofacial development. We can at best speculate about the mechanisms involved, using our interpretation as an indication of useful lines of enquiry for the future. It may be helpful, therefore, to summarise the abnormal events described above in a way which incorporates some of these speculative explanations.

1. When excess vitamin A is administered to rat embryos at the primitive streak stage *in vivo*, or at the early neural plate stage *in vitro*, a characteristic pattern of abnormal development is produced.

2. The first indication of abnormal morphogenesis is that the cranial neural ectoderm is shorter than normal in the antero-posterior plane. As a result, the hindbrain and otocyst are shifted anteriorly relative to the position of the foregut.

3. Preotic neural crest cells (which have been shown in avian embryos to migrate to the mandibular region, Noden, 1975), are retarded in their migration. Since they, too, are in an abnormally anterior position, some of those which would normally reach the mandibular region migrate into the maxillary process.

4. The crest cells which reach the mandibular region form a Meckel's cartilage of abnormal shape, which fails to form normal ear ossicle precursors at its proximal (middle ear) end.

Figure 19 (opposite): Day 18 embryo from day 8-vitamin A-treated dam. Section through region of the "maxillary cartilage" showing membrane bone forming around two sections of cartilage as well as around and lateral to Meckel's cartilage. Mallory trichrome stain, Crooke-Russell modification. T-tongue; MC-Meckel's cartilage; EC-ectopic "maxillary" cartilage. (Bar = 0.05mm).

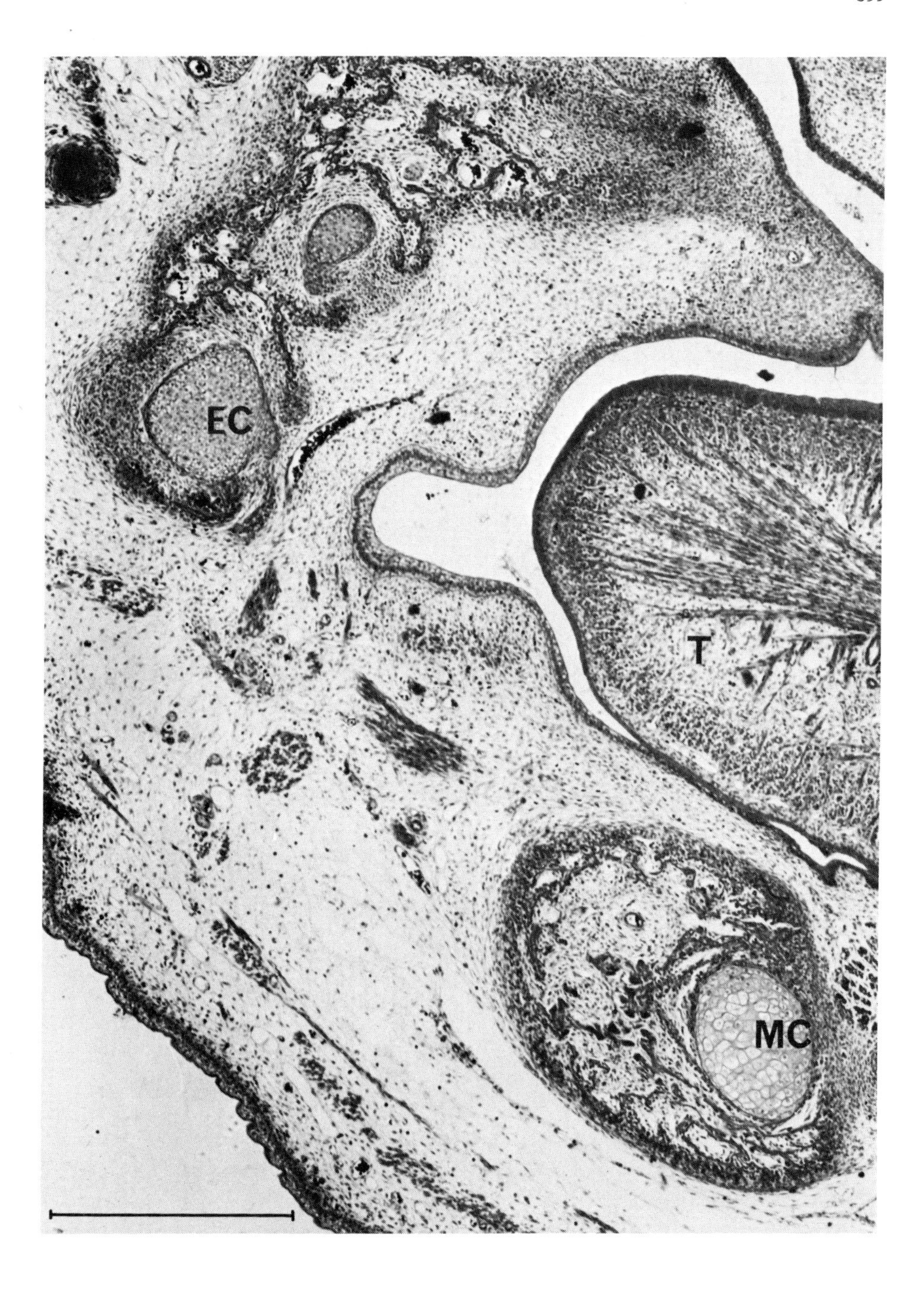
EC
T
MC

5. The crest cells which reach the maxillary region also differentiate as first arch cartilage, possibly involving an interaction with the upper pharyngeal endoderm. Their competence to form ectopic cartilage may also reflect a previously determined restriction of potential of the subectodermal population of preotic neural crest.

6. Membrane bone of the mandible forms in close association with Meckel's cartilage; hence an abnormal mandible develops when Meckel's cartilage is abnormal, and the tympanic bone is absent when the underlying (proximal) part of Meckel's cartilage fails to form. Similarly, when part of the first arch cartilage develops in the maxillary region, ectopic bones form in association with it.

7. The role of pharyngeal endoderm in the differentiation of Meckel's cartilage in normal embryos, and the role of the mandibular epithelium (and possibly also Meckel's cartilage) in the differentiation of membrane bone, remain to be investigated.

POSSIBLE FUTURE DIRECTIONS IN RESEARCH ON MAMMALIAN NEURAL CREST

GENETIC DISSECTION OF NEURAL CREST CELL BEHAVIOUR AND DIFFERENTIATION

The development of coat colour in mammals is due to a large number of factors operating in a sequential fashion (reviewed by Searle, 1968). The demonstration by Rawles (1947, and discussed earlier) that melanocytes in the mouse are NC-derived, and the well-documented range of pigmentation mutants in some species, has allowed analysis of the genetic control of the various steps in the developmental process. Using an experimental design based largely on combinations of mutant and wild-type tissues grafted intra-coelomically into chick embryos and subsequently examined for the presence or absence of pigmented hair in the graft, Mayer (see below) has been able to pinpoint the site of genetic lesions in the 'white spotting' series of mutants. Lack of pigmentation can be due to a number of factors (Searle, 1968), but in white-spotting (which in extreme cases may affect the entire coat but leave the retinal pigment unaffected) the various affected alleles all show a block in the differentiation of melanoblasts that have reached their location in the skin. The loci affected in 'Piebald' (*s*) and in 'Dominant spotting' (W^v) are active at the level of the neural crest itself and thus the early "stem" melanoblasts are affected (Mayer, 1965, 1970, 1977; Mayer and Green, 1968). By contrast, in "steel"

(*Sl*) the melanoblasts are normal but the skin environment into which they migrate is abnormal in some manner and blocks their further differentiation (Mayer 1975; Mayer and Green, 1968). In some of those mutants in which the lesion affects NC cell migration and proliferation directly, as in *s* and W^v, the pleiotropic effects include structures believed to be NC-derived: e.g. the incidence of megacolon in "piebald" (Bielschowsky and Schofield, 1962) and in "lethal spotting" (*ls*) (Mayer and Maltby, 1964), is coincident with a reduction or absence of the intestinal ganglia which, as discussed, have been demonstrated to be NC-derived in birds (Le Douarin and Teillet, 1974) and in mammals are assumed to have a similar origin. Thus those "white-spotting" mutants in which the loci affected operate at the NC cell and stem melanoblast level, apparently have other tissues of NC derivation affected. It has been proposed that the primary effect of mutations at these loci is the NC cell itself (Searle, 1968) and a list of putative 'neural crest' mutants in the mouse has been assembled by Weston (1970).

In spite of the considerable possibilities of genetic dissection of NC cell behaviour and differentiation using this range of material (Weston, 1970, 1972), to date only the development of coat colour has been analyzed in this fashion. Possibly this is a reflection of the fact that melanocytes are the most firmly established NC-derived cell type in mammals. Some of the interactions involved in bird NC cell differentiation have been discussed earlier, particularly those involved in the differentiation of mesenchymal tissues. Irrespective of the precise range of mammalian NC derivatives, it is highly probable that tissue interactions of various types will prove to play a causal role in NC differentiation. By using recombinations of mutant and wild-type tissues, either *in vivo* or *in vitro*, genetic dissection of mammalian NC cell migration, proliferation and differentiation becomes a real and exciting possibility.

THE QUESTION OF PLURIPOTENTIALITY

A number of ostensibly unrelated observations indicate that cells within differentiating or differentiated tissues of NC derivation are able, under certain conditions, to yield a progeny of a cell type other than the one characterizing the tissue in question. Neurogenic cells in embryonic sensory and autonomic ganglia can express a melanogenic potential when explanted *in vitro* (Cowell and Weston, 1970), although this ability is lost later in development.

During fracture repair of mesectoderm-derived membrane bones in the chick (Hall and Jacobsen, 1975) and in mammals (Girgis and Pritchard, 1958), periosteal cells produce a chondrogenic progeny seen in the callus. It has recently been proposed that the phenomenon of 'paradoxical regeneration' of the amphibian limb is due to Schwann cells undergoing metaplasia to produce the range of mesenchymal tissues necessary to rebuild a limb (Wallace, 1972; Maden, 1977). Some classes of tumour provide circumstantial evidence for potential or actual pluripotentiality by cells within a NC-derived tissue. A so-called 'neural crest tumour' has been described and defined on the basis of the unusual combination of neural and mesenchymal cells contained within it (Saxén and Saxén, 1960). The juxtaposition of neurilemma, melanoma, sarcomatous and ganglion cells together with cartilage and bone, a range of cell types exemplifying crest derivatives, led these authors to suggest that the stem cell of the initial primary tumour was NC-derived. Meningiomas contain a range of tissue types similar to that characterizing the 'pluripotential' range of the crest (Ball *et al.*,1975) and although the origin of the meninges was a controversial issue for some years, there is now thought to be a large NC contribution to the pia and arachnoid mater (Weston, 1970).

Can one resolve these phenomena with either the concept of pluripotentiality or that of metaplasia? It is important to remember that both are strictly 'single cell' concepts, although as such they are frequently abused and applied indiscriminately to tissues and populations of cells. None of the systems summarised above has been analysed at the single cell level, so one can only speculate. Do differentiated tissues of NC derivation contain within them cells which are as pluripotential as the NC? If so, could such cells react to a variety of perturbations by expressing some or all of the potential intrinsic to the embryonic crest? Alternatively, is a single differentiated cell in an NC-derived tissue able to respond to a particular perturbation by 'de-differentiating' to a stage of potentiality which is analogous to that of a crest cell, and can it's progeny then 're-express' some or all of that pluripotentiality? Until these systems have been analysed in a more rigorous fashion at the single cell level, these questions cannot be resolved. This rationale can be pushed further back in developmental time to the NC itself. It is usually assumed that the NC is a population of equally pluripotential cells. There is only limited evidence to support this claim and at the present time the possibility that heterogeneity exists amongst the NC cells cannot be discounted. We have already discussed this possibility in relation to the chondrogenic potential of the cranial NC. Viewed in this way the crucial test of pluripotentiality in this system is expressed by the question *'is a single NC cell capable of giving rise to the full range of cell types which characterises the generally accepted definition of NC pluripotentiality?'* Instead the NC population may consist of a collection of covert sub-populations, e.g. some mesenchymal, some melanogenic/neurogenic, and others endocrine: all those groups might be present in the crest at any particular level along the anteroposterior axis. Adjacent crest cells might commence migration simultaneously but the precise fate of cells, within the range of potentiality of each sub-population, could then be determined secondarily by environment. Although the generally held idea of a population of equally pluri-potential cells is a most attractive interpretation, this second explanation, based on limited heterogeneity amongst NC cells, is not totally incompatible with existing evidence. If all sub-populations were represented at any level along the anteroposterior axis, such heterogeneity would not be revealed by heterotopic grafting.

Evidently the need for a clonal approach, whether *in vitro* or *in vivo*, is of paramount importance in examining the crest itself or the differentiation of its progeny. Some systems in the chick embryo have been investigated in this way (e.g. Cohen and Konigsberg, 1975; Newsome, 1976). It is worth stressing that the adoption of a clonal approach in parallel with other methods is essential in future investigations of the differentiation of NC derivatives in mammals (see West, this volume).

MANIPULATION OF EMBRYOS DEVELOPING IN VITRO

Techniques currently available enable rat embryos to be cultured for various periods of time from the stage immediately prior to the onset of neural crest cell migration. Rat embryos utilise the yolk sac placenta in culture; since the chorio-allantoic placenta does not begin to function until the 17-somite stage, it has been possible to perfect the culture technique during the late-presomite to 16-somite period (32 hours), and to reproduce highly favourable conditions for up to 48 hours (New *et al.*,1976a). Less perfect but still useful development can be continued up to 72-95 hours (New *et al.*,1976b). These time periods can be translated in terms of neural crest studies as follows: during the period of migration of cranial and cervical NC cells along their major pathways, normal development can be observed *in vitro;* during the period of tissue interactions and early differentiation of some NC cells (e.g. neuroblasts), satisfactory though imperfect development occurs. Early in this second, less satisfactory stage of development in culture, or late in the first stage, embryonic development has reached the stage where tissue interactions may be studied using existing techniques of organ culture.

The availability of these culture techniques suggests at least three possibilities for studying mammalian neural crest migration and differentiation experimentally:

(i) Chemical intervention: the use of teratogens to study their effects on cell migration, tissue interaction, and differentiation, as described for vitamin A.

(ii) Labelling with ^{3}H-thymidine without surgical intervention, as described previously (Johnston and Krames, section 1). Vital stains might be used in a similar way, though preliminary attempts have not been successful due to spreading of the dye beyond the area to which it was initially applied (Morriss and Steele, unpublished). More permanent localization of the dye may be possible by modifying the technique of application.

(iii) Microsurgical intervention equivalent to that described in section 2 of this paper in avian embryos. This requires two essential tools: a) successful microsurgical techniques; b) a marker which is not diluted, i.e. an equivalent to the chick-quail nucleolar system of Le Douarin (1973).

There is some evidence that microsurgical techniques can be used in day 9 embryos prior to culture. Deuchar (1969), using the older culture technique (New, 1966) transected the axis of day 9 and day 10 embryos at various levels in order to identify the point of initiation of "turning" whereby embryos change their direction of curvature from dorsal-concave to ventral-concave. She found that cuts in day 9 embryos healed well, but that at day 10, possibly due to the presence of the yolk sac around the embryo at this stage, many of the embryos did not survive. Clearly damage to the membranes must be avoided, which means that even at day 9, operations must be carried out from a ventral approach instead of the dorsal approach used in avian embryos. Deuchar's (1969) demonstration that a single cut made in this way will heal, suggests that grafting a whole piece (e.g. of a ^{3}H-thymidine-labelled donor) would be possible with an acceptable rate of success.

The use of a more permanent marker requires that interspecific grafts would be successful. Relative contributions of the two species might subsequently be identified by immunofluorescence and immunoperoxidase techniques in sectioned material. Gardner and Johnson (1973) have used this approach to analyse the relative contributions of rat and mouse cells to primitive streak stage embryos formed after reimplantation of a mouse blastocyst to which rat inner cell mass cells had been added. However, such experiments would be limited by the presently available timescale of the culture technique (see also discussion by West, this volume).

CONCLUSION

At present little direct experimental analysis of mammalian neural crest migration and differentiation has been undertaken. However, as we have shown in this section, many of the techniques for doing so are already available, and others await adventurous and patient exploration. The fruits of this research will be a much clearer understanding of the aetiology of human congenital anomalies involving neural crest. Furthermore, it will be a source of satisfaction and delight to elucidate the mechanisms of organogenesis in mammalian embryos through direct discovery, instead of being confined largely to the somewhat "secondhand" (but nevertheless invaluable) insights derived from studies of non-mammalian vertebrates.

ACKNOWLEDGEMENTS

We wish to thank Beth Crutch and Martin Barker for assistance, Brian Archer and John Heywood for photography, and Chris Graham, John McAvoy, Ginny Papaioannou and Andy Copp for their valuable criticisms of the manuscript at various stages during its preparation. GMM is in receipt of a project grant from the Medical Research Council.

REFERENCES

ABERCROMBIE, M. & HEAYSMAN, J. (1954) Observations on the social behaviour of cells in tissue culture. II "Monolayering" of fibroblasts. Expl. Cell Res. 6, 293-306.

BALINSKY, B.J. (1947) Korrelation in der Entwicklung der Mund- und Kiemenregion und des Darmkanals bei Amphibien. Wilhelm Roux Arch. Entw-Mech. Org. 143, 253-258.

BALL, J., COOK, T.A., LYNCH, P.G. & TIMBERLEY, W.R. (1975) Mixed mesenchymal differentiation in meningiomas. J. Path. 116, 253-258.

BANCROFT, M. & BELLAIRS, R. (1976) The neural crest cells of the trunk region of the chick embryo studied by SEM and TEM. Zoon. 4, 73-85.

BARTELMEZ, G.W. (1960) Neural crest in the forebrain of mammals. Anat. Rec. 138, 269-281.

BIELSCHOWSKY, M. & SCHOFIELD, G.C. (1962) Studies on megacolon in piebald mice. Aust. J. exp. Biol. med. Sci. 40, 395-404.

CHIAKULAS, J.J. (1957) The specificity and differential fusion of cartilage derived from mesendoderm and mesectoderm. J. exp. Zool. 136, 287-300.

CHIBON, P. (1967) Marquage nucléaire par la thymidine tritiée des dérivés de la crête neurale chez l'Amphibien Urodèle *Pleurodeles waltii* Michah. J. Embryol. exp. Morph. 18, 343-358.

COHEN, A.M. (1972) Factors directing the expression of sympathetic nerve traits in cells of neural crest origin. J. exp. Zool. 179, 167-182.

COHEN, A.M. & KONIGSBERG, I.R. (1975) A clonal approach to the problem of neural crest determination. Devl. Biol. 22, 670-697.

COWELL, I. & WESTON, J. (1970) Analysis of melanogenesis in cultured chick embryo spinal ganglia. Devl. Biol. 22, 670-697.

DE BEER, G.R. (1947) The differentiation of neural crest cells into visceral cartilage and odontoblasts in Ambystoma and a re-examination of the germ-layer theory. Proc. Roy. Soc. London B 134, 377-398.

DE LUCA, L. & YUSPA, S.H. (1974) Altered glycoprotein synthesis in mouse epidermal cells treated with retinyl acetate *in vitro*. Expl. Cell Res. 86, 106-110.

DEUCHAR, E.M. (1969) Effects of transecting early rat embryos on axial movements and differentiation in culture. Acta Embryol. Exp. 157-167.

DANIEL, M.R., DINGLE, J.T., GLAUERT, A.M. & LUCY, J.A. (1966) The action of excess vitamin A alcohol on the fine structure of rat dermal fibroblasts. J. Cell Biol. 30, 465-475.

DUNN, G.A. & HEATH, J.P. (1976) A new hypothesis of contact guidance in tissue cells. Expl. Cell Res. 101, 1-14.

EBENDAL, T. (1976) The relative roles of contact inhibition and contact guidance in orientation of axons extending on aligned collagen fibrils in vitro. Expl. Cell Res. 98, 159-169.

EBENDAL, T. (1977) Extracellular matrix fibrils and cell contacts in the chick embryo. Cell Tiss. Res. 175, 439-458.

ELSDALE, T.R. & BARD, J.B.L. (1972) Collagen substrata for studies in cell behavior. J. Cell Biol. 54, 626-637.

EPPERLEIN, H.H. (1974) The ectomesenchymal-endodermal interaction-system (EEIS) of *Triturus alpestris* in tissue culture. I. Observations on attachment, migration and differentiation of neural crest cells. Differentiation 2, 151-168.

EPPERLEIN, H.H. & LEHMANN, R. (1975) The ectomesenchymal-endodermal interaction-system (EEIS) of Triturus alpestris in tissue culture. 2. Observations on the differentiation of visceral cartilage. Differentiation 4, 159-174.

GALBRAITH, D.B. & KOLLAR, E.J. (1976) Procollagen enhancement of mouse tooth-germ development in vitro. Am. Zool. 16, 183 (abstract).

GARDNER, R.L. & JOHNSON, M.H. (1973) Investigation of early mammalian development using interspecific chimaeras between rat and mouse. Nature New Biology 246, 86-89.

GIRGIS, F.G. & PRITCHARD, J.J. (1958) Experimental production of cartilage during the repair of fractures of the skull vault in rats. J. Bone Jt. Surg. 40B, 274-281.

GOODRICH, E.S. (1958) Studies on the structure and development of vertebrates. Vol. I. Dover. N.Y.

GREENBERG, J.H. & PRATT, R.M. (1977) Glycosaminoglycan and glycoprotein synthesis by cranial neural crest cells *in vitro*. Cell Differ. 6, 119-132.

GURR, E. (1973) Biological Staining Methods, Searle Diagnostic, England.

HALL, B.K. & JACOBSON, H.N. (1975) The repair of fractured membrane bones in the newly hatched chick. Anat. Rec. 181, 55-69.

HAMILTON, W.J. & MOSSMAN, H.W. (1972) Human Embryology. Heffers, Cambridge.

HAY, E.D. (1973) Origin and role of collagen in the embryo. Am. Zool. 13, 1085-1107.

HOLTFRETER, J. (1968) Mesenchyme and epithelia in inductive and morphogenetic process. In: P. Fleischmajer and R.E. Billingham (Eds.) Epithelio-Mesenchymal Interactions, Williams and Wilkins Co. Baltimore, pp. 1-30.

HÖRSTADIUS, S. (1950) The Neural Crest. Oxford University Press.

HÖRSTADIUS, S. & SELLMAN, S. (1946) Experimentelle Untersuchungen uber die Determination des Knorpeligen Kopfskelettes bei Urodelen. Nova Acta R. Soc. Scient. Upsal. Ser. IV. 13, 1-167.

JOHNSTON, M.C. (1966) A radioautographic study of the migration and fate of cranial neural crest cells in the chick embryo. Anat. Rec. 156, 143-156.

JOHNSTON, M.C., BHAKDINARONK, A. & REID, Y.C. (1973) An expanded role of the neural crest in oral and pharyngeal development. In: J.F. Bosma, (Ed.), Oral Sensation and Perception, DHEN Publication No.(NIH) 73-546, pp. 31-52.

JOHNSTON, M.C. & LISTGARTEN, M.A. (1972) Observations on the migration, interaction, and early differentiation of orofacial tissues. In: H.C. Slavkin and L.A. Bavetta (Eds.), Academic Press, N.Y. pp. 53-80.

JOHNSTON, M.C., MORRISS, G.M., KUSHNER, D.C. & BINGLE, G.J. (1977) Abnormal organogenesis of facial structures. In: J. G. Wilson and F.C. Fraser (Eds.), Handbook of Teratology, Vol. 2, 421-451.

KALLEN, B. (1953) Notes on the development of the neural crest in the head of *Mus musculus*. J. Embryol. exp. Morph., 1, 393-398.

KOCH, W.E. (1972) Tissue interaction during *in vitro* odontogenesis. In: H.C. Slavkin and L.A. Bavetta (Eds.), Developmental Aspects of Oral Biology, Academic Press, pp. 151-164.

KOLLAR, E.J. (1972) Histogenetic aspects of dermal - epidermal interactions. In: H.C. Slavkin and L.A. Bavetta (Eds.), Developmental Aspects of Oral Biology, Academic Press, pp. 125-149.

KOLLAR, E.J. & BAIRD, G. (1970) Tissue interactions in embryonic mouse tooth germs. II. The inductive role of the dental papilla. J. Embryol. exp. Morph. 24, 173-186.

KWASIGROCH, T.E. & KOCHHAR, D.M. (1976) Locomotory behaviour of limb bud cells. Effects of excess vitamin A in vivo and in vitro. Expl. Cell Res. 95, 269-278.

LE DOUARIN, N.M. (1973) A feulgen-positive nucleolus. Expl. Cell Res. 77, 459-468.

LE DOUARIN, N.M. (1975) The neural crest in the neck and other parts of the body. In: D. Bergsma, A.R. (Ed.) Morphogenesis and Malformation of Face and Brain Birth Defects XI no 7, Liss Inc. pp. 19-50.

LE DOUARIN, N.M. (1976) Cell migration in early vertebrate development studied in interspecific chimaeras. In: Ciba Fnd. Symp. 40, Embryogenesis in Mammals, pp. 71-97.

LE DOUARIN, N.M. & TEILLET, M-A. (1973) The migration of neural crest cells to the wall of the digestive tract in avian embryo. J. Embryol. exp. Morph. 30, 31-48.

LE DOUARIN, N.M. & TEILLET, M-A. (1974) Experimental analysis of the migration and differentiation of neuroblasts of the autonomic nervous system and of neuroectodermal mesenchymal derivatives, using a biological cell marking technique. Devl. Biol. 41, 162-184.

LE LIEVRE, C. (1974) Rôle de cellules mésectodermiques issués des crêtes neurales céphaliques dans la formation des arcs brachiaux et du squelette viscéral. J. Embryol. exp. Morph., 31, 453-477.

LE LIEVRE, C.S. & LE DOUARIN, N.M. (1975) Mesenchymal derivatives of the neural crest: analysis of chimaeric quail and chick embryos. J. Embryol. Exp. Morph. 34, 125-154.

LEWIS, J.H. & WOLPERT, L. (1976) The principle of non-equivalence in development. J. Theor. Biol. 62, 479-490.

LÖFBERG, J. (1976) Scanning and transmission electron microscopy of early neural crest migration and extracellular fiber systems of the amphibian embryo. J. Ultrastruct. Res. 54, 484a.

LÖFBERG, J. (1978) Microarchitecture of contacts between ECM and motile cells in amphibian embryos. Differentiation (In press).

MADEN, M. (1977) The role of Schwann cells in paradoxical regeneration in the axolotl. J. Embryol. exp. Morph. 41, 1-13.

MANGOLD, O. (1936) Experimente zur Analyse der Zusammenarbeit der Keimblatter. Naturwissenschaften 24, 753-760.

MARCHASE, R.B., VOSBECK, K. & ROTH, S. (1976) Intercellular adhesive specificity. Biochim. biophys. Acta 457, 385-416.

MAYER, T.C. (1965) The development of piebald spotting in mice. Devl. Biol. 11, 421-435.

MAYER, T.C. (1970) A comparison of pigment cell development in Albino, Steel, and Dominant-spotting mutant mouse embryos. Devl. Biol. 23, 297-309.

MAYER, T.C. (1975) Tissue environmental influences on the development of melanoblasts in steel mice. In: H.C. Slavkin and R.G. Greulich, (Eds.) Extracellular Matrix Influences on Gene Expression. Academic Press, pp. 555-560.

MAYER, T.C. (1977) Enhancement of melanocyte development from piebald neural crest by a favourable tissue environment. Devl. Biol. 56, 255-262.

MAYER, T.C. & GREEN, M.C. (1968) An experimental analysis of the pigment defect caused by mutations at the *W* and *sl* loci in mice. Devl. Biol. 18, 62-75.

MAYER, T.C. & MALTBY, E.L. (1964) An experimental investigation of pattern development in lethal spotting and belted mouse embryos. Devl. Biol. 9, 269-286.

MEIER, S. & HAY, E.D. (1975) Stimulation of corneal differentiation by interaction between cell surface and extracellular matrix. J. Cell Biol. 66, 275-291.

MILAIRE, J. (1959) Predifferentiation cytochemique des divers ébauches céphaliques chez l'embryon de souris. Arch. Biol. (Liège) 70, 588-724.

MORRISS, G.M. (1972) Morphogenesis of the malformations induced in rat embryos by maternal hypervitaminosis A. J. Anat. 113, 241-250.

MORRISS, G.M. (1973) The ultrastructural effects of excess maternal vitamin A on the primitive streak stage rat embryo. J. Embryol. exp. Morph. 30, 219-242.

MORRISS, G.M. (1975) Abnormal cell migration as a possible factor in the genesis of vitamin A-induced craniofacial anomalies. In: D. Neubert (Ed.) New Approaches to the Evaluation of Abnormal Mammalian Embryonic Development. Geo. Thieme Verlag. Berlin, pp. 678-687.

MORRISS, G.M. & SOLURSH, M. (1978a) The role of primary mesenchyme in normal and abnormal morphogenesis of mammalian neural folds. Differentiation (In press).

MORRISS, G.M. & SOLURSH, M. (1978b) Regional differences in mesenchymal cell morphology and glycosaminoglycans in early neural fold stage rat embryos. J. Embryol. exp. Morph. (In press).

MORRISS, G.M. & STEELE, C.E. (1977) Comparison of the effects of retinol and retinoic acid on postimplantation rat embryos in vitro. Teratology 15, 109-120.

MURRAY, P.D.F. & SMILES, M. (1965) Factors in the evocation of adventitious (secondary) cartilage in the chick embryo. Aust. J. Zool. 13, 351-381.

NEW, D.A.T. (1966) Development of rat embryos cultured in blood sera. J. Reprod. Fert. 12, 509-524.

NEW, D.A.T., COPPOLA, P.T. & COCKCROFT, D.L. (1976a) Comparison of growth in vitro and in vivo of postimplantation rat embryos. J. Embryol. exp. Morph. 36, 133-144.

NEW, D.A.T., COPPOLA, P.T. & COCKCROFT, D.L. (1976b) Improved development of head-fold rat embryos in culture resulting from low oxygen and modifications of the culture serum. J. Reprod. Fert. 48, 219-222.

NEWSOME, D.A. (1972) Cartilage induction by retinal pigmented epithelium of chick embryo. Devl. Biol. 27, 575-579.

NEWSOME, D.A. (1975) Collagen synthesis in cultured neural crest cells, their derivatives and retinal pigmented epithelium: stimulation of $(\alpha I)_3$ collagen production. In: H.C. Slavkin and R.G. Greulich (Eds.) Extracellular Matrix Effects on Gene Expression. Academic Press, pp. 601-607.

NEWSOME, D.A. (1976) In vitro stimulation of cartilage in embryonic chick neural crest cells by products of retinal pigmented epithelium. Devl. Biol. 49, 496-507.

NODEN, D.M. (1973) The migratory behaviour of neural crest cells. In: J.F. Bosma (Ed.) Oral Sensation and Perception. DHEW Publication No.(NIH) 73-546, pp. 9-36.

NODEN, D.M. (1975) An analysis of the migratory behaviour of avian cephalic neural crest cells. Devl. Biol. 42, 106-130.

OKADA, E.W. (1955) Isolationsversuche zur Analyse der Knorpelbildung aus Neurelleistenzellen bei Urodelekein. Men. Coll. Sci. Univ. Kyoto. B22, 23-28.

POSWILLO, D. (1975) The pathogenesis of the Treacher-Collins syndrome (Mandibulofacial dysostosis). Brit. J. oral Surg. 13, 1-26.

PRATT, R.M., LARSEN, M.A. & JOHNSTON, M.C. (1975) Migration of cranial neural crest cells in a cell-free hyaluronate-rich matrix. Devl. Biol. 44, 298-305.

PRATT, R.M., MORRISS, G.M. & JOHNSTON, M.C. (1976) The source, distribution, and possible role of hyaluronate in the migration of chick cranial neural crest cells. J. gen. Physiol. 68, 15a-16a.

PUGIN, M.E. (1973) Sur le comportement des troncons du tube neural et de la corde d'embryon de souris greffés à la place des organes homologues chez l'embryon de poulet. C.R. Acad. Sci. (D) Paris, 276, 3477-3480.

RAVEN, C.P. (1931) Zur Entwicklung der Ganglienleiste. I. Die kinematik der Ganglienleistenentwicklung bei den Urodelen. Roux Arch. Entw. mech. 125, 21-292.

RAWLES, M.E. (1947) Origin of pigment cells from the neural crest in the mouse embryo. Physiol. Zool. 20, 248-270.

REES, D.A., LLOYD, C.W. & THOM, D. (1977) Control of grip and stick in cell adhesion through lateral relationships of membrane glycoproteins. Nature 267, 124-128.

REINBOLD, R. (1968) Rôle du tapetum dans la différentiation de la scléretique chez l'embryon de poulet. J. Embr. exp. Morph. 19, 43-47.

ROMANOFF, A.L. (1960) 'The Avian Embryo'. Macmillan - New York.

SÁXEN, L. & SÁXEN, E. (1960) Malignant embryoma of the Neural Crest. Cancer 13, 899-906.

SCHOWING, J. (1968) Mise en évidence du rôle inducteur de l'encephale dans l'osteogenese du crane embryonnaire du poulet. J. Embryol. exp. Morph. 19, 88-93.

SEARLE, A.G. (1968) Comparative Genetics of Coat Colour in Mammals. Logos Academic Press.

SELLMAN, S. (1946) Some experiments on the determination of the larval teeth in Amblyostoma mexicanum. Odont. T. 54, 1-128.

SHUR, B.D. & ROTH, S. (1975) Cell surface glycosyltransferases. Biochim. biophys. Acta 415, 473-512.

SLAVKIN, H.C. (1972) Intercellular communication during odontogenesis. In: H.C. Slavkin and L.A. Bavetta (Eds.) Developmental Aspects of Oral Biology, Academic Press, pp. 165-199.

SMITH, N.G. Jr., LINSENMAYER, T.F. & NEWSOME, D.A. (1976) Synthesis of type II collagen *in vitro* by embryonic chick neural retina tissue. Proc. Natn. Acad. Sci. USA, 73, 4420-4423.

STEFFEK, A.J., MUJWID, D.K. & JOHNSTON, M.C. (1975) Scanning electron microscopy (SEM) of cranial neural crest migration. J. dent. Res. 54, (Suppl. A), 165.

THESLEFF, I. (1977) Tissue interactions in tooth development in vitro. In: M. Karkinen-Jaaskelainen, L. Saxen, and L. Weiss (Eds.) Cell Interactions in Differentiation, Academic Press, pp. 191-207.

TOOLE, B.P. (1973) Hyaluronate and hyaluronidase in morphogenesis and differentiation. Am. Zool. 13, 1061-1065.

TOOLE, B.P. (1976) Morphogenetic role of glycosaminoglycans, (acid mucopolysaccharides) in brain and other tissues. In: S. Barondes (Ed.), Neuronal Recognition,Chapman and Hall, London, pp. 275-329.

TYLER, M.S. & HALL, B.K. (1977) Epithelial influences on skeletogenesis in the mandible of the embryonic chick. Anat. Rec. 188, 229-239.

VON DER MARK, H., VON DER MARK, M. & GAY, S. (1976) Study of differential collagen synthesis during development of the chick embryo by immunofluorescence. 1. Preparation of collagen Type I and Type II specific antibodies and their application to early stages of the chick embryo. Devl. Biol. 48, 237-249.

WALLACE, M. (1972) The components of regrowing nerves which support the regeneration of irradiated salamander limbs. J. Embryol. exp. Morph. 28, 419-435.

WEISS, P. (1958) Cell Contact. Int. Rev. Cytol. 7, 391-423.

WESSELLS, N.K. (1977) Tissue Interactions and Development. W.A. Benjamin, Inc.

WESTON, J.A. (1963) A radioautographic analysis of the migration and localization of trunk neural crest cells in the chick. Devl. Biol. 6, 279-310.

WESTON, J.A. (1970) The migration and differentiation of neural crest cells. Adv. Morphogen. 8, 41-114.

WESTON, J.A. (1972) Cell interaction in neural crest development. In: L.G. Silvestri (Ed.) Cell Interactions, 3rd Lepetit Colloquium, North Holland, pp. 286-292.

WESTON, J.A. & BUTLER, S.L. (1966) Temporal factors affecting localization of neural crest cells in the chicken embryo. Devl. Biol. 14, 246-266.

WESTON, J.A. & ROTH, S.A. (1969) Contact inhibition: behavioural manifestations of cellular adhesive properties *in vitro*. In: R.T. Smith and R.A. Good (Eds.) Cellular Recognition, Appleton-Century-Crofts, N.Y. pp. 29-37.

ANALYSIS OF CLONAL GROWTH USING CHIMAERAS AND MOSAICS

John D. West

Department of Zoology
University of Oxford
South Parks Road
Oxford OX1 3PS

In the strictest sense, almost any individual is a clone of cells descended from a single cell, the zygote. However, in this chapter we will be concerned with clones that form during development and subdivide each tissue of the organism according to cell lineages. The progeny of any mitotically active cell are clonally related and so belong to the same cell lineage. Early in development a cell will produce a large clone comprising many cells. This will be subdivided into smaller clones by later cell divisions. The extent of cell mixing will determine whether the smaller sub-clones become spatially isolated or whether a larger clone can be identified in the adult as a single group of cells.

The use of genetic cell markers has enabled us to distinguish different populations of cells in a single individual. Where there is little cell mixing during development, cell markers have enabled direct mapping of large clones *in situ*. This has been particularly successful in insects (for example, Stern, 1940; Schneiderman and Bryant, 1971; Garcia-Bellido *et al.*,1973; Lawrence, 1973). Ideally these large clones seen in adult tissues represent the descendant clones initiated at tissue foundation. The time of tissue foundation is defined as the time at which cell mixing between lineages in different tissue primordia becomes minimal. This definition is identical to that defined by McLaren (1976) as the time of "allocation". The construction of mammalian fate maps from adult patterns has proved more difficult than their insect counterparts because there appears to be more cell mixing during mammalian histogenesis. The distribution of cells within a single descendant clone is also affected by cell selection and cell death during development. Thus, a descendant clone will only be seen as a coherent group of cells if there has been no or little mixing of the cells after the initiation of the descendant clone. This may be the situation in insects but is unlikely to be true of mammalian histogenesis. If cell mixing occurs for a few cell generations after tissue foundation and then stops, the number of descendant clones initiated at tissue foundation will be overestimated from the distribution of two marked cell populations seen in the adult tissue. If cell mixing is more extensive and continues until quite late in development the cells of descendant clones initiated at tissue foundation may become dispersed into much smaller isolated groups. However, if these smaller groups of cells remain relatively close together, it may still be possible to estimate the number of descendant clones initiated at tissue foundation from the adult pattern.

It is convenient to distinguish between "coherent clones" and "descendant clones". A coherent clone is that defined by Nesbitt (1974) as "a group of clonally related cells which have remained contiguous throughout the history of the embryo". A coherent clone, therefore, reflects the spatial relationship of clonally related cells in the tissue at the time considered and its size is a function of the degree of cell mixing. Several coherent clones may become adjacent during development and will be seen as a patch. Nesbitt (1974) has defined a patch as "a group of cells of like genotype which are contiguous at the time of consideration". A patch is the unit that can be directly measured in chimaeras and mosaics and, as we will see later, its size is a function of the proportions of the two cell populations in the tissue.

A descendant clone is any group of clonally related cells irrespective of whether they have remained contiguous throughout development. Thus, all the cells of a coherent clone belong to the same descendant clone but the cells of a descendant clone will be separated into different coherent clones unless there is little or no cell mixing during development. Descendant clones reflect the cell lineage relationships within a differentiated tissue, whereas coherent clones indicate the spatial relationships, at the time considered, between cells within a descendant clone. While the size of a coherent clone is dependent on the amount of cell mixing during growth, the size of a descendant clone is dependent only on the number of surviving mitotic progeny. We are normally interested in determining the number of descendant clones initiated at tissue foundation, but since any mitotically active cell initiates a descendant clone, it is important to understand the stage at which the descendant clones we observe were initiated. This problem has been dealt with in considerable detail by McLaren (1972) and Lewis *et al.* (1972), and will be considered further in a later section.

MARKER SYSTEMS

To study clonal growth, we require a system for distinguishing clonally related groups of cells within a single individual. Thus, we need both a means of obtaining two populations of cells early in development and a means of identifying them at later stages. The perfect marker would be stably inherited during cell proliferation, cell autonomous and detectable at the cellular level in all tissues. Ideally such a marker would be used to label a single cell, during development, in such a way that all the descendants of that cell could later be identified. Genetically inherited variants are stable and so fulfil the first of these requirements but few are both detectable at the cellular level, *in situ*, and have a wide tissue distribution. Partly for this reason, it is not yet possible to trace all the descendants of a single cell, although a single cell can be marked early in mouse development by irradiating with X-rays to produce a clone of mutant cells (Russell and Major, 1957) or by mechanically incorporating a single genetically marked cell into a developing mouse embryo (Gardner and Lyon, 1971). However, in most cases several cells are marked simultaneously. Most commonly the two cell populations are obtained by using one of two basic approaches. X-inactivation mosaics, which are functional genetic mosaics and have two populations of cells that express the maternally derived X chromosome or the paternally derived X chromosome respectively, may be used (Lyon, 1961). Alternatively chimaeras may be produced (reviewed by McLaren, 1976) either by aggregating two early embryos (Tarkowski, 1961; Mintz, 1962) or by injecting donor cells into the blastocoel cavity of a host blastocyst (Gardner, 1968). The major differences between X-inactivation mosaics and aggregation chimaeras for producing two marked cell populations are listed in Table 1. The value of the X-inactivation mosaic lies with the ease with which mosaicism can be achieved, simply by producing females heterozygous for the required X-linked genes. The limitation of using X-inactivation mosaics is the small number of suitable X-linked markers at present available in laboratory mammals.

Apart from the experimentally produced chimaeras and the functional mosaicism of X-inactivation mosaics, two cell populations may arise spontaneously in a number of different ways. For example, chromosome mosaics, where the two cell populations differ in karyotype may arise by non-disjunction. Strictly speaking, individuals such as these are mosaics if the two cell populations arise from a single zygote and chimaeras if they arise from more than one zygote (Anderson *et al.*, 1951; Ford, 1969). This distinction between mosaics and chimaeras is widely used by zoologists although botanists normally use the term chimaera to describe plants comprising two cell populations regardless of whether the two populations were derived from the same or different zygotes. Several alternative adjectives have been used to describe chimaeric mice including allophenic, tetraparental, quadriparental and even mosaic, but "chimaeric" has precedence (Tarkowski, 1961).

TABLE 1: COMPARISON OF AGGREGATION CHIMAERAS AND X-INACTIVATION MOSAICS

	Aggregation Chimaeras	X-inactivation mosaics
Number of zygotes	2 (or more)	1
Time of 'marking' of different cell populations	By aggregation 8 + 8 cells (2½ days p.c.* in the mouse)	By X chromosome inactivation** (probably soon after 3½ days p.c. in the mouse)
Spatial relationship between cell populations at time of cell marking	Non-random	Probably random
Maximum genetic difference possible between cell populations	Whole genome	X chromosome
Difference in developmental age between cell populations	Possible	None

* p.c. = post coitum

** The event resulting in two cell populations in X-inactivation mosaics could either be X chromosome inactivation or X chromosome activation.

GENETIC MARKERS

A large number of different genetic markers have been used to examine the distribution of the two cell populations in mammalian chimaeras and X-inactivation mosaics. No attempt will be made to review exhaustively every available marker but many of these are listed in Table 5 and further details are given in the reviews by Mintz (1970a, 1971, 1974), Nesbitt and Gartler (1971) and McLaren (1976). We have already mentioned that the most useful type of genetic marker is one that is cell autonomous and detectable at the cellular level in a wide range of tissues. Unfortunately few markers even approach this ideal, but the most useful markers are those that can be used to visualise directly the different cell populations *in situ*. Chromosome markers and electrophoretic variants of certain enzymes have been widely used to study chimaeras and mosaics. Electrophoretic variants of glucose-6-phosphate dehydrogenase have been particularly useful in the analysis of human X-inactivation mosaics (reviewed by Gartler, 1974; Fialkow, 1976). The recent discovery of an electrophoretic variant of phosphoglycerate kinase in the mouse (Nielsen and Chapman, 1977) will allow a similar analysis of mouse X-inactivation mosaics. However, analysis, using either electrophoretic or chromosome markers, involves the disruption of the tissue and so is unsatisfactory for many types of study and can only be used for indirect analysis of clonal growth. Unless otherwise stated the markers discussed below have been used in mouse chimaeras to look at different cell populations *in situ*.

MORPHOLOGICAL AND HISTOLOGICAL MARKERS

The genes producing pigment cells of the coat and eye were the first genetic markers used to detect experimental chimaerism in mice (Tarkowski, 1964) and since then these systems have been extensively exploited to examine clonal growth. The pigmentation in the coat, inner ear and ciliary body, iris and choroid of the eye is produced by migratory melanocytes derived from the neural crest and at least in the coat the melanin granules are secreted extracellularly (Searle, 1968). The pigmented cells in the retinal pigment epithelium arise from the optic cups and the melanin granules remain *in situ* to provide an excellent cell marker. Variegated phenotypes comprising two cell populations are most clearly seen in mouse chimaeras and X-inactivation mosaics marked with variants at the albino locus. Such variegated pigment patterns may be obtained in mouse X-inactivation mosaics by using Cattanach's insertion which involves the insertion of part of chromosome 7 (including the albino locus) into one X-chromosome (Cattanach, 1961). Other coat colour and hair structure variants have been used in mouse chimaeras and X-inactivation mosaics, although in many cases it is not known whether the variegated phenotype directly reflects the underlying distribution of the two cell populations (McLaren, 1976).

A number of genetic variants producing degeneration of neurological tissues are available as potential cell markers. These include retinal degeneration, *rd*, (Tansley, 1954), Purkinje cell degeneration, *pcd*, and nervous, *nr*, (Mullen and LaVail, 1975; Mullen *et al.*, 1976) in the mouse and retinal dystrophy, *rdy*, (Mullen and LaVail, 1976) in the rat. All of these cause degeneration of the retinal photoreceptor cells and the mouse *rd* and *pcd* genes and the rat *rdy* gene have been incorporated into chimaeras (Mintz and Sanyal, 1970; Wegmann *et al.*,1971; Sanyal and Zeilmaker, 1974, 1976; LaVail and Mullen, 1976; Mullen and LaVail, 1976; West 1976b; Mullen 1977). The rat *rdy* gene appears to act via the retinal pigment epithelium and so is not a cell autonomous marker. The mouse *rd* gene is probably cell autonomous although the spatial organisation of the original two cell populations seems to be lost after the death of the *rd/rd* cells. It is not yet known whether the surviving +/+ cells simply move towards the retinal pigment epithelium to take the place of overlying dead *rd/rd* cells and so compact the remaining layer of photoreceptor cells or whether there is additional movement of cells around the circumference of the retina. Until the extent of cell movement is known, there must be reservations about using these various mutants as cell markers for clonal analysis in chimaeras.

The gene ichthyosis (*ic*) is a potentially more useful histological marker because it does not cause any alteration in the structure of the tissue. Ichthyosis causes abnormal nuclear morphology of leukocytes (Green *et al.*,1975) and its full potential as a cell marker has yet to be explored in other tissues. An elegant cell marker system has been devised for analysis of mouse X-inactivation mosaics by Drews and Alonso-Lozano (1974). This has been used to visualise mosaicism in epididymal cells of sex-reversed phenotypic males. The X-inactivation mosaics have two X chromosomes and are heterozygous for the *Tfm* allele, at the testicular feminisation locus, which causes a deficiency of androgen receptor protein. The androgen responsive tissue (in this case the epididymis) was produced by incorporating the sex reversal gene (*Sxr*) into the genome and thereby producing a phenotypic male with epididymides and two X chromosomes. The epididymal cells expressing the *Tfm*-bearing X chromosome failed to respond to androgens whereas the other cell population responded and appeared enlarged in histological sections.

Sex chromatin has been used as a marker in X-chromosome mosaics where the two cell populations differ in their number of X chromosomes (Klinger and Schwarzacher, 1962). However, sex chromatin is not a very reliable cell marker because other chromatin may interfere with the scoring of the presence or absence of sex chromatin. Thus, some cells in normal XY males will appear as sex chromatin positive and a proportion of the cells in normal XX females will appear to be sex chromatin negative. The reliability of this technique is dependent both on the species and the tissue analysed.

BIOCHEMICAL AND IMMUNOLOGICAL MARKERS

Biochemical and immunological markers offer the advantage of a wide tissue distribution although relatively few of these markers have yet been used to mark individual cells *in situ*. The difference in β-glucuronidase enzyme activity produced by the alleles Gus^h and Gus^b (or Gus^a) was first used by Wegmann (1970) to infer the presence of patches of two cell populations in the livers of chimaeric mice using a biochemical assay. These patches were later visualised histochemically (Condamine *et al.*,1970; West 1976c), as shown in Figure 1A, and Feder (1976) has now refined this technique and extended its use to other tissues. Histochemical techniques have also been used with activity variants of β-galactosidase to study nervous tissue in mouse chimaeras (Dewey *et al.*,1976). Other potential histochemical cell markers include glucose phosphate isomerase and ornithine carbamoyl transferase. The thermolabile variant of glucose phosphate isomerase ($Gpi\text{-}1^c$) recently discovered by Padua *et al.*(1978) will probably prove to be one of the most useful markers for studying clonal growth in mouse chimaeras, as this enzyme has a ubiquitous tissue distribution and a histochemical technique for the visualisation of enzyme activity in tissue sections is already available (Orchardson and McGadey, 1970). Patches of two cell populations have been seen in livers of human X-inactivation mosaics heterozygous for ornithine carbamoyl transferase deficiency (see Figure 1B and Ricciuti *et al.*,1976). An unstable form of this X-linked enzyme is available in laboratory stocks of mice (DeMars *et al.*, 1976) and should prove useful for examining patterns of mosaicism in the mouse liver.

The use of antigen differences coupled with immunofluorescent labelling has been successfully applied in the analysis of interspecific chimaeras comprising both rat and mouse cells (Gardner and Johnson, 1973) and in some cell types of intraspecific mouse chimaeras (Barnes *et al.*,1974; Dewey *et al.*,1976). Unfortunately technical difficulties have dis-

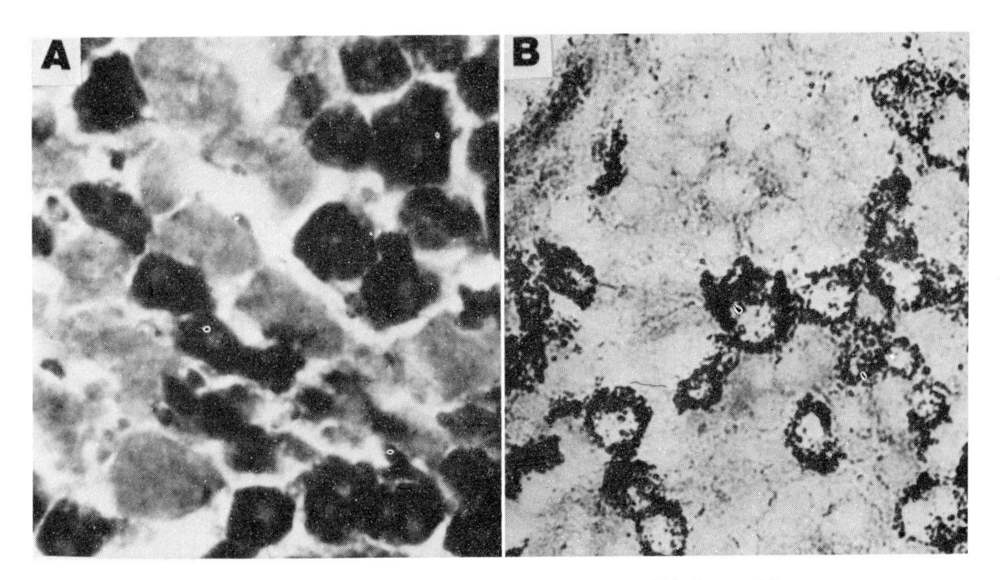

Figure 1: Histochemical cell markers in the liver. (A) Two cell populations in the liver of a chimaeric mouse revealed by histochemical staining for β-glucuronidase. The darkly staining hepatocytes have normal levels of this enzyme whereas the paler staining cells have the lower levels produced by the Gus^h variant. (B) Two cell populations in the liver of a human X-inactivation mosaic revealed by histochemical staining for ornithine carbamoyl transferase by Ricciuti *et al.*(1976). The darkly, staining cells express the X chromosome carrying the normal allele responsible for the synthesis of this enzyme whereas the paler coloured cells express the other X chromosome which carries an allele causing a deficiency of this enzyme. Figure 1B is reproduced by kind permission of the authors.

couraged the wider application of this promising approach to *in situ* labelling of different cell populations. These difficulties include the restricted tissue distribution of suitable antigens, the need for high titres of antibody and the rapid fading of immunofluorescent markers. However, it seems probable that alternative immunological labelling techniques will overcome some of the difficulties of the rather short-lived labelling provided by immunofluorescence.

Genetically determined effects, including effects on cell selection or differential cell adhesion, must be considered when using any genetic marker to study clonal growth. Such effects may be caused by the genetic marker itself or by other genetic differences between the two cell populations. Consistent differences in genetic background associated with a particular marker occur when chimaeras are made between two inbred strains of mice. These differences can be reduced by using X-inactivation mosaics or chimaeras made from congenic stocks, or may be randomised by constructing chimaeras from appropriately marked, randomly-bred animals.

PATTERNS OF CLONAL GROWTH

In most analyses of clonal growth, the distribution of the two cell populations is examined in the adult and attempts are made to interpret events that occurred during morphogenesis. This is obviously unsatisfactory since the arrangement of the two cell populations may be influenced by cell mixing, cell selection and cell death at any time during development. Despite the difficulties in assessing the relative contributions of each of these processes, by extrapolation from adult patterns some useful information has been obtained. We will consider two of the questions that have been posed regarding the development and growth of a particular tissue. First, how many cells are allocated to the tissue at the time of tissue foundation and second, how much cell mixing occurs between the different cell lineages within a tissue?

Before considering these questions we should briefly examine various patterns of growth that could occur during tissue development. In Figure 2 we compare four possible growth patterns of a hypothetical one-dimensional tissue comprising a string of 72 cells. Two genetically marked populations are represented respectively as black and white cells. The inner group of six cells represents the cells at the time of tissue foundation and the outer string of 72 cells represents the adult tissue at the time of analysis. The adult tissue has been divided into six equal parts representing the positions occupied by six descendant clones if no cell mixing occurs, between these six cell lineages, during growth. This situation is illustrated in Figure 2A and has been referred to by Nesbitt (1974) as strict coherent clonal growth. The other extreme, non-clonal growth, is represented by Figure 2B. Here so much cell mixing has occurred during growth that black and white cells are randomly arranged in the adult tissue. There is no correlation between the position of black and white cells in the tissue primordium and the two cell types in different parts of the adult tissue. It is, therefore, impossible to determine the number of descendant clones from the distribution of the two cell types in the adult tissue. Figures 2C and 2D represent intermediate patterns of growth referred to by Nesbitt (1974) as limited coherent clonal growth. In Figure 2C the distribution of cells in the adult is as though groups of three cells (all black or all white) were randomly arranged in a line. As in Figure 2B, there is no relationship between the position of the two cell types in the tissue primordium and their proportions in different regions of the adult tissue. This type of cell distribution could be most easily produced by extensive cell mixing early in development followed by a reduction in cell mixing later. The distribution of cells in Figure 2D differs from that shown in Figure 2C in that there is some correlation between the positions of the black and white cells at tissue foundation and the proportions of the two cell types in different regions of the adult tissue. This situation could arise by coherent clonal growth early in development followed either by alternate periods of cell mixing and coherent clonal growth or by a period of limited cell mixing. Thus, in the adult we can see two superimposed clonal patterns. The positions of the descendant clones are still marked, although distorted, and a period of some cell mixing is implied by the breakdown of the pattern into smaller coherent clones.

The terms "descendant clone", "coherent clone" and "patch" have been defined earlier. In the adult tissue shown in Figure 2A, there are four patches comprising six cell lineages or descendant clones. In Figures 2B and 2C the cells in the adult are arranged such that the six descendant clones initiated at tissue foundation cannot be identified. In Figure 2B the tissue comprises 36 patches and 72 cells arranged randomly rather than in coherent clones. The adult tissue shown in Figure 2C comprises 14 patches and 24 coherent clones of 3 cells each and the adult tissue shown in Figure 2D comprises 12 patches and 24 coherent clones of 3 cells each. However, in Figure 2D the adult pattern retains some of the spatial arrangement of the six descendant clones. Analysis of a number of individuals with this type of pattern could give a reasonable estimate of the number of descendant clones, although this is unlikely from a single individual. Estimation of the number of descendant clones gives us some idea of the number of cell lineages in the tissue whereas the size of the smaller

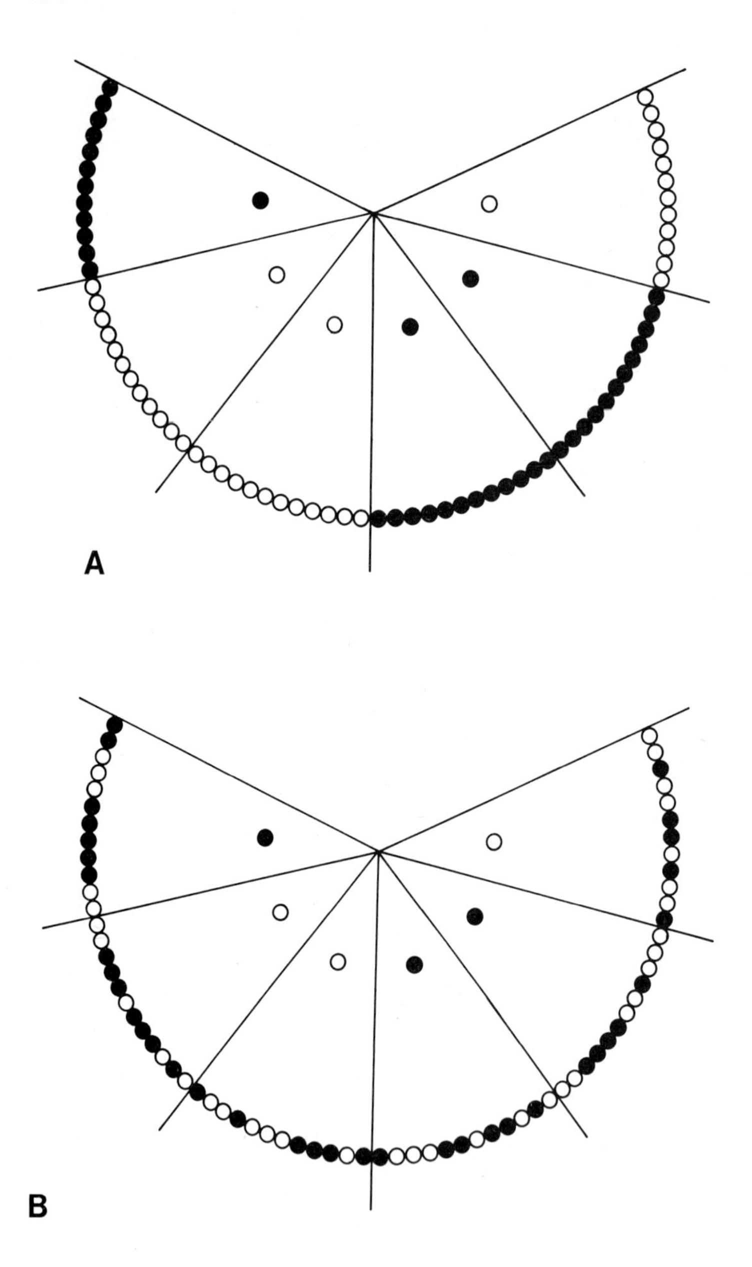

Figure 2: Four possible patterns of tissue growth. A hypothetical one-dimensional tissue, comprising a string of 72 cells, may develop from a row of six progenitor cells by A) strict coherent clonal growth; B) non-clonal growth; C) limited coherent clonal growth with extensive cell mixing between descendant clones; or D) limited coherent clonal growth with local cell mixing between adjacent descendant clones.

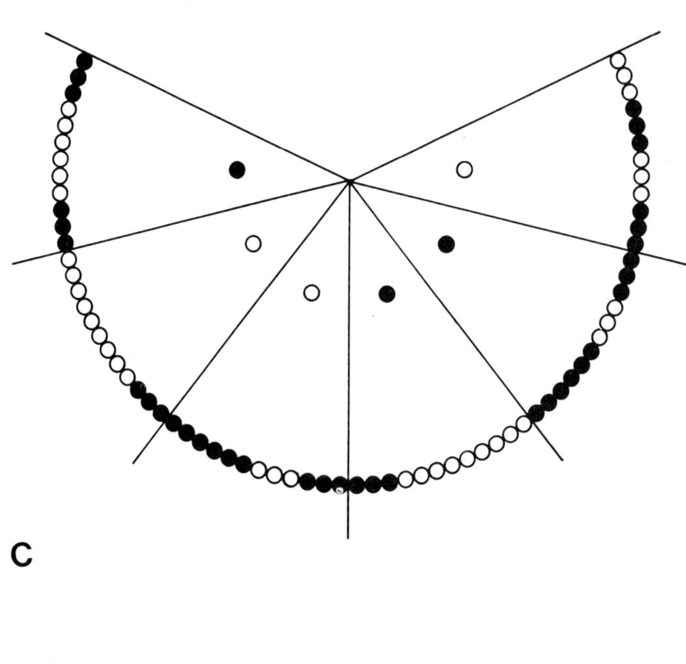

C

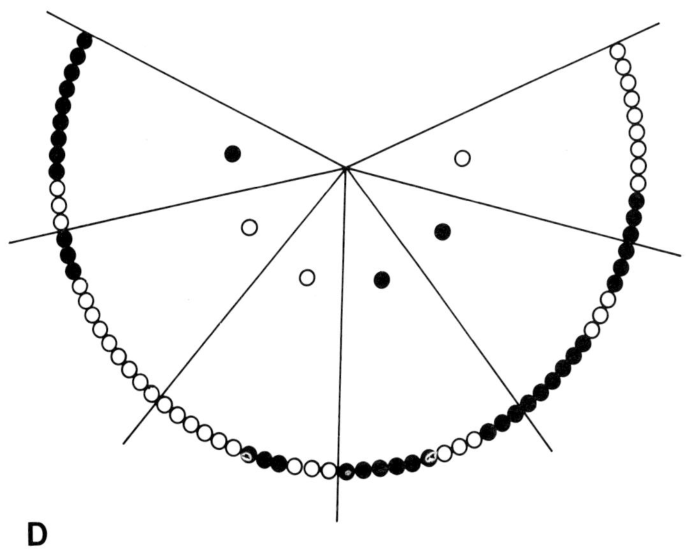

D

coherent clones indicates the amount of cell mixing that has occurred among neighbouring descendant clones.

It should be emphasised that Figure 2 is intended only to illustrate the main differences between four patterns of growth. None of the four adult distributions shown is intended as a realistic representation of the cell distribution in any adult mammalian tissue. The strings of coherent clones, diagrammatically represented in Figures 2C and 2D as each comprising exactly three cells, obviously grossly oversimplify the expected distribution of cells following limited coherent clonal growth. We might expect to see coherent clones, each comprising an identical number of cells, if cell mixing distributed the cells randomly and then abruptly ceased. If all the cells were dividing synchronously, each coherent clone would contain an equal number of cells. (The number of cells per coherent clone might be 2, 4, 8, 16 etc. depending on the number of cell divisions after cell mixing stopped). However, it is improbable that cell mixing would cease abruptly so that, even in the unlikely case of synchronous cell division, some coherent clones in the adult tissue would comprise fewer cells than others. Some cells could be completely isolated from others of the same cell population, even if the mean number of cells per coherent clone was relatively large.

METHODS OF ANALYSING CLONAL GROWTH

The basic method of analysing clonal growth from patterns in chimaeric or mosaic tissues is to measure or count groups or patches of cells of like type and to calculate the size or number of descendant clones or coherent clones from these experimental observations. The direct observation of patches is relatively straightforward when the genetic markers used mark the two cell types *in situ*. However, some cell markers such as chromosomes or electrophoretic variants require the disruption of the tissue during the analysis. Such markers are primarily useful for estimating the proportions of the two cell populations and we have to rely on rather indirect statistical approaches to estimate the size or number of patches, descendant clones or coherent clones.

Our discussion of clonal analysis is divided into two parts. First, we will examine methods used when a genetic difference can be used to directly mark the two cell populations *in situ*. Pigmented and unpigmented cells in the mouse retinal epithelium will be used to illustrate this approach. Second, we will consider some of the statistical approaches used when an indirect genetic cell marker is used to label the two cell populations. The terms "direct" and "indirect" are used to distinguish between markers that can be used to visualise the two cell populations *in situ* and those that can be used only after tissue disruption. Thus, we include some histochemical markers as direct markers even though the two cell populations cannot be seen without a specific histochemical stain.

ANALYSIS OF CLONAL GROWTH USING DIRECT (IN SITU) CELL MARKERS

The retinal pigment epithelium (RPE) is a monolayer of cells between the neural retina and the choroid of the eye (Figure 3). This tissue has several advantages for examining clonal growth by interpreting patterns in a mosaic or chimaeric individual. First, the tissue is a two-dimensional sheet of cells which greatly simplifies the analysis. Second, as we have seen, variants at the albino locus provide excellent *in situ* cell markers in both mouse chimaeras and X-inactivation mosaics. The two cell populations can be easily distinguished in the RPE of such animals from the thirteenth day of gestation to the adult.

Cell lineages and descendant clones

Ideally whole mounts of the entire RPE would be examined to determine the complete pattern of variegation (Mintz, 1971) but unfortunately the curvature of the eye and the dark pigmentation of the overlying choroid present problems to this approach. The heavily pigmented choroid frequently masks most of the underlying RPE and cannot easily be removed without damaging it. For this reason, histological sections have been used in most studies of clonal growth of the RPE, although chimaeras with unpigmented choroids and variegated RPEs can be produced if the pigment-producing cell population is homozygous for the *W* allele (Gordon, 1977). The most complete information on variegation in the RPE has been produced by the reconstruction of the entire tissue from drawings of serial and/or semi-serial sections (Mintz and Sanyal, 1970; Sanyal and Zeilmaker, 1977). Analysis of smaller regions has also been useful, particularly for comparative studies (Deol and Whitten, 1972; West, 1976a).

Sanyal and Zeilmaker (1977) used a computer to produce two-dimensional reconstruction maps of the RPE of chimaeric mice from camera lucida drawings of pigmented and albino patches in every third section. Four of Sanyal and Zeilmaker's reconstruction maps are presented in Figure 4 and will be used to illustrate different methods used to analyse the pattern of patches. In the four examples shown, the pigmented cells,

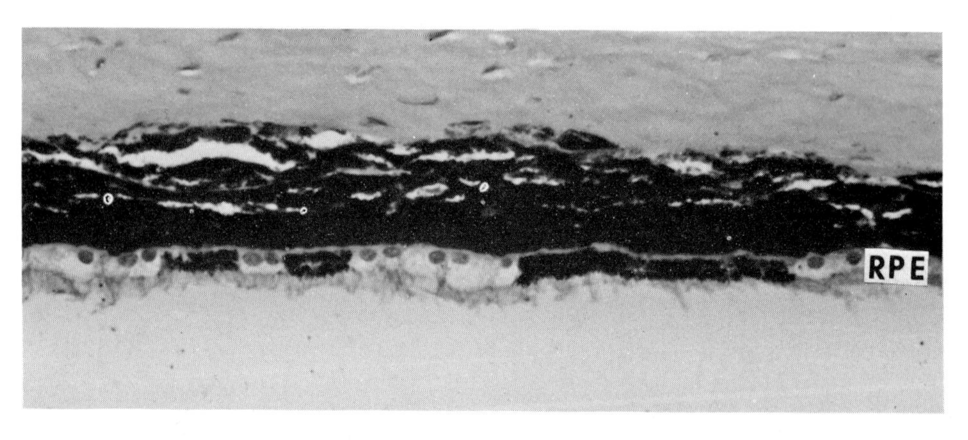

Figure 3: Histological section of part of an eye from a mouse chimaera showing pigmented and non-pigmented cells in the monolayer forming the retinal pigment epithelium (RPE). Above the RPE, towards the outside of the eye, is the heavily pigmented choroid and the sclera. The neural retina lies immediately below the RPE and has been removed from this eye.

represented as black areas, comprise only a minority of the tissue. The distribution of the pigmented cells is clearly not random in the map shown in Figure 4A and most of them are arranged as a sector radiating from the centre. Similarly, most of the pigmented cells in Figure 4B fall within two, or possibly three, such sectors. Mintz and Sanyal (1970) suggested that these radiating patterns represented descendant clones formed by the radial proliferation of a small number of primordial RPE cells. These patterns can be seen in Figure 4 by holding the page at arms length but closer inspection shows that the pigmented cells are arranged in smaller isolated patches. This indicates that local cell mixing occurs during growth and is sufficient to break up the descendant clones into smaller patches but is not extensive enough to disperse these patches randomly over the RPE. This type of limited coherent clonal growth is similar to that represented by Figure 2D. The positions of the sectors is less clearly defined in Figures 4C and 4D where the two populations of cells are more equally represented. In Figure 4C some of the patches of pigmented cells appear to be arranged as radiating sectors although over the rest of the RPE the patches are distributed more randomly. Sanyal and Zeilmaker (1977) interpret the pattern shown in Figure 4D as 10 alternating sectors (5 of pigmented cells and 5 of albino cells) which is similar to the "basic plan" described for the neural retina by Mintz and Sanyal (1970).

Such patterns could be used to estimate the number of descendant clones in the tissue in two ways. First, tissues from a large number of mice could be examined to determine the maximum number of sectors or the smallest possible sector. The former is likely to be more difficult to determine because of the local cell mixing discussed earlier. Among the 38 chimaeric eyes studied by Sanyal and Zeilmaker (1977), the tissue shown in Figure 4A had the lowest proportion of pigmented cells (0.4%). If all of these pigmented cells represented a single descendant clone, we would calculate that there are about 250 such cell lineages in the RPE. However, this is clearly a crude estimate and the total number of descendant clones may be considerably smaller if cell selection or cell death has reduced the proportion of pigmented cells during development. The series of BALB/c↔C3H mouse chimaeras studied by Sanyal and Zeilmaker shows evidence of being "unbalanced" for pigmentation of the RPE as most animals have a high proportion of albino cells in this tissue. (The term "unbalanced" was coined by Mullen and Whitten (1971) to describe a series of chimaeras in which one cell population predominated in most individuals). However, the RPE is predominantly pigmented in a few chimaeras and in one case the RPE comprised 99.6% pigmented cells. This again suggests that there is a large number of descendant clones.

The second approach also uses the proportion of the two cell populations and the total number of sectors, but the analysis is not restricted to the extremes of a series of animals. This approach could be applied to chimaeric or mosaic tissues if the number of sectors could be reliably interpreted. Sanyal and Zeilmaker (1977) suggested that the

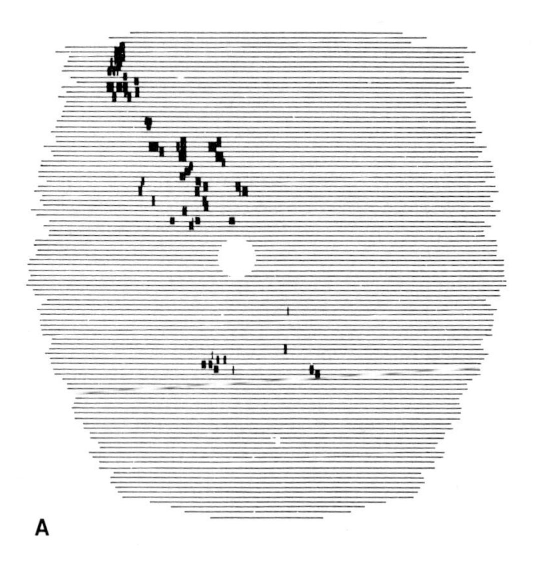

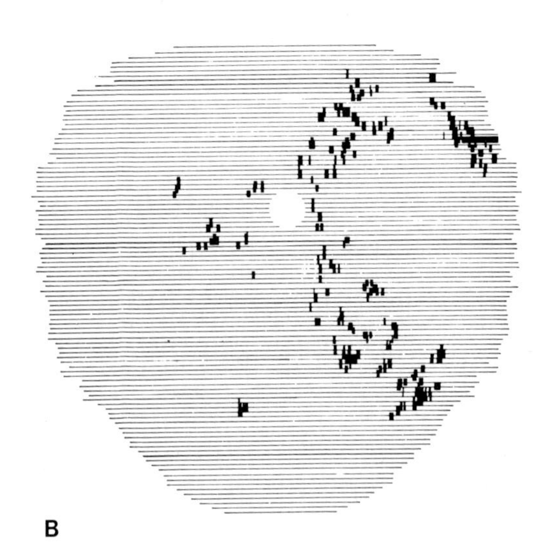

Figure 4: Reconstruction maps of the retinal pigment epithelium from eyes of four chimaeric mice analysed by Sanyal and Zeilmaker (1977). All four reconstruction maps are reproduced by kind permission of the authors.

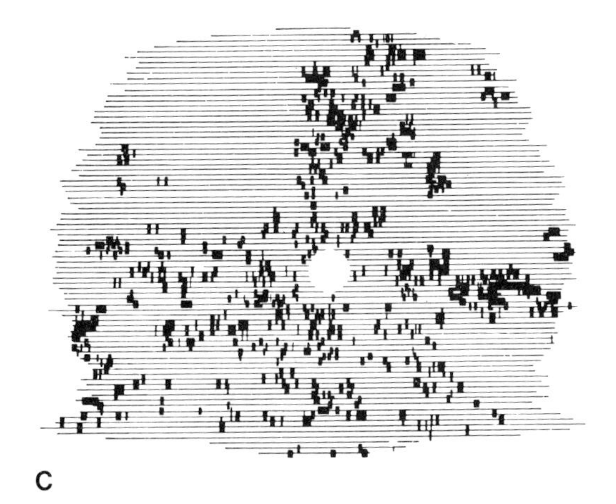

C

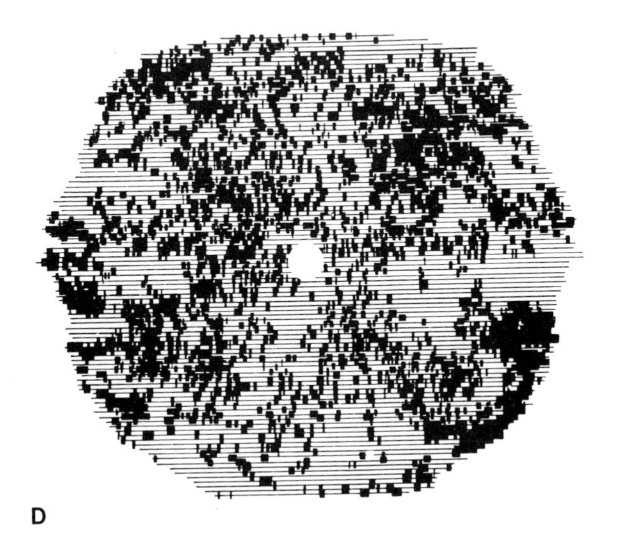

D

tissue shown in Figure 4D comprises 10 sectors. Each sector probably does not represent a single descendant clone because adjacent descendant clones may carry the same marker. If the proportion of pigmented cells is known, the expected mean number of descendant clones per sector can be calculated either statistically or by using a computer simulation. In this case, we have assumed that the sectors arise by radial proliferation of a circle of primordial RPE cells. The counting of sectors and clones is therefore a one-dimensional problem and can be analysed more easily.

The one-dimensional relationship between the number of descendant clones and the number of sectors is a function of the proportions of the two cell types. Considering one cell population at a time the expected mean number of descendant clones per sector is calculated as $1/(1-p)$, where p represents the proportion of that cell population in the tissue (Roach, 1968; West, 1975). The relationship is derived (Roach, 1968) by considering a random one-dimensional sequence of pigmented and albino descendant clones. The number of sectors can be seen to be equal to the number of descendant clones forming the left hand end of a sector. A descendant clone will be the left hand end of a pigmented sector if it is pigmented and its left hand neighbour is albino. For any descendant clone this has the probability of $p(1-p)$ where p is the proportion of pigmented descendant clones. Thus, in a line of N descendant clones the expected number of sectors is $Np(1-p)$. The average number of descendant clones per pigmented sector is the expected number of pigmented descendant clones, Np, divided by the expected number of pigmented sectors, $Np(1-p)$, which reduces to $^{1}/_{(1-p)}$.

The estimated number of descendant clones in the tissues shown in Figures 4A, 4B and 4D is shown in Table 2, assuming the total number of sectors to be 2, 4 and 10 respectively and using the proportions of pigmented cells given by Sanyal and Zeilmaker (1977). The large discrepancy between the results for different eyes partly reflects the difficulty in assessing the true number of sectors in Figure 4D and the dependence of the calculation on a very minor component in Figures 4A and 4B. Clearly, statistical analysis must be applied to a large number of eyes before any conclusions can be drawn.

TABLE 2: ESTIMATION OF THE NUMBER OF DESCENDANT CLONES IN THE RETINAL PIGMENT EPITHELIA SHOWN IN FIGURES 4A, 4B and 4D

Eye	4A		4B		4D	
Cell Population	Pigmented	Albino	Pigmented	Albino	Pigmented	Albino
p*	0.004	0.996	0.012	0.988	0.230	0.770
1/(1-p)	1.0	250	1.0	83	1.3	4.3
Number of sectors	1	1	2	2	5*	5*
Estimated number of descendant clones	1	250	2	166	6.5	21.5
	251		168		28	

* From Sanyal and Zeilmaker (1977)

The estimated number of descendant clones is related to the number of cells set aside to form the RPE primordium of each eye at tissue foundation. However, this relationship is not a direct one as McLaren (1972) and Lewis *et al.* (1972) have made clear. If the cells are randomly arranged at tissue foundation, the estimate will approximate the number of descendant clones initiated at this stage. If, however, cell mixing prior to tissue foundation is incomplete, the cells will already be arranged into coherent clones rather than at random and the estimate will measure the mean number of coherent clones at tissue foundation and not the number of cells sampled. The small coherent clones sampled into the tissue primordium will each initiate a descendant clone which may be detected in the adult tissue. These descendant clones are therefore each initiated by single cells, present before the time of tissue foundation, that contribute at least some of their progeny to the RPE. If there is a period of extensive cell mixing in the RPE before the initiation of the sectoring, the number of descendant clones in the adult will overestimate the number of coherent clones at tissue foundation. Our adult descendant clones, therefore, really estimate the number of coherent clones present at (or after) tissue foundation rather than the number of cells at tissue foundation.

Cell mixing

The small isolated patches of pigmented cells seen in all four eyes shown in Figure 4 indicate that some degree of cell mixing has occurred during the growth of the RPE. Patches have distinct boundaries and so are more easily measured than the sectors discussed above. Crude counts or measurements of these patches, however, tell us little about the amount of cell mixing because, as we have already seen, the patch size depends partly on the proportions of pigmented and albino cells. However, we can estimate the number or size of coherent clones to allow comparison between eyes with different proportions of pigmented cells.

The distribution of coherent clones in the RPE is two-dimensional and, in this respect, the analysis of coherent clones is different from that of the descendant clones discussed above. The two-dimensional relationship between the patch size and the coherent clone size has been simulated using a computer (West, 1975; Ransom, Hill and Kacser, unpublished; Whitten and Rupp, personal communication). Unfortunately, this relationship is dependent on the array size (the total number of coherent clones) and the shape of the coherent clones. However, when the proportions of the two cell populations are very unequal the array size is less important for estimating the mean coherent clone size in two dimensions. Thus, an estimate of the expected number of coherent clones may be made from Figure 4A. The proportion of pigmented cells (p) is 0.004 and the number of isolated patches is approximately 34. The mean number of coherent clones per patch can be estimated as 1.131 from the antilog_{10} of $(1.48p + 0.3708)^3$ which is the empirical relationship derived from a computer simulation study (West, 1975). Thus, statistically we would expect 0.4% of the RPE to contain about 38 coherent clones and the entire tissue to comprise nearly 10,000 such coherent clones. There are about 50,000 cells in the adult RPE (West, 1976a) so an average coherent clone will comprise about 5 cells.

This value is a statistical estimate and represents the mean size of regularly shaped groups of cells that would produce the observed mean patch size if distributed at random. In reality, the coherent clones may be quite variable in size and some cells may be completely isolated from others of the same cell population (see Deol and Whitten, 1972; Figure 3A in

West 1976a). Figure 4A shows that the pigmented cells are not distributed at random but grouped together in a sector, so the mean adult coherent clone size may be smaller than 5 cells. On the other hand, the coherent clone size may be under-estimated if the coherent clones are irregularly shaped. However, the absolute size of the coherent clones is less important for comparative studies where the sizes of coherent clones in equivalent tissues from different groups of chimaeras or mosaics are compared. It is for this type of comparative approach that the analysis of coherent clones is most useful.

In a comparative study of coherent clones in the RPE the author (West, 1976a) measured the lengths of patches of pigmented and unpigmented (not necessarily albino) cells in sections of eyes from embryonic and adult chimaeras and mosaics. The average coherent clone length was calculated by dividing the mean patch length by the expected number of coherent clones per patch for a linear array. This calculation is the same as that already described in the previous section to estimate the number of descendant clones per sector, although the size of a patch can be more accurately measured than the size or number of sectors. The average coherent clone area was taken to be the square of the calculated average clone length and the number of cells per coherent clone was estimated from the calculated cell area. The increase in coherent clone size in the RPE of X-inactivation mosaics during development was roughly parallel to that in a series of chimaeras made between members of the randomly bred Q strain (Table 3). Between the thirteenth day of gestation and the adult there was a 4 fold increase in the number of cells per coherent clone and a 3 fold increase in the number of coherent clones. This suggests that a decrease in cell mixing results in limited coherent clonal growth for part of this period. One group of adult chimaeras shown in Table 4 had larger coherent clones than the other groups. This presumably reflects less cell mixing during the development and growth of the RPE. This could result from a tendency of cells of unlike genotype to form less stable interactions in this particular strain combination. Deol and Whitten (1972) reported a difference in mean patch length between sections of the RPE from a group of X-inactivation mosaics and a series of C57BL/10↔SJL mouse chimaeras. However, since the patch size was measured rather than the coherent clone size, this difference may be largely due to the difference in proportions of the two cell populations between the mosaics and chimaeras.

TABLE 3: ESTIMATED NUMBER OF CELLS PER TWO-DIMENSIONAL COHERENT CLONE IN THE RETINAL PIGMENT EPITHELIUM OF A SERIES OF MOUSE AGGREGATION CHIMAERAS AND X-INACTIVATION MOSAICS AT VARIOUS STAGES OF DEVELOPMENT (FROM WEST, 1976a)

Age	Mean number of cells per coherent clone*	
(days post coitum)	Mosaics	Aggregation chimaeras**
12½	1.20 ± 0.09 (12)	1.33 ± 0.18 (10)
13½	1.82 ± 0.26 (10)	-
14½	1.56 ± 0.21 (10)	1.83 ± 0.14 (2)
15½	2.24 ± 0.18 (10)	4.72 ± 0.35 (2)
18½	2.18 ± 0.23 (10)	-
Birth		
20½	2.79 ± 0.10 (10)	4.36 ± 0.56 (8)
Adult	4.85 ± 0.27 (20)	6.05 ± 0.44 (11)

* Expressed as mean ± standard error (number of eyes)

** Made with pigmented and non-pigmented Q (random bred) strain mouse embryos.

The extremely small coherent clone sizes at 12½ days (Table 3) imply that cell mixing was sufficient to disrupt any tendency towards coherent clonal growth at this stage, unless the coherent clone size was consistently underestimated. The results are compatible with an almost random distribution of cells over the RPE at 12½ days but this seems unlikely in view of the sectoring patterns so clearly demonstrated by Sanyal and Zeilmaker (1977) in adult eyes. From estimates of cell numbers at different ages it seems that there are only about four cell doublings in the RPE between 12½ days and the adult (West, 1976a). Therefore, if the distribution of cells at 12½ days was random, it would be necessary to postulate aggregation of pigmented cells to explain the sectoring seen in Figures 4A and 4B, unless the sectoring is only prominent in chimaeras of certain strain combinations where cell mixing is reduced. It seems more likely that the sectoring was already present at 12½ days in West's mice but was not detected by the analysis of coherent clones. This would be expected if cell mixing was sufficient to disrupt any tendency for daughter cells to remain together but not sufficient to move cells over long distances. (The sectoring should, however, be detectable as variation in the proportion of pigmented cells either among different areas of a section or among different sections, depending on the plane of the sections).

TABLE 4: ESTIMATED NUMBER OF CELLS PER TWO-DIMENSIONAL COHERENT CLONE IN THE RETINAL PIGMENT EPITHELIUM OF ADULT MOUSE X-INACTIVATION MOSAICS AND AGGREGATION CHIMAERAS OF THREE STRAIN COMBINATIONS (FROM WEST, 1976a)

Group	Number of eyes	Mean number of cells per coherent clone ± standard error
X-inactivation mosaics	20	4.85 ± 0.27
Aggregation chimaeras:		
Pigmented-Q↔Unpigmented-Q	11	6.05 ± 0.44
Pigmented-Q↔Recessive	16	5.70 ± 0.76
(C57BL x C3H)F_1↔ Recessive	19	8.82 ± 0.77

From these studies of the RPE it seems that a number of descendant clones are initiated early in the development of the tissue. Local cell mixing appears to mix the cells of neighbouring descendant clones by 12½ days post coitum. Cell mixing is probably not sufficient to mix many cells between widely separated descendant clones, although the scattering of a few patches of pigmented cells away from the main sectors in Figures 4A and 4B suggest that this does sometimes occur. The degree of cell mixing is reduced during later development and limited clonal growth increases the mean size of the coherent clones.

ANALYSIS OF CLONAL GROWTH USING INDIRECT CELL MARKERS

In the above discussion, we have seen the type of information that can be obtained from direct markers where the spatial relationships can be directly seen *in situ*. Even with this type of marker the interpretation of growth patterns from variegation in the adult tissue is far from precise, particularly when we are trying to uncover two- or three-dimensional patterns from one- or two-dimensional information respectively. The use of indirect markers relies more heavily on various statistical approaches to determine the number of descendant clones initiated earlier in development, because these markers only allow us to estimate the proportions of the two cell types.

Binomial expansion and the frequency of non-variegation

The binomial theorem has been used to estimate the number of "tissue progenitor cells" in chimaeras and mosaics, usually from the proportion of individuals that show a uniform, non-variegated phenotype (see Table 5 and Gandini *et al.*,1968). If there are only three progenitor cells then, on average, assuming no differential cell selection, 1/4 would be non-variegated (1/8 would be entirely composed of one cell population and 1/8 would be entirely composed of the other cell population). In turn n progenitor cells would produce $2 \times 1/(2)^n$ non-variegated individuals. McLaren (1972) and Lewis *et al.*(1972) have discussed this approach and conclude: (1) if no cell mixing occurs during development of the individual, the estimate relates to the "stage at which cell heterogeneity arises" by the marking of two cell populations; (2) if cell mixing occurs during development, the estimate relates to the cell generation at which mixing between cell lineages of different tissues becomes minimal. The cells may already be arranged in coherent clones at the time of tissue foundation or, alternatively, extensive cell mixing may continue to randomise the cell distribution in the developing tissue for several more cell generations. However, if we determine the frequency of non-variegation in the entire tissue we are not concerned with the mixing of cell lineages within a tissue and our estimate reflects the number of coherent clones allocated to the tissue primordium at tissue foundation. As we have already discussed, unless the cells are randomly distributed at tissue foundation this estimate approximates the number of descendant clones initiated at an earlier stage.

This method has been used to estimate the number of descendant clones at tissue foundation in several tissues in the mouse (Table 5) but relies heavily on the assumptions that there is no differential cell selection or cell death as well as the assumption that the two cell populations are randomly arranged at tissue foundation (McLaren, 1976). These assumptions are particularly unlikely for mouse chimaeras of certain strain combinations that are unbalanced in terms of the proportions of the two cell populations (Mullen and Whitten, 1971) or those that show a non-random distribution of cell types within an adult tissue (Moore and Mintz, 1972; West and McLaren, 1976).

A similar approach was used by Russell (1964) in order to estimate the number of descendant clones in the germ line of a group of radiation-induced mouse mosaics. The mosaics that were selected for study contained about 50% of each cell population in the coat. However, the distribution of the two cell populations in the germ line was not clustered around 50% but was much broader. Russell suggested that this indicates that the number of descendant clones must be very small, perhaps around 5. This impression was reached by comparing the distribution to the expansion of the binomial $(0.5 \text{ mutant} + 0.5 \text{ non-mutant})^5$. If 5 coherent clones were sampled from a large population comprising 50% of each cell population, we would expect 3% of the mosaics to have no mutant cells in the germ line ($p=0$, where p is the proportion of the mutant germ cells). The expected frequencies of individuals containing mutant germ cells would be 16% with $p=0.2$, 31% with $p=0.4$, 31% with $p=0.6$, 16% with $p=0.8$, and 3% with $p=1.0$. The fit of the results to these six classes is very crude and clearly this method also suffers from all of the problems discussed above.

Variations on these two types of analysis using the binomial theorem have been used by Nance (1964); Linder and Gartler (1965); and Wegmann and Gilman (1970).

Methods based on the variance of the proportions of each cell population

Another approach, also based on the binomial theorem, and therefore also dependent on the above mentioned assumptions, has been used to estimate the number of descendant clones initiated at tissue foundation when only the proportions of the two cell populations can be determined from the tissue analysis. The methods discussed by Wegmann (1970) and Nesbitt (1971) fall into this category. Part of the variation in the proportion of each cell population in a given tissue, among a group of chimaeras or mosaics, results from a sampling of a group of cells at the time of tissue foundation. The smaller the number of cells sampled to form the tissue, the greater is the expected variance. The number of descendant clones (N) can be estimated from the formula $N=[p(1-p)]/s^2$ by assuming binomial sampling of randomly distributed descendant clones into the tissue primordium. The proportion of the two cell populations (p and 1-p) and the variance of p (s^2) can be determined, assuming that differential cell selection or cell death does not significantly alter p during tissue growth. This provides a better method of fitting the results to the appropriate binomial expansion than that discussed in the previous section.

Unfortunately, the variance among the tissues from different individuals will be partly due to the sampling events at tissue foundation and partly due to earlier sampling events that cause variation in the proportions of the two cell populations in the pool of cells from which the tissue progenitor cells are sampled. Two different approaches have been used to overcome this problem. Wegmann (1970) calculated the variance among samples taken from one tissue in a single individual. Nesbitt (1971) analysed several tissues from each of a number of X-inactivation mosaics using a more complex statistical analysis. Nesbitt's statistical analysis was designed to derive both the number of descendant clones in a specific tissue and the number of cells involved in an earlier sampling event, common to all the tissues analysed. (The earlier sampling event was assumed in this case to be the "marking" of the two cell populations in the early embryo due to the differential expression of the two X chromosomes.)

Wegmann (1970) estimated the proportions of the two cell populations in samples of liver from chimaeric mice using biochemical assays of the enzyme β-glucuronidase. In this case s^2 represented the variance among six samples from a single liver and p and 1-p were respectively the mean proportions of each cell population in these six samples. Several livers were analysed in this way and revealed a mean of 17.2 "clones" per sample. As each sample represented about 1/20 of the liver, this is interpreted to suggest a mean of 344 such "clones" per liver. In addition to the assumptions already discussed, this approach assumes that the variance among samples increases as the sample size decreases. This would probably not be true if, as in the case of the retinal pigment epithelium, the distribution of the two cell populations is a result of two or more superimposed patterns. For example, if the patterns shown for the RPE in Figure 4, were analysed by Wegmann's method, we would expect different results depending on the sample size. If the whole RPE could be used as a sample (ignoring problems of earlier sampling events, cell distribution and cell selection) N would be related to the number of coherent clones allocated to the tissue primordium or the number of descendant clones initiated at or before tissue foundation.

However, if a number of small samples were taken from a single sector of one eye, N would more nearly reflect the size of the much smaller coherent clones in the adult tissue. In both of these cases the sample size is much larger than the "clone" size. In his analysis of chimaeric liver, Wegmann used quite large samples (1/20 of the whole liver), but if this sample size is not considerably larger than the descendant clones, each sample will contain a significant proportion of incomplete descendant clones. In this case the variance will probably be attributable partly to the size of the descendant clones and partly to their shape. The local cell mixing that occurs during tissue growth produces the pattern of patches of small coherent clones and so affects the shape of the descendant clones. Therefore, it seems probable that 344 represents an overestimate of the number of descendant clones in the liver. However, it is unlikely that these large "clones" refer directly to coherent clones in the adult as Nesbitt and Gartler (1971) infer because histochemical studies by Condamine *et al.*(1970) and West (1976c) indicate that the two cell populations are quite finely interspersed. Wegmann's method has recently been used in an analysis of the epidermis of mouse chimaeras (Iannaccone *et al.*,1978). In this case, the authors assumed that they were measuring coherent clones in the adult tissue.

The analysis of several tissues in each of a series of X-inactivation mosaics has been undertaken by Nesbitt (1971) and Fialkow (1973). This has also been discussed by Nesbitt and Gartler (1971) and McLaren (1976). Nesbitt (1971) assumes that there are two successive sampling events involving n_1 and n_2 cells (or coherent clones) respectively. The number of cells (or coherent clones) at the early sampling event, common to all tissues, is represented by n_1 and is calculated from the covariance of p from paired tissues. In the context of clonal growth of tissues, we are more concerned with n_2 which is equivalent to the number of descendant clones initiated at or before tissue foundation. This value is calculated separately for each tissue from n_1 and N, which is the number of descendant clones that would have accounted for the variance for that tissue if there was no earlier sampling event. The statistical analysis is complex and rests on the same tenuous assumptions as the other binomial methods. In Nesbitt's experiments, the period of tissue culture before scoring the proportions of the two cell populations adds to the probability of differential cell selection and so further complicates the interpretation of her results.

Fialkow (1973) analysed the proportions of the two cell populations in a number of tissues from human X-inactivation mosaics and found strong correlations between most tissues studied. From this, Failkow argues that the number of cells allocated to each tissue primordia might be quite large. If so, the variance, in the proportions of the two cell populations, between individuals is generated mainly by the earlier of the two sampling events considered by Nesbitt (1971).

Frequency of small samples that contain both cell populations

Hutchison (1973) has developed a method of analysis designed to estimate the size of coherent clones, at the time of analysis, from the frequency of small samples that contain both cell populations. This method is applicable when the sample size is smaller than the coherent clone size. Both cell populations will be detected in a sample when coherent clones from different patches are present. The expected frequency of these mixed samples is a function of the frequency of the two cell populations in the entire tissue and the ratio of the sample size to mean coherent clone size. Hutchison (1973) considered this relationship for various geometrically shaped coherent clones and has shown that, unless the coherent clones are irregular, the relationship is not strongly dependent on their shape.

This model has now been used in a number of studies (Gartler *et al.*,1971; Friedman and Fialkow, 1976; Iannaccone *et al.*,1977). From their analysis of hair follicles in the epidermis of the scalp of adult human X-inactivation mosaics, Gartler *et al.*,(1971) estimated that each coherent clone comprised between three and eleven hair follicles. In this analysis each hair follicle represented a separate experimental sample.

STUDIES OF CLONAL GROWTH OF CHIMAERIC AND MOSAIC TISSUES

Tissues from chimaeras and mosaics have been analysed to try to estimate the number of cells allocated to various tissue primordia and to determine the lineage relationships between various tissues, by use of the methods discussed in the previous sections.

STUDIES OF CELL LINEAGES WITHIN A TISSUE

A number of studies of clonal growth in chimaeras and mosaics have been undertaken in order to try to determine the number of cells allocated to a tissue primordium at the time of tissue foundation. We have defined this as the time at which cell mixing between lineages in different tissue primordia is reduced to a minimum. This may not be accompanied by any biochemical or overt morphological changes and is not necessarily the time at which tissue specific genes are first activated. Most attempts to estimate the number of cells at tissue foundation fall short of their goal. Some of the reasons for this have already been mentioned and this subject is considered in detail by McLaren (1972, 1976) and Lewis *et al.*(1972). Thus, the numbers originally derived as estimates for the number of cells at tissue foundation may more closely represent the number of coherent clones present at this time, the number of descendant clones initiated after tissue foundation (if there is a lot of cell mixing within a tissue primordium) or may be unacceptable estimates based on unsound assumptions. Table 5 lists examples of various types of analysis of descendant clones in the mouse. These estimates were originally intended as estimates of the number of tissue precursor cells but have been re-classified in Table 5 according to the scheme mentioned above. In addition, the numerical values shown in the table are subject to the criticisms mentioned earlier.

A somewhat different approach was used in the study reported by Russell and Major (1957). Two cell populations of coat melanocytes were marked at 10¼ days *post coitum* by somatic mutation, induced by X-rays. This is probably after the time of tissue foundation for this tissue, although, since coat

TABLE 5: ANALYSIS OF CELL LINEAGES IN MOUSE CHIMAERAS AND MOSAICS

Probable class of clone analysed*	Type of experimental mouse	Tissue	Estimated mean clone number*	Marker	Method	Reference
A. Cells at tissue foundation	Aggregation chimaera	Germ Cells	2 - 10	(see method)	Minimum = 2 because germ cell chimaerism occurs. Maximum = 10 by direct observation of 10 cells in 8-day embryo	Mintz, 1968; 1971
B. Descendant clones. Each initiated from a separate coherent clone present at tissue foundation	Aggregation chimaera	Hair follicle	≥ 5	Hair colour (agouti *A*)	Frequency of variegation and binomial expansion	Wegmann & Gilman, 1970
		Blood	≥ 5	Haemoglobin and globulin		
		Hair follicle	3?	Hair colour (agouti *A*)	Frequency of non-variegated coat in chimaeras and binomial expansion	Mintz, 1970a
		Liver	2?	Isozyme (malic enzyme) (*Mod-1*)	Frequency of non-variegated livers in chimaeras and binomial expansion	Mintz, 1970a
		"Embryo"	3?	Coat colour, isozymes and immunological	Frequency of non-chimaeras after aggregation and binomial expansion	Mintz, 1971
		Somite	2 - 5	Isozyme (glucose phosphate isomerase *Gpi-1*)	Frequency of non-variegated eye muscles in chimaeras and binomial expansion	Gearhart & Mintz, 1972

TABLE 5: continued on next page

Somatic mutation mosaic	Germ Cells	5?	Coat colour of off-spring	Expansion of binomial to fit the observed broad distribution of the proportion of each cell population	Russell, 1964
X-inactivation mosaic	Kidney proximal tubule	4 or 5	Testicular feminisation and β-glucuronidase (*Tfm* and *Gus*^{h})	Frequency of non-variegation in mosaics	Tettenborn *et al.*,
	Melano-blasts	> 10	Coat pigmen-tation	Frequency of non-variegation on mosaics	Lyon, 1972
	Melano-blasts	22	karotype and labelling patterns by autoradio-graphy	Statistical analysis based on variance of the proportions of the two cell types and binomial sampling	Nesbitt, 1971
	Spleen	21			
	Thymus	23			
	Right Lung	41			
	Left Lung	43			
	Abdominal fascia	58			

TABLE 5: continued on next page

Probable class of clone analysed*	Type of experimental mouse	Tissue	Estimated mean clone number*	Marker	Method	Reference
C. Descendant clones. Each either iniated from a separate coherent clone present at tissue foundation or initiated after tissue foundation	Aggregation chimaera	Vertebral column	120	Skeletal morphology	Smallest unit = 1/4 vertebra. (30 vertebrae with 4 lineages per vertebra)	Moore & Mintz, 1972
		Melanoblasts:		Coat pigmentation	Maximum number of stripes	Mintz, 1967; 1971
		head	6			
		body	12			
		tail	16			
		Hair follicles (coat and tail)	150 - 200	Hair structure colour (agouti, *A*; Fuzzy, *fz*)	Maximum number of stripes	Mintz, 1969; 1971. (Also see Green *et al.*, 1977)
		Visual retina	10 per eye	Retinal degeneration (*rd*)	"Basic plan" of 10 radiating sectors per eye. (Assumed to be maximum number of sectors)	Mintz & Sanyal, 1970
	X-inactivation mosaic	Melanoblasts:		Coat pigmentation	Maximum number of stripes	Cattanach, 1974
		head	6			
		body	14			

TABLE 5: continued on next page

Probable class of clone analysed*	Type of experimental mouse	Tissue	Estimated mean clone number*	Marker	Method	Reference
D. Probably an over-estimate of (C) above	Aggregation chimaera	Liver	344	β-glucuronidase (*Gus*^{h})	Variation in proportion of the two cell populations between samples. (Claims to measure adult "clone" size.)	Wegmann, 1970
E. Descendant clones initiated at 10½ days	Somatic mutation mosaic	Coat melanoblasts	150 - 200	Coat pigmentation	Mean size of spot of mutant pigment induced at 10½ days	Russell & Major, 1957

* The numerical values listed in B and C were originally intended as estimates of the number of tissue precursor cells. These have been reclassified as described in the text. In addition, the validity of the numerical estimates depends on a number of assumptions which are discussed by McLaren (1976) and in the text.

colour markers were used, the possibility of some descendants of the mutant cell contributing to other tissues could not be properly tested. The "spots" of mutant colour in the coat of each mosaic were either restricted to a single area or separated by a short distance. This indicated that all the mutant pigment was probably produced by a single descendant clone of melanocytes initiated by one cell at 10¼ days. The proportion of the coat occupied by mutant pigment was used to infer the presence of 150-200 such descendant clones in the entire coat, although the diffuse outline of the mutant "spots" made it difficult to accurately estimate this proportion. The estimation of the number of descendant clones from the size of the mutant "spots" assumes that the mutant melanoblasts were not at any selective advantage or disadvantage, and that all the melanoblasts present at 10¼ days colonise a similar area of coat. This second assumption may not be true if the migration of melanoblasts is already further advanced in the anterior part of the embryo at 10¼ days, but since Russell and Major (1957) studied 31 mosaics with "spots" on various parts of the body, the "spots" that they measured are probably representative of the total population.

STUDIES OF SHARED CELL LINEAGES

If the foundation of a certain tissue follows a sequential series of sampling processes, the fate of a group of cells becomes progressively restricted as development proceeds. Thus, some tissues may share a common pool of cells until relatively late in development whereas others diverge much earlier. Mintz and Palm (1969), Wegmann and Gilman (1970) and Mintz and Niece (reported by Mintz, 1971) showed a positive correlation, for the proportion of each cell population, between white blood cells (or cells producing gamma globulin) and red blood cells, in chimaeric mice. In a similar survey, Gornish *et al.*(1972) suggested a common origin for mitotic cells of spleen, bone marrow and thymus, whereas Ford *et al.* (1975) suggested separate origins for a group including bone marrow, thymus and Peyer's patches and a group comprising spleen and lymph nodes. Earlier studies on human X-inactivation mosaics also prompted the conclusion of a common origin of erythrocytes, granulocytes, and perhaps lymphocytes (Gandini *et al.*,1968; Gandini and Gartler, 1969). However, simple correlations between tissues do not provide evidence for shared pools of cells unless other tissues correlate less well and Fialkow (1973) has shown good correlations between granulocytes and both skin and muscle as well as between granulocytes and lymphocytes. Also, correlations could arise either as a result of shared pools of cells or could be due to similar selection pressures. Rapidly dividing tissues would be expected to show the effects of cell selection most clearly. A number of observations suggest that cell selection may occur in the haematopoietic system (Mintz and Palm, 1969; Nyhan, *et al.*,1970;Nesbitt and Gartler, 1971; Tuffrey *et al.*,1973; West, 1977) and possibly other tissues as well (Mintz, 1970b). Warner *et al.*(1977) have also suggested that mechanisms, other than genetically determined cell selection, may also change the proportions of the two cell populations in tissues of the immune system in chimaeric mice, although this has yet to be confirmed.

Similar correlations have been made between tissues less obviously sensitive to selection pressures. Gartler *et al.* (1971) argued for a common origin for the outer sheath and bulb of single hairs in human X-inactivation mosaics, and Sanyal and Zeilmaker (1977) suggested that the retinal pigment epithelia in left and right eyes of chimaeric mice arise from a common pool of cells which divides to produce the two optic cups. Sanyal and Zeilmaker showed a stronger correlation in the proportion of pigmented cells in the RPE between left and right eyes than between the RPE and the choroid of the same eye. The authors suggested that the RPE from left and right eyes is derived from a single pool of cells and that daughter cells may segregate, one into each eye. Sanyal and Zeilmaker measured each patch in the RPE in all slides examined but they used a less sensitive method for scoring the proportion of pigmented cells in the choroid. This could contribute to the poorer correlation between the RPE and choroid. However, on embryological grounds we would expect a closer correlation between left and right RPEs than between the pigmentation in the choroid and the RPE (which are derived from the neural crest and the optic cups respectively) even if the two RPE primordia were eventually sampled from separate pools of cells. Attempts to correlate the proportions of each cell population in the RPE and neural retina have been confused by the problems of using retinal degeneration as a cell marker. Some reports (Mintz and Sanyal, 1970; Sanyal and Zeilmaker, 1974, 1976) suggested that the proportions of the two cell types in the neural retina differed from the proportions in the RPE and most other tissues. However, this may be partly a result of an inadequate method of estimating the proportions in the neural retina (West, 1976b). Despite these problems, Sanyal and Zeilmaker (1974, 1976) and LaVail and Mullen (1976) reported that the proportions of the two cell populations in the neural retina are often similar in the left and right eyes.

Another possible approach is suggested by the work of Russell and Major (1957), Russell (1964), Garcia-Bellido *et al.* (1973) and Lawrence (1973) who all used X-rays to induce somatic mutation at known times in development and traced the descendant clones derived from the initial mutant cells. If a wide range of tissues were surveyed for electrophoretic mutants induced by X-rays at a large number of loci, it might be possible to trace the progressive restriction of the fate of the descendants of a mutant cell by irradiating embryos at different ages. In principle this technique could also be used to estimate the time of tissue foundation and the number of cells present in different tissue primordia. One limiting factor might be the difficulty in detecting a very small proportion of the mutant cell population.

STUDIES OF CELL MIXING

In tissues such as the mouse retinal pigment epithelium (Sanyal and Zeilmaker, 1977) and the melanocytes of the mouse dorsal coat (Russell and Major, 1957; Mintz, 1967; Cattanach, 1974) regions of the tissue can be identified in some chimaeric or mosaic individuals that are predominantly occupied by one cell population. These regions appear as "stripes", "spots" or "sectors" and probably mark the area colonised by one or more descendant clones. A descendant clone is less likely to be completely disrupted by cell mixing once it has reached a certain size (Lewis, 1973). Thus, perhaps the establishment of relatively clearly defined descendant clones (for examples, see Figures 4A and 4B) depends on coherent clonal growth early in development. The patterns of clonal growth may vary among individuals and in some cases cell mixing may disrupt this early clonal growth and result in a more finely dispersed pattern, as shown in Figure 4C. Cell mixing quite early in development probably also accounts for the smaller groups of pigmented cells some distance from the main sectors in Figures 4A and 4B. Even relatively clearly defined descendant clones are broken up into a number of isolated patches of coherent clones. This can be seen directly by close inspection of the sectors shown in Figure 4 and by the diffuse outline of the stripes in the coats of chimaeric and mosaic mice (Russell and Major, 1957; Mintz, 1967). In many cases single hairs contain pigment secreted from melanocytes of both cell populations (McLaren and Bowman, 1969).

The relative amount of cell mixing during histogenesis has been assessed by estimating the coherent clone sizes at different stages of development in the retinal pigment epithelium of mouse chimaeras and X-inactivation mosaics (West, 1976a). The coherent clones were very small at early fetal stages but larger at later stages. This implies that a period of extensive cell mixing is followed by limited coherent clonal growth and a reduction in cell mixing. A similar conclusion was reached by Gartler *et al.*(1971) for the scalp epidermis of human X-inactivation mosaics although experimental observations were only made on adults. A rather different comparative study was reported by Gearhart and Mintz (1972) who examined the electrophoretic patterns of the dimeric enzyme glucose phosphate isomerase (GPI) in somites from chimaeric mouse embryos and in adult eye muscles, each derived from a single somite. Most of the embryonic somites (30/38) contained both forms of GPI and neighbouring somites tended to have similar proportions of the two cell populations. This indicates that individual somites are produced by more than one cell and that the distribution of cells into neighbouring somites is non-random. In other words, some coherent clonal growth has probably occurred before somite formation. All 33 of the adult eye muscles contained both forms of GPI but 11 of these lacked the hybrid isozyme band, formed when two myoblasts of unlike genotype fuse to form a muscle fibre (Mintz and Baker, 1967). Unless there was a tendency against fusion of cells of unlike genotype this suggests that the two cell populations were arranged into a few large patches and that there was little cell mixing before the formation of the muscle fibres.

Garner and McLaren (1974) found evidence for coherent clonal growth rather than cell mixing between the aggregation of two eight-cell mouse embryos and the formation of the blastocyst. Other observations on the sizes of coherent clones or patches *in situ*, at one specific stage of development, give us some indication of the degree of cell mixing but comparative studies between different times of development are more informative and less dependent on the accuracy of numerical estimates. Small patch sizes have been seen in adult mouse chimaeras, both in the liver (Condamine *et al.*,1970; West, 1976c) and the Purkinje cell layer of the brain (Dewey *et al.*, 1976; Mullen, 1977). This indicates that the development of these tissues has been accompanied by considerable cell mixing. Mullen (1977) recently estimated that the coherent clones in the Purkinje cell layer comprised an average of only 1.06 cells. Reconstruction of part of this tissue from serial sections suggested that the cells were distributed almost randomly, over the region examined, rather than in coherent clones. Small patches have also been seen in the livers of human X-inactivation mosaics (Ricciuti *et al.*,1976). However, valid numerical estimates of the mean size or numbers of coherent clones are difficult to obtain for three-dimensional tissues, such as the liver. In contrast to these observations, quite large patches have been reported during embryonic development (Klinger and Schwarzacher, 1962; Gardner and Johnson, 1973). Gardner and Johnson demonstrated large patches at the egg cylinder stage in rat↔mouse chimaeras using immunofluorescent labelling. However, the amount of cell mixing in interspecific chimaeras may not represent the normal situation in either species as genetically similar cells may show a tendency to stay together even in some intraspecific mouse chimaeras (Table 4). In their study of a human XY/XXY chromosome mosaic, Klinger and Schwarzacher (1962) found large "patches" in both the ectodermal epithelium and mesodermal connective tissue of the amnion. These "patches" were marked by regions of high or low frequency of sex chromatin positive cells respectively. Even in normal XX individuals, the frequency of sex chromatin positive cells falls short of 100% so Klinger and Schwarzacher were unable to mark the exact boundaries of the "patches". However, "patches" with a high proportion of sex chromatin positive cells could be identified by dividing the tissue into small areas and classifying the percentage of sex chromatin positive cells to the nearest 20%. "Patches" of sex chromatin positive cells frequently showed gradients ranging from 80-100% sex chromatin positive cells at the centre to lower frequencies at the outside. A "patch" therefore probably included cells from both populations and more nearly reflected the distribution of descendant clones rather than coherent clones. The "patches" of sex chromatin positive cells in the ectoderm of the amnion often occurred over "patches" of sex chromatin negative mesoderm cells, as would be expected from the independent origins of these two layers. Although sex chromatin is not a completely reliable cell marker, the "patches" were considerably larger in the amnion and chorion than in the body tissues of the fetus. This suggests that cell mixing may disrupt the descendant clones less in two-dimensional sheets of tissues than in the three-dimensional tissues studied.

Apart from the histochemical observations of Ricciuti *et al.*, (1976) clonal growth in human X-inactivation mosaics has been analysed using electrophoretic variants of glucose-6-phosphate dehydrogenase (G6PD). The mean coherent clone size has been estimated for a number of human tissues (Gartler *et al.*,1971; Friedman and Fialkow, 1976). These are usually presented in terms of area or volume of tissue but Fialkow (1976) has listed estimates of 10,000 cells for normal adult uterine tissue, 950-3,500 cells for adult scalp and 1,500 cells for coherent clones in the skin epithelium overlying the vulva. The estimate for the adult scalp (Gartler *et al.*,1971) is based partly on an estimate of 3-11 hair follicles per coherent clone and partly on the random distribution of G6PD genotypes in follicles 1mm apart. The second observation indicates that a coherent clone is no larger than 1 mm^2 (about 3,000 cells) but it could be very much smaller than this. The estimate of 3-11 hair follicles per coherent clone was used to estimate the number of cells per coherent clone both at hair follicle initiation and in the adult, assuming that there was no mixing of cells between hair follicles after the time of follicle initiation. The estimation of 12-66 cells per hair follicle at follicle initiation also rests on a rather tentative estimate of the number of progenitor cells for each hair follicle.

From a comparison of the estimates of human coherent clone sizes with estimates derived for the mouse, Linder and Gartler

(1965) and Gartler (1974) suggested that mouse coherent clones were larger than their human counterparts. However, the comparison was between different tissues and the estimates for the mouse more nearly reflected descendant clones. No parallel analyses have been done to compare coherent clone sizes in the same tissues of mice and humans. Iannaccone *et al.*(1978) used electrophoretic variants in chimaeric mice to estimate a mean of 600 cells per coherent clone in the skin epidermis. Since this is unlikely to be an overestimate for reasons given earlier and other analyses of coherent clones in mice suggest that the two cell populations are finely interspersed, it seems unlikely that coherent clones are any larger in mice than in humans.

Local cell mixing, at least, appears to be a normal part of histogenesis although the extent of cell mixing in experimental mouse chimaeras may depend on the genotypes of the two cell populations. The extent of cell mixing also probably changes during development and differs between tissues. The experiments of Garner and McLaren (1974) suggest that there is little cell mixing immediately prior to the blastocyst stage in the mouse. Similarly, the work of Gearhart and Mintz (1972) argues for relatively little cell mixing in the eye muscles before myoblast fusion (in the fetus). In contrast, there is evidence for considerable cell mixing in the retinal pigment epithelium, during early fetal development (West 1976a), and in the Purkinje cell layer, later in development (Mullen, 1977).

CONCLUSIONS

Studies of clonal growth in chimaeras and mosaics have rarely succeeded in reliably estimating the number of cells, or even coherent clones, present at the time of tissue foundation. However, extensive analysis of some adult mouse tissues, including the retinal pigment epithelium (Sanyal and Zeilmaker, 1977) and the melanoblasts of the dorsal coat (Russell and Major, 1957; Mintz, 1967; Cattanach, 1974), strongly suggests that in these tissues descendant clones initiated early in development retain a degree of spatial identity in at least some individuals. Cell mixing tends to disrupt these descendant clones into smaller patches of coherent clones so that the final pattern of variegation may be a product of two superimposed distributions: one of descendant clones and the other of coherent clones. The relative contributions of cell mixing and coherent clonal growth can best be examined using cell autonomous markers detectable at the cellular level. Numerical estimates of various parameters of clonal growth are unlikely to be accurate because statistical models never precisely fit the biological reality. This is a particularly difficult problem when analysing three-dimensional tissues. For these reasons, comparative studies of a tissue at different times in development probably offers the most reliable approach to the analysis of clonal growth in chimaeras and mosaics.

ACKNOWLEDGEMENTS

I am very grateful to Dr. Leon Rosenberg of Yale University, and Dr. Somes Sanyal of Erasmus University, Rotterdam, for kindly sending me the photographs shown in Figures 1B and 4 respectively.

I also thank Drs. Virginia Papaioannou, Richard Gardner and Anne McLaren for much helpful discussion during the preparation of this manuscript.

REFERENCES

ANDERSON, D., BILLINGHAM, R.E., LAMPKIN, G.H. & MEDAWAR, P.B. (1951) The use of skin grafting to distinguish between monozygotic and dizygotic twins in cattle. Heredity, London. 5, 379-397.

BARNES, R.D., HOLLIDAY, J. & TUFFREY, M. (1974) Immunofluorescence and elution studies in tetraparental NZB↔CFW chimaeras and graft-versus-host diseased NZB. Immunology 26, 1195-1206.

CATTANACH, B.M. (1961) A chemically-induced variegated type position effect in the mouse. Z. VererbLehre 92, 165-182.

CATTANACH, B.M. (1974) Position effect variegation in the mouse. Genet. Res. 23, 291-306.

CONDAMINE, H., CUSTER, R.P. & MINTZ, B. (1971) Pure-strain and genetically mosaic liver tumors histochemically identified with the β-glucuronidase marker in allophenic mice. Proc. natn. Acad. Sci. USA 68, 2032-2036.

DeMARS, R., LeVAN, S.L., TREND, B.L., & RUSSELL, L.B. (1976) Abnormal ornithine carbamoyltransferase in mice having the sparse-fur mutation. Proc. natn. Acad. Sci. USA 73, 1693-1697.

DEOL, M.S. & WHITTEN, W.K. (1972) Time of X chromosome inactivation in retinal melanocytes of the mouse. Nature New Biol. 238, 159-160.

DEWEY, M.J., GERVAIS, A.G. & MINTZ, B. (1976) Brain and ganglion development from two genotypic classes of cells in allophenic mice. Devel. Biol. 50, 68-81.

DREWS, U. & ALONSO-LOZANO, V. (1974) X-inactivation pattern in the epididymis of sex-reversed mice heterozygous for testicular feminization. J. Embryol. exp. Morph. 32, 217-225.

FEDER, N. (1976) Solitary cells and enzyme exchange in tetraparental mice. Nature Lond. 263, 67-69.

FIALKOW, P.J. (1973) Primordial cell pool size and lineage relationships of five human cell types. Ann. Hum. Genet. Lond. 37, 39-48.

FIALKOW, P.J. (1976) Clonal origin of human tumors. Biochim. biophys. Acta. 458, 283-321.

FORD, C.E. (1969) Mosaics and chimaeras. Brit. Med. Bull. 25, 104-109.

FORD, C.E., EVANS, E.P. & GARDNER, R.L. (1975) Marker chromosome analysis of two mouse chimaeras. J. Embryol. exp. Morph. 33, 447-457.

FRIEDMAN, J.M. & FIALKOW, P.J. (1976) Viral "tumorigenesis" in man: cell markers in Condylomata acuminata. Int. J. Cancer 17, 57-61.

GANDINI, E. & GARTLER, S.M. (1969) Glucose-6-phosphate dehydrogenase mosaicism for studying development of blood cell precursors. Nature, Lond. 224, 599-600.

GANDINI, E., GARTLER, S.M., ANGIONI, G., ARGIOLAS, N. & DELL' ACCUA, G. (1968) Developmental implications of multiple tissue studies in glucose-6-phosphate dehydrogenase-deficient heterozygotes. Proc. Nat. Acad. Sci. USA 61, 945-948.

GARCIA-BELLIDO, A., RIPOLL, P. & MORATA, G. (1973) Developmental compartmentalisation of the wing disk of Drosophila. Nature New Biol. 245, 251-253.

GARDNER, R.L. (1968) Mouse chimaeras obtained by the injection of cells into the blastocyst. Nature, Lond. 220, 596-597.

GARDNER, R.L. & JOHNSON, M.H. (1973) Investigation of early mammalian development using interspecific chimaeras between rat and mouse. Nature New Biol. 246, 86-89.

GARDNER, R.L. & LYON, M.F. (1971) X chromosome inactivation studied by injection of a single cell into the mouse blastocyst. Nature, Lond. 231, 385-386.

GARNER, W. & McLAREN, A. (1974) Cell distribution in chimaeric mouse embryos before implantation. J. Embryol. exp. Morph. 32, 495-503.

GARTLER, S.M. (1974) Utilization of mosaic systems in the study of the origin and progression of tumors. In: J. German (Ed.), Chromosomes and Cancer, Wiley, New York, pp. 313-334.

GARTLER, S.M., GANDINI, E., HUTCHISON, A.T., CAMPBELL, B. & ZECHHI, G. (1971) Glucose-6-phosphate dehydrogenase mosaicism: Utilization in the study of hair follicle variegation. Ann. hum. Genet. 35, 1-7.

GEARHART, J.D. & MINTZ, B. (1972) Clonal origins of somites and their muscle derivatives: evidence from allophenic mice. Dev. Biol. 29, 27-37.

GORDON, J. (1977) Modification of pigmentation patterns in allophenic mice by the W gene. Differentiation 9, 12-27.

GORNISH, M., WEBSTER, M.P. & WEGMANN, T.G. (1972) Chimaerism in the immune system of tetraparental mice. Nature New Biol. 237, 249-251.

GREEN, M.C., SCHULTZ, L.D. & NEDZI, L.A. (1975) Abnormal nuclear morphology of leukocytes in the mouse mutant ichthyosis. A possible transplantation marker. Transplantation 20, 172-175.

GREEN, M., DURHAM, D., MAYER, T.C. & HOPPE, P.C. (1977) Evidence from chimaeras for the pattern of proliferation of epidermis in the mouse. Genet. Res. Camb. 29, 279-284.

HUTCHISON, H.T. (1973) A model for estimating the extent of variegation in mosaic tissues. J. theor. Biol. 38, 61-79.

IANNACCONE, P.M., GARDNER, R.L. & HARRIS, H. (1978) The cellular origin of chemically induced tumours. J. Cell Sci., 29, 249-269.

KLINGER, H.P. & SCHWARZACHER, H.G. (1962) XY/XXY and sex chromatin positive cell distribution in a 60mm human fetus. Cytogenetics 1, 266-290.

LeVAIL, M.M. & MULLEN, R.J. (1976) Role of the pigment epithelium in inherited degeneration analysed with experimental mouse chimaeras. Exp. eye Res. 23, 227-245.

LAWRENCE, P.A. (1973) Maintenance of boundaries between developing organs in insects. Nature New Biol. 242, 31-32.

LEWIS, J. (1973) The theory of clonal mixing during growth. J. theor. Biol. 39, 47-54.

LEWIS, J.H., SUMMERBELL, D. & WOLPERT, L. (1972) Chimaeras and cell lineage in development. Nature, Lond. 239, 276-278.

LINDER, D. & GARTLER, S.M. (1965) Distribution of glucose-6-phosphate dehydrogenase electrophoretic variants in different tissues of heterozygotes. Amer. J. Hum. Genet. 17, 212-220.

LYON, M.F. (1961) Gene action in the X-chromosome of the mouse (Mus musculus L.) Nature, Lond. 190, 372-373.

LYON, M.F. (1972) X-chromosome inactivation and developmental patterns in mammals. Biol. Rev. 47, 1-35.

McLAREN, A. (1972) Numerology of development. Nature, Lond. 239, 274-276.

McLAREN, A. (1976) Mammalian Chimaeras. Cambridge University Press, Cambridge-London.

McLAREN, A. & BOWMAN, P. (1969) Mouse chimaeras derived from fusion of embryos differing by nine genetic factors. Nature, Lond. 224, 238-240.

MINTZ, B. (1962) Formation of genotypically mosaic mouse embryos. Amer. Zool. 2, 432 (Abstr. 310).

MINTZ, B. (1967) Gene control of mammalian pigmentary differentiation. I. Clonal origin of melanocytes. Proc. Nat. Acad. Sci. USA 58, 344-351.

MINTZ, B. (1968) Hermaphroditism, sex chromosomal mosaicism and germ cell selection in allophenic mice. J. Anim. Sci. (Suppl. 1) 27, 51-60.

MINTZ, B. (1969) Gene control of the mouse pigmentary system. Genetics, 61, (Suppl.) 41 (Abstr.).

MINTZ, B. (1970a) Gene expression in allophenic mice. In: H.A. Padykula (Ed.), Control Mechanisms in the Expression of Cellular Phenotypes, Academic Press, New York and London, pp. 15-42.

MINTZ, B. (1970b) Neoplasia and gene activity in allophenic mice. In: Genetic Concepts and Neoplasia. Annual Symposium on Fundamental Cancer Research, Williams and Wilkins, Baltimore, pp. 477-517.

MINTZ, B. (1971) The clonal basis of mammalian differentiation. In: D.D. Davies and M. Balls (Eds.), Control Mechanisms of Growth and Differentiation, Symposia of the Society for Experimental Biology, 25, Cambridge University Press, London, pp. 345-370.

MINTZ, B. (1974) Gene control of mammalian differentiation. Ann. Rev. Genetics 8, 411-470.

MINTZ, B. & BAKER, W.W. (1967) Normal mammalian muscle differentiation and gene control of isocitrate dehydrogenase synthesis. Proc. Nat. Acad. Sci. USA 58, 592-598.

MINTZ, B. & PALM, J. (1969) Gene control of hematopoiesis. I. Erythrocyte mosaicism and permanent immunological tolerance in allophenic mice. J. exp. Med. 129, 1013-1027.

MINTZ, B. & SANYAL, S. (1970) Clonal origin of the mouse visual retina mapped from genetically mosaic eyes. Genetics 64 (Suppl.), 43-44.

MOORE, W.J. & MINTZ, B. (1972) Clonal model of vertebral column and skull development derived from genetically mosaic skeletons in allophenic mice. Dev. Biol. 27, 55-70.

MULLEN, R.J. (1977) Site of pcd gene action and Purkinje cell mosaicism in the cerebella of chimaeric mice. Nature, Lond. 270, 245-247.

MULLEN, R.J. & LaVAIL, M.M. (1975) Two new types of retinal degeneration in cerebellar mutant mice. Nature, Lond. 258, 528-530.

MULLEN, R.J. & LaVAIL, M.M. (1976) Inherited retinal dystrophy: primary defect in pigment epithelium determined with experimental rat chimaeras. Science 192, 799-801.

MULLEN, R.J. & WHITTEN, W.K. (1971) Relationship of genotype and degree of chimaerism in coat color to sex ratios and gametogenesis in chimaeric mice. J. exp. Zool. 178, 165-176.

MULLEN, R.J., EICHER, E.M. & SIDMAN, R.C. (1976) Purkinje cell degeneration a new neurological mutation in the mouse. Proc. natn. Acad. Sci. USA. 73, 208-212.

NANCE, W.E. (1964) Genetic tests with a sex-linked marker: glucose-6-phosphate dehydrogenase. Cold Spring Harbor Symp. Quant. Biol. 29, 415-425.

NESBITT, M.N. (1971) X-chromosome inactivation mosaicism in the mouse. Dev. Biol. 26, 252-263.

NESBITT, M.N. (1974) Chimaeras vs. X inactivation mosaics: significance of differences in pigment distribution. Dev. Biol. 38, 202-207.

NESBITT, M.N. & GARTLER, S.M. (1971) The applications of genetic mosaicism to developmental problems. Ann. Rev. Genet. 5, 143-162.

NIELSEN, J.T. & CHAPMAN, V.M. (1977) Electrophoretic variation for sex-linked phosphoglycerate kinase (PGK-1) in the mouse. Genetics. 87, 319-325.

NYHAN, W.L., BAKAY, B., CONNOR, J.D., MARKS, J.F. & KEELE, D.K. (1970) Hemizygous expression of glucose-6-phosphate dehydrogenase in erythrocytes of heterozygotes for the Lesch-Nyhan syndrome. Proc. natn. Acad. Sci. USA. 65, 214-218.

ORCHARDSON, R. & McGADEY, J. (1970) The histochemical demonstration of phosphoglucose isomerase. Histochemie 22, 136-139.

PADUA, R.A., BULFIELD, G. & PETERS, J. (1978) Biochemical genetics of a new glucosephosphate isomerase allele ($Gpi-1^c$) from wild mice. Biochem. Genet. 16, 127-143. (In press).

RICCIUTI, F.C., GELEHRTER, T.D. & ROSENBERG, L.E. (1976) X-chromosome inactivation in human liver: confirmation of X-linkage of ornithine transcarbamylase. Am. J. Hum. Genet. 28, 332-338.

ROACH, S.A. (1968) The Theory of Random Clumping. Methuen, London.

RUSSELL, L.B. (1964) Genetic and functional mosaicism in the mouse. In: M. Locke (Ed.) The Role of Chromosomes in Development. Academic Press, New York and London, pp. 153-181.

RUSSELL, L.B. & MAJOR, M.H. (1957) Radiation-induced presumed somatic mutations in the house mouse. Genetics, 42, 161-175.

SANYAL, S. & ZEILMAKER, G.H. (1974) Gene action and cell lineage in retinal development in experimental chimaeric mice. Teratology 10, 322.

SANYAL, S. & ZEILMAKER, G.H. (1976) Comparative analysis of cell distribution in the pigment epithelium and the visual cell layer of chimaeric mice. J. Embryol. exp. Morph. 36, 425-430.

SANYAL, S. & ZEILMAKER, G.H. (1977) Cell lineage in retinal development of mice studied in experimental chimaeras. Nature 265, 731-733.

SCHNEIDERMAN, H.A. & BRYANT, P.J. (1971) Genetic analysis of development mechanisms in Drosophila. Nature, Lond. 234, 167-194.

SEARLE, A.G. (1968) Comparative genetics of coat colour in mammals. Logos Press, Academic Press, London, p. 52.

STERN, C. (1940) The prospective significance of imaginal discs in Drosophila. J. Morph. 67, 107-122.

TANSLEY, K. (1954) An inherited retinal degeneration in the mouse. J. Hered. 45, 123-127.

TARKOWSKI, A.K. (1961) Mouse chimaeras developed from fused eggs. Nature, Lond. 190, 857-860.

TARKOWSKI, A.K. (1964) Patterns of pigmentation in experimentally produced mouse chimaeras. J. Embryol. exp. Morph. 12, 575-585.

TETTENBORN, U., DOFUKU, R. & OHNO, S. (1971) Noninducible phenotype exhibited by a proportion of female mice heterozygous for the X-linked testicular feminization mutation. Nature, Lond. 234, 37-40.

TUFFREY, M., BARNES, R.D., EVANS, E.P. & FORD, C.E. (1973) Dominance of AKR lymphocytes in tetraparental AKR↔CBA-T6T6 chimaeras. Nature New Biol. 243, 207-208.

WARNER, C.M., McIVOR, J.L. & STEPHENS, T.J. (1977) Chimaeric drift in allophenic mice. Differentiation 9, 11-18.

WEGMANN, T.G. (1970) Enzyme patterns in tetraparental mouse liver. Nature, Lond. 225, 462-463.

WEGMANN, T.G. & GILMAN, J.G. (1970) Chimaerism for three genetic systems in tetraparental mice. Dev. Biol. 21, 281-291.

WEGMANN, T.G., LaVAIL, M.M. & SIDMAN, R.L. (1971) Patchy retinal degeneration in tetraparental mice. Nature, Lond., 230, 333-334.

WEST, J.D. (1975) A theoretical approach to the relation between patch size and clone size in chimaeric tissue. J. theor. Biol. 50, 153-160.

WEST, J.D. (1976a) Clonal development of the retinal epithelium of mouse chimaeras and X-inactivation mosaics. J. Embryol. exp. Morph. 35, 455-461.

WEST, J.D. (1976b) Distortion of patches of retinal degeneration in chimaeric mice. J. Embryol. exp. Morph. 36, 145-149.

WEST, J.D. (1976c) Patches in the livers of chimaeric mice. J. Embryol. exp. Morph. 36, 151-161.

WEST, J.D. (1977) Red blood cell selection in chimaeric mice. Exp. Hematol. 5, 1-7.

WEST, J.D. & McLAREN, A. (1976) The distribution of melanocytes in the dorsal coats of a series of chimaeric mice. J. Embryol. exp. Morph. 35, 87-93.

SUBJECT INDEX